# Environmental Science and Engineering

The ultimate goal of this series is to contribute to the protection of our environment, which calls for both profound research and the ongoing development of solutions and measurements by experts in the field. Accordingly, the series promotes not only a deeper understanding of environmental processes and the evaluation of management strategies, but also design and technology aimed at improving environmental quality. Books focusing on the former are published in the subseries Environmental Science, those focusing on the latter in the subseries Environmental Engineering.

More information about this series at http://www.springer.com/series/7487

Faith Ka Shun Chan · Hing Kai Chan ·
Tiantian Zhang · Ming Xu
Editors

# Proceedings of the 2020 International Conference on Resource Sustainability: Sustainable Urbanisation in the BRI Era (icRS Urbanisation 2020)

Springer

*Editors*
Faith Ka Shun Chan
Department of Geographical Sciences
University of Nottingham Ningbo China
Ningbo, China

Hing Kai Chan
Nottingham University Business
School China
University of Nottingham Ningbo China
Ningbo, China

Tiantian Zhang
Nottingham University Business
School China
University of Nottingham Ningbo China
Ningbo, China

Ming Xu
School for Environment and Sustainability
University of Michigan
Ann Arbor, MI, USA

ISSN 1863-5520 ISSN 1863-5539 (electronic)
Environmental Science and Engineering
ISBN 978-981-15-9604-9 ISBN 978-981-15-9605-6 (eBook)
https://doi.org/10.1007/978-981-15-9605-6



This Springer imprint is published by the registered company Springer Nature Singapore Pte Ltd.
The registered company address is: 152 Beach Road, #21-01/04 Gateway East, Singapore 189721, Singapore

# Contents

# Estimation of Metal Values of Obsolete Mobile Phones: A Case of China

**Jingying Li, Tong Xu, Jiangxian Wen, Shuli Gong, and Ni An**

**Abstract** In recent decades, obsolete mobile phones have attracted global concern due to huge production, rich valuable metal content, and potential environmental risk. Printed circuit boards (PCBs) in mobile phones contain various metals. Reusing these metal resources with a reasonable way into industrial production will bring great economic and social benefits to human society. Under the context of circular economy and sustainable development, the recycling of obsolete mobile phones will be of great significance in China. Based on previous studies, this study estimated that the contents of metals in PCBs coming from obsolete mobile phones in China and the forecasted the metals recycling values of obsolete mobile phones in 2030. The future average possession in per 100 inhabitants of cellular phones and the total population are calculated using the logistic model and the new population prediction model, respectively. According to the estimation results, the total weight of obsolete mobile phones generated each year will reach 220.0 Gg and 265.2 Gg in 2020 and 2030 respectively. In addition, 85.13 tons of silver and 31.02 tons of gold will be recovered from obsolete mobile phones in 2020, and the number will reach 102.63 and 37.39 in 2030. Among all the metals, gold has the most recycled value, accounting for 72.348% of the total recycling value, the total economic value of metal recovery will exceed 2149.98 million dollars in 2030. Finally, this article proposes some references and hopes to help establish a comprehensive obsolete mobile phones management system in China.

**Keywords** Obsolete mobile phones · Recycling values · Estimation · China

---

J. Li (✉) · T. Xu · J. Wen · S. Gong · N. An
College of Environment and Safety Engineering, Qingdao University of Science & Technology, Qingdao, Shandong 266042, People's Republic of China
e-mail: lijy741018@qust.edu.cn



F. K. S. Chan et al. (eds.), *Proceedings of the 2020 International Conference on Resource Sustainability: Sustainable Urbanisation in the BRI Era (icRS Urbanisation 2020)*, Environmental Science and Engineering,
https://doi.org/10.1007/978-981-15-9605-6_1

# 1 Introduction

Owing to increasing demand for electronic products, the electronics and electronic equipment (EEE) industry has increased the production scale and the kinds of new electronic products. In recent two decades, the speed of updating and elimination of EEE continues to accelerate, producing a large amount of waste electronics and electronic equipment (WEEE), especially in China. WEEE mainly includes obsolete computers, mobile phones, TVs, refrigerators and washing machines. In recent years, many smart electronic devices have also added such as waste window cleaning robots, VR devices and 3D printers. Due to the various factors mentioned above, WEEE has become the fastest growing solid wastes with the highest volume of discard in the world [1–3]. Moreover, due to the large amount of mobile phones and the fast changing frequency, obsolete mobile phones become special e-wastes.

The challenges following the increase of WEEE are not only limited in its huge amount of generation but also the complexity and harmfulness [4]. The effective management of WEEE and its proper disposal have gained widespread attention around the world. Above all, research on electronic waste (e-waste) recycling has also increased accordingly. On the other hand, in order to promote the management of e-waste recycling, the government has enacted and promulgated relevant laws and regulations.

The mobile phone and particularly smart phones, as the most commonly EEE utilized as integrated telecommunication and information equipment all over the world, has become an essential tool in people's life all over the world. There is also a person who has several mobile phones. In order to meet the needs of consumers, mobile phone manufacturers make the performance and quality of phones better. However, the life span of the mobile phone has similarly diminished. The ideal technical life of a phone is about 10 years [5], but the service life of cellular phones in real life is limited to 3 years owing to various needs, such as the pursuit of fashion and desire of new functions, resulting in the generation of a large quantity of obsolete mobile phones [6–8]. As the number of mobile phone users has grown exponentially over the past two decades, the amount of obsolete mobile phones is growing faster than other EEE. Miao Wei, Minister of the Ministry of Industry and Information, said that by July 2016, the total number of China mobile phone users reached 1.304 billion. Since 2012, China has become the world's largest mobile phone consumer market, with the result of a sharp increase in obsolete mobile phones. In China, consumers change their phones every 15 months, and about 100 million obsolete mobile phones are produced every year [9]. According to the model, Li et al. [10, 11] estimated that the output of obsolete mobile phones in China will be 280,000 and 799 million in 1998 and 2013 respectively. The global recovery rate of obsolete mobile phones is about 3%; while the recovery rate is only 1% in China [12, 13], lower than the average level of recovery rate in the world. Therefore, in China, the recycling of obsolete mobile phones should attract more attention.

A variety of metals and non-metals materials are contained in obsolete mobile phones. Mobile phones are generally made up of housing, liquid crystal display

(LCD), printed circuit boards (PCBs), battery, keyboard, antenna, speaker and accessory (such as headphone and charger) and other components with high resource value [14]. The PCB of an ordinary mobile phone weigh 15–43% by weight [15], which are high in precious metals content and economic value. It consists 28% of metals and 70% of non-metals [16]. Lisinska et al [17] made a more detailed summary that one phone consists of 40% metal, 30% plastic, and 30% ceramics.

Precious metals have been used as contact materials in cellular phones due to their satisfactory physical and chemical properties such as high electrical conductivity, good ductility and chemical stability [18]. Waste printed circuit boards (WPCBs) mainly contain precious metals (such as Au, Ag; small amounts of specific PCBs contain Pt) and base metals (such as Cu, Al, Zn, Fe, et al.). These metal elements are the most valuable material in the WPCBs. Moreover, WPCBs also contain hazardous metals, such as Hg, Pb, Cd, As, Sb [19]. Leaking or disposing of these harmful metals can seriously pollute the soil and groundwater, posing a tremendous threat to human health. For example, cadmium contained in a cellular phone can contaminate 600,000 L of water [20]. The recycling of obsolete mobile phones and the quantifying value of metals from phones has become a hot topic in the world. In this rapidly developing society, collecting and recycling the huge amount of metal resources of obsolete mobile phones has become an urgent task in case that obsolete mobile phones give rise to serious environmental pollution.

In developed countries, the recycling system for obsolete mobile phones has been matured. Significant achievements have been made by Switzerland, United States (US), UK, Germany, Japan and Korea in collecting and recycling obsolete mobile phones [21]. First of all, Switzerland established the producer responsibility organization (PROs) [22]. Meanwhile, obsolete mobile phones were also included in the SWICO system [21, 23, 24], which is one of the most important PROs. Switzerland also enacted ordinance to recognize the legal responsibilities of each stakeholder(de [25], so, *The Return, the Taking Back and the Disposal of Electrical and Electronic Equipment (ORDEE)* was carried out in 1998.

As developed countries, Germany and UK were actively implementing the WEEE Directive and RoHS Directive to control the management of obsolete mobile phones. The recovery rate of obsolete mobile phones is as high as 9% in UK, higher than the average level of recovery rate in the world. Regenersis, which devotes to encouraging the reuse and recycling of the mobile phone has treated over 25 million obsolete mobile phones. In Japan, there are no laws specifically aimed at mobile phone recycling, but a cellular phone recycling association has been established. Major recycling companies include KDDI Co., LTD., NTT Do Co Mo, Yokohama Metal Co., Ltd., etc. Recycling methods were changed from shredding to manual dismantling, and the recycling rate increased from 60 to 100% [26]. While domestic recycling and cleaning processes in developed countries are well implemented, it is also important to note that the e-waste in developed countries is still being shipped to developing countries in large quantities.

Polák and Drápalová [27] estimated the end of life of mobile phones generation in Czech, and pointed out that it will recover about 31 tonnes of gold and 325 tonnes of silver from obsolete mobile phones in the next 10 years. Hira et al. [28] studied characteristics and evaluation of metal in obsolete mobile phone. According to their research, obsolete mobile phones must be considered to be hazardous due to the toxic metals contained. However, mobile phones can be an asset if used properly [28]. The obsolete mobile phone is actually a treasure trove of resources with great economic value. Finding a scientific, reasonable and efficient recycling method for obsolete mobile phones can not only reduce the environmental damage caused by the e-waste pollution, but also reduce the direct harm to the human body and realize the recycling of resources, which is of great practical significance.

So far, domestic and foreign scholars' research on obsolete mobile phones has focused on the estimation of total production and the precious metal content in mobile phones. However, there are few reports on the economic benefits of recycling of obsolete mobile phones. In this paper, we estimate the annual production of obsolete smart phones and the recycling benefits of metals in smart phones (assuming that the prices of various metals will not fluctuate much) from 2019 to 2030 based on the production of obsolete mobile phones and metal content in mobile phones from 1998 to 2018 in China, by using a simplified method.

## 2 Estimation of Obsolete Cellular Phones Generation

### *2.1 Definition and Estimated Range*

All mobile phones mentioned in this study refer to phones that include a shell, screen, battery and PCB, and do not include charger, earphone and other accessories. The estimation of obsolete cellular phones generation from 2019 to 2030 is for smart phones.

There are many definitions of the service life of a product, including "total lifespan" and "service lifespan", among others [3, 29, 30]. Golev et al. [31] has an in-depth description of the lifespan of mobile phones when using the Australian mobile phone as an example to demonstrate product flow analysis. The product lifespan in their study was measured by the following three methods: Average lifespan estimation based on the Leaching model; Average lifespan estimation based on consumer surveys; Lifespan distribution estimation based on the use of the Weibull function. In this paper, average lifespan based on consumer surveys are considered as the mobile phone lifespan, the specific data of service lifespan are derived from the social questionnaire survey results, a study that the research team has done.

The average weight of a mobile phone refers to the average weight of different types of mobile phones on the market in China, including batteries but not including chargers, earphones and other accessories.

The geographical area of statistics is limited to mainland China. Based on the population of mainland China and the service life of mobile phones, this study estimates the annual obsolete mobile phones generated from 1998 to 2018 in China and forecasts the generation amount of obsolete smart phones from 2019 to 2030.

## 2.2 Estimation Methods

Currently, there is a lack of statistical data on the generation of obsolete mobile phones. Many researches on the estimation of the generation of obsolete mobile phones at country level have been conducted [4, 27, 32] by using information about subscribers, sales, lifespan or average lifespan, domestic demand and other parameters of cell phones with the mathematical models [3].

In this paper, the logistic model is used to forecast the future average possession of mobile phones. Based on past possession data (data come from Ministry of industry and information technology of the People's Republic of China, http://www.miit.gov.cn/n1146295/n1146592/n1146764/n4666969/index.html), the average possession amount of mobile phones per 100 inhabitants per year can be calculated by Eq. (1). The annual total number of cellular phones in China can be calculated by Eq. (2).

$$\overline{P}t = \overline{P}\text{max}/\left[1 - a \cdot \exp\{-b(t - t0)\}\right] \quad (1)$$

$$P\text{t} = \overline{P}t \times Nt \quad (2)$$

where $\overline{P}t$ represents the average possession of mobile phones per 100 inhabitants in year t. $\overline{P}$max represents the maximum value of the average possession of mobile phones per 100 inhabitants, and the maximum value was identified as 121 units. $t_0$ represents the first year for logistic regressions. The values of a and b are −33.45 and 0.27 respectively [33]. $N$t represents the number of 100 inhabitants in year t. $P$t represents the total number of mobile phones in China at year t. Figure 1 shows the average possession of mobile phones from 1998 to 2030 in China.

The average possession per 100 inhabitants was only 1.89 units in 1998. With the rapid economic development in China, the number reached 105.1 units in 2018, and the estimated value will reach 118.3 and 120.3 units in 2025 and 2030, respectively.

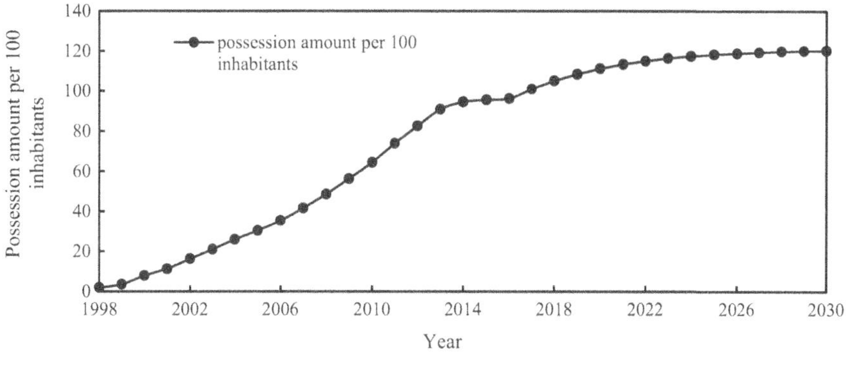


**Fig. 1** The average possession of mobile phones per 100 inhabitants

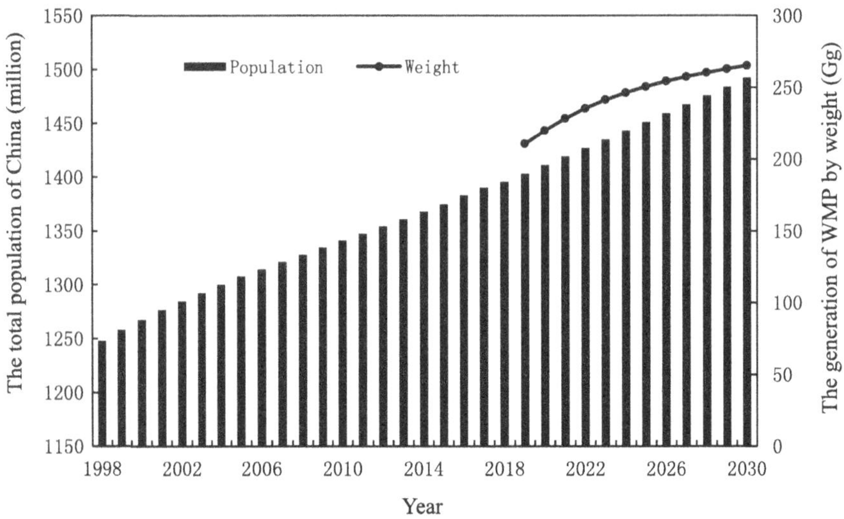


**Fig. 2** Estimation of obsolete mobile phones described by weight and total population of China

Meanwhile, this paper predicted the total population of China by 2030. The new population prediction model is used to forecast the future total population in China, as described below. According to the trend of increasing population in mainland China from 1998 to 2018(data were derived from National Bureau of Statistics of the People's Republic of China, http://www.stats.gov.cn/), the natural population growth rate was assumed to be 5.613‰ to forecast the total population from 2019 to 2030, and the total population can be calculated by Eq. (3). The relationship between $Mt$ and $Nt$ is shown in Eq. (4), and then we can get the value of $Nt$.

$$Mt = Mt'0 \cdot e^{k(t-t'0)} \tag{3}$$

$$Mt = Nt \times 100 \tag{4}$$

where $Mt$ represents the total population at year t. $Mt'0$ represents the total population in 2018, and the value is 1395.38 million. t'0 represents the year of 2018. k represents the natural population growth rate, 5.613‰, the average value of the population growth rate from 1998 to 2018. The predicted population result is shown in Fig. 2.

## 2.3 *Estimation of the Generation Amounts of Obsolete Mobile Phones*

In our previous survey results, 39.63% of the people changed their mobile phones in 1–2 years, and 38.17% replaced their mobile phones in 2–3 years. According to

**Table 1** Weight data of major mobile phone products in the Chinese consumer market

| Brand | Product | Weight (g) | Brand | Product | Weight (g) |
|---|---|---|---|---|---|
| Apple | iPhone 6 | 129 | HUAWEI | honor8 | 153 |
| | iPhone 6P | 172 | | hono9 | 153 |
| | iPhone 7 | 138 | | honorV9 | 184 |
| | iPhone 7P | 188 | | honor10 | 153 |
| | iPhone 8 | 148 | | P9 | 144 |
| | iPhone 8p | 202 | | P10 | 145 |
| | iPhone X | 174 | | Mate10 | 186 |
| OPPO | R9 | 145 | vivo | X9 | 157 |
| | R9s | 145 | | X20 | 159 |
| | R11 | 150 | MI | MI5 | 129–139 |
| | R15 | 175 | | MI4s | 133 |
| | A3 | 159 | | MI note4 | 175 |

calculations, the average time for Chinese consumers to replace a mobile phone is 2.2 years. In this paper, it can be assumed that the average service life of one mobile phone for Chinese consumers is two years generally. In other words, we roughly think that the total number of mobile phones in 2017 will become the number of obsolete mobile phones in 2019, and the amount of obsolete mobile phone was calculated.

The 2017 Mobile Internet User Analysis Report relies on the tens of millions of mobile phone user data in the carrier network, which truly exposed the use status of mobile phones in 2017. According to the "Report" data, the top five brands of mobile phone brands in the network in 2017 are Apple, Huawei, OPPO, vivo, and MI. At present, the weight data of several mobile phones that account for a relatively large number in the mobile phone consumer market are listed in Table 1.

It can be seen from the statistics in the Table 1 that one mobile phone weighs between 130 g and 200 g in the consumer market of the current mobile phone in China. The screen size of these mobile phones is mostly between 4.5 and 6 in., which is a suitable size for human to use. It will not change much in the future development process, nor will its weight. So we can assume the weight of the phone is a fixed value, and the average weight was assumed to be 150 g. The annual estimation of obsolete mobile phones by weight from 2019 to 2030 is shown in Fig. 2.

Figure 2 shows that 210.6 Gg of obsolete mobile phones will be generated in 2019. The total weight of obsolete mobile phones generated every year will reach 220.0 Gg and 265.2 Gg in 2020 and 2030 respectively. In the ten years from 2021 to 2030, the total amount of obsolete mobile phones will reach 16.68 billion units and 2502.1 Gg by unit and weight respectively.

## 3 Metal Content and Value of Obsolete Cellular Phones

After the mobile phone is scrapped, the metal, especially the precious metal, is the main object of recycling and the main value of obsolete mobile phones [26]. Yu et al. [34] and Sugiyama et al. [35] used the methods of the material flow analysis (MFA) to estimate the life cycle impacts of cellular phones in China and Japan respectively. Yu gave the average metal content of typical mobile phones based on the researches [36–38] related to studying of metal concentration, Cu (13%), Al(2%), Fe (5%), Ni(0.1%), Pb(0.3%), Sn(0.5%), Ag(0.14%), Au(0.035%), Pd(0.02%). Vats and Singh [15] have evaluated the content of gold and silver in PCBs of obsolete mobile phones by subjecting the PCB to roasting and acid digestion. They selected 10 different types of cell phones to analyse, and pointed out that the content of gold and silver in the PCBs was in the range of 0.009–0.017% and 0.25–0.79% by weight respectively. Hageluken [39] found that there are 250 tons of silver, 24 tons of gold, 9 tons of palladium and 9000 tons of copper in every 1000 million units of mobile phones. Another study showed that concentration of precious metals in one ton of the mobile phone is 3573 g of silver, 368 g of gold and 287 g of palladium respectively [40]. Zhao et al. [41] studied the metal content of PCBs powder of the cellular phones with different particle sizes and indicated that Au and Ag are mainly enriched in the particle size range of less than 0.106 mm, for <0.075 mm and 0.075–0.106 mm particles, Au (7.82 ppm, 9.75 ppm) and Ag (0.0038%, 0.0025%) respectively. Recently, new research has summarized the chemical composition of mobile phone. Lisinska et al. [17] pointed out in the study of printed circuit boards coming from obsolete mobile phones that content of gold and silver are 140–1000 ppm and 350–2000 ppm, respectively.

It is worth noting that, due to the unskilled manufacturing technology, "Shanzhai" cellphone needs more metal in the process of the manufacture of mobile phones. The metal content in PCBs is higher than the brand mobile phone.

Much work has been devoted to the determination of metal content in PCBs coming from obsolete mobile phones. The weighted percentage of several major metals in a typical PCB is presented in Table 2.

## 4 Value Chain of Obsolete Cellular Phones Recycling

In the next ten years, a lot of obsolete mobile phones will be generated, as shown in Fig. 2. These obsolete mobile phones are huge invisible mineral resources. According to the data already reported, we believe that the weight of PCB accounts for 30% of the total weight of the phone.

According to the average content of metal by weight in PCBs of the mobile phone, we can roughly figure out that we will get 58.05 tons of silver, 21.15 tons of gold, 16419.6 tons of copper, 1082.7 tons of iron, 738.9 tons of nickel, 469.8 tons of lead, 1137.6 tons of tin and 3.6 tons of palladium in every billion units of obsolete mobile phones, if the metal can be recovered fully. There will be 85.13 tons of silver and

**Table 2** Weighted percentage of major metals found in PCBs, by various researches (%)

| Metals | Ag | Au | Cu | Fe | Ni | Pb | Sn | Pd |
|---|---|---|---|---|---|---|---|---|
| [42] | 0.21 | 0 | 34.49 | 10.57 | 2.63 | 1.87 | 3.39 | – |
| [43] | – | 0.045 | 66.00 | – | 2.30 | – | – | – |
| [14] | 0.06 | 0.08 | 38.57 | 4.27 | 2.54 | 1.22 | 2.58 | – |
| [44] | 0.054 | 0.0043 | 39.86 | – | 0.396 | – | – | – |
| Chehade (2012) | 0.156 | 0.039 | 18.448 | – | – | – | – | 0.009 |
| [45] | 0.027 | 0.088 | 37.97 | – | 2.34 | – | 2.05 | – |
| [46] | 0.106 | 0.0065 | 40.8 | 0.28 | 0.39 | 1.36 | 1.60 | 0.005 |
| [47] | 0.22 | 0.04 | 10.45 | 1.06 | 0.61 | 0.58 | 0.98 | 0.01 |
| [5] | 0.2 | 0.12 | 41.8 | 0.66 | 1.93 | 0.19 | 4.57 | – |
| Average | 0.129 | 0.047 | 36.488 | 2.406 | 1.642 | 1.044 | 2.528 | 0.008 |

"–" Not reported in the corresponding research

31.02 tons of gold recovered from obsolete mobile phones in 2020, and the number will reach 102.63 and 37.39 in 2030 respectively. From 2021 to 2030, we can reclaim 968.33 tons of silver, 352.80 tons of gold, 273894.76 tons of copper, 18060.48 tons of iron, 12325.56 tons of nickel, 7836.72 tons of lead, 18976.26 tons of tin and 60.05 tons of palladium from the cumulative amount of obsolete mobile phones this decade.

In a cellular phone, the total weight of several major metals account for less than half of the total weight of the PCB (Fig. 3a). Among the metals, copper accounts for the largest proportion, followed by stannum and iron, while precious metal gold and silver account for only 0.05% and 0.13%, respectively.

According to the data of the metal price from SMM (https://www.smm.cn/), the recovery value of a cellular phone is shown in Table 3. Among these metals, gold has the most recycled value due to the highest unit price, accounting for 72.348% of the total recycling value (Fig. 3b). As a precious metal, the unit price of silver is higher than that of copper, but the content of copper in PCBs of mobile phones is much higher than that of silver, as shown in Fig. 3a, which has higher recycling value, followed by silver.

The total economic value will reach 58.70 million dollars and 1555.47 million dollars by recovering silver and gold from obsolete mobile phones in 2030, and the total economic value of metal recovery will exceed 2149.98 million dollars. It is estimated that there will be an economic benefit of 20285.99 million dollars from 2020 to 2030. More results are shown in Table 4.

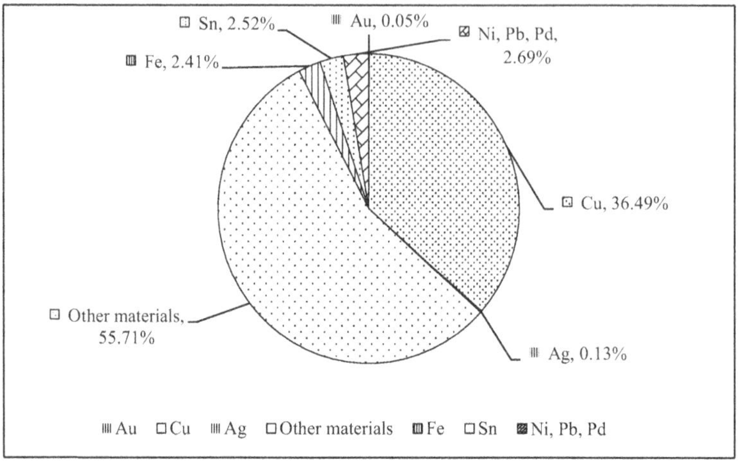


(a) Proportion of weighted age

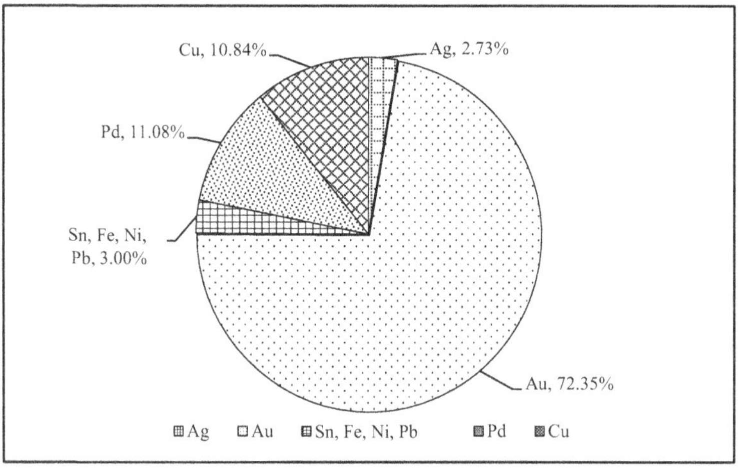


(b) Proportion of recovery value

**Fig. 3** Estimated major metals proportion of a mobile phone PCB: **a** weighted age, **b** proportion of recovery value

## 5 Conclusion and Advice

The estimation results of the generation and metal values of obsolete mobile phones in China show that the generation amount continues to increase and recycling obsolete mobile phones has a considerable economic benefit in the future years. However, the results in this paper will be affected by many factors, such as the population, mobile phone lifespan, the metal price assumed to be fixed, the average quality of mobile phones, etc., which will change slightly with the development process. For example, this paper forecast that the population of China in 2019 will reach 1403.21 million,

**Table 3** Analysis of the recovery value of a mobile phone

| Metals | Weighted age (%) | Metal prices (USD/kg) | Recycle value of one unit (USD) | Proportion of value (%) |
|---|---|---|---|---|
| Ag | 0.129 | 571.97 | 0.033 | 2.730 |
| Au | 0.047 | 41,600 | 0.880 | 72.348 |
| Cu | 36.488 | 8.0300 | 0.132 | 10.842 |
| Fe | 2.406 | 0.5300 | 0.001 | 0.047 |
| Ni | 1.642 | 13.788 | 0.010 | 0.838 |
| Pb | 1.044 | 2.761 | 0.001 | 0.107 |
| Sn | 2.528 | 21.477 | 0.024 | 2.009 |
| Pd | 0.008 | 37.424 | 0.135 | 11.078 |

**Table 4.** The total recovery value of metal in mobile phones

| Year | Value (million dollars) | Year | Value (million dollars) |
|---|---|---|---|
| 2019 | 1707.40 | 2025 | 2031.30 |
| 2020 | 1783.49 | 2026 | 2062.01 |
| 2021 | 1851.52 | 2027 | 2087.70 |
| 2022 | 1908.25 | 2028 | 2110.07 |
| 2023 | 1956.92 | 2029 | 2130.84 |
| 2024 | 1997.41 | 2030 | 2149.98 |

while the actual population of China at the end of 2019 is 1390.08 million according to official data. This change in numbers will affect a series of subsequent data of estimation and prediction. Similarly, changes in metal prices and mobile phone life will also influence our findings.

In this study, the generation of obsolete mobile phones was approximately 1403.98 million units in 2019, and it was projected to be 1670.32 million units in 2025; these estimations are larger than the previous estimations of 877 million units and 937 million units in 2017 and 2025 [3], while the amount of obsolete mobile phones was 719.8 million units by 2025 in the study of [48]. Meanwhile, the calculated results of the metals in obsolete cellular phones indicate that the amounts of silver and gold will reach 96.83 tons and 35.28 tons; these estimations are larger than their estimations of 56.25 tons and 28.13 tons [3], respectively, in 2025. This may be related to the different research methods which we use. They used material flow analysis to define life cycles, etc. We used a simplified population growth model to predict the population in the future.

Recycling the metal in obsolete mobile phones can reduce the pollution in the environment and reduce the harm to the human body. It can also alleviate the shortage of resources. As a large country that produces and consumes mobile phones, China's follow-up management of obsolete mobile phones is not standardized. China should speed up the management pace of its WMPs.

Electronic waste management is an area that needs to be comprehensively considered. It is learned that Japan is actively developing new technologies to reduce the amount of precious metals in mobile phones and developing alternative technologies for precious metals. Therefore, the cellular phone is bound to gradually reduce the content of precious metals. Furthermore, improving metal recycling rate is an important part of solving the problem of obsolete mobile phones.

Considering the current recycling of obsolete mobile phones in China, legislation and laws to clarify the responsibility of consumers, manufacturers and distributors in the mobile phone recycling market is an indispensable part of building a reasonable recycling market in China. Secondly, the government should install and maintain the proper recycling facilities in combination with culture and economic factors and carry out education publicity. Finally, in order to ensure the reasonable recovery and disposal of obsolete mobile phones, it is the government's responsibility to establish legislation and laws which stipulate that WMP treatment processes should be standardized and the environmental impact should be strictly controlled during the dismantling and disposal process. What we expect is that the results of this paper will have positive effects on the environmental protection and management of obsolete mobile phones.

**Acknowledgements** This project was supported by Shandong Provincial Natural Science Foundation, China (ZR2013EEM008), and supported by Key Laboratory for Solid Waste Management and Environment Safety, Ministry of Education of China, Tsinghua University: SWMES 2017–02. The authors are grateful to the reviewers who help us improve the paper by many pertinent comments and suggestions.

## References

1. Wei, L., Liu, Y.: Present status of e-waste disposal and recycling in China. Proc Environ Sci **16**(4), 506–514 (2012)
2. Guo, X., Liu, J., Qin, H., et al.: Recovery of metal values from waste printed circuit boards using an alkali fusion–leaching–separation process. Hydrometallurgy **156**, 199–205 (2015)
3. Guo, X., Yan, K.: Estimation of obsolete cellular phones generation: a case study of China. Sci Total Environ **575**, 321–329 (2017)
4. Rahmani, M., Nabizadeh, R., Yaghmaeian, K., et al.: Estimation of waste from computers and mobile phones in Iran. Resour Conserv Recycl **87**, 21–29 (2014)
5. Tan, Q., Dong, Q., Liu, L., et al.: Potential recycling availability and capacity assessment on typical metals in waste mobile phones: a current research study in China. J Clean Prod **148**, 509–517 (2017)
6. Huang, E.M., Yatani, K., Truong, K.N., et al.: Understanding mobile phone situated sustainability: the influence of local constraints and practices on transferability. IEEE Pervasive Comput **8**(1), 46–53 (2009)
7. Ha, V.H., Lee, J.C., Jeong, J., et al.: Thiosulfate leaching of gold from waste mobile phones. J Hazard Mater **178**(1–3), 1115–1119 (2010)
8. Ongondo, F.O., Williams, I.D.: Mobile phone collection, reuse and recycling in the UK. Waste Manag **31**(6), 1307–1315 (2011)
9. Yin, J., Gao, Y., Xu, H.: Survey and analysis of consumers' behaviour of waste mobile phone recycling in China. J Clean Prod **65**, 517–525 (2014)

10. Li, B., Yang, J., Lu, B., et al.: Estimation of retired mobile phones generation in China: a comparative study on methodology. Waste Manag **35**, 247–254 (2015)
11. Li, B., Yang, J., Lu, B., et al.: Temporal and spatial variations of retired mobile phones in China. Acta Sci Circum **35**(12), 4095–4101 (2015). (in Chinese)
12. Liu, W., Liang, C., Qin, W.Q., et al.: A new technology for recovery of metals from waste printed circuit boards. Appl Mech Mater **675–677**(2), 698–703 (2014)
13. Li, J., Ge, Z., Liang, C., et al.: Present status of recycling waste mobile phones in China: a review. Environ Sci Pollut Res **24**(20), 16578–16591 (2017)
14. Kasper, A.C., Berselli, G.B.T., Freitas, B.D., et al.: Printed wiring boards for mobile phones: characterization and recycling of copper. Waste Manag **31**(12), 2536–2545 (2011)
15. Vats, M.C., Singh, S.K.: Assessment of gold and silver in assorted mobile phone printed circuit boards (PCBs): original article. Waste Manag **45**, 280–288 (2015)
16. Sohaili, J., Muniyandi, S.K., Mohamad, S.S.: A review on printed circuit boards waste recycling technologies and reuse of recovered nonmetallic materials. Int J Sci Eng Res **3**(2), 1–7 (2012)
17. Lisinska, M., Saternus, Willner, J.: Research of leaching of the printed circuit boards coming from waste mobile phones. Arch Metall Mater **63**(1), 143–147 (2018)
18. Cui, J., Zhang, L.: Metallurgical recovery of metals from electronic waste: A review. J Hazard Mater **158**(2–3), 228–256 (2008)
19. Khaliq, A., Rhamdhani, M.A., Brooks, G., et al.: Metal extraction processes for electronic waste and existing industrial routes: a review and Australian perspective. Resources. **3**(1), 152–179 (2014)
20. Manivannan, S.V.: Environmental and health aspects of mobile phone production and use: suggestions for innovation and policy. Environ Innov Soc Trans **21**, 69–79 (2016)
21. Xu, C., Zhang, W., He, W., et al.: The situation of waste mobile phone management in developed countries and development status in China. Waste Manag **58**, 341–347 (2016)
22. Wath, S.B., Dutt, P.S., Chakrabarti, T.: E-waste scenario in India, its management and implications. Environ Monit Assess **172**(1–4), 249–262 (2011)
23. Nnorom, I.C., Osibanjo, O.: Overview of electronic waste (e-waste) management practices and legislations, and their poor applications in the developing countries. Resour Conserv Recycl **52**(6), 843–858 (2008)
24. Ongondo, F.O., Williams, I.D., Cherrett, T.J.: How are WEEE doing? A global review of the management of electrical and electronic wastes. Waste Manag **31**(4), 714–730 (2011)
25. Oliveira, C.R.D., Bernardes, A.M., Gerbase, A.E.: Collection and recycling of electronic scrap: a worldwide overview and comparison with the Brazilian situation. Waste Manag **32**(8), 1592–1610 (2012)
26. Zheng, X.J., Wang, J.W.: Used mobile phones' resourcing model in the concept of cyclic economy. Ecol Econ **30**, 30–142 (2014). (in Chinese)
27. Polák, M., Drápalová, L.: Estimation of end of life mobile phones generation: the case study of the Czech Republic. Waste Manag **32**(8), 1583–1591 (2012)
28. Hira, M., Yadav, S., Morthekai, P., et al.: Mobile phones-an asset or a liability: a study based on characterization and assessment of metals in waste mobile phone components using leaching tests. J Hazard Mater **342**, 29–40 (2018)
29. Murakami, S., Oguchi, M., Tasaki, T., et al.: Lifespan of commodities, Part I: the creation of a database and its review. J Ind Ecol **14**(4), 598–612 (2010)
30. Oguchi, M., Murakami, S., Tasaki, T., et al.: Lifespan of commodities, Part II: methodologies for estimating lifespan distribution of commodities. J Ind Ecol **14**(4), 613–626 (2010)
31. Artem, Golev, Werner Tim, T., Xuan, Zhu, et al.: Product flow analysis using trade statistics and consumer survey data: a case study of mobile phones in Australia. J Clean Prod **133**, 262–271 (2016)
32. Jang, Y.C., Kim, M.: Management of used and end-of-life mobile phones in Korea: a review. Resour Conserv Recycl **55**(1), 11–19 (2011)
33. Kim, S., Oguchi, M., Yoshida, A., et al.: Estimating the amount of WEEE generated in South Korea by using the population balance model. Waste Manag **33**(2), 474–483 (2013)

34. Yu, J., Williams, E., Ju, M.: Analysis of material and energy consumption of mobile phones in China. Energy Policy **38**(8), 4135–4141 (2010)
35. Sugiyama, K., Honma, O., Mishima, N.: Quantitative analysis of material flow of used mobile phones in Japan. Proc CIRP **40**, 79–84 (2016)
36. Wu, B.Y., Chan, Y.C., Middendorf, A., et al.: Assessment of toxicity potential of metallic elements in discarded electronics: a case study of mobile phones in China. J Environ Sci **20**(11), 1403–1408 (2008)
37. Lindholm ME (2003) Toward environmentally conscious product design-a comprehensive DfE implementation in new generation cellular phones. In: IEEE international symposium on electronics and the environment, pp 251–254
38. Tan KCN (2005) Life cycle assessment of a mobile phone. Communications Technologies, pp 1–111
39. Hageluken, C.: OECD-UNEP conference paris-resource recovery from e-scrap. OECD-UNEP Conf Resour Eff **24**, 1–22 (2008)
40. Chancerel, P., Meskers, C.E.M., Hagelüken, C., et al.: Assessment of precious metal flows during preprocessing of waste electrical and electronic equipment. J Ind Ecol **13**(5), 791–810 (2009)
41. Zhao, C., Zhang, X., Ding, J., et al.: Study on recovery of valuable metals from waste mobile phone PCB particles using liquid-solid fluidization technique. Chem Eng J **311**, 217–226 (2017)
42. Yamane, L.H., de Moraes, V.T., Espinosa, D.C., et al.: Recycling of WEEE: characterization of spent printed circuit boards from mobile phones and computers. Waste Manag **31**(12), 2553–2558 (2011)
43. Kim, E.Y., Kim, M.S., Lee, J.C., et al.: Selective recovery of gold from waste mobile phone PCBs by hydrometallurgical process. J Hazard Mater **198**(2), 206–215 (2011)
44. Li, J.Y., Xu, X.L., Liu, W.Q.: Thiourea leaching gold and silver from the printed circuit boards of waste mobile phones. Waste Manag **32**(6), 1209–1212 (2012)
45. Petter, P.M., Veit, H.M., Bernardes, A.M.: Evaluation of gold and silver leaching from printed circuit board of cellphones. Waste Manag **34**(2), 475–482 (2014)
46. Xiu, F.R., Qi, Y., Zhang, F.S.: Leaching of Au, Ag, and Pd from waste printed circuit boards of mobile phone by iodide lixiviant after supercritical water pre-treatment. Waste Manag **41**, 134–141 (2015)
47. Bian, J., Bai, H., Li, W., et al.: Comparative environmental life cycle assessment of waste mobile phone recycling in China. J Clean Prod **131**, 209–218 (2016)
48. He, P., Wang, C., Zuo, L.: The present and future availability of high-tech minerals in waste mobile phones: evidence from China. J Clean Prod **192**, 940–949 (2018)

# Robust Optimizing a Multi-period Multi-objective Closed-Loop Supply Chain Network for Perishable Goods Using Hybrid Heuristic Algorithm Under Uncertainty

**Zhongjia Sun, Jianquan Guo, Chengji Liang, and Mitsuo Gen**

**Abstract** Considering the high demand volatility, short shelf life and sales period of perishable goods, it is necessary to analyze the application of multi-period distribution in closed-loop supply chain network (CLSC) system. At the same time, considering the environmental pollution caused by improper handling, sales and reputation loss caused by untimely delivery, uncertain demand and recovery of perishable goods in different periods, a mixed integer programming (MIP) model and its corresponding robust optimization model of multi-period and multi-objective CLSC network for perishable goods are established aiming at the minimum environmental impact, the minimum economic cost and the maximum on-time delivery rate. The validity and feasibility of robust optimization model were verified by a hybrid heuristic algorithm with a case study of cross-region CLSC network covering cities of Shanghai, Suzhou, Wuxi, Changzhou, Jiaxing and Huzhou in Eastern China. The numerical results show that, compared with the single period system, the multi-period system has the advantages of good dynamic performance and low system cost, and the multi-objective optimization is better than the single-objective optimization on the whole; meanwhile, the importance of uncertain factors in the decision-making of robust optimization model is verified. This research provides important managerial insights about network design and optimal operation of CLSC for perishable goods under uncertainty.

Z. Sun (✉) · J. Guo
University of Shanghai for Science and Technology, Shanghai, P. R. China
e-mail: sunzhongjiausst@126.com

J. Guo
e-mail: jasonguousst@126.com

C. Liang
Shanghai Maritime University, Shanghai, P. R. China
e-mail: liangcj@shmtu.edu.cn

M. Gen
Fuzzy Logic Systems Institute, Tokyo University of Science, Tokyo, Japan
e-mail: gen@flsi.or.jp



F. K. S. Chan et al. (eds.), *Proceedings of the 2020 International Conference on Resource Sustainability: Sustainable Urbanisation in the BRI Era (icRS Urbanisation 2020)*, Environmental Science and Engineering,
https://doi.org/10.1007/978-981-15-9605-6_2

**Keywords** Perishable goods · Multi-period · Multi-objective · Closed-loop supply chain network · Robust optimization model · Hybrid heuristic algorithm

# 1 Introduction

With the development of social economy and the aggravation of environmental pollution, government and consumers have put forward higher requirements for environmental effect and efficiency of Supply chain (Rahimi and Ghezavati 2018; Alshamsi and Diabat 2015). Study of closed-loop supply chain (CLSC) network is generated in accordance with this social development trend. This paper structures a multi-period multi-objective CLSC network model with perishable product as research object to provide a systematic solution to the problem of model design and decision-making.

## *1.1 Literature Review*

Currently, scholars have conducted a lot of work in the field of CLSC network. Most papers optimize the research of CLSC network by using improved meta-heuristic algorithms such as GA (genetic algorithm) and PSO (particle swarm optimization algorithm). GA has strong robustness and strong search ability, but this method tends to fall into local convergence when solving complex problems (Gou et al. 2020). PSO can converge to the global optimal solution with high probability, but it cannot effectively maintain the high-quality solution of the population (Gretel et al. 2020). Based on the complementary characteristics of GA and PSO, a hybrid GA-PSO algorithm is proposed in this paper.

Perishable product is an important research object in CLSC network. Rafie-Majd et al. (2018) classified perishable products into two general categories: A group of these products such as vegetables, human blood and negatives that have the maximum usable lifetime; the other type, such as alcohol, gasoline and radioactive materials that don't have a shelf lifetime. Considering the characteristics of perishable products, a scientific and efficient design of CLSC network can be beneficial to reduce the logistics cost, improve customer satisfaction and optimize the utilization of resources (Asim et al. 2019).

Most papers adopt stochastic programming, fuzzy programming and robust programming to conduct research on the design of CLSC network. However, the stochastic programming approach requires an accurate distribution of the product (Kannan and Cheng 2018; Cheng et al. 2017); moreover, fuzzy theory focuses more on defining uncertain parameters and its constraints are difficult to weigh (Joonrak et al. 2018). In view of the shortcomings of the above programming approaches, this paper tends to construct robust optimization model of CLSC network.

Multi-period multi-objective planning has a great influence on CLSC network design (Guo et al. 2017; Wu et al. 2015). Rafie-Majd et al. (2018) modeled a multi-period supply chain integrated inventory-location-routing problem. Mohammed et al. (2019) designed a multi-objective CLSC network to optimize both supply chain total cost and carbon emissions. In this paper, a multi period CLSC network model is established to minimize total cost, minimize environmental impact and maximize on-time delivery rate.

### *1.2 Aim and Main Work*

Based on above analysis, a robust optimization model of cross-regional and multi-period CLSC network for perishable products is established and optimized under the objectives of minimizing economic cost, minimizing environmental effect and maximizing on-time delivery. This paper would help perishable products enterprises to obtain strategy of optimal facility location and vehicle route under uncertain environment.

The remainder of the paper is organized as follows: Sect. 2 describes the problem and network structure. A MIP formulation for multi-period multi-objective CLSC network and its robust optimization model are established in Sect. 3. In Sect. 4, the steps of hybrid GA-PSO are introduced. In Sect. 5, a case study based on 6-city network in China and its calculating results are analyzed. Finally, Sect. 6 presents the conclusions and vision of the future.

## 2 Problem Statement and Structure Design

Considering the characteristics of perishable products, the establishment of a cross-regional multi-period multi-objective CLSC network system can accurately grasp the dynamic demand of consumers, location of facilities and optimization of vehicle routes.

The network structure of the CLSC network for perishable products is shown in Fig. 1, which is composed of six parts: pick-up points of consumers, perishable product enterprises, distribution centers, perishable product suppliers, food processing plants and landfills.

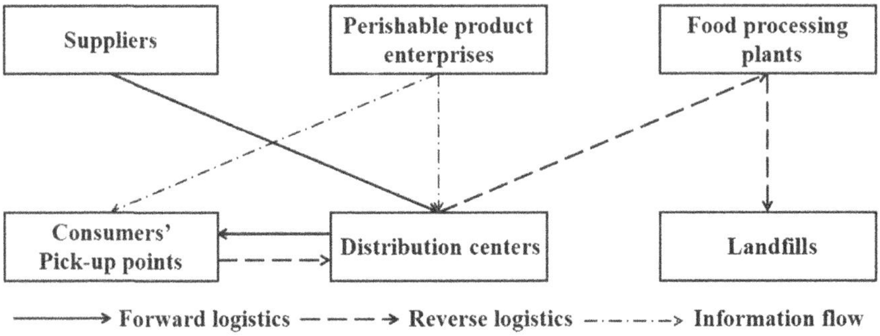


**Fig. 1** Structure of perishable products CLSC network

# 3 Mip Model and Robust Transformation

## 3.1 Assumptions

- The candidate locations and quantities of various facilities are known, and the processing capacity of different nodes of the same facility are similar.
- The processing plants and landfills of returned perishable products are third-party factories, so fixed construction costs of facilities are not necessary to consider.
- In the case of multi-period, the construction cost of facilities is only considered in period 1, and maintenance cost is considered in other periods.
- Each region has at least one distribution center.
- The transportation cost from a certain distribution center to suppliers, processing plants and landfills is linearly related to the transportation volume and distance; the transportation cost from a certain distribution center to a pick-up point and from a pick-up point to another pick-up point is only linearly related to the transportation distance.

## 3.2 Mathematical Formulation

Notations.

| Sets | |
|---|---|
| $i/i'$ | Set of regions,$i/i' \in \{1, 2, \ldots, I\}$ |
| $j/j'$ | Set of candidate consumers' pick-up points,$j \in \{1, 2, \ldots, J\}$ |
| $d$ | Set of distribution centers,$d \in \{1, 2, \ldots, D\}$ |
| $s$ | Set of perishable products suppliers, $s$ $s \in \{1, 2, \ldots, S\}$ |
| $e$ | Set of perishable products enterprises,$e \in \{1, 2, \ldots, \text{E}\}$ |

(continued)

(continued)

| Sets | |
|---|---|
| $i/i'$ | Set of regions, $i/i' \in \{1, 2, \ldots, I\}$ |
| $p$ | Set of food processing plants, $p \in \{1, 2, \ldots, P\}$ |
| $l$ | Set of landfills, $l \in \{1, 2, \ldots, L\}$ |
| $t$ | Set of delivery periods, $t \in \{1, 2, \ldots, T\}$ |
| $v$ | Set of transport vehicles, $v \in V$ |
| $R$ | Set of transport routes, $R = 1, 2, \ldots, R$ |
| *Parameters* | |
| $d_i^t$ | Demand quantity of the region $i$ in the $t$th period |
| $r_i^t$ | Returned quantity of the region $i$ in the $t$th period |
| $Fc_d$ | Fixed construction cost of a certain distribution center |
| $Fc_{ij}$ | Fixed construction cost of the pick-up point $j$ in the region $i$ |
| $Fm_{ij}^t$ | Facility maintenance cost of the pick-up point $j$ in the region $i$ under the $t$th period |
| $Fm_d^t$ | Facility maintenance cost of a certain distribution center in the $t$th period |
| $tc_p$ | Third-party cooperation cost of plants |
| $tc_l$ | Third-party cooperation cost of landfills |
| $op_{ij}^t$ | Operating cost of the pick-up point $j$ in the region $i$ under the $t$th period |
| $op_d^t$ | Operating cost of a certain distribution center in the $t$th period |
| $uip_e$ | Unit information processing costs of the enterprise |
| $ut_{sd}$ | Unit transport cost from the supplier to a certain distribution center |
| $ut_{dij}$ | Unit transport cost from a certain distribution center to the pick-up point $j$ in the region $i$ |
| $ut_{ijd}$ | Unit transport cost from the pick-up point $j$ in the region $i$ to a certain distribution center |
| $ut_{iji'j'}$ | Unit transport cost from a pick-up point to another pick-up point |
| $ut_{dp}$ | Unit transport cost from a certain distribution center to a plant |
| $ut_{pl}$ | Unit transport cost from a certain distribution center to a landfill |
| $dis_{sd}$ | Distance between the supplier and a distribution center |
| $dis_{dij}$ | Distance between a distribution center and the pick-up point $j$ in the region $i$ |
| $dis_{iji'j'}$ | Distance between a pick-up point and another pick-up point |
| $dis_{dp}$ | Distance between a distribution center and a plant |
| $dis_{pl}$ | Distance between a distribution center and a landfill |
| $mc_d$ | Maximum capacity of a distribution center |
| $mc_{ij}$ | Maximum capacity of the pick-up point $j$ in the region $i$ |
| $mc_p$ | Maximum capacity of a plant |

(continued)

(continued)

| Sets | |
|---|---|
| $i/i'$ | Set of regions, $i/i' \in \{1, 2, \ldots, I\}$ |
| $mc_l$<br>$q_{co2}$<br>$C^t_{sd}$ | Maximum capacity of a landfill<br>$co_2$ emission per unit distance of perishable goods with vehicle load per unit weight<br>The percentage of on-time delivery for products which are sent from the supplier $s$ to a distribution center under the $t^{th}$ period |
| $W$ | Maximum cargo capacity of per vehicle |
| $\gamma$ | Reprocessing rate of returned products |
| *Decision variables* | |
| $X^t_{ij}$ | 1 if the pick-up point $j$ in the region $i$ under the $t$th period is open; 0, otherwise |
| $X^t_d$ | 1 if a distribution center in the $t$th period is open, 0, otherwise |
| $X^t_p$ | 1 if a plant in the $t$th period is open, 0, otherwise |
| $X^t_l$ | 1 if a landfill in the $t$th period is open, 0, otherwise |
| $Y^R_{vt}$ | 1 If the route $R$ was transported by vehicle in the $t$th period, 0, otherwise |
| $Y_{dijt}$ | 1 if products are shipped from a distribution center to the pick-up point $j$ in<br>the region $i$ under the $t$th period, 0, otherwise |
| $Y_{iji'j't}$ | 1 if products are shipped from a pick-up point to another pick-up point in the $t$th period, 0, otherwise |
| $Y_{ijdt}$ | 1 if products are shipped from the pick-up point $j$ in the region $i$ to a distribution center under the $t$th period, 0, otherwise |
| $Q^t_{sd}$ | Quantity of products shipped from the supplier to a distribution center in the $t$th period |
| $Q^t_{di}$ | Quantity of products shipped from a distribution center to the region $i$ in the $t$th period |
| $Q^t_{id}$ | Quantity of products shipped from the region $i$ to a distribution center in the $t$th period |
| $Q^t_{dp}$ | Quantity of products shipped from a distribution center to a plant in the $t$th period |
| $Q^t_{dl}$ | Quantity of products shipped from a distribution center to a landfill in the $t$th period |

Minimum total economic cost in single period:

$$\min E_{1\sin} = u_1(\sin) + u_2 + u_3 \tag{1}$$

Minimum total economic cost in multi-period:

$$\min E_{1multi} = u_1(multi) + u_2 + u_3 \tag{2}$$

The fixed cost and facility maintenance cost in single period:

$$u_1(\sin) = \sum_{d=1}^{D} Fc_d X_d + \sum_{j=1}^{J}\sum_{i=1}^{I} Fc_{ij} X_{ij} + \sum_{d=1}^{D}\sum_{s=1}^{S} Fm_d X_d Q_{sd} + \sum_{j=1}^{J}\sum_{i=1}^{I}\sum_{d=1}^{D} Fm^t_{ij} X_{ij}(Q_{di} + Q_{di}) \quad (3)$$

The fixed cost and facility maintenance cost in multi-period:

$$u_1(multi) = \begin{cases} \sum_{d=1}^{D} Fc_d X_d + \sum_{j=1}^{J}\sum_{i=1}^{I} Fc_{ij} X_{ij} & (t = 1) \\ \sum_{d=1}^{D}\sum_{s=1}^{S} Fm_d X_d^{t-1} X_d^t Q_{sd} + \sum_{j=1}^{J}\sum_{i=1}^{I}\sum_{d=1}^{D} Fm^t_{ij} X_{ij}^{t-1} X^t_{ij}(Q_{di} + Q_{di}) & (t \geq 2) \end{cases} \quad (4)$$

Transport cost (both single period and multi-period):

$$u_2 = \sum_{t=1}^{T}\sum_{s=1}^{S}\sum_{d=1}^{D} X^t_d dis_{sd} ut_{sd} Q^t_{sd} + \sum_{t=1}^{T}\sum_{d=1}^{D}\sum_{p=1}^{P} X^t_d X^t_p dis_{dp} ut_{dp} Q^t_{dp} + \sum_{t=1}^{T}\sum_{d=1}^{D}\sum_{p=1}^{P}\sum_{l=1}^{L} X^t_d X^t_p X^t_l dis_{pl} ut_{pl} Q^t_{dl} + \sum_{t=1}^{T}\sum_{d=1}^{D}\sum_{i=1}^{I}\sum_{j=1}^{J}\sum_{v \in V} Y^R_{v \subset t} Y_{dijt} dis_{dij} ut_{dij} + \sum_{t=1}^{T}\sum_{j=1}^{J}\sum_{i=1}^{I}\sum_{v \in V} Y^R_{v \in t} Y_{iji'j't} dis_{iji'j'} ut_{iji'j'} + \sum_{t=1}^{T}\sum_{d=1}^{D}\sum_{i=1}^{I}\sum_{j=1}^{J}\sum_{v \in V} Y^R_{v \in t} Y_{ijdt} dis_{ijd} ut_{ijd} \quad (5)$$

Operating cost, third-party cooperation cost and information processing cost (both single period and multi-period):

$$u_3 = \sum_{t=1}^{T}\sum_{d=1}^{D}\sum_{p=1}^{P} X^t_p tc_p Q^t_{dp} + \sum_{t=1}^{T}\sum_{d=1}^{D}\sum_{l=1}^{L} X^t_l tc_l Q^t_{dl} + \sum_{t=1}^{T}\sum_{s=1}^{S}\sum_{d=1}^{D} X^t_d op^t_d Q^t_{sd} + \sum_{t=1}^{T}\sum_{d=1}^{D}\sum_{j=1}^{J}\sum_{i=1}^{I} X^t_{ij} op^t_{ij} \left(Q^t_{di} + Q^t_{id}\right) + \sum_{t=1}^{T}\sum_{d=1}^{D}\sum_{p=1}^{P} uip_e \left(Q^t_{di} + Q^t_{id}\right) \quad (6)$$

Minimum environmental impact (both single period and multi-period):

$$\min E_2 = \sum_{t=1}^{T}\sum_{s=1}^{S}\sum_{d=1}^{D} X_d^t dis_{sd} q_{co2} Q_{sd}^t + \sum_{t=1}^{T}\sum_{d=1}^{D}\sum_{p=1}^{P} X_d^t X_p^t dis_{dp} q_{co2} Q_{dp}^t$$
$$+ \sum_{t=1}^{T}\sum_{d=1}^{D}\sum_{p=1}^{P}\sum_{l=1}^{L} X_d^t X_p^t X_l^t dis_{pl} q_{co2} Q_{dl}^t$$
$$+ \sum_{t=1}^{T}\sum_{d=1}^{D}\sum_{i=1}^{I}\sum_{j=1}^{J}\sum_{v\in V} Y_{v\in t}^R Y_{dijt} dis_{dij} q_{co2} + \sum_{t=1}^{T}\sum_{j=1}^{J}\sum_{i=1}^{I}\sum_{v\in V} Y_{v\in t}^R Y_{iji'j't} dis_{iji'j'} q_{co2}$$
$$+ \sum_{t=1}^{T}\sum_{d=1}^{D}\sum_{i=1}^{I}\sum_{j=1}^{J}\sum_{v\in V} Y_{v\in t}^R Y_{ijdt} dis_{ijd} q_{co2} \tag{7}$$

Maximum on-time delivery rate (both single period and multi-period):

$$\max E_3 = \sum_{s=1}^{S}\sum_{d=1}^{D} C_{sd}^t Q_{sd}^t \tag{8}$$

s. t.

$$Q_{di}^t \geq \sum_{j=1}^{J} \mathrm{d}_i^t Y_{dijt}, \quad \forall d, i, j, t; \tag{9}$$

$$Q_{id}^t \geq \sum_{j=1}^{J} r_i^t Y_{ijdt}, \quad \forall d, i, j, t; \tag{10}$$

$$\sum_{d}^{D}\sum_{i=1}^{I} Q_{di}^t = \sum_{s=1}^{S}\sum_{d=1}^{D} Q_{sd}^t, \forall t; \tag{11}$$

$$\sum_{i=1}^{I}\sum_{d=1}^{D} \gamma * Q_{id}^t = \sum_{d=1}^{D}\sum_{p=1}^{P} Q_{dp}^t, \forall t; \tag{12}$$

$$\sum_{d=1}^{D}\sum_{l=1}^{L} Q_{dl}^t = \sum_{i=1}^{I}\sum_{d=1}^{D} Q_{id}^t - \sum_{d=1}^{D}\sum_{p=1}^{P} Q_{dp}^t, \forall t; \tag{13}$$

Constraints (9) and (10) guarantee that demand and returned quantities in each region do not exceed the respective freight volume. Constraints (11)–(13) are flow balance constraints.

$$\sum_{j=1}^{J}\left(Y_{iji'j't} + Y_{i'j'ijt}\right) \leq 1, \forall t; \tag{14}$$

$$Y_{ijijt} = 0, \forall t; \tag{15}$$

$$\sum_{j=1}^{J}\sum_{j'=1}^{J'} Y_{vt}^{R} Y_{i'j'ijt} = j, \forall \text{v}, t, R; \tag{16}$$

Constraints (14) and (15) ensure that there is only one route between two different pick-up points. Besides, constraint (16) requires that one pick-up point only belongs to one route in each region.

$$\sum_{v \in V} Y_{vt}^{R} \le 1, \forall t, R; \tag{17}$$

$$\sum_{d=1}^{D}\sum_{j=1}^{J} Y_{vt}^{R} Y_{i'j'ijt} Q_{di}^{t} \le W, \forall \text{v}, i, t, R; \tag{18}$$

Constraints (17) and (18) state that there is only one vehicle for each route, and the total freight volume of each route does not exceed the vehicle cargo capacity.

$$\sum_{d=1}^{D}(Q_{di}^{t} + Q_{id}^{t}) \le \sum_{j=1}^{J} mc_{ij}, \forall \text{v}, t; \tag{19}$$

$$\gamma \sum_{i=1}^{I}\sum_{d=1}^{D} Q_{id}^{t} \le \sum_{p=1}^{P} mc_{p}, \forall t; \tag{20}$$

$$(1-\gamma)\sum_{i=1}^{I}\sum_{d=1}^{D} Q_{id}^{t} \le \sum_{l=1}^{L} mc_{l}, \forall t; \tag{21}$$

$$\sum_{i=1}^{I}(Q_{di}^{t} + Q_{id}^{t}) \le mc_{d}, \forall t; \tag{22}$$

Constraints (19)–(22) concern the limitations in the capacities of pick-up points, distribution centers, plants and landfills.

$$\sum_{j=1}^{J} X_{ij}^{t} \ge 1, \forall t; \tag{23}$$

$$\sum_{d=1}^{D} X_{d}^{t} \ge 1, \forall t; \tag{24}$$

$$\sum_{p=1}^{P} X_p^t \geq 1, \forall t; \tag{25}$$

$$\sum_{l=1}^{L} X_l^t \geq 1, \forall t; \tag{26}$$

$$\sum_{R} \sum_{v \in V} Y_{vt}^R \geq 1, \forall t; \tag{27}$$

Constraints (23)–(26) guarantee that there are at least one pick-up point, one distribution center, one plant and one landfill are selected in per region; Constraint (27) indicates that there is at least one route for the distribution of the whole closed-loop supply chain network.

## 3.3 Robust Optimization Model Design

To effectively avoid the influence of uncertain parameters, the cardinality constrained robust approach is adopted to convert deterministic constraints into robust correspondences.

Assume that the demand $d_i^t$ of each node is evaluated on the uncertain set of box which is equivalent to a bounded set of symmetric intervals. $\widetilde{d_i^t} \in \left[d_i^t - \widehat{d_i^t}, d_i^t + \widehat{d_i^t}\right]$, where $d_i^t$ represents the regular value of the demand for the $t$th period,$\widehat{d_i^t}$ indicates the maximum deviation of the demand for the $t$th period. In constraint (9), the uncertainty budget parameter $\Gamma_{it}^{\mathrm{d}}$ is introduced to adjust the conservativeness of the result, where $\Gamma_{it}^{\mathrm{d}} \in [0, m]$, $m$ represents the number of demand uncertainties of pick-up points in each period. In the actual closed-loop supply chain of perishable product, not all demand is uncertain. The robust optimization method constructed in this paper is to guarantee that there are at most $\Gamma_{it}^d * \widetilde{d_i^t}$ fluctuations within the interval. The deviation from demand quantity at region $i$ is $\left(\Gamma_{it}^d - \Gamma_{it}^d\right)\widehat{d_i^t}$. Thus, constraint (9) can be converted to:

$$Q_{di}^t \geq \sum_{j=1}^{J} d_i^t Y_{dijt} + u\left(d_i^{t*}, \Gamma_{it}^d\right), \forall d, i, t; \tag{28}$$

where,

$$u\left(d_i^{t*}, \Gamma_{it}^d\right) = \max_{\{S_{it} \cup \{g_{it}\} | S_{it} \subseteq J_{it}, |S_{it}| \leq \lfloor \Gamma_{it}^d \rfloor, o_{it} \in J_{it} \setminus S_{it}\}} \left\{ \sum_{j \subseteq S_{it}}^{J} \widehat{d_i^t} Y_{dijt} + \left(\Gamma_{it}^d - \Gamma_{it}^d\right) \widehat{d_i^t} Y_{dio_{it}t}^R \right\}.$$

$u\left(\mathrm{d}_i^{t*}, \Gamma_{it}^d\right)$ can be linearly transformed into the following:

$$u\left(d_i^{t*}, \Gamma_{it}^d\right) = \max\left\{\widehat{d_i^t} Y_{dijt} \beta_i^d : \sum_{i=1}^{I} \beta_i^d \le \Gamma_{it}^d, 0 \le \beta_i^d \le 1, \forall i \in I\right\} \tag{29}$$

Note: here $\lfloor * \rfloor$ indicates that the value $*$ is rounded down, $S$ indicates the set of nodes deviating from the mean, $u\left(d_i^{t*}, \Gamma_{it}^d\right)$ is protection function, and $\beta_i^d$ is the auxiliary variable.

Enterprises can adjust the robustness of the objective function values by adopting the value of $\Gamma_{it}^{\mathrm{d}}$. Since constraint (24) is bounded, according to the dual theory, the dual problem (Eq. (26)) is bounded too and the results are also consistent.

$$\min\{\rho_i^t \Gamma_{it}^d + \sum_{i=1}^{I} \sigma_i^t : \rho_i^t + \sigma_i^t \ge \widehat{d_i^t} Y_{dijt}, \rho_i^t \ge 0, \sigma_i^t \ge 0, \forall d, j, t \tag{30}$$

According to constraint (28)–(30), the robust equivalences of constraint (9) related to demand quantity are as follows:

$$Q_{di}^t \ge \sum_{j=1}^{J} d_i^t Y_{dijt} + \rho_i^t \Gamma_{it}^d + \sum_{i=1}^{I} \sigma_i^t, \forall i, d, t \tag{31}$$

$$\rho_i^t + \sigma_i^t \ge \widehat{d_i^t} Y_{dijt}, \forall i, j, d, t \tag{32}$$

$$\rho_i^t \ge 0, \sigma_i^t \ge 0, \forall i, t \tag{33}$$

Similarly, constraint (10) related to returned quantity is transformed into the following robust correspondences.

$$Q_{id}^t \ge \sum_{j=1}^{J} \mathrm{r}_i^t Y_{ijdt} + \varphi_i^t \Gamma_{it}^{\mathrm{d}} + \sum_{i=1}^{I} \omega_i^t, \forall i, d, t \tag{34}$$

$$\varphi_i^t + \omega_i^t \ge \widehat{r_i^t} Y_{ijdt}, \forall i, j, d, t \tag{35}$$

$$\varphi_i^t \ge 0, \omega_i^t \ge 0, \forall i, t \tag{36}$$

# 4 Hybrid Heuristic Algorithm

Due to the various candidate nodes and constraints, uncertain parameters, multi-period and other factors in the perishable CLSC network, this paper is no longer a simple MIP model, but an NP-hard problem. Hybrid heuristic algorithms such as GA-PSO are generally recommended in solving this kind of model.

The main calculation process is as following:

Step 1: Coding.

In this paper, the representation of a chromosome is *(IJ + D + P + L + V(R) + DIJ(R) + IJI'J'(R) + IJD(R) + SD + DIJ + IJI'J' + IJD + DP + DL)*. The first four parts of chromosome are binary string (0 or 1), representing the open of various facilities. The fifth to eighth parts of chromosome are also 0–1 binary string, which shows that vehicle V and logistics distribution (DIJ + IJI'J' + IJD) are all transported on route R. And the last six parts refer to the quantities transported between nodes.

Step 2: Fitness assessment.

The fitness of chromosome reflects the probability of selection. In the case of satisfying the formulas (9–27) and (28–33), the fitness value of individual is calculated by using the objective functions (1–8), and can directly reflect the merits and demerits of the individual.

Step 3: Selection operation.

The selection operation is to pick up the respective individuals to the next generation. Using roulette strategy for selection can avoid the omission of local optimal solution, and improve the global convergence.

Step 4: Optimization operation.

Set $M$ as particle size, randomly initialize the particles for searching, and each particle represents an optimal solution. Assume that the position of the particle after $T$th iterations is $X_i = (x_{i1}, x_{i2}, \ldots, x_{in})^{\mathrm{T}}$, the velocity is $V_i = (v_{i1}, v_{i2}, \ldots, v_{in})^{\mathrm{T}}$. Take the position of the particles after iteration into the objective functions (1–8), then select the personal optimal location ($P_{best}$) and global optimal position ($G_{best}$), that is, $P_i = (p_{i1}, p_{i2}, \ldots, p_{in})^{\mathrm{T}}$ and $P_g = \left(p_{g1}, p_{g2}, \ldots, p_{gn}\right)^{\mathrm{T}}$. Constraints on update position and velocity of particles are as follows:

$$v_{in}^{k+1} = w*v_{in}^{k+1} + c_1 r_1*\left(p_{in} - x_{in}^{k}\right) + c_2 r_2*\left(p_{gn} - x_{in}^{k}\right) \tag{37}$$

$$x_{in}^{k+1} = x_{in}^{k} + v_{in}^{k+1} \tag{38}$$

$$1 \le i \le n;\ 1 \le n \le N \tag{39}$$

Step 5: Terminal condition.

Terminates the algorithm when particles reach the optimal fitness value, otherwise, turns to step 2.

# 5 Case Study

## 5.1 Data Description

The data of this model comes from a case study of vegetable enterprise in Shanghai, P.R. China. The distribution area covers Shanghai, Suzhou, Wuxi, Changzhou, Jiaxing and Huzhou. It is assumed that the single period length of vehicle distribution is one week, while the multi- period is divided into four cycles.

According to the investigation of the enterprise, the coordinates obtained by google map of the candidate distribution centers are respectively (145.1, 60.5), (160.9, 69.4) and (153.0, 46.8); the coordinates of selected vegetable suppliers is (188.2, 30.5); the coordinates of candidate food processing plants are (192.7, 43.4) and (179.4, 34.8); the coordinates of candidate landfills are (183.6, 23.9) and (35.6,203.1); the coordinates of 11 candidate consumers' pick-up points, the predicted demand and returned quantities for each period are shown in Table 1. The values of uncertain budgets ($\Gamma^{\mathrm{d}}_{it}$ and $\Gamma^{\mathrm{r}}_{it}$) are assigned 0, 4, and 8, respectively. Other parameter values are shown in Table 2.

**Table 1** Pick-up point coordinates, demanded and returned quantities at the certain region in each period

| | Coordinates (x, y) | | $d_i^1$ | $d_i^2$ | $d_i^3$ | $d_i^4$ | $r_i^1$ | $r_i^2$ | $r_i^3$ | $r_i^4$ |
|---|---|---|---|---|---|---|---|---|---|---|
| | $J_1$ | $J_2$ | | | | | | | | |
| $I_1$(Shanghai) | (129.0, 66.9) | (132.6, 33.6) | 91.8 | 47 | 51 | 63.9 | 0 | 11.8 | 8.4 | 8.8 |
| $I_2$(Jiaxing) | (88.9, 6.4) | (111.9, 7.4) | 62.8 | 89.6 | 0 | 74.4 | 0 | 10.7 | 0 | 7.2 |
| $I_3$(Huzhou) | (22.1, 20.8) | / | 58.5 | 0 | 69.0 | 36.0 | 0 | 0 | 3.7 | 5.9 |
| $I_4$(Changzhou) | (12.1, 107.4) | (29.5, 120.6) | 64.2 | 71 | 56 | 0 | 0 | 10.9 | 9.2 | 0 |
| $I_5$(Wuxi) | (44.6, 89.5) | (61.1, 97.8) | 69.6 | 92.6 | 0 | 85.2 | 0 | 11.9 | 0 | 6.6 |
| $I_6$(Suzhou) | (60.8, 62.7) | (88.3, 71.4) | 59.4 | 73.4 | 56.5 | 69.2 | 0 | 6.1 | 10.7 | 8.5 |

**Table 2** The relevant parameter data

| Parameters | Value | Parameters | Value | Parameters | Value |
|---|---|---|---|---|---|
| $Fc_d$ | 100,000 CNY | $op_d^t$ | 60 CNY.t$^{-1}$ | $uip_e$ | 20 CNY ·t$^{-1}$ |
| $Fc_{ij}$ | 50000 CNY | $mc_d$ | 800 t | $ut_{dij}$ | 50 CNY ·km$^{-1}$ |
| $Fm_d^t$ | 5 CNY ·t$^{-1}$ | $mc_{ij}$ | 150 t | $ut_{ijd}$ | 60 CNY ·km$^{-1}$ |
| $Fm_{ij}^t$ | 2 CNY ·t$^{-1}$ | $mc_p$ | 200 t | $ut_{iji'j'}$ | 60 CNY ·km$^{-1}$ |
| $tc_p$ | 15 CNY ·t$^{-1}$ | $mc_l$ | 100 t | $ut_{sd}$ | 4 CNY ·(km · t)$^{-1}$ |
| $tc_l$ | 12 CNY ·t$^{-1}$ | $W$ | 250 t | $ut_{dp}$ | 5 CNY ·(km · t)$^{-1}$ |
| $op_{ij}^t$ | 20 CNY ·t$^{-1}$ | $\gamma$ | 60% | $ut_{pl}$ | 5 CNY ·(km · t)$^{-1}$ |
| $q_{co2}$ | 150 g ·km$^{-1}$ | $C_{sd}^t$ | 85% | | |

**Table 3** The comparison between multi-period and single period under different robust levels

| | Single period | Total multi-period |
|---|---|---|
| case1($\Gamma_{it}^{d}=0, \Gamma_{it}^{r}=0$) | (1042879,241866,235.6) | (864764,221866,270.2) |
| case2($\Gamma_{it}^{d}=0, \Gamma_{it}^{r}=4$) | (1055148,246599,234.9) | (881847,226599,268.8) |
| case3($\Gamma_{it}^{d}=0, \Gamma_{it}^{r}=8$) | (1067436,258224,233.5) | (908252,230104,267.3) |
| case4($\Gamma_{it}^{d}=4, \Gamma_{it}^{r}=0$) | (1094083,265490,230.8) | (978491,233613,261.1) |
| case5($\Gamma_{it}^{d}=4, \Gamma_{it}^{r}=4$) | (1100003,273561,229.3) | (993893,236636,260.3) |
| case6($\Gamma_{it}^{d}=4, \Gamma_{it}^{r}=8$) | (1109936,281417,227.8) | (1004505,240121,259.2) |
| case7($\Gamma_{it}^{d}=8, \Gamma_{it}^{r}=0$) | (1143895,289097,218.4) | (1044856,242853,248.4) |
| case8($\Gamma_{it}^{d}=8, \Gamma_{it}^{r}=4$) | (1150006,296422,217.9) | (1055508,246535,243.9) |
| case9($\Gamma_{it}^{d}=8, \Gamma_{it}^{r}=8$) | (1163994,309898,216.6) | (1068497,251177,237.4) |

## 5.2 *Results and Analyses*

The hybrid GA-PSO are coded by software MATLAB 2018b. Then, the comparison between multi-period and single period under different robust levels are shown in Table 3.

(1) Analysis of variation of robustness
Observing the vertical comparison of 9 kinds of robust levels, with the increase of the uncertain budget $\Gamma_{it}^{d}$ and $\Gamma_{it}^{r}$, the deviations of demand and return of both multi-period and single period are increasing, thus, the total system cost is increasing too.
In Fig. 2, the quantities of returned perishable products are only 5%-25% of the demand, so the amount of demand deviation is greater than return deviation, which makes the robustness and rising trend of demand more significant, as shown in the case3–4 and case6–7.

(2) Analysis of multi period optimization
As can be seen from Table 4, Figs. 3 and 4, the demand and returned quantities of subdivided unit period are more targeted for facilities location, thereby the multi-period total cost is less than single period.
In addition, in the most conservative case (case9), the number of multi-period facilities and the delivery vehicles selected are less than the single period, as shown in Figs. 3 and 4.

(3) Analysis of multi objective optimization

From the perspective of multi-objective analysis, it can be concluded that when the three sub objectives are optimized at the same time, the performance of each sub objective has made some concessions compared with the single objective optimization, but the overall optimization result is better than the single objective optimization.

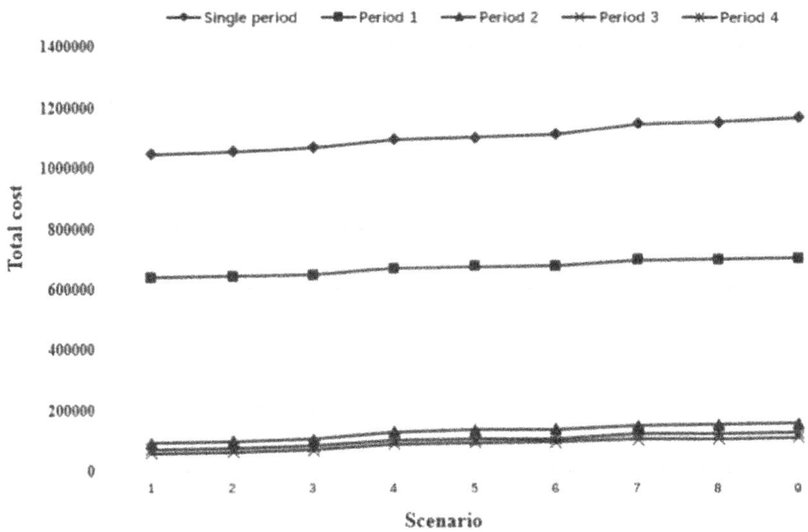


**Fig. 2** Total cost variation at different robust levels

**Table 4** The location at different robust levels under the calculation of GA-PSO

| | Single period | Multiple period | | | |
|---|---|---|---|---|---|
| | | Period 1 | Period 2 | Period 3 | Period 4 |
| Case 1 | $D_1, D_3, P_2, L_1$ | $D_3, P_2, L_1$ | $D_3, P_2, L_1$ | $D_3, P_2, L_1$ | $D_3, P_2, L_1$ |
| Case 2 | $D_1, D_3, P_2, L_1$ | $D_3, P_2, L_1$ | $D_3, P_2, L_1$ | $D_3, P_2, L_1$ | $D_3, P_2, L_1$ |
| Case 3 | $D_1, D_3, P_2, L_1$ | $D_3, P_2, L_1$ | $D_3, P_2, L_1$ | $D_3, P_2, L_1$ | $D_3, P_2, L_1$ |
| Case 4 | $D_1, D_3, P_2, L_1$ | $D_1, P_1, L_2$ | $D_3, P_2, L_1$ | $D_3, P_2, L_1$ | $D_3, P_2, L_1$ |
| Case 5 | $D_1, D_3, P_2, L_1$ | $D_1, P_1, L_2$ | $D_3, P_2, L_1$ | $D_3, P_2, L_1$ | $D_3, P_2, L_1$ |
| Case 6 | $D_1, D_3, P_2, L_1$ | $D_1, P_1, L_2$ | $D_1, P_1, L_2$ | $D_1, P_1, L_2$ | $D_1, P_1, L_2$ |
| Case 7 | $D_1, D_3, P_1, P_2, L_1, L_2$ | $D_1, P_1, L_2$ | $D_1, P_1, L_2$ | $D_1, P_1, L_2$ | $D_1, P_1, L_2$ |
| Case 8 | $D_1, D_3, P_1, P_2, L_1, L_2$ | $D_1, P_1, L_2$ | $D_1, P_1, L_2$ | $D_1, P_1, L_2$ | $D_1, P_1, L_2$ |
| Case 9 | $D_1, D_3, P_1, P_2, L_1, L_2$ | $D_1, P_1, L_2$ | $D_1, P_1, L_2$ | $D_1, P_1, L_2$ | $D_1, P_1, L_2$ |

## 6 Conclusions

In this paper, perishable products are set as research objects. In order to obtain the decision-making of the optimal facilities locations and vehicle routes, a multi-period multi-objective CLSC network and its corresponding MIP model, and robust optimization plan are established. The feasibility of the model is verified by a hybrid GA-PSO with a calculating case study of a cross-region multi-period multi-objective CLSC network in Eastern China. Finally issues such as the quality uncertainty

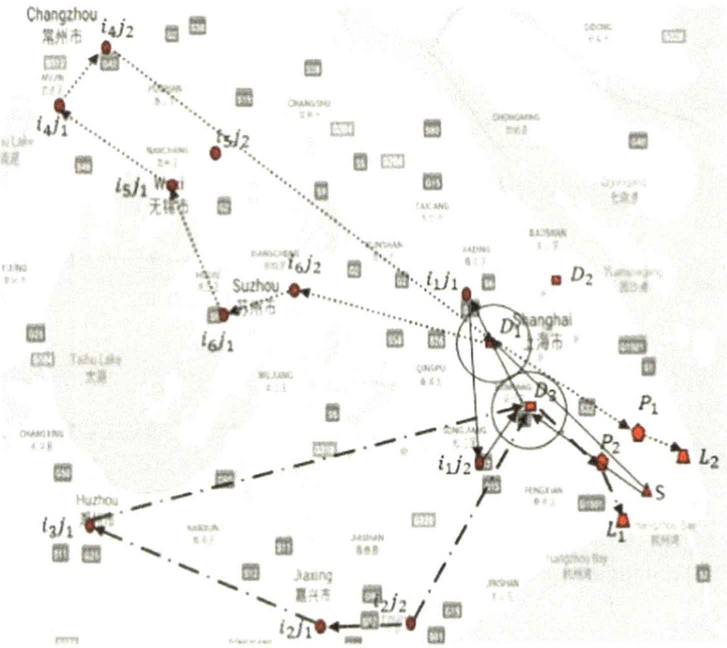


**Fig. 3** Single period route diagram at case 9 (GA)

of returned product and uncertainty of network response time, which are more appropriate to real-world situations, are left for further exploration.

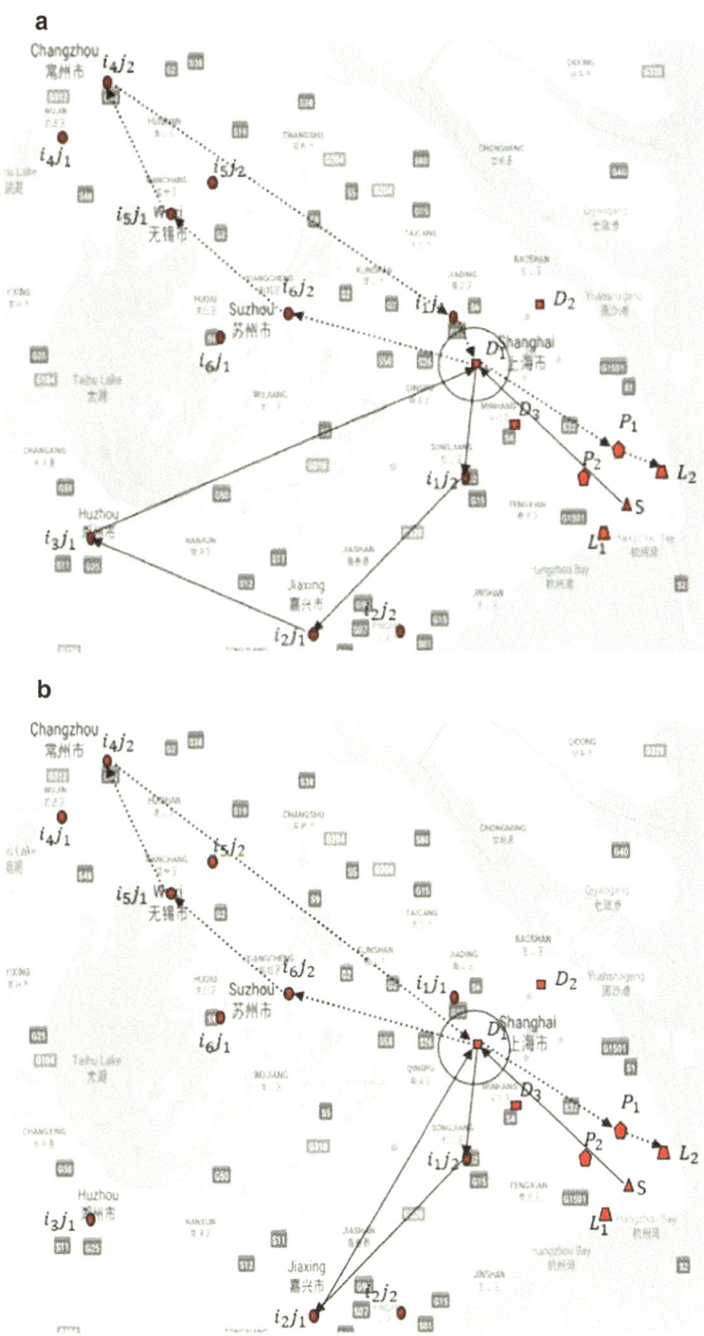


**Fig. 4** Multi-period route diagram at case 9 (GA-PSO) (a-d)

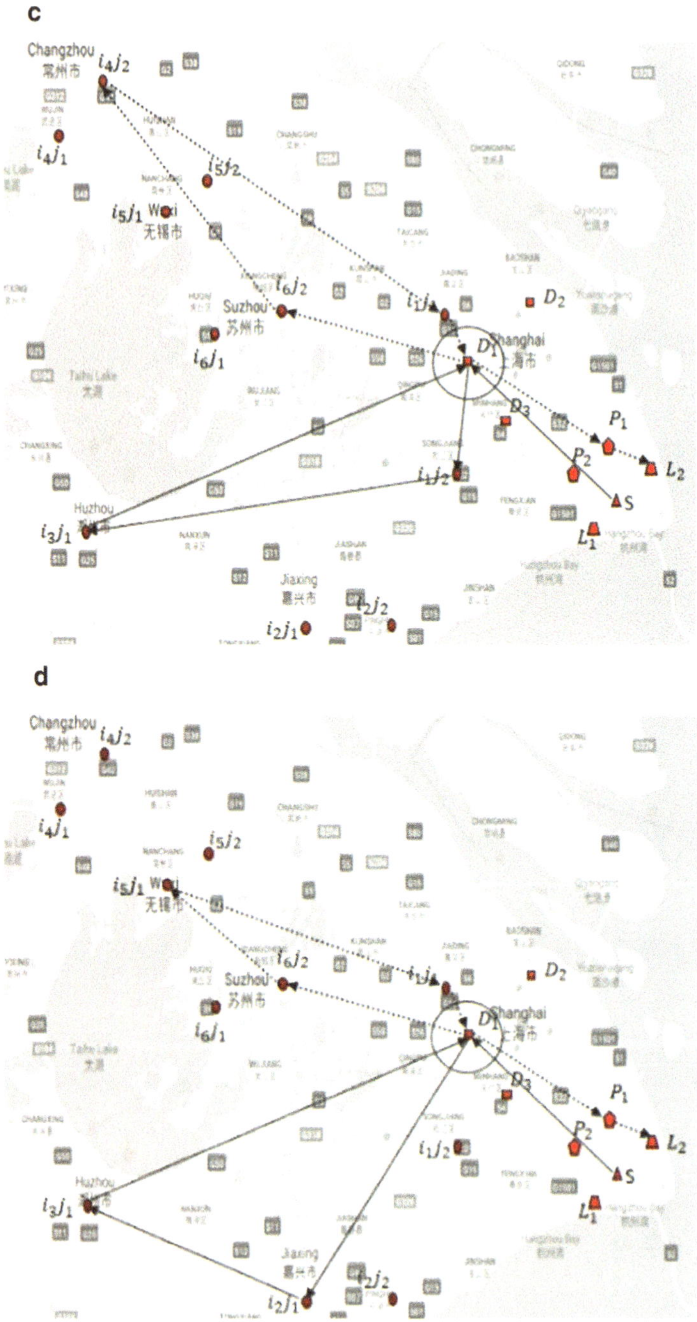


**Fig. 4** (continued)

## References

1. Accorsi, R., Gallo, A., Manzini, R.: A climate driven decision-support model for the distribution of perishable products. J Clean Prod **165**, 917–929 (2017)
2. Adenso-Díaz, B., Lozano, S., Palacio, A.: Effects of dynamic pricing of perishable products on revenue and waste. Appl Math Model **45**, 148–164 (2017)
3. Gou, J., Yang, T.: Formation mechanism of tetragonal nanoprecipitates in Fe–Ga alloys that dominate the material's large magnetostriction. Scripta Mater. **185**, 129–133 (2020)
4. Gretel, B., José, F.: A PSO-based algorithm for mining association rules using a guided exploration strategy. Patt Recogn Lett **138**, 8–15 (2020)
5. Alshamsi, A., Diabat, A.: A reverse logistics network design. Journal of Manufacturing Systems **37**(3), 589–598 (2015)
6. Amin, S.H., Zhang, G.: A multi-objective facility location model for closed-loop supply chain network under uncertain demand and return. Appl Math Model **37**(6), 4165–4176 (2013)
7. Amin, S.H., Zhang, G., Akhtar, P.: Effects of uncertainty on a tire closed-loop supply chain network. Expert. Syst. Appl. **73**, 82–91 (2017)
8. Asim, Z., Jalil, S.A., Javaid, S.: An uncertain model for integrated production-transportation closed-loop supply chain network with cost reliability. Sustain Prod Consump **17**, 298–310 (2019)
9. Azadeh, A., Elahi, S., Farahani, M.H., Nasirian, B.: A genetic algorithm-Taguchi based approach to inventory routing problem of a single perishable product with transshipment. Comput Ind Eng **104**, 124–133 (2017)
10. Cheng, J., Liu, Z., Tang, M., Tan, J.: Robust optimization of uncertain structures based on normalized violation degree of interval constraint. Comput Struct **182**, 41–54 (2017)
11. Cui, Y., Guan, Z., Saif, U., Zhang, L., Zhang, F., Mirza, J.: Close loop supply chain network problem with uncertainty in demand and returned products: genetic artificial bee colony algorithm approach. J Clean Prod **162**, 717–742 (2017)
12. Li, Y., Zobel, C.W.: Exploring supply chain network resilience in the presence of the ripple effect. Int J Prod Econ **228**, 107693 (2020)
13. Huang, Y.S., Fang, C.C., Lin, Y.A.: Inventory management in supply chains with consideration of Logistics, green investment and different carbon emissions policies. Comput Ind Eng **139**, 106207 (2020)
14. Guo, J., Liu, X., Jo, J.: Dynamic joint construction and optimal operation strategy of multi-period reverse logistics network: a case study of shanghai apparel e-commerce enterprises. J Intell Manuf **28**(3), 819–831 (2017)
15. Guo, J., Wang, X., Fan, S., Gen, M.: Forward and reverse logistics network and route planning under the environment of low-carbon emissions: a case study of shanghai fresh food e-commerce enterprises. Comput Ind Eng **106**, 351–360 (2017)

# City Water Resources Vulnerability: The Case of Jinan and Qingdao in Shandong Province, China

**Menglu Sun and Takaaki Kato**

**Abstract** In this paper, I want to describe the water resource vulnerability index of the two biggest cities in Shandong Province in China, Qingdao and Jinan, which represent the coastal and the inland areas of Shandong respectively. The period we study from 2008 to 2018. The Water Resources Vulnerability Index (WRVI) is according to the DPSIR (Drive, Pressure, State, Impact, and Response) framework, an index is established to qualify water resource vulnerability through water scarcity, water stress, water pollution, and water production efficiency. The result indicates that Climate change has a bigger impact on WRVI. Qingdao is more vulnerable than Jinan.

**Keywords** Water resources vulnerability index (WRVI) · Qingdao and Jinan city · Comparison

## 1 Introduction

In view of industrial and agricultural economy, Shandong Province holds an important position in China. Shandong, as a big province of industry and agriculture, ranks the third in China's GDP and first in the North China. With developed industry and agriculture, Shandong has a large population. During the primary stage of industrial development, it consumes a large amount of natural resource and damages the environment greatly. Water resources in Shandong Province are scarcely compared with other regions, for example up to 12% of the fruits and 13% of the vegetables in China were produced by Shandong Province using only 1% of China's water resources.

M. Sun (✉)
Graduate School of Environmental Engineering, University of Kitakyushu, 808-0135 Kitakyushu, Japan
e-mail: a9dac401@eng.kitakyu-u.ac.jp

T. Kato
Faculty of Environmental Engineering, University of Kitakyushu, 808-0135 Kitakyushu, Japan



F. K. S. Chan et al. (eds.), *Proceedings of the 2020 International Conference on Resource Sustainability: Sustainable Urbanisation in the BRI Era (icRS Urbanisation 2020)*, Environmental Science and Engineering,
https://doi.org/10.1007/978-981-15-9605-6_3

Economic development imposes great pressure on the water environment. Comprehensive studies on city water resources vulnerability are important to sustainable development in China, especially for Shandong Province.

There are huge local characters and climatic differences in Shandong. It is divided into two area, coastal areas and inland areas. As for moisture, Coastal areas is relative moisture, Evaporation is less than that in inland areas. As for large rivers that can store water, coastal area is lack of. Yellow river, the second large river in China is located in inland areas. Precipitation is relative same in Shandong. Relative unbalanced economic development between regions appears in Shandong. In coastal area, industry is more developed, more scarce water resources, more water resources contradiction.

To make this study more inclusive, we study the two biggest cities in Shandong Province as case study, Qingdao and Jinan, which represents the coastal and the inland areas of Shandong respectively. Water resources are vulnerable to climate change, especially precipitation. Since 2010, Shandong has suffered from drought. As a result, water resources in Qingdao decreased dramatically from 20 billion $m^3$ in 2011 to 2.87 billion $m^3$ in 2015. However, water consumption in Qingdao remains at 9 billion $m^3$ and has changed little over recent years. In contrast, water consumption in Jinan is almost 16 billion $m^3$ annually, which is much larger than that of Qingdao.

Due to the complexity of its internal mechanisms and diversity of regional environmental systems (including natural, social and economic environments), the water is a kind of environmental properties as an integral part of ecological environment, we consider vulnerability with respect to sustainable water resources only. In general, the vulnerability of a natural and socio-economic system can be determined by the character, magnitude, rate of a threat's development, and by the system's sensitivity and adaptive capacity on the other. Water resource vulnerability can be defined as the weaknesses and flaws of water resource systems that make it easy for the system to collapse [1], or total annual withdrawals as a percentage of annual available water resource [2].

Considering social factors, there are many indexes to evaluate the water environment, for example: the Water Poverty Index [3], the Canadian Water Sustainability Index [4], and the Watershed Sustainability Index [5]. Adoption of indicators is useful for the evaluation and monitoring of progress towards sustainable development. [6]. There are many indicators that not only measure natural indicators, but also indicators that take social factors into account are more powerful for measuring sustainable development, because the relationship between human and nature is interactive and coordinated.

Huang and Cai [1] propose a new index system named the Vulnerability Assessment of Water Resources (VAWR), which contains a limited number of indicators to assess. It is a systematic and analytic approach based on an indicator-component-index system for evaluating the sensitivity of the water resource system to its potential threats. Cai et al. (2017) studied China's water resources vulnerability during the period of 2003–2013. They found that water resources in the northern and central coasts are more vulnerable [7]. Natural environmental resources are fragile economic development has an inevitable influence on it.

We use the Driver Pressure State Impact and Response (DPSIR) model [8]. The model can capture cause and effect relationships between the social economic and environmental systems. It is a sustainability is and evaluation framework. In addition, the indicator-component index system is also used to find water resources in two main cities in Shandong Province. In the stable equilibrium of a subsystem, a small but a gradual change might rapidly lead to the collapse of the whole system.

We think indicators can be used to effectively measure the state of the water scarcity and its factors. The application of indicators of water use and management can contribute to a better allocation of this limited resource and have important impact on policy making. The relevance of indicators for the decision-making process is one of the most important features of the indicators in relation to other forms of information. In the paper we calculate indicators for two water scarce cities in Shandong Province in order to measure the environmental situation and provide information for a response to make social and environmentally sustainable development.

## 2 Methodology and Data

### *2.1 Rational*

We chose water environmental vulnerability because it is important to measure the balance between economic and environmental development. Vulnerability is the degree to which a system is susceptible to, or unable to recover with, the adverse effects of environmental changes [9]. Water environmental capacity and water resources constitute the foundation of the water vulnerability. It represents the self-sustaining, self-regulating and self-purification ability of water environment system. Water self-purification capacity is quantified in the form of water environmental capacity, which is the basis of the carrying capacity of water environment. Climate change is why assessments on water resources and water supply vulnerability, estimation of water scarcity, and analyses of droughts are necessary tools

As for framework can be developed and organized for the evaluation of sustainability. The DPSIR Driving Forces–Pressure–State–Impact–Response approach is the most widely used framework applied for environmental indicators [10, 11]. DPSIR framework designed to provide a more comprehensive approach to the analysis of environmental problems.

DPSIR is based on the pressure-state-response (PSR) introduced by the OECD (1994) [12]. The DPSIR framework organizes the indicators according to the cause–effect schema under these: Drive Forces, Pressure, State, Impact and Response. The adoption of a framework is especially important in the case of indicators related to sustainable development, which encompasses many subjects and dimensions [13].

This system's view of analysis is that economic and social development, which is common driving forces (D). It is social, demographic and economic developments in societies and the corresponding changes in lifestyles which is overall levels of consumption and production. Patterns exert pressure (P). It is developments bring pressure in using of resources and the use of land on the environment, changes in the state (S). It represents indicator of condition of different environmental compartments and systems. Changes have impacts (I) Impacts on human beings, ecosystems and man-made capital resulting from changes in environmental quality. Because of these impacts, society responds (R) to these driving forces, or acts directly upon the pressure, state or impacts through preventive, adaptive, or curative solutions. Responses by groups (and individuals) and government in society attempts to prevent, compensate and adapt to changes in the state of the environment policy response options.

In the paper, we focus on S state and use indicators to measure the water situation, impacts of economic development, and the pressure of environment. Economy growth is driving forces. At the same time due to economy and population growth, environment suffer more and more pressure (P). Understanding present situation is essential to Government and company to respond (R).

## 2.2 *Indicators and Data*

According to the DPSIR framework and indicator-component index system, an index is established to qualify water resource vulnerability through water scarcity, water stress, water pollution, and water production efficiency. The different elements are combined for calculation water scarcity, water stress, water pollution, and water production efficiency and the Water Resources Vulnerability Index (WRVI), like: annual per capita water resources total water supply; annual total water resources total wastewater discharge annual gross domestic product (GDP) The Water Resources Vulnerability Index (WRVI) can be expressed as [7]:

$\mathrm{WRVI} = (\mathrm{WSc} + \mathrm{WSt} + \mathrm{WSt} + \mathrm{WPr})/4$

WSc is Water scarcity. WPo is Water pollution Water stress (WSt) is urban water consumption as a proportion of the total amount of water resources. Water pollution (WPo) is the proportion of sewage emissions of total water resources. Water productivity (WPr) is water demand per unit of GDP

### 2.2.1 Water Scarcity

Water scarcity (WSc) is Indicators to measure per capita water resources. its equation can be presented as

$\mathrm{WSc} = (1700 - \mathrm{WRP})/1700\ (\mathrm{WRP} \leq 1700)$

$\mathrm{WSc} = 0\ (\mathrm{WRP} > 1700)$

As for WSc With regard to the value of 1700, it is a threshold value proposed by Falkenmark (1989). It indicates that there is no vulnerability in water scarcity if the

WRP value reaches above 1700 $m^3$/person. WRP—annual per capita water resources ($m^3$/person).

### 2.2.2 Water Stress

Water stress (WSt) is Urban water consumption as a proportion of the total amount of water resources.

WSt = Wsu/WR

WSu—annual total water supply ($m^3$). WR—annual total water resources ($m^3$).

### 2.2.3 Water Pollution

WPo = WW/WR/0.1 (WW < 0.1 × WR).

WPo = 1 (WW ≥ 0.1 × WR)

WR—annual total water resources ($m^3$); WW—annual total wastewater discharge ($m^3$); As for WPo where WW is annual total wastewater discharge ($m^3$). Regarding the value of 0.1, 1 unit of wastewater can make approximately 10 unit of unpolluted water totally unusable.

### 2.2.4 Water Productivity

WPr = (40 − GDPWW)/40(GDPWW ≤ 40)

WPr = 0(GDPWW > 40)

GDPWW—annual gross domestic product (GDP) in constant prices divided by annual total water withdrawal (RMB (yuan)/$m^3$); GDPWW—the global average WP (RMB(yuan)/$m^3$).

Under DPSIR framework, we studied the period from 2008 to 2018 to calculate the water resource vulnerability index. The water resource data come from Shandong Water Resource Bulletin 2015–2018, Jinan Water Resource Bulletin 2011–2014, and Shandong Statistic Yearbook 2008–2011. Water consumption data, population data, and GDP data come from the Jinan Statistic Yearbooks 2009–2019 and the Qingdao Statistic Yearbooks 2009–2019.

## 3 Results and Discussion

To facilitate the harmonious development of the water ecological environment and social economy the current management system for domestic basin resources and the environment aim to manage the resources and environment within the watershed boundary as a unit.

## 3.1 Water Resource Vulnerability Situation in Qingdao

Table 1 and Fig. 1 are water resource vulnerability situation in Qingdao. Water scarcity (WSc) keep a high level. Average is 0.91693. Water stress (WSt) is unstable and varies with precipitation. Water pollution (WPo) is unstable, it also varies with precipitation. Water productivity (WPr) keep high level.

**Table 1** Water resource vulnerability index in Qingdao

| Year | WSc | WSt1 | Wpo | Wpr | WRVI |
|---|---|---|---|---|---|
| 2008 | 0.815935 | 0.123488 | 0.125369 | 0.778942 | 0.460934 |
| 2009 | 0.935642 | 0.360108 | 0.383293 | 0.796863 | 0.618976 |
| 2010 | 0.955701 | 0.531893 | 0.581361 | 0.828493 | 0.724362 |
| 2011 | 0.860049 | 0.163006 | 0.196688 | 0.853399 | 0.518286 |
| 2012 | 0.918422 | 0.316649 | 0.347968 | 0.867867 | 0.612727 |
| 2013 | 0.907271 | 0.271257 | 0.33402 | 0.870019 | 0.595642 |
| 2014 | 0.918263 | 0.371114 | 0.404694 | 0.879103 | 0.643293 |
| 2015 | 0.981442 | 1 | 1 | 0.907521 | 0.972241 |
| 2016 | 0.95833 | 0.660015 | 0.79043 | 0.915458 | 0.831058 |
| 2017 | 0.935481 | 0.451335 | 0.524253 | 0.921343 | 0.708103 |
| 2018 | 0.899694 | 0.312819 | 0.366555 | 0.92226 | 0.625332 |
| Average | 0.91693 | 0.414699 | 0.459512 | 0.867388 | 0.664632 |

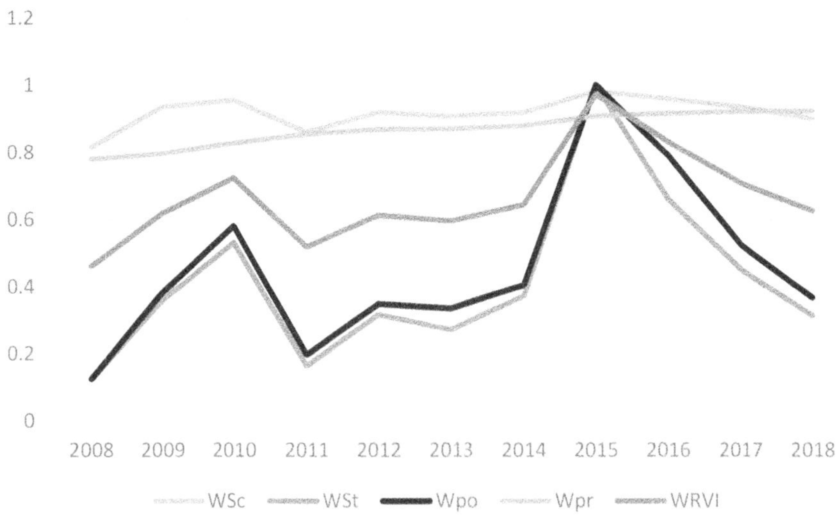


**Fig. 1** Trend of water resource vulnerability situation in Qingdao between 2008 and 2018

Water scarcity(WSc) and Water stress(WSt) represent natural water scarcity. Water scarcity (WSc) keep high level. The water pressure range is larger, which is easily affected by the climate. As for water quality, the change of Water pollution range is bigger, because Water pollution emissions remain basically remain unchanged, but due to climate factors, the low rainfall, Water purification, poor ability of Water pollution to Water vulnerability has a greater impact. Water production remains stable and water consumption required for economic development remains constant.

### 3.2 *Water Resource Vulnerability Situation in Jinan*

Table 2 and Fig. 2 are water resource vulnerability situation in Jinan. In water vulnerability, Water scarcity (WSc) keep a high level. Water stress (WSt) is unstable. It varies with precipitation. Water Stress (WSt) and Water Pollution (WPo) share the same trend and are affected by natural precipitation. Water Productivity (WPr) shows a linear growth trend, indicating that the Water consumption in social economy is increasing. Water pollution (WPo) is unstable, it also varies with precipitation. Water productivity (WPr) in low level. There are 2 peak in Jinan, the first one appear on 2014 and the second one appear on 2017. The curve of WRVI is relative stable.

The obtained result indicates that drought has a huge impact on water resource vulnerability. Additionally, Qingdao is more vulnerable than Jinan. The variance and Mean WRVI of Qingdao are larger than Jinan, and water scarcity is 7.8% higher than Jinan. Water stress is 91.4% higher. Water pollution is 107.99% higher. Water productivity is 30.7% higher. In summary, Mean WRVI in Qingdao is 36.2% more vulnerable than Jinan. Variance of WRVI is 204.31% higher.

**Table 2** Water resource vulnerability index in Jinan

| Year | WSc | WSt | Wpo | Wpr | WRVI |
|---|---|---|---|---|---|
| 2008 | 0.874949 | 0.213551 | 0.188816 | 0.453175 | 0.432623 |
| 2009 | 0.767887 | 0.106618 | 0.104917 | 0.503429 | 0.370713 |
| 2010 | 0.769151 | 0.099238 | 0.120502 | 0.575761 | 0.391163 |
| 2011 | 0.810334 | 0.150184 | 0.152321 | 0.623493 | 0.434083 |
| 2012 | 0.852311 | 0.204513 | 0.222253 | 0.635071 | 0.478537 |
| 2013 | 0.846729 | 0.192865 | 0.243977 | 0.70502 | 0.497148 |
| 2014 | 0.925138 | 0.393859 | 0.491774 | 0.707039 | 0.629453 |
| 2015 | 0.888789 | 0.269033 | 0.333508 | 0.736895 | 0.557056 |
| 2016 | 0.842259 | 0.195587 | 0.203477 | 0.751229 | 0.498138 |
| 2017 | 0.907965 | 0.356158 | 0.344518 | 0.785753 | 0.598599 |
| 2018 | 0.870566 | 0.201677 | 0.024129 | 0.824761 | 0.480283 |
| Average | 0.850553 | 0.216662 | 0.220927 | 0.663784 | 0.487981 |

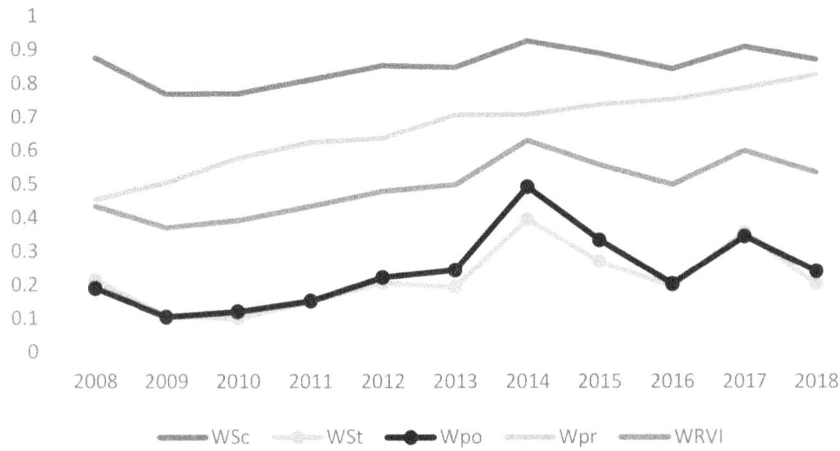


**Fig. 2** Trend of water resource vulnerability situation in Jinan between 2008 and 2018

The highest WRVI point of Qingdao appeared as 0.972 in 2015, and the highest WRVI point of Jinan appeared as 0.629 in 2014. Natural endowment has a greater impact on WRVI. In 2015, Qingdao had 30.9% of the average water resources. In 2014, Jinan had 40.4% of the average water resources. In 2015 both cities finish reform of the laddered water price.

Due to the limitations of available data, the present study mainly considers water resource factors, water environment factors, and some economic factors, which correlates with the ecological environment changes caused by human activities. At present, the water resource environment is fragile. It is necessary for government and enterprise to respond to protect the fragile water environment and keep sustainable development

## 4 Conclusion

The paper assesses the vulnerability of water resource. Under the situation of the climate characteristics and climate changes, Shandong province as a whole is in a state of water shortage. Water resources constraints bring huge effect on water environmental, especially the Water scarcity (WSc) is close to 1 and Water Pollution (WPo) is recorded as 1 in 2015 Water Stress (WSt) in Qingdao, due to the limited precipitation in Qingdao in 2015. Precipitation is very important to water vulnerability, not only providing resources but also purifying water resources.

Climate change has a bigger impact on WRVI. For a long time, water environment is fragile and water vulnerability is poor. In Shandong, water resources cannot support economic and social development, and inter-basin water transfer and other water-saving and water pollution policies are necessary. WRVI can provide basis information for policy making.

The government should take measures to improve the water environment. At present, the main measures are inter-basin water transfer and water-saving policies Inter-basin water transfer will directly improve the water resources environment. Water-saving policies can indirectly regulate water demand.

## References

1. Huang Y, Cai MT (2009) Methodological guidelines—vulnerability assessment of freshwater resources to environmental change. United Nations Environmental Programme, Nairobi (2009)
2. Raskin P, Gleick P, Kirshen P, Pontius G, Strzepek K, Water futures: assessment of long-range patterns 3993 and problems. Stockholm Environment Institute, Stockholm (1997)
3. Sullivan, C.A., Meigh, J.R., Giacomello, A.M., Fediw, T., Lawrence, P., Samad, M., et al.: The water poverty index: development and application at the community scale. Nat Resour Forum **27**(3), 189–199 (2003)
4. Juwana, I., Muttil, N., Perera, B.J.C.: Indicator-based water sustainability assessment—a review. Total Environ Sci **438**, 357–371 (2012)
5. Chaves, H.M.L., Alipaz, S.: An integrated indicator based on basin hydrology, environment, life, and policy: the watershed sustainability index. Water Resour. Manag. **21**(5), 883–895 (2007). https://doi.org/10.1007/s11269-006-9107-2
6. Pires, A., Morato, J., Peixoto, H., Botero, V., Zuluaga, L., Figueroa, A.: Sustainability assessment of indicators for integrated water resources management. Total Environ Sci **578**, 139–147 (2017)
7. Cai, J., Varis, O., Yin, H.: China's water resources vulnerability: a spatio-temporal analysis during 2003–2013 (Part 4). Clean Prod **142**, 2901–2929 (2017)
8. Kelble, C.R., Loomis, D.K., Lovelace, S., Nuttle, W.K., Ortner, P.B., Fletcher, P., Boyer, J.N.: The EBM-DPSER conceptual model: integrating ecosystem services into the DPSIR framework. PLoS ONE **8**(8), e70766 (2013)
9. IPCC TAR WG1Houghton JT, Ding Y, Griggs DJ, Noguer M, van der Linden PJ, Dai X, Maskell K, Johnson CA (eds) Climate Change 2001: The scientific basis. Contribution of Working Group I to the Third Assessment Report of the Intergovernmental Panel on Climate Change. Cambridge University Press (2001)
10. Spangenberg, J.H., Douguet, J.-M., Settele, J., Heong, K.L.: Escaping the lock-in of continuous insecticide spraying in rice: developing an integrated ecological and socio political DPSIR analysis. Ecol. Model. **295**, 188–195 (2015)
11. WWAP—World Water Assessment Programme (2003)The United Nations World Water Development Report: Water for People Water for life Paris: UNESCO, and London: Earthscan
12. OECD—Organization for Economic Cooperation and Development (1994) Environmental indicators: OECD core set Paris
13. WWAP—World Water Assessment Programme (2006) The United Nations World Water Development Report 2: Water—A Shared Responsibility. Paris: UNESCO, and London: Earthscan

# Comparison of Solar Glazing Performance of Semi-transparent Amorphous-Silicon (a-Si) and Crystalline-Silicon (c-Si) Photovoltaic Panels: A Case Study for Typical Office Building in Hong Kong

**Bing-Nan Li, Chuan-Rui Yu, Qian-Cheng Wang, and Hui-Yuan Chi**

**Abstract** Due to resource depletion and environmental pollution, research on renewable energy has been conducted for a few years in various professional communities. Building-Integrated Photovoltaics (BIPV) presents tremendous growth potential for building energy conservation. This paper compares the performance of two kinds of solar PV window in BIPV system: namely crystalline-silicon (c-Si) PV glazing and amorphous-silicon (a-Si) PV glazing. The study employs a typical office floor in Hong Kong as a case study. Based on the laboratory test and simulation results, this study reports the performance of two solar PV glazing systems in terms of the daylighting, thermal, electrical and total energy consumption. The analysis results suggest that the amorphous-silicon PV glazing system presents better-daylighting performance and thermal performance, while crystalline-silicon PV glazing performs better in the electrical part. Besides, the study considers the influence of dimming control being applied regarding energy performance. The study found no significant difference in the annual energy performance of two systems without dimming control, while crystalline-silicon PV glazing saves more energy when dimming control is integrated. The results suggest that the energy-saving performance of BIPV systems might be impacted by the type of solar PV system. The findings would provide an informative reference for BIPV technology applications in Hong Kong.

---

B.-N. Li (✉)
Department of Civil and Environmental Engineering, University of California, Berkeley, CA, USA
e-mail: Bingnan_li@berkeley.edu

C.-R. Yu
The Bartlett School of Environment, Energy and Resources, University College London (UCL), London WC1E 6BS, UK

Q.-C. Wang
Department of Land Economy, University of Cambridge, Cambridge CB3 9EP, UK

H.-Y. Chi
School of Marine Science and Technology, Harbin Institute of Technology, Weihai 264200, China



F. K. S. Chan et al. (eds.), *Proceedings of the 2020 International Conference on Resource Sustainability: Sustainable Urbanisation in the BRI Era (icRS Urbanisation 2020)*, Environmental Science and Engineering,
https://doi.org/10.1007/978-981-15-9605-6_4

**Keywords** BIPV · Photovoltaics · Semi-Transparent Amorphous-Silicon Crystalline-Silicon · Energy conservation · Dimming control · Office · Hong Kong

## 1 Introduction

Due to energy shortages and a series of environmental problems, many studies have made an effort on energy-efficient building technologies and their feasibility [1]. Especially, governments and researchers have paid considerable attention to renewable energy like solar energy. Solar photovoltaic (PV) is one of the most widely employed methods to harvest solar energy [2]. By the end of 2015, solar PV power generation has taken 1.2% share in global electricity production, and world solar PV capacity has reached around 200 GW [3]. PV system generates electric current upon exposure to sunlight then supply electricity to appliances. Benefit from its flexible shape and size, the PV system can be employed on many occasions. Building-Integrated Photovoltaics (BIPV) is a specific way to adopt the solar PV system as part of the building design (e.g., rooftop PV, façade PV), aiming to offset building power use and achieve sustainability with the building. BIPV has tremendous growth potential because of its advantages like requiring less space, reducing overall system cost and reducing building power load [4]. There are two common types of solar cells in the BIPV market: Crystalline Silicon (c-Si) and Amorphous silicon (a-Si). The c-Si type has a higher market share (65% of global annual solar energy production in 2019) [5]. Compared with the c-Si solar cells, the a-Si solar cells use less toxic heavy metal materials. Besides, the a-Si type has a better light transmission [6], which makes it possible to integrate into building façade without scarifying daylighting.

Integrating solar PV into building façade system is a been a potential solution to reduce the energy demand of the building sector [7, 8]. Several studies have revealed the energy performance including thermal performance, power generation of a-Si and c-Si PV under different building conditions. For example, Kapsis et al. [4] examined the lighting performance of facades configured c-Si, opaque spaced cells and a thin film PV module with an office case in Toronto. Liao and Xu [5] also compared the energy performance of see-through a-Si PV and traditional glazing under different room structures. Besides, Martellotta et al. [6] analyzed the energy performance of a-Si semi-transparent PV cells and perovskite-based semi-transparent PV cells applied to the office and the residential building based on the double-panel glazing system. Do et al. [9] investigated the energy benefits of different transparencies with DOE-2 simulation system. They stated that south-facing PV window produced the largest amount of electricity for a year, while the east- and west- facing windows had more potential to generate electricity in summer. The above research shows that the semi-transparent a-Si PV has the significant advantages of simultaneously daylighting and energy production on building façade. However, there is only limited research on the difference between c-Si PV glazing and a-Si PV glazing, i.e. the difference of solar PV panel performance, overall glazing performance and building performance when integrated c-Si and a-Si with building glazing.

The two types of solar cells are difference in the optical characteristic, power generation etc. Thus, an actual case scenario is necessary to discuss their actual performance with whole building simulation considering HVAC, daylighting, artificial lighting, power generation etc. This research project mainly aims to investigate and compare the performance of c-Si and a-Si PV modules applied in a building facade. The study first comprehensively evaluates the electrical, thermal and optical characteristics of selected c-Si and a-Si PV modules as well as normal window materials. Then, the research investigate the actual building performance with a building façade glazing integrating the c-Si and a-Si solar cells.

## 2 Methodology

This research project combined laboratory tests and software simulations. Laboratory tests are used in Stage I and software simulations are used in Stage II. The methodology is illustrated in Fig. 1.

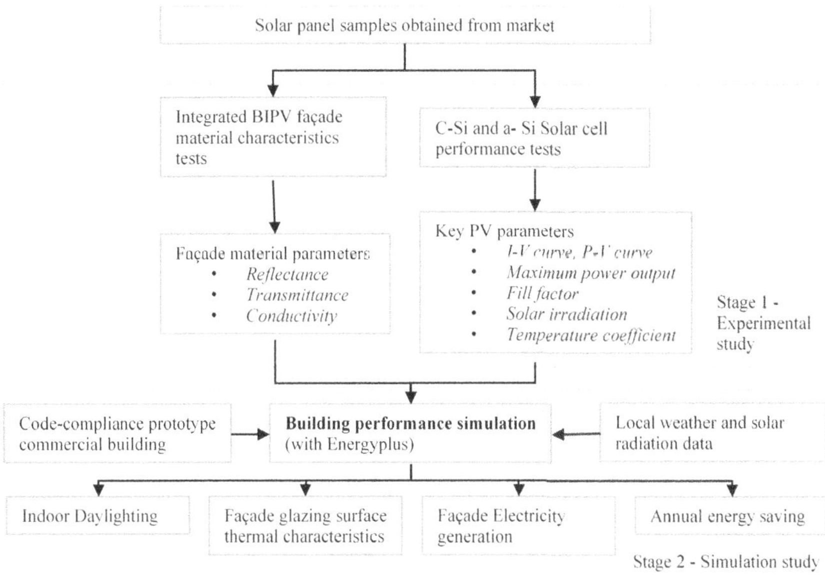


**Fig. 1** Methodology framework. The arrows represent the dataflows through each test

**Table 1** Laboratory test subjects, equipment and outputs

| Test name | Test subject | Key characteristics | Equipment |
|---|---|---|---|
| PV performance | c-Si PV module<br>a-Si PV module | I-V and P-V curve | IV-tracer, data logger, pyranometer, thermal couple, |
| | | Solar irradiation | |
| | | Temperature coefficient | |
| Material characteristics | c-Si PV module<br>a-Si PV module<br>Normal glass | Reflectance | Spectrometer |
| | | Transmittance | |
| | | Conductivity | Thermal conductivity meter |

**Table 2** Standard test condition applied in the PV performance experimental study

| Parameter | Unit | Value |
|---|---|---|
| PV device temperature | °C | 25 |
| Solar irradiance | W/m$^2$ | 1000 |
| Air mass | – | 1.5 |
| Parameter | Unit | Value |

## 2.1 *Stage I—Experimental Study*

To obtain actual performance characteristics of PV modules for accurate simulation, the study conducts a series of lab tests and reports the materials characteristics as well as the related equipment for testing in Table 1. All tests related to PV modules are carried out under Standard Test Conditions (STC) (see Table 2) [10]. Due to the equipment limitation, the research employs the emissivity data from Peng et al. [11] for simulation and analysis.

## 2.2 *Stage II—Simulation Study*

To compare the daylighting impact, thermal impacts, power generation and annual energy consumption between these two types of PV modules, building simulation technique was adopted as a comprehensive set of data can be outputted from the simulation under an identical testing condition. EnergyPlus was developed by National Renewable Energy Laboratory (NREL) of the United State is widely used for building simulation. It simulates occupant, indoor and outdoor heat gain, system use, renewable energy generation etc. for a building on a dynamic basis, providing a accurate and time-based insight for the performance of tested PV modules [1]. Laboratory data about PV panels from stage I was inputted into the EnergyPlus program for more accurate replication for more accurate solar PV performance. The PV power generation was simulated with Equivalent One-Diode Model which

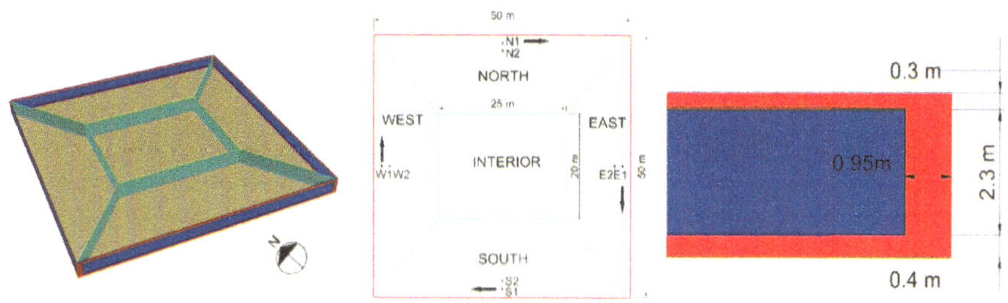


**Fig. 2** Illustration for the typical office energy model. Left—3D isometric view for the model. Middle—layout and dimensions for the thermal zones. Right—Detailed dimensions for the glazing settings on the corners

**Table 3** Key energy simulation model data

| Model parameters | Unit | Value |
|---|---|---|
| Total floor area | $m^2$ | 2500 |
| Room height | m | 3.0 |
| Window-to-wall ratio | – | 0.53 |

predict the current-voltage characteristics and power output of PV system as a module [12, 13].

As for the room settings, this simulation model replicated a code-compliant high-rise office building in Hong Kong with a typical core and shell layout (as shown in Fig. 2) (Table 3). The room parameters such as envelope insulation, window area, and system efficiency were defined according to local basic requirement [14]. The floorplan was separated into five thermal zones (east, west, south, north, and interior) for more accurate simulation result. The impacts on indoor environment were reported with the in-built test points in EnergyPlus simulation.

In office building, the artificial lighting may be equipped with automatic dimming system to save electrical energy when natural light is sufficient. The dimming is based on the illuminance measured 3 m away from the façade center point. The percentage of dimming control is limited to 20% of the total floor area considering the day-long artificial lighting demand in deeper and less daylight spaces. Totally six sets of configurations were tested to compare in lighting environment and energy demands, i.e. normal glass with/without dimming, a-Si PV glazing with/without dimming and c-Si PV glazing with/without dimming.

Daylight was simulated to investigate the lighting and thermal influence on occupants with the test modules. Daylight Autonomy (DA), Useful Daylight Illuminance (UDI) and Unified Glare Rating (UGR) of reference points are adopted to study the lighting performance of the modules for indoor human lighting perception [15, 16]:

- DA indicates whether daylight in a space is sufficient enough to allow to people work by daylight alone. It is calculated by the hours of illuminance over 500 lx divided by the hours of illuminance larger than 0 lx.

- UDI suggests the useful level of daylight, which refers to the daylighting quality. UDI is calculated as the hours of illuminance between 100 lx and 2000 lx divided by the hours of illuminance greater than 0 lx.
- UGR shows the glare level of a given plane. 19 is set as the maximum acceptable glare level for office work. The Glare Exceeding Tolerance Time (GETT) percentage in regarding the time of the glare level which is larger than 0 is calculated to indicate the visual comfort of reference points.

The thermal, electricity performance and annual energy performance unveiled the electricity and energy saving potential of a-Si and c-Si PV panels when integrated into a office building. The thermal performance compared the surface temperature, Solar Heat Gain Coefficient (SHGC) for a general energy saving performance. The electricity performance showed the energy production of the two types of solar PV in a real case scenario. The annual energy performance considered the energy demand reduction from the SHGC as well as the energy production from the two type, which would suggest a more energy efficient façade solar PV solution for actual building use.

# 3 Results & Discussion

## 3.1 Stage 1: Experimental Results

Tables 4 and 5 summarized the physical and electrical characteristics of the a-Si and c-Si PV panels from laboratory tests. A-Si PV panel is lighting transparent with 19% transmittance for visible light (24.5% of normal clear glazing). C-Si PV panels is not transparent, which makes it impossible to apply c-Si panel for 100% area of a glass panel. It might be more preferable to consider c-Si as a shading device fixed to certain areas where view out is not a primary concern on a facade.

The two tested a-Si and c-Si panels performed similarly in terms of electrical characteristics (Figs. 3 and 4). The tested a-Si panel is slightly better performing with higher maximum power and higher efficiency (converting 3.65% of received solar energy into electricity). The data was based on the test samples obtained from markets only, thus cannot represent the general performance characteristics of the

**Table 4** Comparison of optical and thermal characteristics for a-Si PV module, c-Si PV module and normal glass from laboratory tests

| Material | Transmittance | Front reflectance | Back reflectance | Conductivity (W/mK) |
|---|---|---|---|---|
| a-Si PV module | 0.19 | 0.126 | 0.096 | 0.49 |
| c-Si PV module | 0 | 0.093 | 0.080 | 0.1 |
| Normal glass | 0.775 | 0.071 | 0.071 | 0.9 |

**Table 5** Basic characteristics of tested modules for simulation for only one panel sized one square meter

| Name of parameters | Abbreviation | Unit | Type of PV panels | |
|---|---|---|---|---|
| | | | a-Si | c-Si |
| Number of panels | – | – | 1 | 1 |
| Test panel size | – | $m^2$ | 1 | 1 |
| Short circuit current | Isc | A | 2.80 | 2.63 |
| Open circuit voltage | Voc | V | 18.69 | 18.00 |
| Maximum power | Pm | W | 36.45 | 32.26 |
| Maximum current | Imp | A | 2.45 | 2.29 |
| Maximum voltage | Vmp | V | 14.90 | 14.12 |
| Efficiency | η | % | 3.65 | 3.23 |
| Fill factor | FF | – | 0.6965 | 0.6804 |

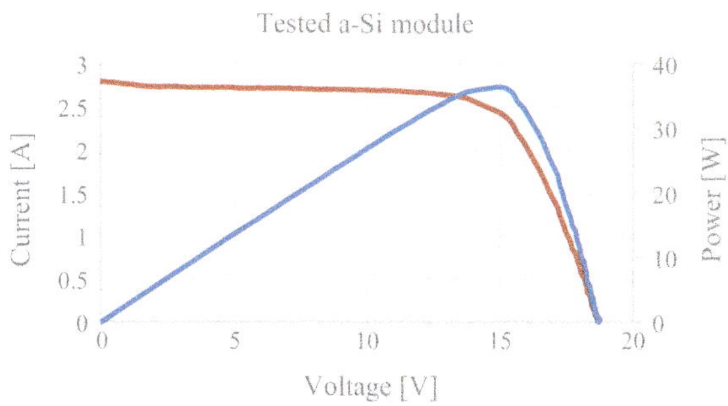


**Fig. 3** p-V and I-V curves for the tested a-Si solar panels

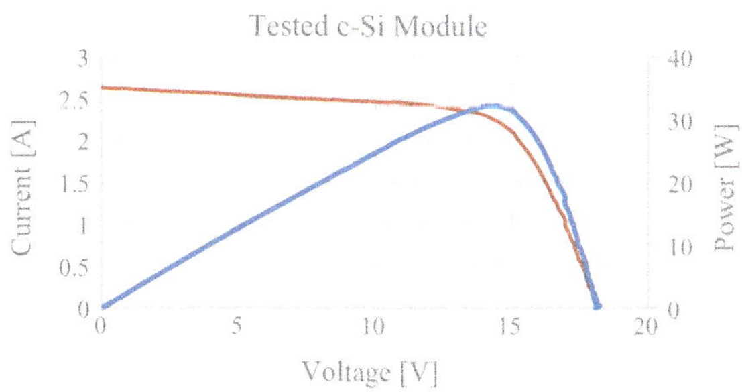


**Fig. 4** p-V and I-V curves for the tested c-Si solar panels

two types of solar PV panels. The information related to PV system will be given and discussed in the electrical performance section.

### 3.2 Stage 2 Simulation Result

**Daylighting and glare performance**. Dimming control was not relevant to the daylighting performance. Therefore, only the three simulation test without dimming control were compared. Figure 5 summarizes the daylighting and glare performance at measuring points by UDI, DA, and GETT index. A-Si PV glazing has the lowest DA. However, its UDI value is twice that of normal glass and its GETT index is also the lowest. The low transmittance blocked more sunlight than the other two types but actually created more time with useful illuminance in a year by reducing the glare problem from glazing façade, making it more preferable for façade and artificial lighting saving.

Glazing material was also found more significant than orientation for daylighting performance. North zone generally shows a less intensive daylighting environment, while south zone and west zones may face more glare problems. The difference in the daylighting index among orientations depends largely by the glazing material. The difference for normal glazing is marginal while difference for a-Si PV glazing is substantial.

A-Si PV glazing has better performance in daylighting, because a-Si PV modules can covers the fenestration area uniformly and reduce direct sunlight glare with its lower transmittance by restricting highly intensive sunlight. C-Si Pv glazing, however, is a combination of clear glass (60%) and c-Si PV modules (40%). The overall glazing transmittance is higher and the gaps between modules allowing high intensive sunlight to still cause glare issues indoor. Thus, the overall glazing performance for c-Si glazing is similar to normal clear glazing panels.

**Thermal Performance**. In Hong Kong, the climate is hot and humid most time of a year thus the cooling energy use dominates the HVAC energy consumption. The discussion for thermal performance here will focus on indoor cooling energy reduction.

Applying a-Si and c-Si glazing changes U-value and Solar Heat Gain Coefficient (SHGC) properties for facade glazing thus changes thermal properties of facade.. As shown in Table 6, both PV glazing systems have lower U-value and SHGC value. With the improved window property, the window surface heat gain rate can be reduced. In Fig. 6, a-Si PV glazing generally helps to reduce 44% of the heat gain rates, and the north zone has the largest value, which is 46%. C-Si PV glazing reduces heat gain rate at a level of 33%, and the north zone has the lowest value, which is 30%.

As in Figs. 7 and 8 PV glazing reduced more peak cooling load in the west and east zone than the north and south zones. Figure 8 indicates that cooling energy saved with different glazing system and dimming control conditions. The most cooling energy saving method is to apply A-Si PV glazing without dimming control, equivalent to 7.7% saving than using normal glass. Other three cases have the similar cooling energy saving. As to the peak cooling demand, as shown in Fig. 9, c-Si PV glazing performances better no matter whether the dimming control is applied or not. The reducing of peak cooling demand can play important roles in system sizing and may help to save lots of investment and operation cost.

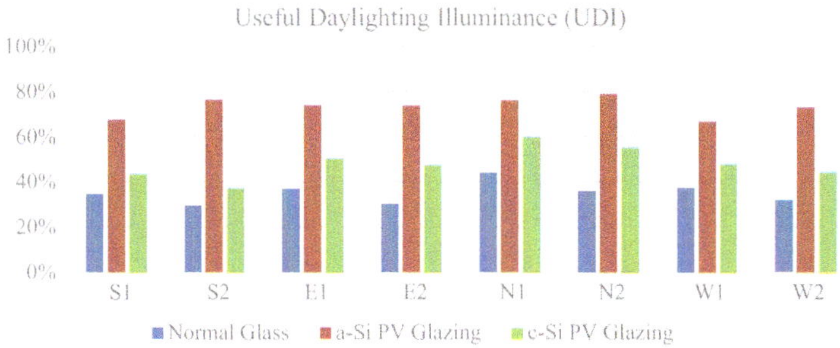


(a)

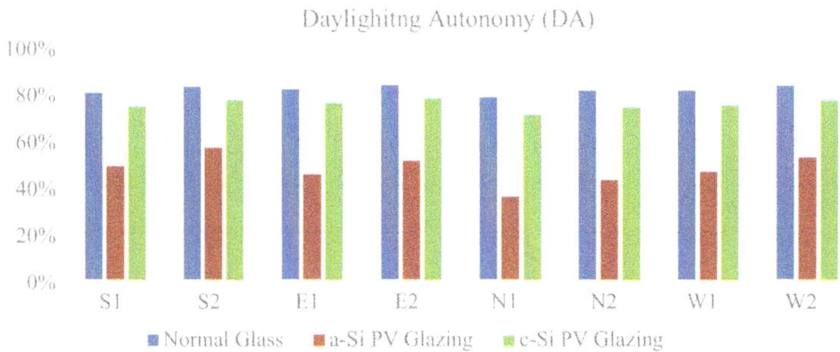


(b)

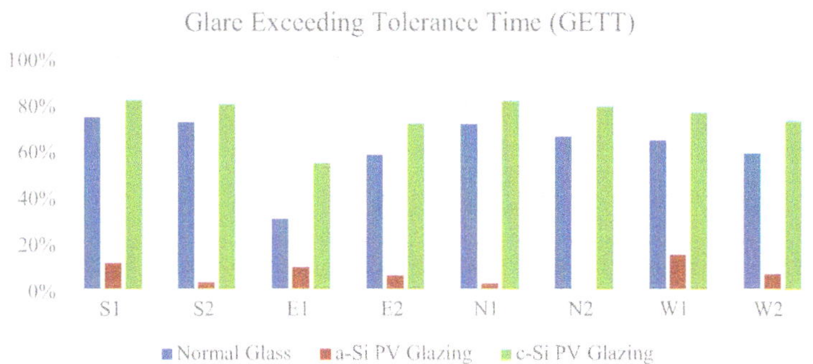


(c)

**Fig. 5** Simulated daylighting performance data on different measurement points for normal, a-Si and c-Si glazing, **a** presents useful daylighting illuminance (UDI), **b** presents daylighting autonomy (DA), **c** presents glare exceeding tolerance time (GETT). S1—measurement point 1.5 m away from the middle of South façade at 1.2 m height. S2—Measurement point at 3 m away from the middle of South façade at 1.2 m height. Similar for E1, E2 N1, N2, W1 and W2 where E, N and W representing East façade, North façade, and West façade respectively

**Table 6** Comparison of window properties between normal glass and façade glazing with a-Si PV panels or c-Si PV panels calculated by simulation program

| Properties | Unit | Normal Glass | a-Si PV Glazing | c-Si PV Glazing |
|---|---|---|---|---|
| U-value | $W/m^2$ K | 5.78 | 5.49 | 5.76 |
| SHGC | – | 0.82 | 0.46 | 0.60 |

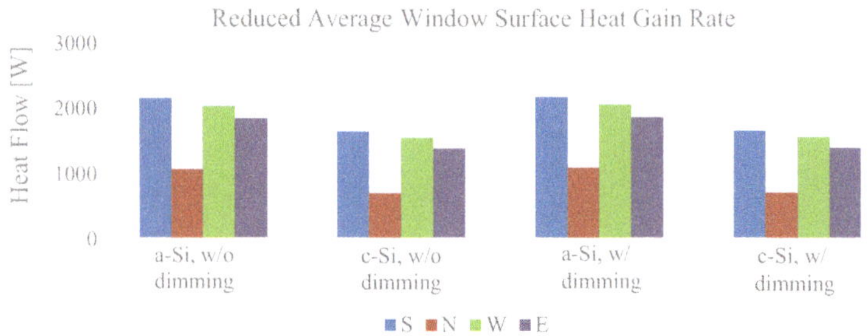


**Fig. 6** Reduced average window surface heat gain rate comparing to normal clear glass façade window. S—South façade window, N—North façade window, W—West façade window, E—East façade window

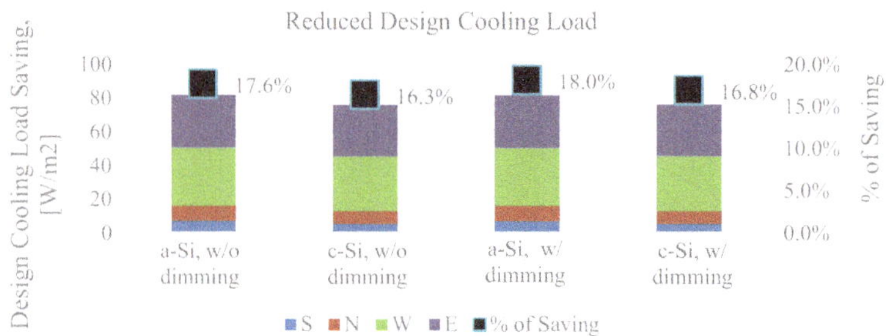


**Fig. 7.** Reduced design cooling load. S—South façade window, N—North façade window, W—West façade window, E—East façade window, % of saving—the percentage of reduction in design cooling load from that of simulated building with normal glazing

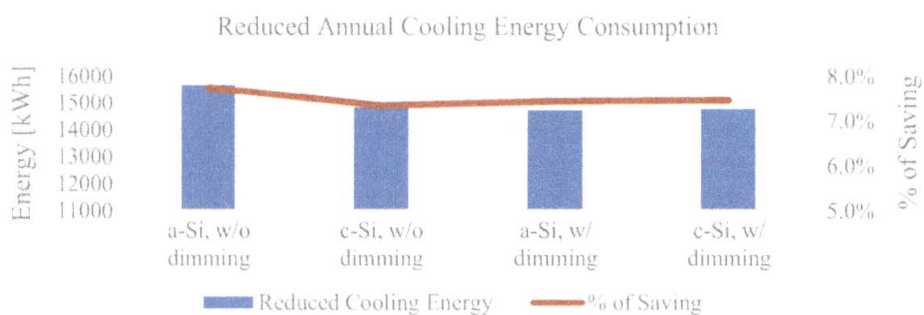


**Fig. 8** Cooling energy saving for a-Si and c-Si PV glazing with or without dimming control

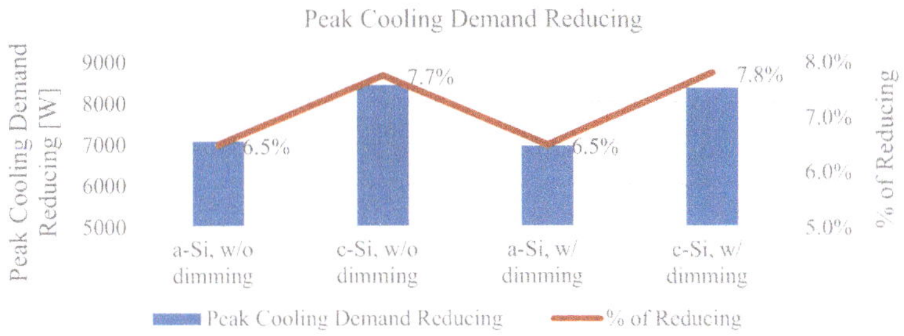


**Fig. 9** Peak cooling demand reducing

**Electrical Performance**. The saving in electricity consumption can be divided into the saving from dimming and the saving from solar PV power generation, considering both shading effect and power generation from the PV modules. The key system characteristics of the two solar façade systems with a-Si PV modules and c-Si PV modules was summarized in Table 7. Figures 10 and 11 describe the electrical performance of different glazing systems.

Figure 10 shows that all two glazing systems saved energy by dimming control. However, with their low transmittance, the saving from dimming control was less than that of normal glass. Figure 11. indicates the power generation of two PVglazing. C-Si PV glazing has higher annual power production and monthly maximum power than

**Table 7** Key Characteristics of two PV systems

| | PV model efficiency (%) | PV cells area ($m^2$) | Total rated power (kW) | Total power generated (kWh) |
|---|---|---|---|---|
| a-Si PV solar façade system | 5.9 | 402.52 | 25.6 | 3686 |
| c-Si PV solar façade system | 22.2 | 173.16 | 38.4 | 6552 |

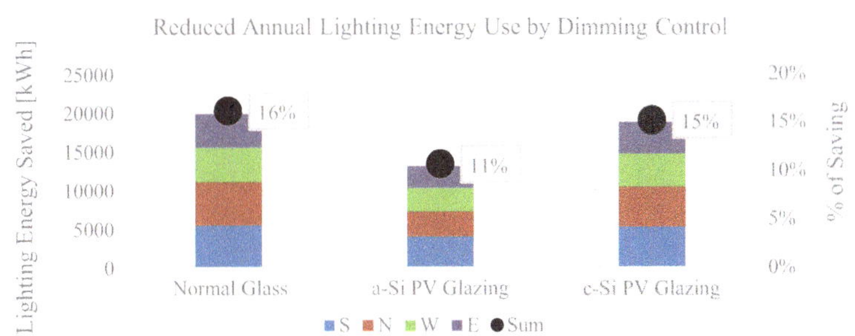


**Fig. 10** Saved annual lighting energy use by dimming control. S—South façade window, N—North façade window, W—West façade window, E—East façade window, % of saving—the percentage of reduction in design cooling load from that of simulated building with normal glazing

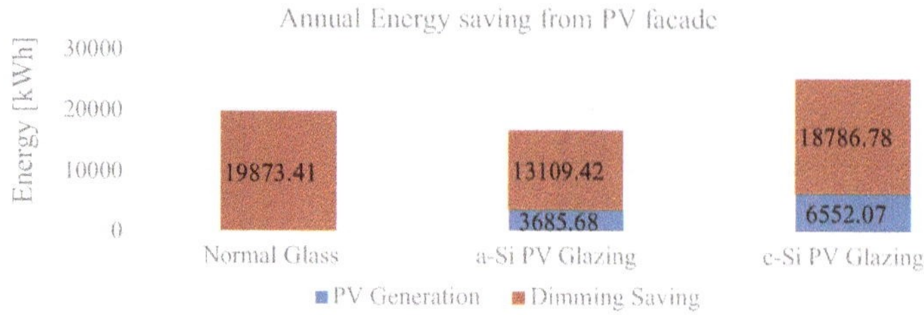


**Fig. 11** Annual power saved in electrical part

a-Si PV glazing. Such difference was closely related to the tested PV characteristics: a-Si PV module generally has a lower efficiency compared to c-Si PV module. Although covering more glazing area, a-Si PV glazing still produce less electricity annually than c-Si PV glazing.

As in Fig. 11 c-Si PV glazing system has a surplus of 5465.44 kWh while the a-Si PV system has a −3078.31 kWh impact on the electricity saving in this part. The power generated by PV takes up 1/4 in the total energy saved by the c-Si PV glazing while 1/5 for the a-Si PV glazing.

**Annual Energy Performance**. The annual energy performance of two PV glazing systems was analyzed by comparing total site energy and net site energy. The net site energy means the total energy consumed after subtracting the total energy generated on site, which is PV power for this project (Table 8). Apart from saving by dimming control, cooling demand reducing and PV power generation, the net site energy analysis also considered the energy use by fans, pumps, and other building system.

As presented in Figs. 12 and 13, without dimming control, the two PV glazing perform similarly in overall energy saving. C-Si PV glazing performs better with dimming control. The less daylight transmittance reduced the energy saving potential by dimming control. Similar phenomenon was also observed in simulation for A-Si PV glazing.

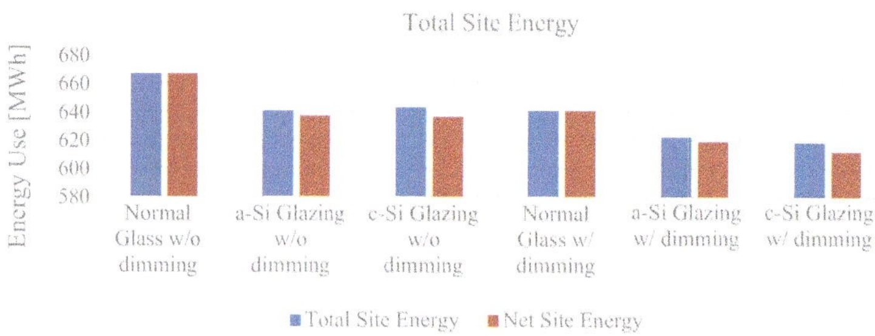


**Fig. 12** Overall energy performance

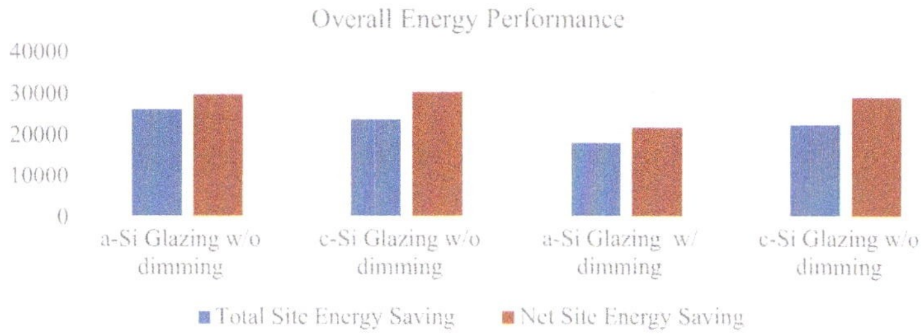


**Fig. 13** Overall energy performance

**Table 8** Annual PV power generated from the façade glazing equipped with different types of PV panels in energy simulation

| | Unit | a-SI PV glazing | c-Si PV glazing |
|---|---|---|---|
| Sum | kWh | 3685.68 | 6552.07 |
| Minimum in a months | kWh | 210.16 | 419.06 |
| Maximum in a months | kWh | 451.4 | 725.21 |

# 4 Conclusion

With the laboratory test and simulation study, the performance of a-Si PV glazing and c-Si PV glazing has been investigated. Their performance can be summarized as follows:

- *Daylighting performance*: a-Si PV glazing can contribute to a more comfortable indoor visual environment with high UDI and low GETT value because of its optical characteristics. C-Si PV glazing has less preferred;
- *Thermal performance*: a-Si PV glazing reduces around 44% of heat gain from the outdoor environment and saves 7.6% cooling energy. C-Si PV glazing preforms better in the peak cooling load reducing, which is about 8400 W peak load reducing;
- *Electrical performance*: with the high efficiency PV module, c-Si PV glazing generates 6500 kWh per year. The energy saved by dimming control of c-Si PV glazing cases is also 4% larger than that of a-Si PV glazing;
- *Annual energy consumption*: when dimming control is not applied, the two materials have the similar energy saving, around 4% of the site energy. When dimming control is applied, c-Si PV glazing has nearly 1% more energy saving than a-Si PV glazing.

Although this project is a case study based on one office floor in Hong Kong, it provides a reference in PV glazing performance and building façade selection. Further study could focus on the whole building performance or performance under different climate. Test chamber is also suggested to have more practical results.

## References


1. Yu, C.-R., Guo, H.-S., Wang, Q.-C., Chang, R.-D.: Revealing the Impacts of passive cooling techniques on building energy performance: a residential case in Hong Kong. Appl Sci **10**(12), 4188 (2020)
2. Chang, S., Wang, Q., Hu, H., Ding, Z., Guo, H.: An NNwC MPPT-based energy supply solution for sensor nodes in buildings and its feasibility study. Energies **12**(1), 101 (2019)
3. Raturi AK (2016) Renewables 2016 Global status report
4. M. E. Mackay, *Solar energy: An introduction*. OUP UK, 2015
5. Green, M.A., Dunlop, E.D., Levi, D.H., Hohl-Ebinger, J., Yoshita, M., Ho-Baillie, A.W.: Solar cell efficiency tables (version 54). Prog Photovoltaics Res Appl **27**(7), 565–575 (2019)
6. Husain, A.A., Hasan, W.Z.W., Shafie, S., Hamidon, M.N., Pandey, S.S.: A review of transparent solar photovoltaic technologies. Renew Sustain Energy Rev **94**, 779–791 (2018)
7. Shukla, A.K., Sudhakar, K., Baredar, P.: Recent advancement in BIPV product technologies: a review. Energy Build **140**, 188–195 (2017)
8. Chang R, Wang Q, Ding Z (2019) How is the energy performance of buildings assessed in Australia?—A comparison between four evaluation systems
9. Do, S.L., Shin, M., Baltazar, J.-C., Kim, J.: Energy benefits from semi-transparent BIPV window and daylight-dimming systems for IECC code-compliance residential buildings in hot and humid climates. Sol Energy **155**, 291–303 (2017)
10. Standard F (2010) Test method for photovoltaic module power rating
11. Peng, J., Curcija, D.C., Lu, L., Selkowitz, S.E., Yang, H., Mitchell, R.: Developing a method and simulation model for evaluating the overall energy performance of a ventilated semi-transparent photovoltaic double-skin facade. Prog Photovoltaics Res Appl **24**(6), 781–799 (2016)
12. Qiu, C., Yi, Y.K., Wang, M., Yang, H.: Coupling an artificial neuron network daylighting model and building energy simulation for vacuum photovoltaic glazing. Appl Energy **263**, 114624 (2020)
13. Griffith BT, Ellis PG (2004) Photovoltaic and solar thermal modeling with the EnergyPlus calculation engine: preprint. National Renewable Energy Lab
14. Guidelines on performance-based building energy code, EMSD, HKSAR (2007)
15. Boyce PRP (2009) SLL Lighting Handbook
16. Xue, P., Mak, C., Huang, Y.: Quantification of luminous comfort with dynamic daylight metrics in residential buildings. Energy Build **117**, 99–108 (2016)

# The Impacts of Big Five Personality Traits on Household Energy Conservation Behavior: A Preliminary Study in Xi'an China

**Liu Xuan, Jian Izzy Yi, Wang Qian-Cheng, Zhou Long-Li, and Xie Qiao-Peng**

**Abstract** The residential sector has become a significant energy consumer and carbon emitter in China. Behavior-driven solutions are widely used for building energy conservation. Research gap exists to understand the impacts of personality traits on the individual differences in energy conservation behavior. This study aims to explore the roles of Big Five personality traits in household energy conservation behavioral process. The research connects the Big Five personality traits with attitude in the theory of planned behavior. Through a detailed questionnaire-based survey, the study collected demographics, energy conservation intentions, and personality characteristics from households in two typical communities in Xi'an China. The sample size of the survey consists of 213 households. The structural equation modelling results show that two personality traits play critical roles in household energy conservation behavioral process. Agreeableness positively contributes to energy conservation attitude while Extroversion presents a negative relationship with the attitude towards household energy conservation. The research offers a shred of an empirical demonstration on the connection between personality traits and pro-environmental behaviors. The findings would contribute to the energy conservation campaigns in China.

---

L. Xuan
Department of the Built Environment, Eindhoven University of Technology, Eindhoven, The Netherlands

J. I. Yi (✉) · X. Qiao-Peng
Department of Building and Real Estate, The Hong Kong Polytechnic University, Hong Kong, China
e-mail: yi.jian@connect.polyu.hk

J. I. Yi
Research Institute of Sustainable Urban Development, the Hong Kong Polytechnic University, Hong Kong, China

W. Qian-Cheng
Department of Land Economy, University of Cambridge, Cambridge CB3 9EP, UK

Z. Long-Li
Faculty of Environment, University of Leeds, Leeds LS2 9JT, UK



F. K. S. Chan et al. (eds.), *Proceedings of the 2020 International Conference on Resource Sustainability: Sustainable Urbanisation in the BRI Era (icRS Urbanisation 2020)*, Environmental Science and Engineering,
https://doi.org/10.1007/978-981-15-9605-6_5

**Keywords** Household · Energy conservation · Pro-environment · Five factor model · Personality traits · Theory of planned behavior · Attitude · China

## 1 Introduction

Energy has played an important role in the development of human society. However, the rapid economic development has also caused energy depletion and a series of environmental problems. The building sector has become one of the major energy consumers in China [1]. In the operation and maintenance stage, behavior of building occupants has a significant impact on building energy efficiency [2, 3]. Behavior-driven building energy conservation programs have become a promising environmental protection strategy. Previous studies have found that behavior-driven methods such as information intervention can reduce building energy consumption by 5–20% [4, 5]. Compared with retrofitting building service systems, the behavior-driven solutions are less likely to require a high initial investment to achieve energy-saving goals in a relatively short period [6].

Understanding the driving factors of energy saving behavior is important in building energy consumption analysis and promotion of energy efficient behavior. Some studies employ psychological models to explain building energy efficiency behavior. For example, Zhang et al. [7] adopted the Norm Activation Model (NAM) to predict the energy-saving behavior of employees in commercial buildings. Liu et al. [6] also used the theory of planned behavior (TPB) to explain energy-saving behavior in residential buildings. These studies have well discussed some critical psychological factors in building energy-saving behavior and picked out many factors that play important roles in the process of energy-saving behavior. However, several empirical cases reported the stable individual differences in building energy conservation behavior [8]. The psychological models in previous studies can hardly explain such individual differences.

Personality traits are habitual patterns of behavior, thought and emotion [9], which are also considered to be the essence of individual differences in behavior. Previous studies have discussed the impacts of personality traits on some pro-environmental behaviors, such as visiting green hotels [10], recycling garbage [11], and taking public transportation [12]. However, only a few studies have focused on the influence of personality traits on building energy efficiency behavior. Besides, previous studies have mostly focused on developed countries and cities in Europe and North America. There are very few studies paying attention to energy conservation and environmental protection in underdeveloped areas. However, the shortage of resources is one of the main reasons for restricting the development of these areas [13, 14]. Therefore, it is necessary to understand the psychological factors that influence the energy-saving behavior of residents in underdeveloped areas.

This article mainly explores the role of personality traits in household energy-saving behaviors. This article first reviews the related literature and then proposes a theoretical framework linking personality traits and TPB factors. Subsequently, the paper analyzes a case in Xi'an, a city in Northwest China. The researchers obtained self-reported data by issuing questionnaires in two typical communities in Xi'an and report the structural equation models (SEM) data analysis results in this paper. Finally, the article further discusses the implications of the findings and the limitations in the research process.

## 2 Literature Review

### *2.1 TPB*

TPB is one of the most commonly used psychological models to explain pro-environmental behavior. The TPB model uses three factors to explain an individual's intention for a specific behavior: attitude, subjective norms and performed behavioral control (PBC). Attitude is an individual's evaluation of whether the outcome of the behavior is positive. For example, residents may choose household energy-saving behaviors because they value environmental protection [6]. Subjective norms reflect the influence of social and external pressure on behavioral intention, and this influence mainly comes from essential people. For example, an employee might refuse to set the air conditioner temperature higher because of concerns about the feelings of colleagues [15, 16]. Students may also engage in energy-saving behaviors because of the expectations of their teachers or parents [17]. PBC reflects the individual's subjective judgment on the difficulty or convenience of behavior. For example, low convenience is one important reason to not engage in household energy-saving behaviors [6, 18].

Most studies have proved that the three TPB factors can explain the intention well. However, the analysis of some research has reached the opposite conclusion. Some studies (e.g., [19]) have found that under certain circumstances, the influence of attitude on intention is not significant. There is also evidence supporting that the effect of subjective norms is insignificant [20, 21]. In addition, Ru et al. [18] also found that the role of subjective norms may become smaller or disappear when additional factors are considered. In addition, some studies believe that PBC has no significant effect on behavior [22]. Thus, this study put forwards the following hypotheses:

$H_1$: *Attitude positively impact energy-saving intention of residents;*
$H_2$: *Subjective Norms positively impact energy-saving intention of residents;*
$H_3$: *PBC positively impact energy-saving intention of residents.*

## 2.2 Personality Traits and Energy-Saving Behavior

Personality traits reflect people's stable feelings and emotional patterns [23]. Therefore, personality traits can explain the different choices made by different people in the same situation and the stable individual differences in similar behaviors. However, how personality traits affect behavior or behavioral intentions has not been discussed in depth. In TPB, attitude plays a vital role in linking personal characteristics with behaviors and personality traits reflect the individual psychological difference in patterns of thought, feeling and action. Some studies, therefore, link personality trait theory with attitude to explain the influence of personality traits on pro-environmental intention (e.g., [24–26]). Especially, Conner and Abraham [27] provide substantial evidence presenting the strong mediating effect of attitude between personality and intention. Also, Pavalache-Ilie and Cazan [25] also supported that the significant correlation between personality traits and attitude. Based on the above viewpoints, the study makes the following hypotheses:

$H_4$: *Personality Traits impact attitude towards energy-saving behavior;*

The Big Five model (also called Five Factor Model, FFM) is the most widely used personality trait model. There are five labelled personality traits in the Big Five model: Agreeableness (A), Openness-to-experience (O), Conscientiousness (C), Extraversion (E), and Neuroticism (N). Agreeableness is the tendency to value social harmony and to get along with others. Openness reflects rich, abstract thinking and an appreciation for the variety and unusual experiences. Conscientiousness is indicated by high levels of self-discipline, respect for duty, and desire for achievement. Extraversion is characterized by an energetic engagement with the world, sociability, and breadth of activities. Finally, Neuroticism is the tendency to experience negative emotions, such as anger, anxiety, and depression.

Openness reflects the degree of intellectual curiosity, creativity and a preference for novelty and variety of an individual. People with higher Openness are more likely to change the world with mew methods positively. Some studies have found a strong correlation between openness and pro-environmental attitude. For example, Ma and Bateson [28] and Brick and Lewis [24] indicate that environmental attitude mediates Openness and pro-environmental behaviors. However, other studies such as Hirsh [29] and Milfont and Sibley [30] failed to link Openness but some sub-traits with behavioral process. The current research put the following hypothesis forwards:

$H_{4\text{-}1}$: *Openness positively impact attitude towards energy-saving behavior;*

Agreeableness refers to motivation to be helpful, cooperative and sympathetic towards others [31]. Majority of studies note that Agreeableness affect attitude towards pro-environmental behaviors (e.g., [29, 32]). Pavalache-Ilie and Cazan [25] show that Agreeableness is associated with attitude and this relationship is stronger than its contribution to environmental concern. However, other research works found this correlation not significant (e.g. [24, 33]). Based on the above studies, the current research put the following hypothesis forwards:

*$H_{4\text{-}2}$: Agreeableness positively impact attitude towards energy-saving behavior;*

Conscientiousness is another important trait. Conscientiousness reflects one's self-discipline, respect for duty, and desire for achievement. Some studies found the correlation between Conscientiousness and pro-environmental attitude is strong (e.g. [25, 34]). Brick and Lewis [24] reported a strong positive link between Conscientiousness and attitude towards low-carbon emission behaviors. Milfont and Sibley [30] found Conscientiousness well predict the electricity saving behaviors. On the other hand, other studies only found this effect inconsistent (e.g. [29, 35, 36]). The current research gives the following hypothesis:

*$H_{4\text{-}3}$: Conscientiousness positively impact attitude towards energy-saving behavior;*

Extraversion refers to an individual's level of outgoing, energetic engagement, sociability, and breadth of activities. Moderate evidence supports that Extraversion contributes to sustainable behaviors and investments (e.g., [24, 25, 34, 37]). However, there are also a few studies failing to find the link significant (e.g., [29, 35]). Accordingly, the current research gives the following hypothesis:

*$H_{4\text{-}4}$: Extraversion positively impact attitude towards energy-saving behavior;*

Neuroticism refers to "the tendency to experience negative emotions, such as anger, anxiety, and depression" [38]. Busic-Sontic et al. [37] found a negative impact of Neuroticism on sustainable investment attitude. However, results in other studies showed that this link is not significant (e.g., [25, 35, 36]). Accordingly, the current research gives the following hypothesis:

*$H_{4\text{-}5}$: Neuroticism positively impact attitude towards energy-saving behavior;*

The research model including $H_1$ to $H_4$ is shown in Fig. 1.

# 3 Methodology

## *3.1 Data Collection*

This study distributed questionnaires in three target communities in Xi'an, from January 2019 to February 2019. The study first invited a small group of residents (N = 15) to conduct a pilot test before the large-scale experiment. Then, the researcher refined some questions according to the feedbacks from pilot study participants. 243 feedbacks were returned out of 300 revised questionnaires sent out, where the study found 213 households (with 579 residents) provided valid responses for the empirical analysis after excluding incomplete or invalid questionnaires. Table 1 summarizes the

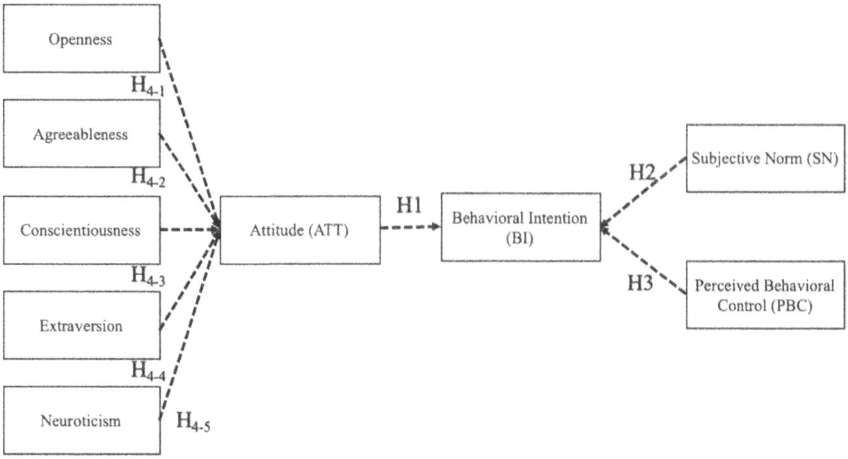


**Fig. 1** The research model

socio-demographic characteristics of the sample participating in the current study. Although this sample is not strictly statistically representative in the gender category, it generally matches the current conditions in Xi'an. Therefore, the data fulfills the requirements for theoretical and practical implications in this study.

## 3.2 Measure

The study adopts uniform online questionnaires with three sections to collect self-reported data from participants. The study provides the definition of household energy-saving behavior and listed 18 typical household energy-saving behavior in Liu et al. [6] ay the cover page. The first section adopts 17 items to measure four TPB factors, including attitude, subjective norms, PBC and behavioral intention. In this section, participants are required to choose one number from 1 (indicating the most negative view) to 5 (meaning the most favorable opinion). This study employs some items from Han et al. [39] and Wan et al. [40]. To minimize the impact of social desirability, the instructions that "there are no right or wrong answer; only your personal opinions matter" was presented at both cover letter and each section. All participants were identified by reference codes rather than their names to ensure the information confidential.

The second part focuses on questions about personality traits of interviewees. The Mini-international Personality Item Pool (Mini-IPIP) Scale has been utilized in this section. This section also applies the five-point scale. Each personality trait is accessed by four items, including two reversed questions. Mini-IPIP has been developed base on the IPIP-Five Factor Model (IPIP-FFM) [41], a 50-item Big Five personality traits scale. The IPIP-FFM scale was commonly-used in previous studies

**Table 1** Respondent profile

| Demographic variable | | N | % | Mean | Var |
|---|---|---|---|---|---|
| Number of family Members | 1 | 35 | 16.432% | 2.718 | 1.207 |
| | 2 | 57 | 26.761% | | |
| | 3 | 78 | 36.620% | | |
| | 4 | 26 | 12.207% | | |
| | 5 | 10 | 4.695% | | |
| | 6 | 7 | 3.286% | | |
| Number of female members | 0 | 36 | 16.901% | 1.347 | 0.967 |
| | 1 | 92 | 43.192% | | |
| | 2 | 68 | 31.925% | | |
| | 3 | 13 | 6.103% | | |
| | 4 | 2 | 0.939% | | |
| | 5 | 0 | 0 | | |
| | 6 | 2 | 0.939% | | |
| Number of male members | 0 | 13 | 6.103% | 1.372 | 0.745 |
| | 1 | 126 | 59.155% | | |
| | 2 | 59 | 27.670% | | |
| | 3 | 12 | 5.634% | | |
| | 4 | 3 | 1.408% | | |
| Average age | N/A | 213 | N/A | 37.540 | 10.027 |
| Highest education | Ph.D. degree or above (6) | 18 | 8.451% | N/A | N/A |
| | Master's degree or equivalent (5) | 52 | 24.413% | | |
| | Bachelor's degree or equivalent (4) | 71 | 33.333% | | |
| | Polytechnic degree or equivalent (3) | 39 | 18.310% | | |
| | Highschool or equivalent (2) | 21 | 9.859% | | |
| | Junior school and below (1) | 12 | 5.634% | | |
| Household Income (per month) | More than ¥30 k ($4412) | 11 | 4.8% | N/A | N/A |
| | ¥20 k to ¥30 k ($2942-$4412) | 26 | 12.6% | | |
| | ¥12 k to ¥20 k ($1765-$2942) | 54 | 25.1% | | |
| | ¥8 k to ¥12 k ($1176-$1765 | 49 | 23.2% | | |

(continued)

**Table 1** (continued)

| Demographic variable | | N | % | Mean | Var |
|---|---|---|---|---|---|
| | ¥4 k to ¥8 k ($588-$1176) | 52 | 25.1% | | |
| | Less than ¥4 k ($588) | 8 | 3.4% | | |
| | Not Applicable | 13 | 5.8% | | |
| Length of residence (year) | N/A | 213 | N/A | 10.58 | 88.477 |
| Ownership | Rent | 57 | 26.761% | N/A | N/A |
| | Self-owned | 156 | 73.240% | | |

because of its exceptional reliability. However, some researchers proposed that the 50-item IPIP-FFM scale has too many items which lower participants' patience in the experiment of their research. Donnellan et al. [42] developed a 15-item short form scale, namely Mini-IPIP, based on the IPIP-FFM with more straightforward questions and fewer questions. The Mini-IPIP scale presents a comparable pattern of convergent, discriminant, and validity with other Big Five based scales [42]. Mini-IPIP is widely applied in linking personality traits and predictors of pro-environmental behaviors. The items in section two are also on a five-point scale.

The third part focuses on socio-demographic details of participating people, where the first four questions record the necessary information, including their genders, ages, education levels and native places. There are also five questions about their families, such as the number of family members, house ownership, household income, and length of residence.

## *3.3 Data Analysis*

Structural equation modelling (SEM) has gained popularity in behavioral sciences [43] and has been widely-used in pro-environmental behavioral modelling (e.g., [6, 18, 40]). The statistical modelling technique integrates factor and path analysis and expresses a strong performance in accounting for measurement errors while estimating the modelled path coefficients. The SEM approaches can be explained in two widely-employed approaches, namely covariance-based SEM (CB-SEM) and partial least squares SEM (PLS-SEM; Wong, 2013). PLS-SEM makes it reliable to estimate and to analyze complex models with only a few observations or without imposing distributional assumptions [44]. CB-SEM is generally used for accurate theoretical confirmation with a larger sample size [44]. The data for this study has two characteristics: (1) sample size is small (i.e., 213); (2) there are eight relationships to be discussed, suggesting that the model is relatively complex. Therefore, PLS-SEM is adopted for qualitative analysis in this study. The study employs SmartPLS 3.0 for SEM analysis.

# 4 Results

## 4.1 Measurement Modelling Analysis

Before structural modelling, this study conducts measurement modelling to ensure all measurement items and constructs are statistically reliable and valid for modelling. The measurement modelling includes two steps: (1) convergent validity examination (CV), and; (2) discriminant validity examination (DV). The CV assessment consists of four criteria to examine the correlation between items measuring one constructs:

(1) The factor loading (FL) of each item should be no less than 0.5;
(2) The average variance extracted (AVE) of each construct should be no less than 0.5;
(3) The composite reliability of each construct should be no less than 0.7;
(4) The Cronbach's alpha ($\alpha$) of each construct should be no less than 0.6.

The result of the assessment (Table 2) indicated that all constructs in the current study adequately met the statistical requirements for CV.

The study then investigates the discriminant validity (DV) of the measurement. The study evaluates the DV through the Heterotrait-Monotrait (HTMT) Ratio (reported in Table 3), which is expected to be less than 0.9.

## 4.2 Structural Modelling Analysis

Table 4 present the structural modelling analysis results.

The research model connects personality traits with the TPB model. The structural modelling analysis result shows that attitude is the most significant factor associated with behavioral intention ($\beta = 0.288$, $p < 0.001$), followed by perceived behavioral control ($\beta = 0.178$, $p = 0.089$ and $p < 0.1$). The contribution of subjective norm to behavioral intention towards household energy saving is relatively small ($\beta = 0.095$, $p < 0.001$). The results support H1, H2 and H3. In addition, the results revealed that some personality traits may directly affect attitude towards household energy saving. Especially, data indicated that Agreeableness ($\beta = 0.224$, $p = 0.096$ and $p < 0.1$) may positively contribute to attitude towards energy saving behaviors at home and Extraversion are likely to be associated with attitude as well ($\beta = -0.111$, $p = 0.088$ and $p < 0.1$). However, other Big Five personality traits do not present significant correlations towards household energy-saving attitude, including Openness ($\beta = 0.021$, $p = 0.389$), Conscientiousness ($\beta = -0.086$, $p = 0.222$) and Neuroticism ($\beta = -0.077$, $p = 0.246$). In addition to the standard TPB factors, the analysis results suggest that household income ($\beta = 0.056$, $p = 0.026$ and $p < 0.1$) may positively contribute to energy-saving behavioral intention. However, the relationship between income and intention is weaker than three TPB factors' contribution. The correlation between intention and other socio-demographic factors, such as building ownership

**Table 2** CV assessment results

| Constructs | Item | FL | AVE | Composite reliability | Cronbach's Alpha (α) |
|---|---|---|---|---|---|
| Attitude | ATT-1 | 0.858 | 0.608 | 0.839 | 0.713 |
| | ATT-2 | 0.936 | | | |
| | ATT-3 | 0.887 | | | |
| | ATT-4 | 0.784 | | | |
| Subjective norms | SN-1 | 0.900 | 0.630 | 0.871 | 0.818 |
| | SN-2 | 0.725 | | | |
| | SN-3 | 0.783 | | | |
| | SN-4 | 0.755 | | | |
| PBC | PBC-1 | 0.800 | 0.523 | 0.773 | 0.856 |
| | PBC-2 | 0.577 | | | |
| | PBC-3 | 0.709 | | | |
| | PBC-4 | 0.625 | | | |
| | PBC-5 | 0.706 | | | |
| Intention | I-1 | 0.779 | 0.505 | 0.748 | 0.722 |
| | I-2 | 0.793 | | | |
| | I-3 | 0.862 | | | |
| | I-4 | 0.802 | | | |
| Extraversion (E) | E-1 | 0.853 | 0.834 | 0.734 | 0.721 |
| | E-2 | 0.579 | | | |
| | E-3 | 0.802 | | | |
| | E-4 | 0.684 | | | |
| Openness (O) | O-1 | 0.610 | 0.622 | 0.838 | 0.772 |
| | O-2 | 0.839 | | | |
| | O-3 | 0.763 | | | |
| | O-4 | 0.767 | | | |
| Agreeableness (A) | A-1 | 0.783 | 0.749 | 0.832 | 0.950 |
| | A-2 | 0.722 | | | |
| | A-3 | 0.669 | | | |
| | A-4 | 0.761 | | | |
| Conscientiousness (C) | C-1 | 0.755 | 0.739 | 0.722 | 0.833 |
| | C-2 | 0.674 | | | |
| | C-3 | 0.713 | | | |
| | C-4 | 0.853 | | | |
| Neuroticism (N) | N-1 | 0.571 | 0.884 | 0.815 | 0.865 |
| | N-2 | 0.739 | | | |
| | N-3 | 0.578 | | | |
| | N-4 | 0.915 | | | |

**Table 3** DV assessment results (heterotrait-monotrait ratio)

| | ATT | A | BI | BO | C | EDU | N | E | FOC | HI | O | PBC |
|---|---|---|---|---|---|---|---|---|---|---|---|---|
| ATT | | | | | | | | | | | | |
| A | 0.343 | | | | | | | | | | | |
| BI | 0.367 | 0.388 | | | | | | | | | | |
| BO | 0.032 | 0.106 | 0.133 | | | | | | | | | |
| C | 0.213 | 0.429 | 0.265 | 0.254 | | | | | | | | |
| EDU | 0.064 | 0.112 | 0.507 | 0.007 | 0.144 | | | | | | | |
| N | 0.114 | 0.234 | 0.144 | 0.255 | 0.318 | 0.129 | | | | | | |
| E | 0.168 | 0.456 | 0.227 | 0.114 | 0.262 | 0.312 | 0.726 | | | | | |
| FOC | 0.071 | 0.140 | 0.048 | 0.115 | 0.123 | 0.402 | 0.112 | 0.112 | | | | |
| HI | 0.087 | 0.113 | 0.132 | 0.605 | 0.516 | 0.801 | 0.602 | 0.208 | 0.153 | | | |
| O | 0.183 | 0.610 | 0.221 | 0.222 | 0.341 | 0.500 | 0.212 | 0.476 | 0.074 | 0.266 | | |
| PBC | 0.392 | 0.223 | 0.453 | 0.196 | 0.249 | 0.708 | 0.414 | 0.312 | 0.079 | 0.102 | 0.243 | |
| SN | 0.495 | 0.238 | 0.616 | 0.144 | 0.209 | 0.650 | 0.800 | 0.127 | 0.126 | 0.022 | 0.080 | 0.304 |

**Table 4** Structural modelling analysis results

| | β | Sample mean | Standard deviation (STDEV) | p-value |
|---|---|---|---|---|
| H1: ATT -> BI | 0.288 | 0.874 | 0.019 | *** |
| H2: SN -> BI | 0.095 | 0.096 | 0.025 | *** |
| H3: PBC -> BI | 0.178 | 0.097 | 0.132 | 0.089# |
| H4-1: O -> ATT | 0.021 | −0.011 | 0.075 | 0.389 |
| H4-2: A ->ATT | 0.224 | 0.167 | 0.172 | 0.096# |
| H4-3: C -> ATT | −0.086 | −0.068 | 0.112 | 0.222 |
| H4-4: E -> ATT | −0.111 | −0.128 | 0.082 | 0.088# |
| H4-5: N -> ATT | −0.077 | −0.068 | 0.113 | 0.246 |
| Building ownership -> BI | −0.044 | −0.047 | 0.026 | 0.045 |
| Education -> BI | −0.004 | −0.005 | 0.024 | 0.439 |
| FOC -> BI | −0.021 | −0.025 | 0.025 | 0.202 |
| Household income -> BI | 0.056 | 0.057 | 0.029 | 0.026* |

*Note* #: $p < 0.10$, *: $p < 0.05$, ***: $p < 0.001$

($\beta = -0.044$, $p = 0.045$), education ($\beta = -0.004$, $p = 0.439$) and frequency of cooking ($\beta = -0.021$, $p = 0.202$) are less significant.

# 5 Discussion

## *5.1 The Personality Bases of Energy-Saving Behaviors*

The result indicated that TPB is an adequate framework for explaining the household energy saving intention. The result presents that attitude is the most significant factor affecting household energy saving behavioral intention and is followed by PBC and subjective norm. The significance between subjective norms and intention is lower. Although the result is in line with another two studies in China [45, 46], the lower significance of the subjective norm is contrary to the expectation. There are two possible reasons. First, the study pays more attention to energy conservation behaviors at home. Although the expectation of other important people, such as superiors, co-workers, and environmental organizations, may influence one's energy-saving intention in public areas, the home environment may weaken this positive influence. Besides, the education level of the participants is higher than the average level in Xi'an City. Previous studies found that well-educated individuals tend to perform eco-friendly behaviors because of their environmental concerns rather than the evaluation of others [47]. Therefore, the subjective norm may play a less significant role in this process.

Agreeableness is significantly associated with attitude toward household energy saving behavior and its contribution is the largest among the Big Five personality variables. The result is also supported by Poškus and Žukauskienė [11], Yu and Yu [32] and Komatsu and Nishio [48]. People with high Agreeableness are likely to have higher levels of selflessness, morality and empathy. Especially, Agreeableness is related to people's concern for the interest of others, the welfare of environment and social development [29, 30, 49]. Thus, considerate individuals tend to extend their concerns for others to the energy saving for its positive impact on the environment. Milfont and Sibley [30] noted that environmental engagement is significantly related to higher levels of empathy and connectedness with though and cognitive ability, which explains the empirical finding that the individuals, who are selflessness-oriented and are cooperative towards others, are more likely to hold strong attitude towards reducing energy consumption and carbon emission in their households.

Extroversion is another significant factor. There is abundant evidence supporting that Extraversion contributes to some pro-environment behaviors and investments [11, 33, 50–52]. For example, Kvasova [52] reported a positive relationship between Extraversion and pro-environmental tourist behavior in Cyprus. Terrier et al. [33] also suggest that extroverted people tend to have higher eco-civic engagement. However, our SEM result shows that Extroversion plays a negative role in the household energy-saving attitude (consistent with [49] and the second case in [26]). The intense sociability of extraverted people may partly explain the adverse results between previous studies on other pro-environmental behaviors and this research. Extraversion reflects the cooperation and breadth of activities of an individual: extraverted people are more likely to consider the conditions and feelings of the surrounding people [53]. Some external factors related to the social environment, such as community attachment and opinions of neighbors, may strongly influence the household energy-saving behaviors of extraverted residents. Extraverted residents may present a higher attitude towards pro-environmental behaviors at public circumstances for going along with others and avoiding losing face, while their pro-environmental attitude might be lower at private places. Extroverted people might be energized and gregarious by the process of organizing an eco-civic event or participating in eco-friendly scheme rather than environmental protection itself.

The current study found the correlation between Conscientiousness and energy saving attitude is less significant. Although some existing studies suggest its positive impact on pro-environmental attitude, intention or behavior (e.g., [24, 27, 30]), the other research (e.g., [29, 34, 36]) supports the current result that household energy saving behaviors have indistinctive link with occupants' feeling of reasonability and adherence to social rules. The analysis also reflects an insignificant relationship between Openness and energy-saving behavioral attitude, which suggests that the degree of creativity and intellectual curiosity might not influence household energy-saving behavioral process. Many previous works argue that Openness to experience may positively contribute to pro-environmental behaviors (e.g., [24]) since people with higher openness-to-experience may accept environmental protection concepts easily. However, some studies (e.g., [52]) even made adverse conclusion. Therefore, further research is required to explore the truth. Similar to most existing studies (e.g., [35, 36]), the current research concludes that the relationship between Neuroticism and energy-saving behavioral attitude is less significant.

## 5.2 Limitations and Further Studies

This study has some limitations. Firstly, the data analysis of this study depends on self-reported behavioral data. Although respondents may have a different standard for their answers to the items, some studies indicated a slight tendency toward over-reporting (e.g., [54]). It is believed that participants may overvalue the items in the study because of social pressure and desirability, so-called social desirability bias. As the effect of social desirability changes with the nature of behavior [55], research in the future would benefit from considering social desirability bias into the experiment design process. For example, further studies would employ measures for social desirability and assess the objective energy-saving behaviors or energy consumption data of participants.

Secondly, the survey in this study can be further improved. The study employed Mini-IPIP scale, a brief measure of personality traits. Most personality-based studies tend to employ longer instruments, such as 240-item Big Five Scale and 50-item Big Five Inventory Scale. However, it may take much time to complete those longer questionnaires. Thus, when time and resources are limited, personality, researchers are often confronted with the choice between a brief or no measure of personality at all [31]. Besides, the shorter questionnaire means less information obtained: The employment of brief measures also limits the analysis on the relationship between the sub-traits and household energy-saving behaviors. Further studies would conduct a more detailed discussion on this topic.

## 6 Conclusion

This paper combines Big Five personality traits and the TPB model to explain household energy-saving behavior. With data collected from 213 households in three communities in Xi'an, the study first examined the TPB model. The results indicated that all three standard TPB predictors are strongly associated with behavioral intention, which well predicts household energy conservation behavior. The study found that attitude is the most significant factor influencing attitude towards energy-saving at home. Notably, residents may make decisions towards household energy-saving behavior based on their interests, such as saving money and comfort. Besides, many residents may perform energy-saving actions at home for environmental protection. Then, the result suggests that attitude play a mediating role between some Big Five personality traits and behavioral intention. The results suggested that people with higher Agreeableness tend to have a stronger attitude towards energy-conservation behaviors at home because of their empathic concerns to other people and the environment. On the contrary, Extroversion is likely to have negative impact on household energy conservation process. The high sociability and connection with others may explain this finding. However, the results indicate that the correlations between other three Big Five personality traits and attitude are insignificant. Besides, the study also demonstrates that household income may have a small but positive effect on energy-saving intention. This paper contributes to the pro-environmental behavior literatures by expounding the nature and the role of personality traits in predicting energy-saving intention. The model will not only provide a reference for further studies but also be conducive to governments and environmental protection organizations in the areas facing severe energy overuse problems in Chinese building sector.

## References


1. Zhang, Y., He, C.-Q., Tang, B.-J., Wei, Y.-M.: China's energy consumption in the building sector: a life cycle approach. Energy Build **94**, 240–251 (2015)
2. Yu, Z., Fung, B.C., Haghighat, F., Yoshino, H., Morofsky, E.: A systematic procedure to study the influence of occupant behavior on building energy consumption. Energy Build **43**(6), 1409–1417 (2011)
3. Klein, L., et al.: Coordinating occupant behavior for building energy and comfort management using multi-agent systems. Autom Constr **22**, 525–536 (2012)
4. Anderson, K., Lee, S., Menassa, C.: Impact of social network type and structure on modeling normative energy use behavior interventions. J Comput Civil Eng **28**(1), 30–39 (2014)
5. Kosonen HK, Kim AA (2017) Advancement of behavioral energy interventions in commercial buildings. *Facilities*

6. Liu, X., Wang, Q., Wei, H.-H., Chi, H.-L., Ma, Y., Jian, I.Y.: Psychological and demographic factors affecting household energy-saving intentions: a TPB-based study in Northwest China. Sustainability **12**(3), 836 (2020)
7. Zhang, Y., Wang, Z., Zhou, G.: Antecedents of employee electricity saving behavior in organizations: an empirical study based on norm activation model. Energy Pol **62**, 1120–1127 (2013)
8. Hong, T., Chen, C.-F., Wang, Z., Xu, X.: Linking human-building interactions in shared offices with personality traits. Build Environ **170**, 106602 (2020)
9. Kassin SM (2003) Essentials of psychology. Prentice Hall
10. Tang CMF, Lam D (2017) The role of extraversion and agreeableness traits on Gen Y's attitudes and willingness to pay for green hotels. Int J Contemp Hospit Manag
11. Poškus, M.S., Žukauskienė, R.: Predicting adolescents' recycling behavior among different big five personality types. J Environ Psychol **54**, 57–64 (2017)
12. Yazdanpanah M, Hadji Hosseinlou M (2017) The role of personality traits through habit and intention on determining future preferences of public transport use. Behav Sci 7(1):8
13. Munyaneza, J., Wakeel, M., Chen, B.: Overview of Rwanda energy sector: from energy shortage to sufficiency. Energy Proc **104**, 215–220 (2016)
14. Kaygusuz, K.: Energy services and energy poverty for sustainable rural development. Renew Sustain Energy Rev **15**(2), 936–947 (2011)
15. Lo, Siu Hing, Peters, Gjalt-Jorn Y., van Breukelen, Gerard J.P., Kok, Gerjo: Only reasoned action? An interorganizational study of energy-saving behaviors in office buildings. Energ Effi **7**(5), 761–775 (2014). https://doi.org/10.1007/s12053-014-9254-x
16. Obaidellah, U.H., Danaee, M., Mamun, M.A.A., Hasanuzzaman, M., Rahim, N.A.: An application of TPB constructs on energy-saving behavioural intention among university office building occupants: a pilot study in Malaysian tropical climate. J Hous Built Environ **34**(2), 533–569 (2019). https://doi.org/10.1007/s10901-018-9637-y
17. De Leeuw, A., Valois, P., Ajzen, I., Schmidt, P.: Using the theory of planned behavior to identify key beliefs underlying pro-environmental behavior in high-school students: Implications for educational interventions. J Environ Psychol **42**, 128–138 (2015)
18. Ru, X., Wang, S., Yan, S.: Exploring the effects of normative factors and perceived behavioral control on individual's energy-saving intention: an empirical study in eastern China. Resour Conserv Recycl **134**, 91–99 (2018)
19. Chiou, J.-S.: The effects of attitude, subjective norm, and perceived behavioral control on consumers' purchase intentions: the moderating effects of product knowledge and attention to social comparison information. Proc Natl Sci Counc ROC (C) **9**(2), 298–308 (1998)
20. Conner, M., Armitage, C.J.: Extending the theory of planned behavior: a review and avenues for further research. J Appl Soc Psychol **28**(15), 1429–1464 (1998)
21. Lim, Y.J., Osman, A., Salahuddin, S.N., Romle, A.R., Abdullah, S.: Factors influencing online shopping behavior: the mediating role of purchase intention. Proc Econ Fin **35**(5), 401–410 (2016)
22. Paul, J., Modi, A., Patel, J.: Predicting green product consumption using theory of planned behavior and reasoned action. J Retail Consum Serv **29**, 123–134 (2016)
23. Matthews G, Deary IJ, Whiteman MC (2003) Personality traits. Cambridge University Press
24. Brick, C., Lewis, G.J.: Unearthing the "green" personality: core traits predict environmentally friendly behavior. Environ Behav **48**(5), 635–658 (2016)
25. Pavalache-Ilie, M., Cazan, A.-M.: Personality correlates of pro-environmental attitudes. Int J Environ Health Res **28**(1), 71–78 (2018)
26. Wang, Q.-C., Wang, Y.-X., Jian, I.Y., Wei, H.-H., Liu, X., Ma, Y.-T.: Exploring the "energy-saving personality traits" in the office and household situation: an empirical study. Energies **13**(14), 3535 (2020)
27. Conner, M., Abraham, C.: Conscientiousness and the theory of planned behavior: toward a more complete model of the antecedents of intentions and behavior. Pers Soc Psychol Bull **27**(11), 1547–1561 (2001)

28. Ma, X., Bateson, D.J.: A multivariate analysis of the relationship between attitude toward science and attitude toward the environment. J Environ Educ **31**(1), 27–32 (1999)
29. Hirsh, J.B.: Personality and environmental concern. J Environ Psychol **30**(2), 245–248 (2010)
30. Milfont, T.L., Sibley, C.G.: The big five personality traits and environmental engagement: associations at the individual and societal level. J Environ Psychol **32**(2), 187–195 (2012)
31. Gosling, S.D., Rentfrow, P.J., Swann, W.B.: A very brief measure of the big-five personality domains. J Res Person **19**(21), 139–152 (2003)
32. Yu, T.-Y., Yu, T.-K.: The moderating effects of students' personality traits on pro-environmental behavioral intentions in response to climate change. Int J Environ Res Publ Health **14**(12), 1472 (2017)
33. Terrier, L., Kim, S., Fernandez, S.: Who are the good organizational citizens for the environment? An examination of the predictive validity of personality traits. J Environ Psychol **48**, 185–190 (2016)
34. Hilbig, B.E., Zettler, I., Moshagen, M., Heydasch, T.: Tracing the path from personality—via cooperativeness—to conservation. Eur J Pers **27**(4), 319–327 (2013)
35. Wuertz TR (2015) Personality traits associated with environmental concern
36. Markowitz, E.M., Goldberg, L.R., Ashton, M.C., Lee, K.: Profiling the "pro-environmental individual": a personality perspective. J Pers **80**(1), 81–111 (2012)
37. A. Busic-Sontic and C. Brick, "Personality trait effects on green household installations," 2018
38. Costa, P.T., McCrae, R.R.: Influence of extraversion and neuroticism on subjective well-being: happy and unhappy people. J Pers Soc Psychol **38**(4), 668 (1980)
39. Han, H., Kim, Y.: An investigation of green hotel customers' decision formation: developing an extended model of the theory of planned behavior. Int J Hosp Manag **29**(4), 659–668 (2010)
40. Wan, C., Shen, G.Q., Choi, S.: Experiential and instrumental attitudes: interaction effect of attitude and subjective norm on recycling intention. J Environ Psychol **50**, 69–79 (2017)
41. Goldberg, L.R.: The development of markers for the Big-Five factor structure. Psychol Assess **4**(1), 26 (1992)
42. Donnellan, M.B., Oswald, F.L., Baird, B.M., Lucas, R.E.: The mini-IPIP scales: tiny-yet-effective measures of the big five factors of personality. Psychol Assess **18**(2), 192 (2006)
43. Hooper D, Coughlan J, Mullen M (2008) Structural equation modelling: guidelines for determining model fit. Articles 2
44. Hair, J.F., Ringle, C.M., Sarstedt, M.: PLS-SEM: indeed a silver bullet. J Market Theor Pract **19**(2), 139–152 (2011)
45. Zhang, Y., Wang, Z., Zhou, G.: Determinants of employee electricity saving: the role of social benefits, personal benefits and organizational electricity saving climate. J Clean Prod **66**, 280–287 (2014)
46. Gao, L., Wang, S., Li, J., Li, H.: Application of the extended theory of planned behavior to understand individual's energy saving behavior in workplaces. Resour Conserv Recycl **127**, 107–113 (2017)
47. Prud'homme B, Raymond L (2013) Sustainable development practices in the hospitality industry: an empirical study of their impact on customer satisfaction and intentions. Int J Hospit Manag 34:116–126
48. Komatsu, H., Nishio, K.-I.: An experimental study on motivational change for electricity conservation by normative messages. Appl Energy **158**, 35–43 (2015)
49. Busic-Sontic, A., Czap, N.V., Fuerst, F.: The role of personality traits in green decision-making. J Econ Psychol **62**, 313–328 (2017)
50. von der Ohe H, Martins N (2016) Reducing the carbon footprint of research by recycling item level data. In ECRM2016-Proceedings of the 15th European Conference on Research Methodology for Business Management": ECRM2016, Academic Conferences and publishing limited, p 322
51. Ribeiro JA, Veiga RT, Higuchi AK (2016) Personality traits and sustainable consumption. Revista Brasileira De Market 15(3)
52. Kvasova, O.: The big five personality traits as antecedents of eco-friendly tourist behavior. Personal Individ Differ **83**, 111–116 (2015)

53. Watson D, Clark LA (1997) Extraversion and its positive emotional core. In: Handbook of personality psychology, Elsevier, pp 767–793
54. Hebert, J.R., Clemow, L., Pbert, L., Ockene, I.S., Ockene, J.K.: Social desirability bias in dietary self-report may compromise the validity of dietary intake measures. Int J Epidemiol **24**(2), 389–398 (1995)
55. Strahan R, Gerbasi KC (1972) Short, homogeneous versions of the Marlowe-Crowne social desirability scale. J Clin Psychol

# Assessing the Transition of Municipal Solid Waste Management Using Combined Material Flow Analysis and Life Cycle Assessment

**Dan Wang, Jun He, and Yu-Ting Tang**

**Abstract** Faced with the challenges to deal with increasingly growing and ever diversified municipal solid waste (MSW), a series of waste directives have been published by European Commission to divert MSW from landfills to more sustainable management options. The presented study assessed the transition of MSW management in Nottingham, UK, since the enforcement of the EU Landfill Directive using a tool of combined materials flow analysis (MFA) and life cycle assessment (LCA). The results show that the MSW management system in Nottingham changed from a relatively simple landfill and energy from waste (EfW) mode to a complex, multi-technology mode. Improvements in waste reduction, material recycling, energy recovery, and landfill prevention have been made. As a positive result, the global warming potential (GWP) of the MSW management system reduced from 1076.0 kg $CO_2$-eq./t of MSW in 2001/02 to 211.3 kg $CO_2$-eq./t of MSW in 2016/17. Based on the results of MFA and LCA, recommendations on separating food waste and textile at source and updating treatment technologies are made for future improvement.

**Keywords** Municipal solid waste · Material flow analysis · Life cycle assessment · Global warming potential · Future improvement

---

D. Wang (✉)
School of Life Science, Institute of Soil Ecology and Remediation, Taizhou University, Taizhou, Zhejiang, P.R. China
e-mail: dan.wang@nottingham.edu.cn

D. Wang · J. He (✉)
Department of Chemical and Environmental Engineering, International Doctoral Innovation Centre, University of Nottingham Ningbo China, Ningbo, Zhejiang, P.R. China
e-mail: jun.he@nottingham.edu.cn

Y.-T. Tang
School of Geographical Sciences, University of Nottingham Ningbo China, Ningbo, Zhejiang, P.R. China

 
F. K. S. Chan et al. (eds.), *Proceedings of the 2020 International Conference on Resource Sustainability: Sustainable Urbanisation in the BRI Era (icRS Urbanisation 2020)*, Environmental Science and Engineering,
https://doi.org/10.1007/978-981-15-9605-6_6

# 1 Introduction

Landfill used to be the main option for disposal of municipal solid waste (MSW) in Europe and is still the most widely adopted MSW management method worldwide. But it is the least sustainable MSW treatment due to its high contamination potentials, such as high greenhouse gas (GHG) emission resulting from the decomposition of biodegradable fraction, water and soil pollution as a result of leachate emission, and resources depletion resulting from unrecycled valuable materials [1–5]. To mitigate the environmental impacts of landfills and to deal with the increased quantity and ever diversified composition of MSW, the EU Landfill Directive was introduced in 1999 (EU Directive 99/31/EC). This directive emphasizes the reduction of landfilled biodegradable municipal waste (BMW). Since then, as illustrated in Fig. 1, waste directives have been successively introduced by European Commission to improve the sustainability of MSW management in Europe by diverting waste from landfills to more environmentally friendly management options at the upper layers of waste management hierarchy such as recycling and energy recovery, thus to facilitate the development of circular economy. Management targets were set in these directives. As a response, waste management policies, regulations and targets have been developed by England and Nottingham City Council (Fig. 1).

Since the implementation of the EU Landfill Directive, studies have been conducted to identify the gaps and difficulties of achieving the MSW management goals [3, 5, 6], to analyze the development of waste management legislations and practices [2, 5, 7, 8], and to evaluate the environmental impacts of waste management strategies [8–11]. However, the performance of MSW management from a transitional perspective under the guidance of EU waste directives has seldom been investigated and assessed.

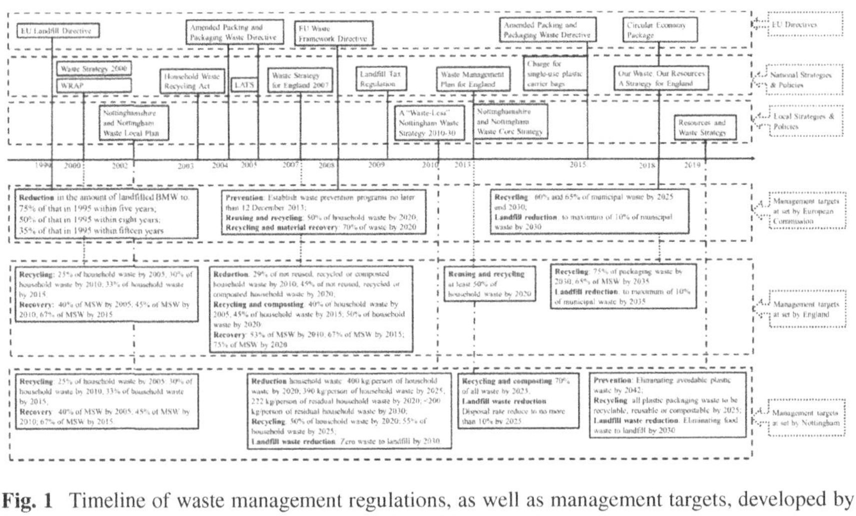


**Fig. 1** Timeline of waste management regulations, as well as management targets, developed by European Commission, England and Nottingham City Council

Material flow analysis (MFA) and life cycle assessment (LCA) are often used tools to assess the performance of MSW management, but they were often separately and independently applied. MFA is a robust, transparent, and useful tool in measuring the performance of an MSW management system by identifying and analyzing the pathways of waste streams, but it alone cannot sufficiently and comprehensively assess or support an MSW management strategy in view of certain goals, such as protection of human health and mitigation of global impact [12]. Even though, MFA provides well-grounded inventory for LCA [12, 13]. A Combination of MFA and LCA could identify the minor changes but might have long-term and/or significant damage and the most promising processes and flows for improvements.

Therefore, the present study investigates and assesses the transitioning MSW management in Nottingham, UK via identifying and quantifying the MSW flows and the associated global warming potential (GWP) using a tool of combined MFA and LCA. Three historical MSW management situations in 2001/02, 2006/07 and 2016/17 corresponding to three transitional stages in Nottingham in response to the EU waste directives were investigated and assessed. The novelty and contributions of this study can be summarized as follows:

- Application of a combined MFA and LCA approach to evaluate the performance of MSW management system at the meso level.
- Assessment of an MSW management from the vision of development.
- An insight into the effectiveness of waste regulations and policies.
- Assistance to local government in planning and decision making.
- Experiences for cities alike.

## *1.1 Case Study*

Nottingham is located in the central UK (52° 57′ N and 1° 09′ W). It was chosen as the study city because its MSW management strategy has been changed for several times in response to the EU and national waste regulations since 2000, and ambitious MSW management targets (e.g., recycling 55% of household waste by 2025, and achieving "zero waste to landfill" by 2030) have been set by the local authority (Fig. 1). Techniques and technologies including kerbside collection which separately collect recyclable materials and garden waste at source, material recovery facility (MRF) which sort and process recyclable materials from mixed recyclables or residual waste, and production of refuse-derived fuel (RDF) have been successively introduced into Nottingham for improving the sustainability of MSW management in the city.

## 2 Methodologies

A combination of MFA and LCA was applied in this study to quantitatively assess the transition of the MSW management system in Nottingham together with its environmental performance throughout the period from 2001/02 to 2016/17. MFA approach was applied to analyze the waste flows and stocks into, within and from the MSW management system. The performing of MFA is based on the first law of the thermodynamics entailing conservation of matter and energy [14, 15]. It is expressed as: mass in = mass out + stocks [13]. Through material balance, the sources, flows, accumulations and changes of wastes become visible [13]. MFA was performed using the free software STAN v2.6 (https://stan2web.net/). The associated GWPs of the waste streams described and quantified by the MFA were assessed using LCA.

### 2.1 *Goal and Scope*

The goals of this study were threefold: (1) to identify and quantify the waste flows in the MSW management situations at different stages of the transition; (2) to quantitatively assess the GWP of each situation; (3) to identify the successes and failures in the transition, and potential improvements. For consistency with targets set in waste regulations and available data, MSW was conceptualized as household waste which includes all waste collected from household sources and street cleaning. Separately collected commercial waste, industrial waste and healthcare waste were excluded from the scope of assessment. Functional unit of LCA was defined as the treatment of one ton of MSW, to ensure the situations were comparable to each other.

### 2.2 *System Boundaries*

Setting appropriate boundaries is critical as it affects the data collection and assessment results. In this study, the system boundaries were set to identify the waste flows and to analyze the GWP from the MSW management system in Nottingham. The spatial boundary was the administrative boundary of Nottingham City Council. The temporal boundary was the statistical year from April to March of the next year; for example, April 2001—March 2002, so that the years to our MSW management situations were expressed to cross two years, i.e. 2001/02. Detailed assumptions included in the assessment boundaries are summarized as follows:

(1) The MSW management processes include waste generation, collection, transfer, transport, treatment and disposal. Waste treatment facilities were identified from the database of WasteDataFlow (www.wastedataflow.org).

(2) GHG emissions include the direct emissions from MSW and indirect emissions from energy consumption for transport and operation of treatment and disposal facilities.
(3) Heat recovered from MSW was assumed to displace the same quantity of heat from natural gas which was the main energy source for home heating in the UK [16].
(4) Due to the variation of energy mix, the emission factors of electricity production were 0.45 kg $CO_2$-eq./kWh in 2002 [17], 0.47 $CO_2$-eq./kWh in 2007 [18] and 0.35 $CO_2$-eq./kWh in 2017 [19].
(5) GHG emissions from the operation of the Civic Amenity (CA) site and bring sites were not included because their data was unavailable.
(6) Bottom ash from incinerator (BAI) was not considered as a source of GHGs.

## 2.3 MSW Management Situations

In total, three historical MSW management situations (S1–S3) had been evaluated and compared in this study.

**S1.** The MSW management in 2001/02. 2001/02 is the earliest year when the MSW management data started to be documented. It is regarded as the beginning of the transition when the Nottinghamshire and Nottingham Waste Local Plan published in response to the EU Landfill Directive. In S1, weekly door-to-door collection was assumed to be provided by the local authority [20]. Source separation was unavailable. Landfilled waste was stored and transferred at transfer station. MSW was either directly disposed in landfills or incinerated in the Eastcroft EfW for energy recovery without pretreatment or material recovery [21]. An CA site and dozens of bring sites were set to collect recyclable materials including paper, glass and metal [21, 22]. BAI, as well as metal in it, was landfilled. Methane collection system at the landfills was not applied.

**S2.** The MSW management in 2006/07 before the enforcement of the Waste Framework Directive. 2006/07 is the earliest year when the waste flow was recorded. In S2, kerbside collection and MRF had been introduced, but kerbside collection had not been provided to all households. Transfer station was used to store and transfer waste to MRF. Residual waste was either disposed in landfills or incinerated in the Eastcroft EfW for energy recovery without pretreatment. Metal from BAI was recycled. Separately collected garden waste was treated via open windrow composting.

**S3.** The MSW management in 2016/17. It was the latest year with available data at the time for analysis. Kerbside collection was further strengthened to serve all households in Nottingham. RDF was produced at MRF. Only residual waste from MRF and fly ash from incinerator were landfilled. BAI was recycled for aggregates.

## 2.4 Life Cycle Inventory

**Collection, Transfer and Transport.** Estimated travel distances and life cycle inventories for collection and transport are presented in Table 1. GHG emission factor for each vehicle type was taken from Ecoinvent v3 database. Distances travelled in kerbside collection, bulky waste collection and door-to-door collection were modelled based on the length of accessible street within Lower Layer Super Output Areas (LSOA) Google Earth, the average population within LSOA and MSW generation per capita in Nottingham. Electricity and diesel consumed for the operation of transfer station were 4 kWh/t and 0.84 kg/t, respectively [23]. Distances between LSOA and waste management facilities and distances between waste management facilities were estimated based on their locations using Google Earth and Google map.

**Landfill.** Diesel and electricity consumptions for the operation of landfill were 1.8 kg/t and 8 kWh/t, respectively [23]. Methane emitted from landfill was estimated using the method and equations reported by Fong et al. [25]. They estimate the total potentially generated methane based on the mass and composition of landfilled waste. The waste compositions of landfilled waste in S1–S2 were shown in Table 2.

**Incineration with Energy Recovery.** The quantity of $CO_2$ generated from incinerated waste was calculated based on the mass and composition of it (Table 3) using the method and equations provided by the IPCC [26]. Heat recovered from waste was assumed to substitute the equivalent heat generated from gas boilers with an efficiency of 89%. The efficiency of the Eastcroft EfW was 15.3% for electricity and 28.2% for heat of the lower heating value (LHV) of MSW [27]. Electricity and fuel

**Table 1** Travel distances of waste collection and transport, and emission factors applied for each vehicle type

| | Transport distance (km) | Vehicle type | Emission factors (kg $CO_2$-eq./tkm) |
|---|---|---|---|
| Kerbside collection | 14 | Road, lorry 16–32 metric ton | 0.177 |
| Bulky waste collection | 14 | Road, lorry 16–32 metric ton | 0.177 |
| CA site collection | 0 | Road, lorry | 0.135 |
| Street cleaning | 20 [23] | Road, lorry 3.5 metric ton [24] | 0.555 |
| Bring sites collection | 20 [23] | Road, lorry | 0.135 |
| Door-to-door collection | 14 | Road, lorry 16–32 metric ton | 0.177 |
| Transport | Distance between waste management facilities | Road, lorry<br>Rail, freight<br>Ocean, ship | 0.135<br>0.0431<br>0.0112 |

**Table 2** Composition of landfilled waste (%)

| Composition category | S1 | S2 | S3 | Degradable organic carbon (DOC) content in wet waste |
|---|---|---|---|---|
| Putrescible | 21.0 | 37.6 | 2.3 | 15 |
| Paper and card | 32.0 | 21.1 | 19.3 | 40 |
| Plastics | 11.0 | 3.0 | 2.4 | 0 |
| Wood | – | 11.5 | 29.6 | 43 |
| Textiles | 2.0 | 4.5 | 1.1 | 24 |
| Glass | 9.0 | 1.5 | 10.6 | 0 |
| Metals | 8.0 | 3.8 | 1.5 | 0 |
| Other | 17.0 | 17.0 | 33.2 | 0 |
| Total | 100 | 100 | 100 | – |

**Table 3** Composition of incinerated waste (%)

| Composition category | S1 | S2 | S3 | Dry matter content of wet weight | Total carbon content in dry weight | Fossil carbon fraction of total carbon |
|---|---|---|---|---|---|---|
| Putrescible | 21.0 | 25.8 | 34.9 | 40 | 38 | – |
| Paper and card | 32.0 | 20.8 | 10.2 | 90 | 46 | 1 |
| Textiles | 2.0 | 3.3 | 9.0 | 80 | 50 | 20 |
| Dense plastics | 6.0 | 8.0 | 7.2 | 100 | 75 | 100 |
| Plastic film | 5.0 | 8.1 | 4.0 | 100 | 75 | 100 |
| Fines (<10 mm) | 7.0 | 3.4 | 0.4 | 90 | 3 | 100 |
| Miscellaneous combustibles | 8.0 | 10.9 | 19.2 | 40 | 70 | 10 |
| Miscellaneous non-combustibles | 2.0 | 3.2 | 4.7 | 100 | – | – |
| Non-ferrous metal | 2.0 | 1.3 | 0.9 | 100 | – | – |
| Glass | 9.0 | 9.4 | 3.2 | 100 | – | – |
| Ferrous metal | 6.0 | 3.3 | 2.6 | 100 | – | – |
| Others | 0 | 2.7 | 3.7 | - | – | – |
| LHV (MJ/kg) | 9.6 | 8.8 | 6.8 | - | – | – |

oil consumed to operate incineration plant were 62 kWh/t and 3.76 kg/t, respectively [28].

**Recycling.** Avoided emission by material recycling was estimated based on the England Carbon Metric Report [29]. This report gives a summary about GHG avoided by materials recycling and reusing compared to landfilling of these materials.

**Composting.** Details of life cycle inventories (LCI) for composting are presented in Table 4. 36% of garden waste was assumed to be non-compostable material [30].

**Table 4** LCI for composting

| | Unit | Value | Reference |
|---|---|---|---|
| *Pre-treatment input* | | | |
| Diesel | kg/t | 0.1 | [23] |
| Electricity | kWh/t | 1.1 | [23] |
| *Composting input* | | | |
| Diesel | kg/t | 3.07 | [31] |
| Electricity | kWh | 0.51 | [31] |
| *Process emission* | | | |
| $CH_4$ | kg/t | 4 | [26] |
| $N_2O$ | kg/t | 0.24 | [26] |
| *GHG emission from production of inorganic fertilizer* | | | |
| N fertilizer | kg $CO_2$-eq./kg | 6.8 | [32] |
| P fertilizer | kg $CO_2$-eq./kg | 1.2 | [32] |
| K fertilizer | kg $CO_2$-eq./kg | 0.5 | [32] |
| *Avoided fertilizer product by applying compost* | | | |
| N fertilizer | kg/t | 3.4 | [33] |
| P fertilizer | kg/t | 2.8 | [33] |
| K fertilizer | kg/t | 9.7 | [33] |

**Material Recovery Facility.** There were two types of MRF in Nottingham. One was designed to sort and process mixed recyclable materials. It consumed 2 kg/t diesel and 35 kWh/t electricity for its operation [23]. Another MRF was designed to recover materials from bulky waste, street waste and residual waste. Diesel and electricity consumption in this MRF were 2 kg/t and 44 kWh/t, respectively [23, 24].

**Production and Incineration of RDF with Energy Recovery.** 40 kWh/t electricity was consumed to produce RDF [34] with standard LHV of 25 MJ/kg and fossil carbon content of 32% by weight [26, 35]. RDF was assumed to be incinerated to only generate electricity with an efficiency of 25% based on the thermal energy production estimated using corresponding LHV of RDF [34]. GHG emission from the incineration of RDF could be calculated based on the mass and fossil carbon content of RDF.

## 2.5 Life Cycle Impact Assessment

The life cycle impact assessment was characterized by GWP at a 100-year time period, with characterization factors taken from IPCC. The GHGs calculated in this study covered carbon dioxide ($CO_2$), methane ($CH_4$) (GWP factor: 25) and nitrous oxide ($N_2O$) (GWP factor: 298). The total GWP of the MSW management was the sum of GWPs of all GHGs, and expressed as $GWP_{100}$.

# 3 Results and Discussions

## 3.1 *Results of MFA*

Comprehensive material flow diagrams for MSW management situations are presented in Figs. 2, 3 and 4. Residual waste was the dominant waste stream for all situations, but its disposal varied in situations. Majority of residual waste (67,617 t) was landfilled in S1 but it was incinerated for energy recovery in S2 and S3. Due to the implementation of kerbside collection and MRF, the recycling (and composting) rate was significantly increased from 3.4% (3.6%) in S1 to 31.4% (44.9%) in S3 (Table 5). But there is still a big gap to the 2020 and 2025 targets. Even though, Nottingham has big achievements on waste prevention, energy recovery from waste and prevention of landfilled waste. The reduction target to 390 kg per person per year of household waste had been achieved in the latest situation (Table 5). In S3, majority of waste (61.9%) was incinerated for energy recovery and the landfill rate reduced to as low as 7.3% (Table 5). As such, MSW management in Nottingham transformed from a relatively simple model (S1) combining landfilling and incineration with energy recovery into a more complex model (S3) integrating source separation, recycling, composting, pre-treating landfilled waste and incineration with energy recovery.

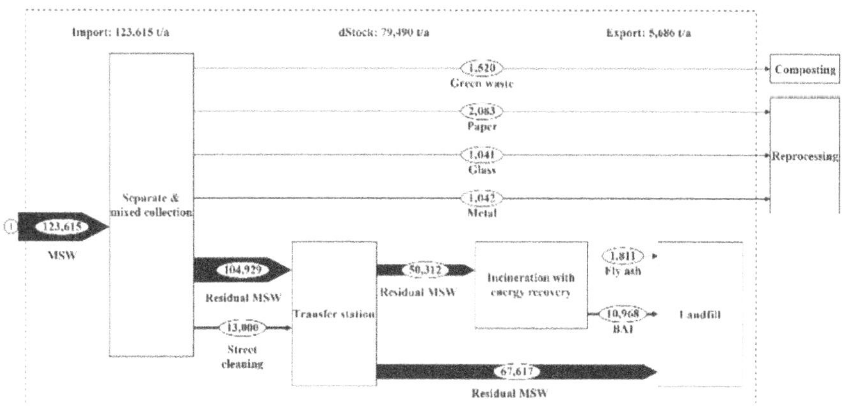


**Fig. 2** Material flow diagram for S1

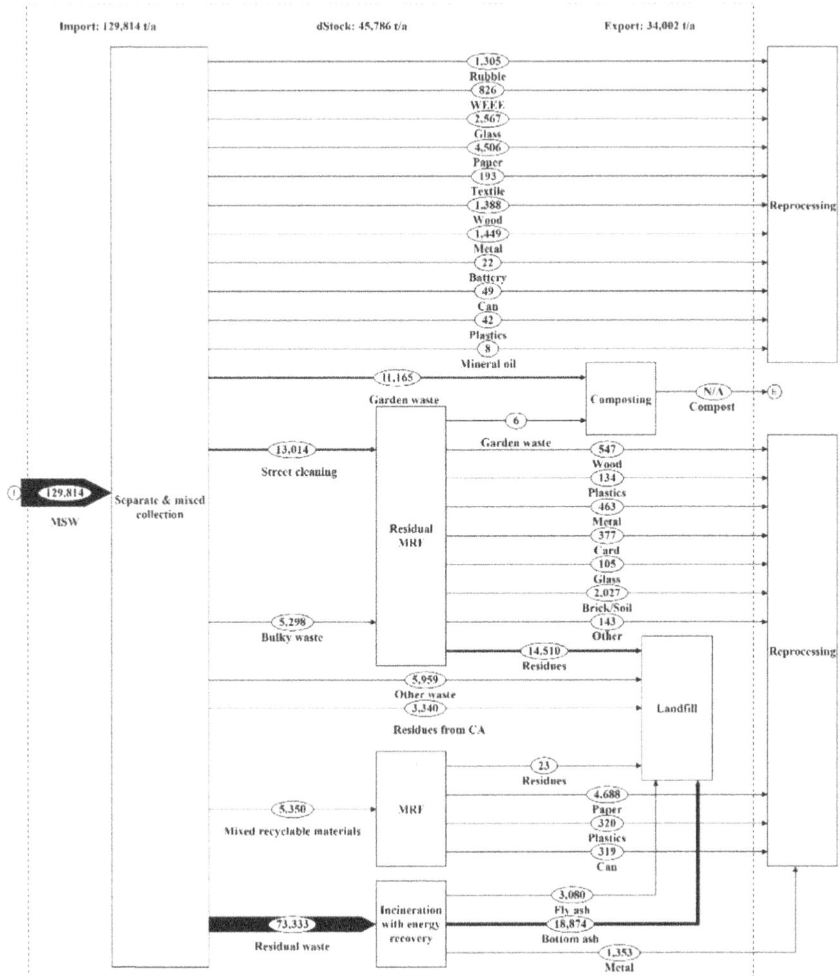


**Fig. 3** Material flow diagram for S2

## 3.2 Life Cycle Interpretation

Based on the results of MFA, LCA has been performed for all MSW management situations. The LCA results are presented in Fig. 5. $GWP_{100}$ of MSW management in Nottingham significantly decreased from 1076.0 kg $CO_2$-eq./t of MSW in S1 to 487.9 kg $CO_2$-eq./t of MSW in S2, and further to 211.3 kg $CO_2$-eq./t of MSW in S3. This is mainly due to the diversion of waste from landfill to more sustainable management options such as recycling, composting and incineration with energy recovery. Followed by incineration, landfills were the main source of GHG in S1 and S2, but their roles were reversed in S3. GHG emission from landfills kept falling

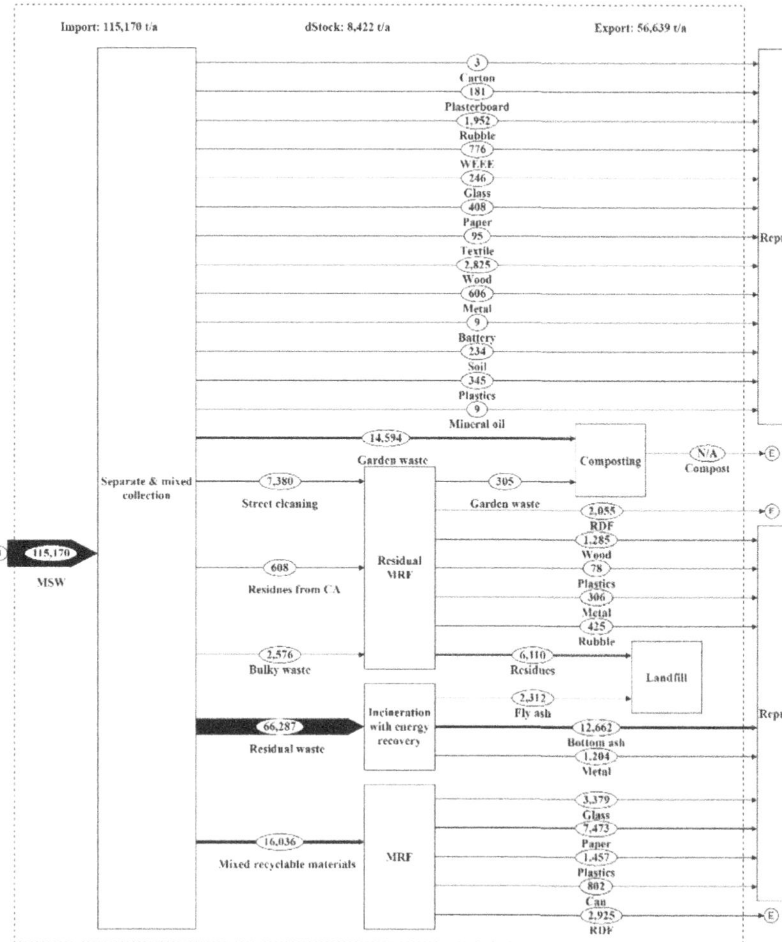


**Fig. 4** Material flow diagram for S3

during the study period because of the reduction of landfilled waste. GHG emission from incineration increased from 210.3 kg $CO_2$-eq./t of MSW in S1 to 294.7 kg $CO_2$-eq./t of MSW in S2, and then decreased to 174.5 kg $CO_2$-eq./t of MSW in S3 even though the share of incinerated waste kept increasing during the study period. The main reason of decreased $GWP_{100}$ of incineration from S2 to S3 is the changed composition of incinerated waste. The shares of materials having high content of carbon, such as plastic film and dense plastics, were lower in S3 than those in S2. Due to the improved separation of garden waste, GHG produced by composting kept growing from 1.3 kg $CO_2$-eq./t of MSW in S1 to 13.2 kg $CO_2$-eq./t of MSW in S3.

However, materials recycling was the only waste management practice that consistently saved GHG emission in all scenarios, while composting of garden waste and incineration with energy recovery added $GWP_{100}$ because of the limitations

**Table 5** Comparison of MFA indicators for the three situations

| Indicator | Description | Situation | | |
|---|---|---|---|---|
| | | S1 | S2 | S3 |
| Waste generation per capita (kg/y) | The MSW generated by each resident in a specific place (in this case is Nottingham) in a statistical year | 463.0 | 465.8 | 361.2 |
| Recycling rate (%) | The ratio between the amount of waste prepared for recycling and the total amount of waste generated | 3.4 | 17.6 | 31.9 |
| Recycling and composting rate (%) | The ratio between the amount of waste pre-pared for recycling and composting and the total amount of waste generated | 4.6 | 26.2 | 44.9 |
| Recovery rate (%) | The ratio between the amounts of waste used for recovery options and the total amount of waste generated | 40.7 | 56.5 | 61.9 |
| Landfill rate (%) | The ratio between the amount of waste disposed in landfill and the total amount of waste generated | 54.7 | 35.3 | 7.3 |

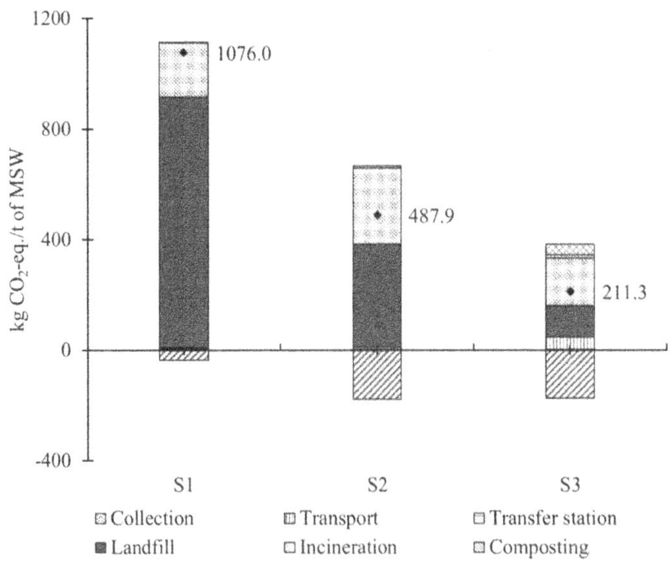


**Fig. 5** Change of total $GWP_{100}$ and $GWP_{100}$ added by MSW management processes

Table 6 Comparison of GHG emissions (unit: kg $CO_2$-eq.) from collection and transport

| | S1 | S2 | S3 |
|---|---|---|---|
| Collection | 3.4 | 3.1 | 2.8 |
| Transport to of recyclable materials to reprocessing facilities | 1.1 | 4.7 | 42.2 |
| Transport between treatment facilities | 3.5 | 2.5 | 2.0 |
| Total | 8.0 | 10.3 | 47.0 |

of outdated technologies and low efficiencies of the facilities. Technology adopted for composting was open windrow composting which had high potential of methane generation. Production of organic fertilizer from each ton of garden waste avoided 20.4 kg $CO_2$-eq./t GHG emission, but the GHG emissions (122.5 kg $CO_2$-eq./t of garden waste) from decomposition of garden waste and facility operation were much more than the emission saved by avoiding landfilling. The energy recovery efficiency of Eastcroft EfW appeared to be lower than other cases reported in the literature, e.g. 26.1% of electricity production, 77.2% of heat production and 52.1% of combined heat and power reported by Reimann [36], and 28% for electricity and 85% for heat reported by Habib et al. [37].

GHG emissions from collection and transport were usually ignored when performing the life cycle impact of MSW management because it only took a small part of the overall GHG emission. However, we noticed that the quantity and share of $GWP_{100}$ contributed by transport of recyclable materials to reprocessing facilities showed an increasing trend. It increased from 1.1 kg $CO_2$-eq./t of MSW in S1 to 42.2 kg $CO_2$-eq./t of MSW in S3 (Table 6). This growth was the result of the combination of growing recycling rate and the distant locations of reprocessing facilities. Some of them were located in overseas.

## 3.3 Recommendations for Future Improvements

According to the results of MFA and LCA, some recommendations have been made for the future improvement of MSW management in Nottingham.

Separating food waste at source and applying biological treatment for it are recommended. First, this measure could help the local authority to meet the recycling and composting target. Second, separating food waste improves the LHV of incinerated waste, thus improves efficiency of incineration and reduces the potentials of toxic gas generation resulting from incomplete combustion. Third, applying biological treatment on separately collected food waste provides additional organic fertilizer and/or renewable energy. Furthermore, separating food waste at source could reduce the potential of contamination of recyclable materials.

The results of MFA show that textile waste has been used more for the energy recovery. Textile was excluded by kerbside collection. Studies indicate that recycling textile generates less environmental burden compared to using virgin materials [38].

Therefore, we recommend to take textile into the scope of kerbside collection and recycling textile materials rather than incinerating it. In this way, the recycling rate in Nottingham could be further improved.

We also recommend the local authority to use advanced biological treatment technology, such as anaerobic digestion, to treat BMW (garden waste and food waste in this case) and to upgrade the EfW facility to improve the energy recovery efficiency. As food waste is recommended to be separately collected and treated, more methane will be produced from it. Adopting anaerobic digestion can not only reduce GHG emission but also improve the energy recovery ability of the MSW management system.

Furthermore, the significantly increased GHG emission from transport indicates that improving domestic reprocessing of secondary materials also needs to be enhanced in the future. Improving energy consumption of vehicles also helps to reduce the GHG emission from transport.

## 4 Conclusions

The enforcement of EU waste directives and the associated targets have stimulated the update of national and local waste management regulations, plans and strategies. To assess the transition of MSW management under the guidance of these waste directives, this study assessed the performance of three historical MSW management situations at different transitional stage in Nottingham using the combined MFA and LCA. Both the MFA and LCA result confirmed the improved performance of the MSW management in Nottingham which changed from a simple model of combined landfilling and incineration with energy recovery to a combination of source separation, recycling, pre-treatment before landfilling, production of RDF, composting, incineration with energy recovery and landfilling with complex waste flows. The transition resulted in a constant reduction of $GWP_{100}$. All these improvements contribute to the development of circular economy in the UK by providing renewable material and energy resources from waste. However, more efforts need to be made to meet national and local management targets. The future improvements can be made by separating food waste and textile waste at sources, replacing open windrow composting by anaerobic digestion, updating the EfW facility and enhancing domestic reprocessing of secondary materials.

## References

1. Pan, J., Voulvoulis, N.: The role of mechanical and biological treatment in reducing methane emissions from landfill disposal of municipal solid waste in the United Kingdom. J. Air Waste Manag. Assoc. **57**(2), 155–163 (2007)

2. Taşeli, B.K.: The impact of the European Landfill Directive on waste management strategy and current legislation in Turkey's specially protected areas. Resour. Conserv. Recycl. **52**(1), 119–135 (2007)
3. Lasaridi, K.: Implementing the landfill directive in Greece: problems, perspectives and lessons to be learned. Geogr. J. **175**(4), 261–273 (2009)
4. Apostol, L., Mihai, F.: The process of closing down rural landfills case study: Neamţ county. Present Environ. Sustain. Dev. **5**(2), 167–174 (2011)
5. Stanic-Maruna, I., Fellner, J.: Solid waste management in Croatia in response to the European landfill directive. Waste Manage. Res. **30**(8), 825–838 (2012)
6. Price, J.L.: The landfill directive and the challenge ahead: demands and pressures on the UK householder. Resour. Conserv. Recycl. **32**(3–4), 333–348 (2001)
7. Scharff, H.: Landfill reduction experience in The Netherlands. Waste Manage. **34**(11), 2218–2224 (2014)
8. Závodská, A., Benešová, L., Smyth, B., Morrissey, A.J.: A comparison of biodegradable municipal waste (BMW) management strategies in Ireland and the Czech Republic and the lessons learned. Resour. Conserv. Recycl. **92**, 136–144 (2014)
9. Pires, A., Martinho, M.G., Silveira, A.: Could MBT plants be the solution of fulfil landfill directive targets in Portugal. In International Symposium MBT 63–72 (2007).
10. Emery, A., Davies, A., Griffiths, A., Williams, K.: Environmental and economic modelling: a case study of municipal solid waste management scenarios in Wales. Resour. Conserv. Recycl. **49**(3), 244–263 (2007)
11. Ionescu, G., Rada, E.C., Ragazzi, M., Mărculescu, C., Badea, A., Apostol, T.: Integrated municipal solid waste scenario model using advanced pretreatment and waste to energy processes. Energy Convers. Manage. **76**, 1083–1092 (2013)
12. Allesch, A., Brunner, P.H.: Material flow analysis as a decision support tool for waste management: a literature review. J. Ind. Ecol. **19**(5), 753–764 (2015)
13. Brunner, P.H., Rechberger, H.: Practical Handbook of Material Flow Analysis: For Environmental, Resource, and Waste Engineers. CRC press (2016).
14. Rotter, V.S., Kost, T., Winkler, J., Bilitewski, B.: Material flow analysis of RDF-production processes. Waste Manage. **24**(10), 1005–1021 (2004)
15. Makarichi, L., Techato, K.A., Jutidamrongphan, W.: Material flow analysis as a support tool for multi-criteria analysis in solid waste management decision-making. Resour. Conserv. Recycl. **139**, 351–365 (2018)
16. Department for Business (DB), Energy & Industrial Strategy (EIS). https://www.gov.uk/government/collections/electricity-statistics, last accessed 2017/11/25.
17. Department for Environment, Food & Rural Affairs (DEFRA), Department for Business, Energy & Industial Strategy (DBEIS). https://www.gov.uk/government/publications/greenhouse-gas-reporting-conversion-factors-2017, last accessed 2019/04/18.
18. Department for Environment, Food & Rural Affairs (DEFRA), Department for Energy & Climate Change (DECC). https://www.gov.uk/government/publications/greenhouse-gas-reporting-conversion-factors-2007, last accessed 2019/04/19.
19. Department for Environment, Food & Rural Affairs (DEFRA), Department for Business, Energy & Industial Strategy (DBEIS). https://www.gov.uk/government/publications/greenhouse-gas-reporting-conversion-factors-2017, last accessed 2019/04/19.
20. Parfitt, J.P., Lovett, A.A., Sünnenberg, G.: A classification of local authority waste collection and recycling strategies in England and Wales. Resour. Conserv. Recycl. **32**(3–4), 239–257 (2001)
21. Nottingham City Council (NCC): Best Value Performance Plan 2005–2006 (2005).
22. Data.Gov. https://data.gov.uk/dataset/c9a3d775-6e00-4b8f-9f80-7f28fea7d944/household-recycling-by-material-and-region-england, last accessed 2018/11/12.
23. Turner, D.A., Williams, I.D., Kemp, S.: Combined material flow analysis and life cycle assessment as a support tool for solid waste management decision making. J. Cleaner Prod. **129**, 234–248 (2016)

24. Nottingham City Council (NCC): Annual population survey—households by combined economic activity status (2017).
25. Fong, W.K., Sotos, M., Doust, M., Schultz, S., Marques, A., Deng-Beck, C.: Global Protocol for Community-Scale Greenhouse Gas Emission Inventories (GPC). New York, NY, USA, World Resources Institute (2015)
26. International Panel on Climate Change (IPCC): 2006 IPCC Guidelines for National Greenhouse Gas Inventories Volume 5 Waste (2006).
27. FCC Environment: Annual Performance Report for Eastcroft Energy from Waste Facility Year-2014 (2015).
28. WRG: Annual Performance Report for Eastcroft Energy from Waste Facility Year-2007 (2008).
29. Department for Environment, Food & Rural Affairs (DEFRA). https://www.gov.uk/government/uploads/system/uploads/attachment_data/file/142019/Carbon_Metric_-_final_published.xls, last accessed 2018/04/10.
30. Nottingham City Council (NCC): Nottingham City Waste Kerbside Composition Analysis (2013).
31. Fisher, K.: Impact of Energy from Waste and Recycling Policy on UK Greenhouse Gas Emissions. Environment Resource Management for Department of Environment, Food and Rural Affairs (2006)
32. Hill, N., Walker, H., Beevor, J., James, K.: Guidelines to DEFRA/DECC's GHG Conversion Factors for Company Reporting: Methodology Paper for Emission Factors. Published by DEFRA (Department of Environment, Food and Rural Affairs) PB13625 (2011).
33. Boldrin, A., Andersen, J.K., Møller, J., Christensen, T.H., Favoino, E.: Composting and compost utilization: accounting of greenhouse gases and global warming contributions. Waste Manage. Res. **27**(8), 800–812 (2009)
34. Burnley, S., Phillips, R., Coleman, T., Rampling, T.: Energy implications of the thermal recovery of biodegradable municipal waste materials in the United Kingdom. Waste Manage. **31**(9–10), 1949–1959 (2011)
35. Materazzi, M., Lettieri, P., Mazzei, L., Taylor, R., Chapman, C.: Fate and behavior of inorganic constituents of RDF in a two stage fluid bed-plasma gasification plant. Fuel **150**, 473–485 (2015)
36. Reimann, D.O. https://www.cewep.eu/wp-content/uploads/2017/09/13_01_15_cewep_energy_report_iii.pdf, last accessed 2019/05/12.
37. Habib, K., Schmidt, J.H., Christensen, P.: A historical perspective of global warming potential from municipal solid waste management. Waste Manage. **33**(9), 1926–1933 (2013)
38. Woolridge, A.C., Ward, G.D., Phillips, P.S., Collins, M., Gandy, S.: Life cycle assessment for reuse/recycling of donated waste textiles compared to use of virgin material: an UK energy saving perspective. Resour. Conserv. Recycl. **46**(1), 94–103 (2006)

# Green Supply Chain in Construction Sector, Government Intervention, Partnerships and Green Practices

**Ying Xie, YiQing Zhao, and YaHui Chen**

**Abstract** Rapid urbanization and investment in infrastructure projects have promoted significant growth in the construction industry, which has led to a series environmental issues and threats. While the contribution of green supply chain management (GSCM) is receiving increasing attention in the construction sector, the role of environmental regulations, their influences on government support and public-private partnerships (PPPs), and the resulted impacts on GSCM practices implementations have not thoroughly investigated. Using an 18-item instrument and drawing from the survey results from 229 Chinese construction companies, this study investigates the causal relationships between constructs using partial least squared based structural equation modelling (PLS-SEM). The results indicated that environmental regulations had a significant positive impact on the GSCM practices implementation, through the sequential mediation effects of government support and PPPs. Firm facing stronger market pressures were found to have stronger desire in seeking for government support and establish PPPs, so as to engage more actively in GSCM practices implementation.

**Keywords** Urbanisation · Green supply chain management · Construction · Government support · Public-private partnerships · Market pressures

## 1 Introduction

Building and construction sectors together are responsible for 39% of all carbon emissions in the world. There has been growing global recognition of the urgent need to tackle the carbon emission from construction industry [4]. European Union introduced directives on waste reduction,the UK Government's strategy for construction,

Y. Xie (✉) · Y. Chen
Faculty of Business and Law, Anglia Ruskin University, Cambridge CM1 1SQ, UK
e-mail: kariarxy@hotmail.com

Y. Zhao
Logistics School, Linyi University, Linyi 276005, China

 
F. K. S. Chan et al. (eds.), *Proceedings of the 2020 International Conference on Resource Sustainability: Sustainable Urbanisation in the BRI Era (icRS Urbanisation 2020)*, Environmental Science and Engineering,
https://doi.org/10.1007/978-981-15-9605-6_7

is to achieve a 50% carbon reduction from the built environment by 2025, compared to a 1990 baseline; the Chinese government also aimed to increase the proportion of newly constructed green buildings in urban areas to be 50% by 2020. These targets impose huge pressures on the construction industry to green their regular practices, i.e., integrate environmental concerns into their operations and business strategies [6]. However, industries in construction sector also need to maintain a level of economic development, and there is also a paucity of sustainable knowledge, which may impede the implementation of environmental management practices [25].

Green Supply Chain Management (GSCM) has emerged as an environmentally sustainable organisational technological innovation, integrating environmental concerns into organisational supply chain (SC) management activities [37], to help organization gain both environmental performances and economic profitability [44]. GSCM has been extensively investigated by researchers as a systematic approach to facilitate transformation of the construction sector through operational changes [11] and strategic partnering and alliance [36, 40].

However, the construction SC exhibits unique characteristics that differentiate from other industries. Construction industry is characterised by the project-based nature, the large number of actors involved (potentially thousands), short term and ad-hoc project relationships, lack of continuity and relatively unstable SC [31]. There is often a lack of trust, or a high level of conflict among SC stockholders [18]. The unique nature of construction SC means that the traditional GSCM activities may not be applicable or become less effective [4]. In addition, enterprises in construction SC may vary dramatically in their efforts to green the SC, influenced by the potential benefits of taking such an initiative. A systematic literature review of GSCM in construction reveals that there is a need to identify a mechanism to resolve conflicts among SC actors, to faster the development of relationships among them, and to manage the transition to green SC processes and activities [4].

The scale and the number of actors in a construction SC make it challenging for a single actor, such as developer or contractor, to manage the transition to GSCM. Government intervention, implemented through hierarchical structure, in the form of regulations, incentives, and support, backed by real enforcement measures, has been identified as an effective mechanism to influence organisations adopting green practices [42], and improve firms' environmental and innovation capabilities [26]. But there has been very limited research as to whether government intervention can act as an effective mechanism to resolve conflicts among supply chain actors or to enable the establishment of a collaborative paradigm, especially in a very complex SC context such as construction SC, which is the first research objective of our study.

There have been consistent needs for enhancing collaboration across the player across the construction SC, to expedite progress on tackling global carbon emission. As a form of collaborative relationship, partnership has emerged as a popular contracting concept to respond to the adversarial culture and high levels of conflict associated with the construction SC [9].

Under a collaborative paradigm, innovation in SC is argued to be a collaborative and systematic mechanism that seeks to achieve a radical breakthrough changes on all three dimensions of sustainability, economic, environmental and social performances

[20]. Zhu et al. [44] defined GSCM as an emergent environmentally sustainable technological innovation. GSCM could offer a system-level approach to facilitate the transformation of construction industry by reducing the environmental impacts of the SC. The second research objective is to gain insight into inter-organisation partnership and knowledge transfer of innovative skills and practices between SC actors, and to analyse the impacts of the partnership on GSCM adoption.

Our study is the first to provide empirical evidence of associations between government support and partnerships. The empirical findings demonstrate the positive serial mediation effects of government support and partnerships in fostering collaborative GSCM practices.

Literature on the construction SC and GSCM is reviewed in Sect. 2. In Sect. 3, the GSCM framework is developed. In Sect. 4, statistical analysis is undertaken to identify the relationships between GSCM constructs. Discussion and conclusions are presented in Sect. 5.

## 2 Literature Review

### *2.1 Construction Supply Chain*

The construction SC covers a wide and complex range of activities, with three main subdivisions, construction, services and materials. Construction refers to project management and building activities carried out by contractors or subcontractors. Services relate to architectural design, engineering, survey and wholesale activities. Materials cover a wide range of construction industry products and materials, including construction, installation, engineering, building decorations, etc.

The number of organizations involved in a large construction project can reach into hundreds or thousands and these organizations tend to have low trust or adversarial trading relations between them [2]. A simplified construction SC encompassing the materials developer, architect/consultants, contractors and suppliers is depicted in Fig. 1. The developer initiates the construction project as a conceptual design and hires the architect and other consultants to prepare drawings and the project specification. Contractors are appointed to execute the project, covering a wide range of activities including extraction of materials, manufacture of parts, engineering, assembly of elements and final construction on site. The contractor is responsible for the employment of subcontractors and procurement of materials from suppliers.

### *2.2 Green Supply Chain Management*

Ofori et al. [31] was one of the earliest publications in this field, exploring the potential of applying GSCM to integrate construction processes in Singapore. Competitor

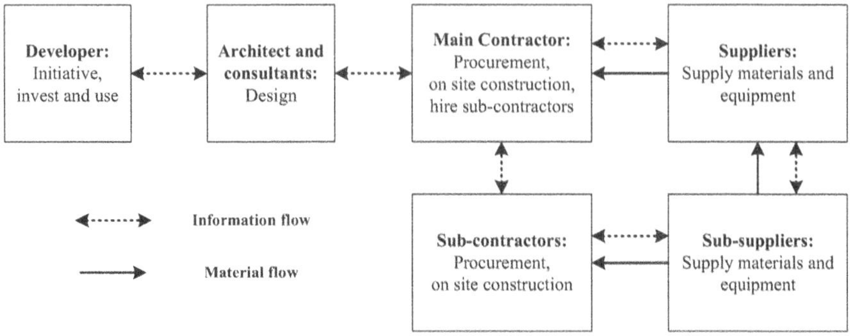


**Fig. 1** A representative construction supply chain

pressure was found to be one of the main drivers of green practices in the Singapore construction industry. Following this research, Balasubramanian and Shukla [6] developed, validated and applied a multi-dimensional GSCM framework for the construction industry in the UAE. Although a comprehensive list of relevant GSCM constructs were identified for the construction industry, Balasubramanian and Shukla's work [6] does not capture the important roles of government intervention and the partnerships.

The buy in from participants in the construction SC required to boost the development of green buildings necessitates constant stimulation from government policy/regulations and standardization of systems. Wang et al. [41] proposed an empirical method to estimate carbon emissions generated from highway construction projects in China. Their results show that over 80% were generated from raw material production,onsite construction and material transportation only accounted for 10% and 3% respectively of the total. It is clear that eco-design of materials will play a key role in reducing carbon emissions generated from materials supply, but such eco-friendly materials usually come with higher costs. If the developer's emphasis is commercial profit, contractors or sub-contractors will choose low cost materials over environmentally friendly ones. To reduce carbon emissions from the construction industry and the buildings, participants in the construction SC need to work collaboratively to engage with continuous technology development, to implement innovative green practices, and make continuous improvements in performance [14], which is defined as GSCM. The lack of uptake of GSCM across the construction SC is mainly due to the lack of inter-organisational collaboration [6].

## 3 Theoretical Model and Hypotheses Development

Greening a construction SC not only requires individual firms to implement green practices, but also involves institutionalizing environmental principles in SC development, management and operation. In addition to building environmentally friendly

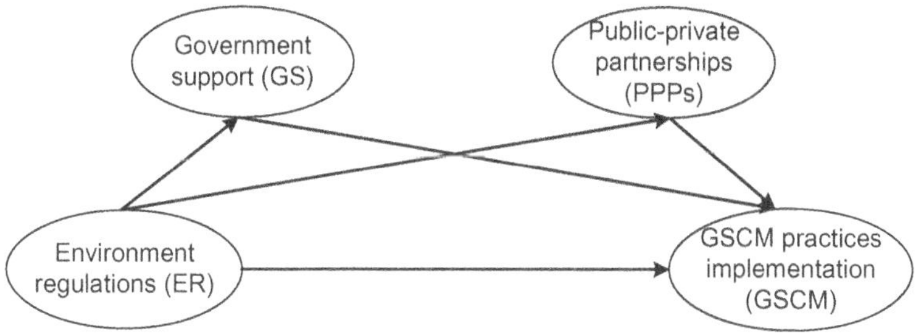


**Fig. 2** Proposed model

firms, we need to understand how the stakeholders in the construction SC collectively react to the external and internal drivers/barriers relating to green practices. Taking Suppliers and Constructors as representatives of a construction SC, we place emphasis on the three stages of the product and service life cycle [39], i.e., from the production of construction materials to their consumption during construction and, finally, waste disposal.

In conjunction with a comprehensive and in-depth synthesis of the literature of green related work in construction industry, we configure the framework in Fig. 2, with the constructs and the underlying measurement items summarized in Table 1.

## *3.1 Environmental Regulations and GSCM Practices*

Government agencies are power groups that may influence the actions of an organization by exerting regulations and policies [34]. The scarcity of resources, degradation of the living environment, and the increasing pressures from the public to protect environment and resources, have promoted governments to introduce environmental regulations, to impose pressures and constraints on industries and businesses, to regulate their practices [5]. Environmental regulations are defined as coercive pressures driving the implementation of GSCM to improve environmental performances [45]. Enterprises also need guidance and regulations from national governments in their environmental efforts, to transform to green production and operation [23]. In construction sector, building energy standards were introduced to regulate construction industry actions, to certify green buildings, and to promote top technologies for sustainable construction [30]. Therefore,

**H1** Environmental regulations (ER) are positively related to GSCM practices implementation (GSCM).

### 3.2 Government Support

Central and regional governments are regarded as key role players in promoting environmental protection to create sustainable society, whilst safeguarding economic development, which is essentially the concept of Triple Win [17]. Stakeholders across a SC, if located in different regions, may have fragmented and disjointed interpretations of green regulations, which further emphasize the important role played by local government in supporting the stakeholders to fully engage in implementing GSCM practices. In addition to exerting pressures through regulations, local governmental subsidies and financial support, such as levying taxes or grants, are significant motivators for many organizations adopt GSCM practices [24].

Government support, underpinned by regional regulations, is identified as a key enabler for the environmental development in construction industry. In China, state-centred top-down approach has resulted in a remarkable growth of green building projects, as the state provides the conditions and simulates self-regulation at regional level via economic mechanisms or via public interest [13]. In the construction SC, stakeholders are more successful when national government builds support for action at the ministerial level, and equips them with the necessary financial and technical support. Therefore,

**H2** Government support (GS) mediate the relationship between environmental regulations (ER) and GSCM practice implementation (GSCM).

### 3.3 Partnerships

The overall GSCM performance depends on the performance of stakeholders (partners) in a SC, and regular performance improvement at each stage of the SC would improve the performance of partners the next stage [38]. Collaboration in SC is identified as an important determinant of $CO_2$ emission reduction across the SC [29]. Under the collaborative paradigm, partnerships have been introduced as a policy instrument to advance development of green policies and green infrastructure, and act as an effective platform for knowledge and practice diffusion [28]. Partnerships are self-organising and non-hierarchical alliances between multiple levels of government, businesses, research organisations and civil societies [8].

In this research, we focus on public-private partnerships (PPPs) where private parties participate in and provide support for, the provision of infrastructure, and deliver public infrastructure-based services. On one hand, PPPs offer ways to governments to simultaneously manage rising costs and reducing governmental budgets [10, 12]. On the other hand, PPPs allow public and private actors to share resources, expertise, and skills together so that they can achieve broader and mutual beneficial objectives [35].

There might a distinct disconnect between knowledge, capital and projects among SC players, and they may have different levels of resources available to fulfil the responsibilities of reducing carbon reduction [22]. PPPs prove to be effective in mediating between governmental levels, fostering networking between public and private actors, sharing resources and eventually facilitating implementation of green practices across levels of government and societal domains [8]. Therefore,

**H3** PPPs mediate the relationship between environmental regulations and GSCM practices implementation.

# 4 Research Methodology

## *4.1 Instrument Development*

A pre-tested questionnaire was used to collect data to test hypotheses. The questionnaire was tested and revised through a pilot study with 8 mid-level managers in construction sector. They were asked to review the survey instrument in terms of structure, readability, ambiguity, and completeness. Based on suggestions from respondents, we modified the questionnaire by removing repeated items and clarifying questions.

The constructs used in the model were operationalised as reflective constructs. The dimensions (indicators) of each construct used in this study were developed based on a comprehensive review of industrial practices, expert opinions and the relevant literature, as shown in Table 1. The indicators were measured on a five-point Likert type scale, with 1 indicating the strongly disagree, and 5 indicating strongly agree.

## *4.2 Data Collection*

The survey was conducted with Chinese contractors and construction materials suppliers. The contractors have headquarters in different cities in China, but they undertake construction projects across China or export construction services to other countries. A total of 408 surveys were distributed through the subsidies of China State Construction Engineering Corporation (CSCECS). Structured questionnaires along with a cover letter were distributed in three group meetings of the CSCECS, explaining the aim and methodologies of the research. The surveys were distributed in two waves, with 5 weeks apart. 210 questionnaire were distributed in the first wave, and 103 complete and valid responses were received. In the second wave of distribution, the 198 questionnaires were sent, and 126 valid responses were received. This design is suitable for research in Chinese unique social and cultural context where business activities are largely influenced by personal relationship and network [7]. Personal relationships and support from CSCECS significantly

**Table 1** Measurement model (factor loadings, SCR, AVE)

| Factor | Items | Factor loadings | SCR | AVE |
|---|---|---|---|---|
| Environmental regulations (ER) [45] | The national environmental regulations and green building industrial standards are important for GSCM adoption. (ER1) | 0.720 | 0.772 | 0.532 |
| | The regional environmental regulations and green building standards are important for GSCM adoption. (ER2) | 0.758 | | |
| | The regional resource and conservation regulations are important for GSCM adoption? (ER3) | 0.705 | | |
| Government support (GS) [1] | There is strong level of government support in developing infrastructure for green practices. (GS1) | 0.774 | 0.841 | 0.571 |
| | There is strong level of government support in fostering collaborations between organisations (including both private and public sectors). (GS2) | 0.706 | | |
| | There is strong level of government incentive and financial support for green practices. (GS3) | 0.761 | | |
| | There is strong level of government support in providing training an education on green technologies and practices. (GS4) | 0.776 | | |
| Public private partnerships (PPP) (Zhou 2015) | The company has knowledge transfer partnerships with other organizations. (PPP1) | 0.649 | 0.772 | 0.533 |
| | The company receives training and education on green innovations through partnerships. (PPP2) | 0.783 | | |

(continued)

**Table 1** (continued)

| Factor | Items | Factor loadings | SCR | AVE |
|---|---|---|---|---|
| | The company learns good practices or new skills through partnerships. (PPP3) | 0.748 | | |
| GSCM [6, 45, 46] | The company cooperates with designers and customers for eco-design. (GSCM1) | 0.701 | 0.822 | 0.502 |
| | The company adopts clean technology or green materials in production/construction. (GSCM2) | 0.650 | | |
| | The company reduces the use of hazardous materials. (GSCM3) | 0.709 | | |
| | The company complies with environmental regulations and green building standards. (GSM4) | 0.670 | | |
| | The company corporates with contractors and sub-contractors for environmental objectives. (GSCM5) | 0.705 | | |
| | The company reduces transportation volume and distance. (GSCM6) | 0.627 | | |

improved the response rate. In total 229 completed questionnaires were received, and the characteristics of the survey participants are presented in Table 2.

## 4.3 *Statistical Analysis and Findings*

Partial Least Square Structural equation modeling (PLS_SEM) is used to analyse the empirical data and test the hypothesized relationships as shown in Fig. 2. The data in this research consist primarily of perceptual measures, and our hypotheses represent a series of relationships that include exogenous and endogenous variables. Different from covariance based (CB) SEM, PLS_SEM tend to focus primarily on the predictive power of the hypothesized-model [15], and it is a preferred method to exploratory research. In this study, the relationships proposed between constructs are guided by distinct theories but these relationships are not examined or tested

**Table 2** Profile of survey respondents

| | Responses | Percentage |
|---|---|---|
| *Ownership* | | |
| Foreign or joint venture | 145 | 0.63 |
| Private sector | 40 | 0.17 |
| State owned | 44 | 0.20 |
| *Industry sector* | | |
| Contractors | 116 | 0.51 |
| Suppliers | 113 | 0.49 |
| *Firm size* | | |
| Small (<250) | 36 | 0.16 |
| Medium (250–2000) | 147 | 0.51 |
| Large (>2000) | 46 | 0.20 |
| Total | 229 | 100 |

in the literature, which means this research bears an exploratory nature. Therefore, PLS_SEM is chosen as a suitable technique for data analysis [16]. We followed Anderson and Gerbing's [3] recommended two-step approach to test our hypotheses, (1) we test the measurement model to examine validity and reliability of the scales used in our analysis, and (2) testing and analysing the structural relationships between constructs.

## 4.4 Measurement Model Reliability and Validity

To test the measures satisfied all the requirements of reliability and validity, we use, (1) Scale composite reliability (SCR) to check internal consistency; (2) factor loadings to evaluate inter-item reliability [21], (3) the average variance extracted (AVE) to assess convergent validity; and (4) the Heterotrait-Monotrait ratio of the correlations (HTMT) between these constructs was calculated to evaluate discriminant validity [27]. The results of these tests show that the SCRs were greater than 0.7, and the AVE for each construct was higher than 0.5. The HTMT ratios between paired constructs were lower than 0.90. All measures indicate adequate reliability and validity.

## 4.5 Structural Model Assessment

**Collinearity,** Having confirmed the construct measures as reliable and valid, we are going to assess the structural model. Step 1 is to examine the collinearity between environment regulation (ER), government support (GS), and public-private partnership (PPP), as these serve as predictors of GSCM practices implementation (GSCM).

**Table 3** Path coefficients and significance

| Hypothesis | Relationships | Beta | *t*-values | *p*-values | Results | Mediation |
|---|---|---|---|---|---|---|
| 1 | ER -> GSCM | 0.355 | 5.123 | 0.000*** | supported | |
| 2 | ER -> GS -> GSCM | 0.112 | 3.345 | 0.001*** | supported | Partial |
| 3 | ER -> PPPs -> GSCM | 0.071 | 2.862 | 0.004** | supported | Partial |

VIF values of these variables ranged from 1.449–1.802, well below the threshold of 10, providing confidence that the structural model results are not negatively affected by collinearity.

**Explanatory and predictive power**, The coefficient of determination of $R^2$ was examined to determine the explanatory power of the proposed model in Fig. 2. The values of $R^2$ achieved for the endogenous variables GSCM, GS and PPP are above the suggested threshold of 0.10. Falk and Miller [19] showing moderate strength in explaining these variables.

## 5 Hypotheses Testing

We used Smart PLS to test hypotheses and assess the significance of path coefficients [3]. Bootstrapping with 5000 samples was performed to determine the structural equation model and the significance level of the parameter estimates [21]. The results of hypothesis testing, i.e., the beta ($\beta$) values of the paths and their corresponding *p*-values are summarized in Table 3.

The H1 posits that there is a positive relationship between environmental regulations (ER) and the GSCM practices implementation (GSCM). The results in Table 3 fully support this hypothesis. The path coefficient of 5.123 highlights the important role of ER in driving GSCM.

H2 and H3 speculate the mediation effects of GS and PPP. To conduct mediation analysis, we followed recommendations suggested by [32] and tested indirect effects caused by the mediators by performing bootstrapping. Results support hypotheses H2 and H3, showing that GS mediates relationship between ER and GSCM (H2), and PPPs mediate the relationship between ER and GSCM (H3).

## 6 Discussions and Conclusions

The current study further elaborates the arguments from [43, 45] that environmental regulations enforce organizations to implement GSCM. Organizations often make their environmental related decisions under the pressures from environmental regulations, markets, and competitors. The parameters affecting the effectiveness of these pressures, such as government intervention, have been rarely investigated. From a research perspective, understanding the roles of government and partnerships in terms of GSCM adoption creates new research avenues to develop theories on GSCM and more broadly on agreen practices. Building upon the arguments of [8, 33, 43], our empirical findings highlight government support and partnerships as important mediators in partially mediating the impacts of environmental regulations on GSCM implementations. Based on recent debates surrounding the valuable contributions made by PPPs to complex projects, there exists a clear opportunity for organizations to develop partnerships with multi-levels of governments, to engage in decision making, to share resource, disseminate knowledge and build up capacities.

The insights obtained from this study can help practitioners and decision markers take necessary action or make interventions to support the construction industry in minimizing its environmental impact. The GSCM discussed in this research will need to be evaluated and assessed by construction SC stakeholders to best suit their own ambitions and their clients' needs.

## References


1. Adetunji, I., Price, A.D.F., Fleming, P.: Achieving sustainability in the construction supply chain. Eng. Sustain. **161**(3), 161–172 (2008)
2. Akintoye, A., McIntosh, G., Fitzgerald, E.: A survey of supply chain collaboration and management in the UK construction industry. Eur. J. Purchasing Supply Manage. **6**(3–4), 159–168 (2000)
3. Anderson, J.C., Gerbing, D.W.: Structural equation modeling in practice, a review and recommended two-step approach. Psychol. Bull. **103**(3), 411 (1988)
4. Badi, S., Murtagh, N.: Green supply chain management in construction, a systematic literature review and future research agenda. J. Cleaner Prod. **223**, 312–322 (2019)
5. Bai, X., Imura, H.: A comparative study of urban environment in East Asia, stage model of urban environmental evolution. Int. Rev. Environ. Strateg. **1**(1), 135–158 (2000)
6. Balasubramanian, S., Shukla, V.: Green supply chain management, an empirical investigation on the construction sector. Supply Chain Manage. Int. J. **22**(1), 58–81 (2017)
7. Baruch, Y., Holtom, B.C.: Survey response rate levels and trends in organizational research. Human Relat. **61**(8), 1139–1160 (2008)
8. Bauer, A., Steurer, R.: Multi-level governance of climate change adaptation through regional partnerships in Canada and England. Geoforum **51**, 121–129 (2014)
9. Black, C., Akintoye, A., Fitzgerald, E.: An analysis of success factors and benefits of partnering in construction. Int. J. Project Manage. **18**(6), 423–434 (2000)
10. Blanken, A., Dewulf, G.: PPPs in health, static or dynamic? Aus. J. Public Adm. **69**, S35–S47 (2010)
11. Bohari, A.A.M., Skitmore, M., Xia, B., Teo, M.: Green oriented procurement for building projects, Preliminary findings from Malaysia. J. Cleaner Prod. **148**, 690–700 (2017)
12. Bosakova, L., Madarasova Geckova, A., van Dijk, J.P., Reijneveld, S.A.: Increased employment for segregated Roma may improve their health, outcomes of a public–private partnership project. Int. J. Environ. Res. Public Health **16**(16), 2889 (2019)
13. Buttel, F.H.: Ecological modernization as social theory. Geoforum **31**(1), 57–65 (2000)
14. Chen, W., Chen, J., Xu, D., Liu, J., Niu, N.: Assessment of the practices and contributions of China's green industry to the socio-economic development. J. Cleaner Prod. **153**, 648–656 (2017)
15. Davcik, N.S.: The use and misuse of structural equation modeling in management research. J. Adv. Manage. Res. **11**(1), 47–81 (2014)

16. Dubey, R., Gunasekaran, A., Childe, S.J., Papadopoulos, T., Hazen, B.T., Roubaud, D.: Examining top management commitment to TQM diffusion using institutional and upper echelon theories. Int. J. Prod. Res. **56**(8), 2988–3006 (2018)
17. Elkington, J.: Towards the sustainable corporation: win-win-win business strategies for sustainable development. Calif. Manage. Rev. **36**(2), 90–100 (1994)
18. Eriksson, P.E.: Procurement effects on coopetition in client-contractor relationships. J. Constr. Eng. Manage. **134**(2), 103–111 (2008)
19. Falk, R.F., Miller, N.B.: A Primer for Soft Modeling, University of Akron Press (1992).
20. Gao, D., Xu, Z., Ruan, Y.Z., Lu, H.: From a systematic literature review to integrated definition for sustainable supply chain innovation (SSCI). J. Cleaner Prod. **142**, 1518–1538 (2017)
21. Hair Jr, J.F., Hult, G.T.M., Ringle, C., Sarstedt, M.: A primer on partial least squares structural equation modeling (PLS-SEM), Sage Publications (2016)
22. Hasan, M.S., Zhang, R.J.: Critical barriers and challenges in implementation of green construction in China. Int. J. Curr. Eng. Technol. **6**, 435 (2016)
23. He, S., Yin, J., Zhang, B., Wang, Z.: How to upgrade an enterprise's low-carbon technologies under a carbon tax, the trade-off between tax and upgrade fee. Appl. Energy **227**, 564–573 (2018)
24. Hsu, C.C., Tan, K.C., Mohamad Zailani, S.H., Jayaraman, V.: Supply chain drivers that foster the development of green initiatives in an emerging economy. Int. J. Oper. Prod. Manage. **33**(6), 656–688 (2013)
25. Hwang, B.G., Tan, J.S.: Green building project management, obstacles and solutions for sustainable development. Sustain. Dev. **20**(5), 335–349 (2012)
26. Joo, H.Y., Seo, Y.W., Min, H.: Examining the effects of government intervention on the firm's environmental and technological innovation capabilities and export performance. Int. J. Prod. Res. **56**(18), 6090–6111 (2018)
27. Kline, R.B.: Principles and practice of structural equation modeling, 2nd edn. Guilford, New York (2005)
28. Lee, N. Public-Private Partnerships as a policy instrument to advance green growth. Режим доступу. www.greengrowthknowledge.org/blog/public-private-partnerships-policy-instrumentadvance-green-growth (2014).
29. Liu, J., Feng, Y., Zhu, Q., Sarkis, J.: Green supply chain management and the circular economy. Int. J. Phys. Distrib. Logis. Manage. **48**(8), 794–817 (2018)
30. Liu, J.Y., Low, S.P., He, X.: Green practices in the Chinese building industry, drivers and impediments. J. Technol. Manage. China **7**(1), 50–63 (2012)
31. Ofori, G.: Greening the construction supply chain in Singapore. Eur. J. Purchasing Supply Manage. **6**(3), 195–206 (2000)
32. Preacher, K.J., Hayes, A.F.: SPSS and SAS procedures for estimating indirect effects in simple mediation models. Behav. Res. Methods Instrum. Computers **36**(4), 717–731 (2004)
33. Regeczi, D.: Limited partnership, The lack of sustainable development in relation to participation in Hungarian public-private partnerships. Bus. Strat. Environ. **14**(4), 205–215 (2005)
34. Rivera, W.M., Alex, G.: The continuing role of government in pluralistic extension systems. J. Int. Agric. Ext. Educ. **11**(3), 41–52 (2004)
35. Roehrich, J.K., Lewis, M.A., George, G.: Are public–private partnerships a healthy option? A systematic literature review. Soc. Sci. Med. **113**, 110–119 (2014)
36. Seuring, S., Gold, S.: Sustainability management beyond corporate boundaries, from stakeholders to performance. J. Cleaner Prod. **56**, 1–6 (2013)
37. Seuring, S., Müller, M.: From a literature review to a conceptual framework for sustainable supply chain management. J. Cleaner Prod. **16**(15), 1699–1710 (2008)
38. Singh, R., Centobelli, P., Cerchione, R.: Evaluating partnerships in sustainability-oriented food supply chain, a five-stage performance measurement model. Energies **11**(12), 3473 (2018)
39. Tsoulfas, G.T., Pappis, C.P.: Environmental principles applicable to supply chains design and operation. J. Cleaner Prod. **14**(18), 1593–1602 (2006)

40. Vachon, S., Klassen, R.D.: Extending green practices across the supply chain, the impact of upstream and downstream integration. Int. J. Oper. Prod. Manage. **26**(7), 795–821 (2006)
41. Wang, X., Duan, Z., Wu, L., Yang, D.: Estimation of carbon dioxide emission in highway construction, a case study in southwest region of China. J. Cleaner Prod. **103**, 705–714 (2015)
42. Zhang, S., Yu, Y., Zhu, Q., Qiu, C.M., Tian, A.: Green innovation mode under carbon tax and innovation subsidy, an evolutionary game analysis for portfolio policies. Sustainability **12**(4), 1385 (2020)
43. Zhu, Q., Sarkis, J., Cordeiro, J.J., Lai, K.H.: Firm-level correlates of emergent green supply chain management practices in the Chinese context. Omega **36**(4), 577–591 (2008)
44. Zhu, Q., Sarkis, J., Lai, K.H.: Green supply chain management innovation diffusion and its relationship to organizational improvement, an ecological modernization perspective. J. Eng. Technol. Manage. **29**(1), 168–185 (2012)
45. Zhu, Q., Sarkis, J., Lai, K.H.: Institutional-based antecedents and performance outcomes of internal and external green supply chain management practices. J. Purchasing Supply Manage. **19**(2), 106–117 (2013)
46. Zsidisin, G.A., Hendrick, T.E.: Purchasing's involvement in environmental issues, a multi-country perspective. Ind. Manage. Data Syst. **7**, 313–320 (1998)

# Performance, Environmental Benefit and Economic Analysis of Constructed Wetland Using Construction Waste as Substrate

**Lu Zhou, Zhi Cao, and Zhaojun Huang**

**Abstract** Constructed wetland is a kind of environment-friendly surface water treatment technology, but the traditional constructed wetland with gravel as substrate has weak phosphorus removal capacity. Using construction waste as the substrate of constructed wetland can not only reduce the environmental impact of gravel production, but also improve the phosphorus removal efficiency. This study evaluates the environmental impact and costing of traditional constructed wetland, and how much environmental impact and costing will be reduced when traditional constructed wetland's substrate is replaced by construction waste. The result shows that compared with the traditional constructed wetland, the environmental impact and costing of constructed wet-land filled with construction waste decreased by 77.81% and 14.72% respectively. Natural Land Transformation, Climate Change Ecosystem, Human Toxicity and Fossil Depletion are the main categories of the reduced environmental impact of constructed wetland after the replacement of substrate. The decrease in production and landfill of gravel are the main reasons for the reduction of environmental impact. The paper studies the important factors that affect the environmental impact of constructed wetland and pro-vides a theoretical basis for improving its environmental and economic performance.

**Keywords** Constructed wetland · Phosphorus · Construction waste · Polluted surface water

## 1 Introduction

Surface water is one of the essential resources for industry production and daily life, which has a huge impact on the national economy. With the population growth, social progress, and rapid development of industry and agriculture, people's living standards have increased rapidly. As a result, a large amount of wastewater is discharged into

L. Zhou (✉) · Z. Cao · Z. Huang
School of Environment, Tsinghua University, 100084 Beijing, People's Republic of China
e-mail: zhoulu@mail.tsinghua.edu.cn



F. K. S. Chan et al. (eds.), *Proceedings of the 2020 International Conference on Resource Sustainability: Sustainable Urbanisation in the BRI Era (icRS Urbanisation 2020)*, Environmental Science and Engineering,
https://doi.org/10.1007/978-981-15-9605-6_8

the river, resulting in serious surface water pollution. Excessive phosphorus loading in river and lakes has long been a key factor leading to eutrophication, algal blooms, and other serious environmental risks.

Constructed wetland (CW) is an artificially constructed sink with an impervious barrier at the bottom [1]. The sink is filled with substrate and macrophytes are planted on the surface [2]. It utilizes the synergistic action of substrate, plant and microorganism to purify wastewater. CW is considered as a promising in-situ purification technology for polluted surface water due to their low cost and low maintenance requirements [3]. Although CW performs well in removing suspended matter and organic matter, it has low phosphorus removal capacity.

Phosphorus removal in CW is complex. The mechanisms include substrate adsorption, plant uptake, and microbial effects. In previous studies, it was generally believed that substrate contributed significantly to phosphorus removal.

The first consideration when choosing substrate is whether it can provide a good growth environment for plants and microorganisms, that is, it can ensure that the roots of plants can grow normally downward and provide habitat for microorganisms. In addition, the removal of contaminants from the wastewater by the substrate needs to be considered. Gravel is the most commonly used substrate because it is cheap and provides good support for plants and microorganisms [4].

At present, slag, ceramsite, and some porous media with good permeability and large specific surface area, such as zeolite, fly ash, coal slag, peat, limestone, and activated carbon, have also been used as substrate in CWs, greatly improving the removal of the pollutants, especially nitrogen and phosphorus.

Construction waste is a byproduct of urbanization. Construction waste is not only generated in large quantities, but also has low recycling rates in China, which has caused serious environmental pollution and resources waste [5]. Concrete brick (CB), red brick (RB) and fly ash foamed concrete block (FAFCB) are three of the most typical construction wastes. They are porous and rich in calcium, iron, and aluminum, which are ideal substrates for phosphorus removal in CWs. Shi found that construction waste has a stronger ability to adsorb phosphorus than gravel [6]. However, the current research mainly focuses on experimental research, which explores the feasibility of construction waste as constructed wet-land substrate and the advantages of construction waste compared with tradition-al gravel substrate in the ability of phosphorus removal. However, the positive environmental and economic benefits of replacing traditional substrate with construction waste have not been quantified.

Life cycle assessment (LCA) has been widely used to characterize and quantify the potential environmental impacts of wastewater treatment systems [7, 8]. Life cycle costing (LCC) is a method to calculate the total cost in an engineering project's life cycle. Construction costs, maintenance costs, equipment replacement costs, and waste disposal costs can be taken into account [9].

In this paper, phosphorus adsorption capacity of three kinds of typical construction waste (CB, RB, FAFCB) and gravel were studied. And the phosphorus removal efficiency in synthetic surface water of constructed wetland with construction waste and gravel was studied in lab-scale CW tests. Finally, LCA and LCC were used to study the environmental impact and costing of constructed wetland system filled with construction waste or gravel. The purpose of this study is to evaluate the component of environmental impact and costing in traditional constructed wetland, and how much environmental impact and costing will be reduced when traditional CW's substrate is replaced by construction waste.

## 2 Materials and Method

### 2.1 *Substrate*

CB, RB, FAFCB and gravel were obtained from Beijing, China. They were analyzed with X-ray Fluorescence Spectrometer (XRF-1700, SHIMAZU CORP, Japan) and AutoPore IV 9510 Mercury Injection Instrument (MICROMERITICS INSTRUMENT CORP, America). The major chemical composition is $SiO_2$ (40.54, 57.03, 50.13, 57.70%). The total content of $Al_2O_3$, $Fe_2O_3$ and CaO is 32.55, 33.25, 45.16 and 19.21% of each substrate respectively. The porosity of CB, RB, FAFCB is 71.72, 47.92, 35.39%.

### 2.2 *P Adsorption Experiments*

Each substrate was dried at 105 °C to eliminate moisture interference before adsorption test. 0.5, 1.0, 2.0, 5.0, 10, 20 and 25 g of dried substrate (sieved to 1–2 mm) were added into Erlenmeyer flask containing P solutions of 250 mL with different concentrations of 0.5, 1, 2 mg/L. The Erlenmeyer flasks were shaken on a rotating shaker at 25 °C and 180–190 r/min for 48 h. The phosphorus concentration of supernatant after adsorption saturation was measured using Mo-Sb anti-spectrophotometer method. Freundlich adsorption isotherm equation was used to represent the adsorption equilibrium since it is more suitable when adsorbing phosphorus from aqueous with low TP concentration.

### 2.3 *Lab-Scale CW Tests*

4 lab-scale CWs filled with CB, RB, FAFCB and gravel were installed. Cylinder was used as the reactor (H = 60 cm, R = 20 cm). The size grading of each substrate was between 5–30 mm. Dracaena Sanderiana was chosen as the wetland macrophyte. Different amounts of $KH_2PO_4$ were added into the tap water to adjust TP concentration. TP in the influent was controlled at 0.5 and 1.5 mg/L in two tests respectively. TP concentration in the effluent was detected.

## 2.4 Life Cycle Assessment

**Brief Description of the Assessed Systems**. $S_1$ is a real CW engineering project treating polluted surface water with capacity of 7000 t/d. Influent TP and TN concentration ranges from 0.5–2 and 5–20 mg/L respectively. It covers 70,000 $m^2$ area, consisting of 6 subsystems, $A_1$–$A_6$. $A_1$ is made of 8 chambers (80 m long and 15 m wide) while $A_2$–$A_6$ are made of 7 chambers (101 m long and 16 m wide). $S_1$ uses gravel as substrate, removing 20–60% TP and 20–50% TN from the influent. $S_2$ was supposed to have the same construction structure as $S_1$, but it uses FAFCB as substrate. According to the lab-scale CW experiment result, it is assumed that $S_2$ removes 45–65% TP from the influent. Since FAFCB is not better than gravel in the removal of TN, it is assumed that $S_2$ removes 20–50% TN from the influent, the same as $S_1$.

**System Boundary**. This paper studies $S_1$ and $S_2$'s life-cycle environmental impact, mainly comprising of 5 parts, construction, substrate, transportation, operation and decommission. System boundaries are shown in Fig. 1. It is assumed that both 2 systems will be decommissioned after 50 year of operation.

Concrete, reinforcing steel, timber, geomembrane and UPVC pipes are needed to construct a CW. The consumption of each material was calculated based on $S_1'$ design drawings.

Previous studies have demonstrated that the substrate must be renewed every 1.5–8.5 year to maintain high TP removal capacity. It is assumed that the lifetime of $S_1$ and $S_2'$s substrate conforms to the normal distribution, with an average of 5 year and a probability of 99% falling within the range of 1.5–8.5 year. The environmental impact of gravel production was included, while that of FAFCB production was excluded in this study. It is assumed that the average transportation distance of all the construction materials and substrate is 20 km.

During operation, constructed wetland can remove the nutrients from the wastewater. However, it will discharge $CH_4$ and $N_2O$ into the atmosphere. According to the assumption, Inlet TP concentration (M), inlet TN concentration (N), $S_1$'s TP removal efficiency ($K_1$), $S_2$'s TP removal efficiency ($K_2$), $S_1$'s TN removal efficiency ($L_1$),

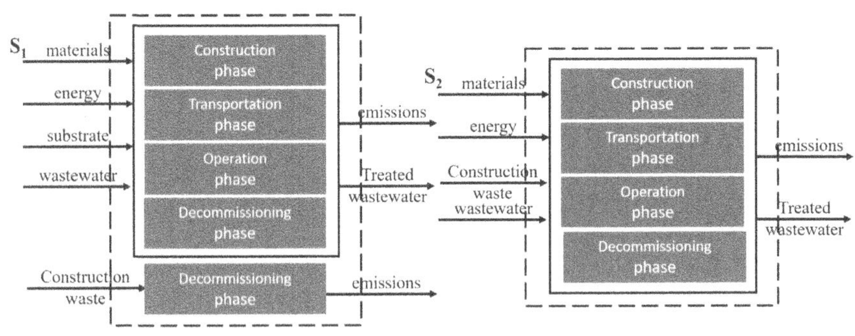


**Fig. 1** LCA System boundary of each system

$S_2$'s TN removal efficiency ($L_2$) subject to the following distribution: M ~ N (1.25, 0.09); N ~ N (12.5, 9); $K_1$ ~ N (0.4, 0.006); $L_1$ ~ N (0.35, 0.0034); $K_2$ ~ N (0.55, 0.0018); $L_2$ ~ N (0.35, 0.0034).

Direct $CH_4$ and $N_2O$ emissions were calculated based on algorithms from the International Panel on Climate Change (IPCC) reports. Biogenic $CO_2$ direct emissions from the systems were not considered. Electricity consumption mainly comes from water pump lifting. $S_1$'s average actual water pump lifting height is 10 m. So is $S_2$.

During decommission, it is assumed that concrete, timber and substrate will be landfilled, while reinforcing steel, UPVC pipe and geomembrane will be recycled, regardless of the environmental impact caused by construction waste transportation and structures demolition. The environmental impact caused by the landfill of FAFCB was taken into account in $S_1$'s system boundary.

**Wastewater Treatment System's LCA**. The functional unit was one cubic meter of treated wastewater. LCA was conducted based on ISO 14,040 and ISO 14,044. The background data was retrieved from ecoinvent 3.6 database. Among the impact categories considered were Agricultural Land Occupation (ALO), Climate Change Ecosystems (CCE), Freshwater Ecotoxicity (FET), Freshwater Eutrophication (FE), Marine Ecotoxicity (MET), Natural Land Transformation (NLT), Terrestrial Acidification (TA), Terrestrial Ecotoxicity (TET), Urban Land Occupation (ULO), Climate Change Human Health (CCHH), Human Toxicity (HT), Ionizing Radiation (IR), Ozone Depletion (OD), Particulate Matter Formation (PMF), Photochemical Oxidant Formation (POF), Fossil Depletion (FRD), Metal Depletion (MRD), which were evaluated based on the ReCiPe Endpoint method. The impact aggregation follows the ReCiPe method guideline. Monte Carlo simulation was applied to perform the uncertainty analysis (100 times).

## 2.5 *Life Cycle Costing*

LCC has included construction costs, substrate costs, operation costs, transportation costs and decommission costs. Construction costs comprise of labor costs, material costs and machinery costs; operation costs comprise of wage costs and electricity costs; decommission costs comprise of demolition costs and landfill costs. Labor costs and machinery costs were estimated based on several similar engineering projects. Material costs, substrate costs, labor costs, wage costs, demolition costs and landfill costs were calculated based on quantity and unit price obtained according to field survey result. All the costs are based on 2020. The discount rate was assumed to be 10%.

**Table 1** Parameters related with the phosphorus adsorption isotherm equation of each studied substrate

| Substrate | $C_0$ | k | 1/n | $R^2$ |
|---|---|---|---|---|
| Concrete block | 0.5 | $7.48 \times 10^3$ | 1.72 | 0.99 |
| | 1.0 | $9.93 \times 10^4$ | 1.31 | 0.94 |
| | 2.0 | $5.85 \times 10^6$ | 0.86 | 0.96 |
| Red brick | 0.5 | $8.80 \times 10^2$ | 1.38 | 0.95 |
| | 1.0 | $9.74 \times 10^2$ | 0.87 | 0.90 |
| | 2.0 | $8.80 \times 10^3$ | 0.66 | 0.91 |
| Fly ash foamed concrete block | 0.5 | $1.48 \times 10^0$ | 0.81 | 0.99 |
| | 1.0 | $3.33 \times 10^3$ | 0.63 | 0.94 |
| | 2.0 | $1.44 \times 10^0$ | 0.51 | 0.95 |
| Gravel | 0.5 | $2.21 \times 10^{-11}$ | 6.71 | 0.89 |
| | 1.0 | $1.97 \times 10^{-10}$ | 6.05 | 0.94 |
| | 2.0 | $5.39 \times 10^{-9}$ | 5.52 | 0.96 |

# 3 Results and Discussion

## 3.1 *P Adsorption Isotherm Experiments*

The parameters of each substrate's adsorption isotherm equation at different $C_0$ are presented in Table 1. Our research shows that at all the three phosphorus concentrations of the aqueous, gravel' 1/n is greater than 5 while three kinds of construction waste's 1/n is between 0.5 and 1.5. From high to low, three kinds of construction waste's 1/n are as follows: FAFCB, CB, RB. According to the meaning of Freundlich isothermal equation, material's adsorption capacity increases as 1/n decreases. Therefore, the adsorption capacity of gravel for phosphorus is poor. In contrast, construction waste has strong adsorption capacity for phosphorus. Since construction waste's adsorption performance shows a positive correlation with its total content of Al, Fe and Ca, it is assumed that P adsorption from aqueous at low phosphorus concentration in three substrates are dominated by chemical adsorption.

## 3.2 *Lab-Scale CW Tests*

The effluent TP concentrations as well as TP removal efficiency under different influent TP concentrations of CB, RB, FAFCB and gravel are shown in Fig. 2. Compared with construction waste, gravel has little use in TP removal. TP removal efficiency of 3 kinds of construction waste was FAFCB, RB and CB from high to low, which was consistent with the results of the adsorption test. Based on the experiment

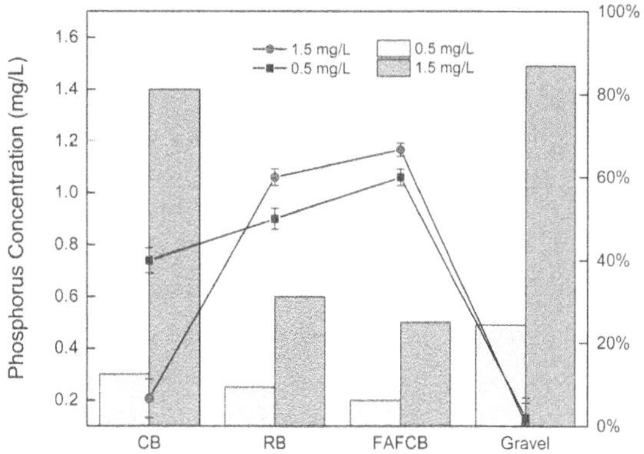


**Fig. 2** Effluent TP concentration and TP removal efficiency of lab-scale CW filled with CB, RB, FAFCB and gravel under different influent TP concentrations

results, it can be concluded that construction waste is more efficient than gravel in TP removal, and FAFCB has the best TP removal performance among three kinds of selected construction waste in this study.

## 3.3 *Life Cycle Assessment*

**LCI result**. Life cycle inventory and ecoinvent process was summarized in Table 2.

**LCA result**. The environmental impact of $S_2$ decreased by 77.81% compared with $S_1$, among which Ecosystem decreased by 88.58%, Human health decreased by 71.05%, and Resources decreased by 64.58%. By analyzing various environmental impacts of $S_1$, it can be found that Ecosystem accounts for 46.02% of the total environmental impacts, mainly composed of NLT and CCE (NLT accounts for 36.34%; CCE accounts for 12.99%). Human health accounts for 33.76% of the total environmental impact, among which HT accounts for 19.07%, CCHH accounts for 10.27% and PMF accounts for 4.38%. Resources accounts for 20.22% of the total environmental impact, among which FRD accounts for 14.46% and MRD accounts for 5.76%. By analyzing various environmental impacts of $S_2$, it can be found that Ecosystem accounts for 23.69% of the total environmental impact. Human health accounts for 33.76% of the total environmental impact. Resources accounts for 21.55% of the total environmental impact (Fig. 3).

**Contributor Analysis**. Substrate accounts for 87.02% of $S_1$'s total environmental impact related with NLT, which is mainly caused by the stone mining in mountainous areas during the process of gravel production. Decommission accounts for 8.92%, which is mainly caused by the occupation of natural land by landfill. Compared with $S_1$, $S_2$ does not use gravel as substrate, which can eliminate the environmental impact caused by gravel production and landfill.

**Table 2** Summary of life cycle inventory of S1 and S2

| Item | Value ($S_1$) | Value ($S_2$) |
|---|---|---|
| Concrete ($m^3$) | 1.92E−05 | 1.92E−05 |
| Cement (kg) | 0.00E+00 | 0.00E+00 |
| Steel (kg) | 2.87E−03 | 2.87E−03 |
| Timber ($m^3$) | 9.99E−08 | 9.99E−08 |
| Sand (kg) | 0.00E+00 | 0.00E+00 |
| Gravel (pool) (kg) | 0.00E+00 | 0.00E+00 |
| Geomembrane (kg) | 2.47E−04 | 2.47E−04 |
| UPVC pipes (kg) | 3.16E−05 | 3.16E−05 |
| Gravel (substrate) (kg) | 8.49E+00 | 0.00E+00 |
| FB (substrate) (kg) | 0.00E+00 | 4.11E+00 |
| Transportation (tkm) | 4.27E−01 | 2.08E−01 |
| Electricity (kWh) | 2.78E−03 | 2.78E−03 |
| Removed TP (kg) | −5.00E−04 | −6.88E−04 |
| Removed TN (kg) | −4.38E−03 | −4.38E−03 |
| $CH_4$ (kg) | 4.34E−04 | 4.34E−04 |
| $N_2O$ (kg) | 3.44E−05 | 3.44E−05 |
| Concrete (kg) | 4.52E−02 | 4.52E−02 |
| Cement (kg) | 0.00E+00 | 0.00E+00 |
| Steel (kg) | 2.87E−03 | 2.87E−03 |
| Timber (kg) | 4.99E−05 | 4.99E−05 |
| Sand (kg) | 0.00E+00 | 0.00E+00 |
| Gravel (pool) (kg) | 0.00E+00 | 0.00E+00 |
| Gravel (substrate) (kg) | 8.49E+00 | 0.00E+00 |
| FB (substrate) (kg) | 4.11E+00 | 4.11E+00 |
| Geomembrane (kg) | 2.47E−04 | 2.47E−04 |
| UPVC pipes (kg) | 3.16E−05 | 3.16E−05 |

Decommission phase accounts for about 50% of $S_1$'s total impact related with CCE, mainly caused by greenhouse gas emissions from landfill process. Without the greenhouse gas emissions from gravel production, transportation, and landfill, $S_2$'s CCE is reduced by 64.39%.

Due to the pollutants generated by diesel combustion during gravel mining and crush process, substrate contributes the highest to the environmental impact related with HT, accounting for around 46.86%. Decommission accounts for 27.04% of the total environmental impact related with HT, mainly due to the carbon particles generated during landfill. Besides, incomplete combustion of fuel during steel and plastic recycling process will discharge toxic chemicals into the atmosphere. Even implementing the European Union 4 motor vehicle emission standard, transportation still accounts for 21.35% of the total environmental impact.

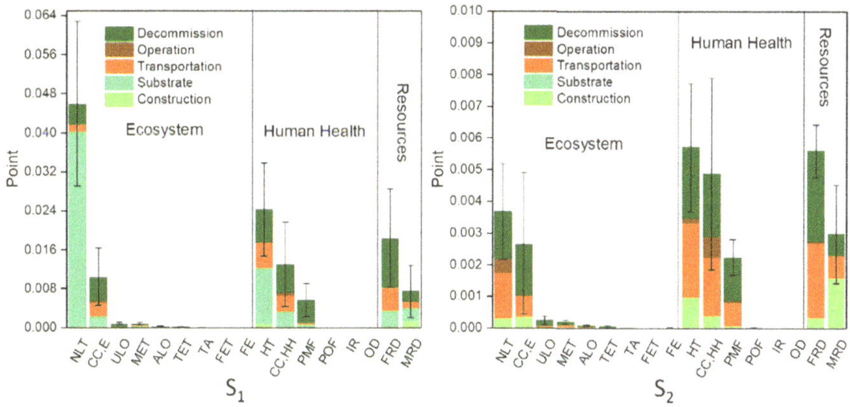


**Fig. 3** LCA result of $S_1$ and $S_2$

The highest contributor of FRD is decommissioning phase, accounting for 54.77%, which mainly comes from the emission of carbon particles generated by landfill into the atmosphere, leading to energy waste. Transportation is followed, accounting for 25.97% (Fig. 4).

### *3.4 Life Cycle Costing Analysis*

Not only the environmental performance of $S_2$ was superior to the environmental performance of $S_1$, it was cheaper. The economic assessment of the systems is presented in Table 3. It can be seen from the table that the total LCC of $S_2$ is reduced by 14.72% compared with $S_1$. The total LCC of $S_1$ is mainly composed of construction costs (60.00%), substrate costs (17.79%) and operation costs (16.32%). The construction and operation cost of $S_2$ and $S_1$ is the same. But $S_2$'s substrate cost is reduced by 67.74% which is the core reason for the decrease of $S_2$'s LCC.

## 4 Conclusion

The paper studies the important factors that affect the environmental impact and costing of constructed wetland, and provides a theoretical basis for improving its environmental and economic benefits.

Constructed wetland that uses construction waste as substrate has stronger dephosphorization capability than constructed wetland that uses gravel as substrate. In this paper, FAFCB is the best of the three kinds of construction waste to remove phosphorus.

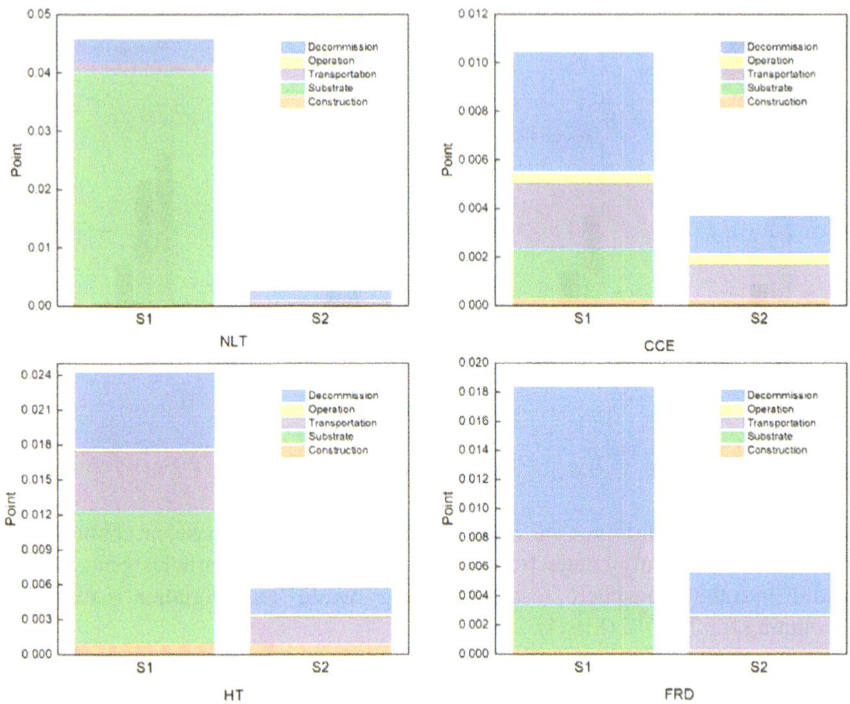


**Fig. 4** Contributor analysis of $S_1$ and $S_2$'s environmental impact related with NLT, CCE, HT and FRD

**Table 3** LCCA result of $S_1$ and $S_2$

| Item | | Amount ($\times 10^6$ USD) | |
|---|---|---|---|
| | | S1 | S2 |
| Construction | Labor cost | 19.92 | 19.92 |
| | Material cost | 9.16 | 9.16 |
| | Machinery cost | 4.51 | 4.51 |
| | Construction cost | 33.59 | 33.59 |
| Substrate | | 19.92 | 6.43 |
| Transportation | | 1.80 | 0.88 |
| Operation | Wage | 18.16 | 18.16 |
| | Electricity | 0.11 | 0.11 |
| Decommission | Demolishing the structure | 0.78 | 0.78 |
| | Landfill | 4.24 | 2.16 |
| | Recycle (steel) | −0.20 | −0.20 |
| | Recycle (plastic) | −0.02 | −0.02 |
| Total | | 111.97 | |

The environmental impact of the substrate accounts for a large proportion of the total environmental impact in traditional constructed wetlands. If the construction waste is used to replace the gravel substrate, the environmental impact of constructed wetland treating 1 $m^3$ wastewater will be reduced by 77.81%.

Analyzing the reduced environmental impact, if the gravel substrate cannot be replaced with construction waste, people can also consider using a cleaner gravel production method or increase the reuse rate of the gravel to reduce the environmental impact.

Finally, the costing of constructed wetland mainly comes from the construction phase, which accounts for 70% of the total costing. Therefore, how to save money in the construction phase, such as lower labor costs, material costs and machinery costs, is a reasonable direction.

## References

1. Nguyen, X.C., Nguyen, D.D., Tran, Q.B., Nguyen, T.T.H., Tran, T.K.A., Tran, T.C.P., Nguyen, T.H.G., Tran, T.N.T., La, D.D., Chang, S.W., Balasubramani, R., Chung, W.J., Yoon, Y.S., Nguyen, V.K: Two-step system consisting of novel vertical flow and free water surface constructed wetland for effective sewage treatment and reuse. Biores. Technol. **306**(12), 2098–2106 (2020)
2. He, S., Wang, Y., Li, C., Li, Y., Zhou, J.: The nitrogen removal performance and microbial communities in a two-stage deep sequencing constructed wetland for advanced treatment of secondary effluent. Biores. Technol. **248**(16), 82–88 (2018)
3. Wu, H., Fan, J., Zhang, J., Ngo, H.H., Guo, W.: Large-scale multi-stage constructed wetlands for secondary effluents treatment in northern China: carbon dynamics. Environ. Pollut. **233**(18), 933–942 (2018)
4. Wu, H., Zhang, J., Guo, W., Liang, S., Fan, J.: Secondary effluent purification by a large-scale multi-stage surface-flow constructed wetland: a case study in northern China. Biores. Technol. **249**(17), 1092–1096 (2018)
5. Li, X.: Recycling and reuse of waste concrete in China. Resour. Conserv. Recycl. **53**(1–2), 36–44 (2008)
6. Shi, X., Fan, J., Zhang, J., Shen, Y.: Enhanced phosphorus removal in intermittently aerated constructed wetlands filled with various construction wastes. Environ. Sci. Pollut. Res. **24**(28), 22524–22534 (2017)
7. Lam, K.L., Zlatanovic, L., Hoek, J.P: Life cycle assessment of nutrient recycling from wastewater: a critical review. Water Res. **173**(12), 115519–115524 (2020)
8. Zhang, Y., Zhang, C., Qiu, Y., Li, B., Pang, H., Xue, Y., Liu, Y., Yuan, Z., Huang, X.: Wastewater treatment technology selection under various influent conditions and effluent standards based on life cycle assessment. Resour. Conserv. Recycl. **154**(20), 1140–1147 (2020)
9. Rawal, N., Duggal, S.K: Life cycle costing assessment-based approach for selection of wastewater treatment units. Nat. Acad. Sci. Lett. **39**(2), 103–107 (2016)

# Achieving Single-Stage Partial Nitritation and Anammox (PN/A) Using a Submerged Dynamic Membrane Sequencing Batch Reactor (DM-SBR)

**Xiaohuan Yang and Qian Li**

**Abstract** Single-stage partial nitration and anammox (PN/A) is achieved using a submerged dynamic membrane (DM-SBR) in this study. The DM-SBR was stably operated for 200 days, and the nitrogen removal efficiency (NRE) was sustained at 70.3 ± 7.2% at a nitrogen loading rate (NLR) ranging from 0.1 to 0.3kgN/m$^3$/d with a hydraulic retention time (HRT) of 24 h. When the NLR was 0.2 kgN/m$^3$/d, the NRE achieved was high as 80% with a low concentration of dissolved oxygen (DO) of 0.13 mg/L. In addition, the specific activity of anammox bacteria (AnAOB) and ammonia-oxidizing bacteria (AOB) reached was 2.72 and 16.80 gN/gVSS/d, respectively. The dynamic membrane (DM) intercepted the biomass due to the lamellar, intact, dense biofilm self-generated on the surface of the supporting material, which had an effluent turbidity of 10 NTU. The enriched anammox functional bacteria were *Candidatus Jettenia*, and the relative abundance was 11.06%. The AOB-like functional bacteria consisted primarily of *Nitrosomonas*, with a relative abundance of 2.76%, which ensured the nitrogen removal process reaction. This study provides a novel reactor configuration of the single-stage PN/A process in the view of practical applications.

**Keywords** Sequencing batch reactor · Partial nitritation · Anammox · Submerged dynamic membrane

---

X. Yang · Q. Li (✉)
School of Environmental and Municipal Engineering, Xi'an University of Architecture and Technology, No. 13, Yanta Road, 710055 Xi'an, China
e-mail: Qian.LI@xauat.edu.cn

X. Yang
e-mail: 1090633365@qq.com



F. K. S. Chan et al. (eds.), *Proceedings of the 2020 International Conference on Resource Sustainability: Sustainable Urbanisation in the BRI Era (icRS Urbanisation 2020)*, Environmental Science and Engineering,
https://doi.org/10.1007/978-981-15-9605-6_9

# 1 Introduction

Compared with the nitrification-denitrification which is conventional biological nitrogen removal process, anaerobic ammonium oxidation (anammox) is sustainable and increasingly popular processes as it has less energy consumption for oxygen demand, does not require the addition of organic carbon sources, and produces less excess sludge, resulting in low operation costs. Single-stage PN/A processes have been developed in the basis of anammox technology for a wide variety of practical applications. In the process, ammonia is first partially converted to nitrite by ammonia-oxidizing bacteria (AOB) under aerobic conditions, then the rest of the ammonia produces nitrogen gas due to anammox bacteria (AnAOB) under anaerobic conditions with the converted nitrite. The single-stage PN/A process, which combines two reactions in a single unit, has prominent advantages for the reduction of the footprint and infrastructure and operation costs. It also produces lower nitrate and nitrous oxide gas emissions.

The single-stage partial nitritation and anammox (PN/A) system is a typical multi-culture and multi-substrate process. In this system, anammox bacteria have a low growth rate with a doubling time of 10–14d at 30–40 °C and a low cell yield of 0.11gVSS/$NH_4^+$-N. In addition, anammox and AOB are easily washed out with effluent. Biomass retention can be realized by sludge granulation, immobilization technologies, and excellent biomass interception dominated by membrane bioreactors. Moreover, dynamic membrane can provide more unique merits, such as high flux, low membrane costs and easy to clean compared to conventional membrane bioreactors.

In addition to biomass retention, the influence of other factors on reactor operation should be considered. AnAOB are highly sensitive to changes in environmental conditions, such as changes in the amounts of free ammonia (FA), free nitrous acid (FNA), organic matter, salinity, dissolved oxygen (DO). In addition, NOB will compete with anammox for converted $NO_2^-$-N, which requires a strict strategy of DO control, such as adjusting the aeration rate, and the allocation of SBR cycle. Therefore, the primary aim of this study is to investigate the operational performance, formation and biomass retention ability of DM and the characteristics of the bulk sludge and DM microbial communities in a DM-SBR. In this study, a submerged DM-SBR is shown to achieve the single-stage PN/A process of treating ammonia wastewater at 35 °C. The main objective is to evaluate the feasibility of this system and provide a control strategy for the DM-SBRs for practical applications.

# 2 Materials and Methods

As shown in Fig. 1a, the single-stage PN/A was performed in a lab-scale SBR with an internal submerged membrane module. It was made of cylindrical transparent PVC plastic with a working volume of 8 L. The interior was reaction and the external

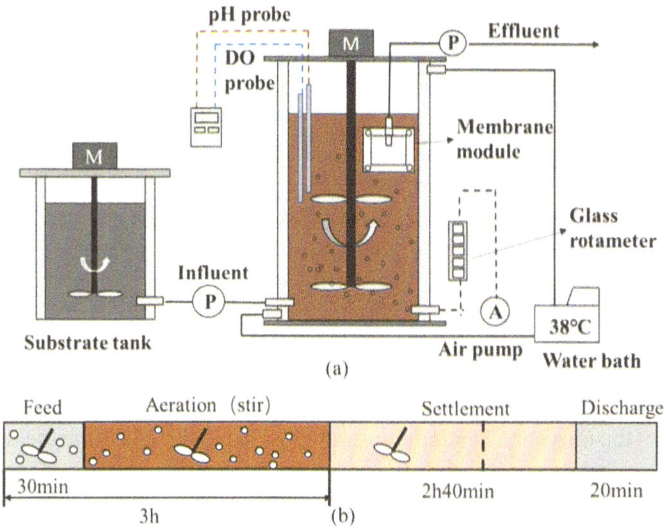


**Fig. 1** Diagram (**a**) and operational model (**b**) of the single-stage partial nitritation and anammox process using the dynamic membrane

was water bath. An agitation device was set in the center to ensure sufficient mixing. The influent was introduced into the reactor and the effluent was drained from the reactor using peristaltic pumps. Sensors for temperature (a thermometer), pH, and DO (Hach probes) were installed in the reactor to monitor operational conditions in real time. The temperature was maintained in the range of $35 \pm 2$ °C using a thermostatic water bath. The DO was supplied using a gas pump, and the aeration rate was adjusted using a glass rotameter in pace with operational conditions. The supporting material of the membrane module was made of nylon net with a surface area of 16 $cm^2$ (two parallel filtering surfaces of $40 \times 40$mm) and a pore size of 30–50μm. A digital pressure meter (SIN-P300.Sinomeasure, China) was installed in the tubes between the effluent peristaltic pump and the membrane module to record the trans-membrane pressure.

The stable operation of the reactor was realized by SBR cycles shown in Fig. 1b. An SBR cycle of 6 h consisted of four phases with no idle time: feeding, aeration, settlement and discharge phase. The specific times of each phase were 30 min, 3 h, 3 h and 20 min, respectively. The feeding phase was accompanied by aeration and stirring, and the settlement phase began to operate without aeration but with a constant stir rate for an hour. Intermittent aeration was selected to supply dissolved oxygen, thereby avoiding excessive aeration and suppressing NOB. The DO concentration depended on the length of time of the aeration phase in an SBR cycle (3 h to 3 h 20 min), the aeration interval in the aeration phase (10s), and the aeration rate adjusted using a glass rotameter (2 L/min). The time of every phase varied with operational conditions and was adjusted by logical timers.

Seed sludge was collected from the $A^2O$ tanks of the Xi'an wastewater treatment plant. The influent had synthetic wastewater added that had no organic carbon and was in the form of 100–300 mg/L of $NH_4HCO_3$, inorganic salt, and trace element at 4 °C with pH of 8.0–8.2.

To monitor the operation of the reactor, samples were collected every two days to analyze the N concentration. A specific SBR cycle was selected to monitor the variation of the N concentrations. The bulk sludge was removed from the reactor to evaluate the specific activity of AOB, NOB, and anammox bacteria and to further understand the characteristics of the microbial communities by the throughput sequencing technique at different NLRs. In a complete cycle of specific SBR, samples were collected from operating reactor to monitor real-time nitrogen conversion. Some operating performances of dynamic membrane are evaluated including turbidity, transfer-membrane pressure and surface morphology. To further understand the characteristics of the microbial communities, samples were collected from the bulk sludge of the reactor and they were analyzed by the throughput sequencing technique.

## 3 Results and Discussion

In Fig. 2, the DM-SBR was stably operated for 200 days, and the nitrogen removal efficiency (NRE) was sustained at 70.3 ± 7.2% at a nitrogen loading rate (NLR) ranging from 0.1 to 0.3 kgN/m$^3$/d with a hydraulic retention time (HRT) of 24 h and the concentration of influent $NH_4^+$ from 100 to 300mg/L. When the NLR was 0.2 kgN/m$^3$/d, the NRE achieved was high as 80% with a low concentration of dissolved oxygen (DO) of 0.13 mg/L. As shown in Fig. 2c, the concentration of FA in the influent increased from 12.67 to 38.4 mg/L in stage II, and 47.6 mg/L in Stage IV, and concentration of FA and FNA in the effluent was maintained at less than 10 mg/L and 0.0005 mg/L, respectively. As shown in Fig. 2d, it was found that the reactor operated under stable operation when the $NO_3^-$-$N_{eff}$ /$\Delta NH_4^+$-N was maintained within a relatively narrow range of 0.1–0.3 in accordance with a theoretical value of 0.13 of the CANON. The ratio was near 0.11 (Fig. 2d) in Stage II with optimum NRE (Fig. 2b).

The DM module, with a stable filtering layer as shown in Fig. 3, displayed that a complete biofilm formed on the nylon mesh, an additional filter of the membrane module. When a new membrane was operated in the reactor, the effluent turbidity was up to 150 contributed to from 300NTU.The dynamic membrane gradually formed after approximately 2 days and 8 cycles, the effluent turbidity decreased to 15.5 NTU and was stably maintained approximately 10 NTU. This indicated that the dynamic membrane had a good retaining sludge ability. As shown in Fig. 3b, once the effluent pump was running, the trans-membrane pressure (TMP) increased to 4.0 kPa. After 20 days of operation, the TMP jumped to approximately 40 kPa, then DM needed to be cleaned or was replaced with a new membrane. The distribution of the particle diameters of the bulk sludge and the DM at an NLR of 0.2 kgN/m$^3$/d are shown in Fig. 3c. The mean size and the median size of the bulk sludge in the reactor were

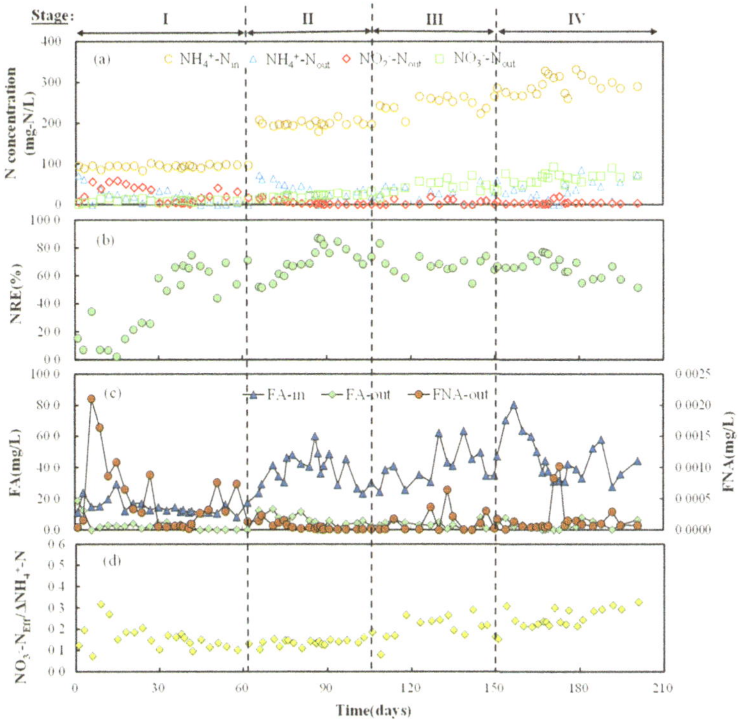

**Fig. 2** Reactor performance: (**a**) concentration of $NH_4^+$-N in the influent and $NH_4^+$-N, $NO_2^-$-N, and $NO_3^-$-N in the effluent; (**b**) the nitrogen removal efficiency; (**c**) the concentration of FA and FNA; and (**d**) the ratio of $NO_3^-$-N/$NH_4^+$-N

139.2 and 86.64 μm, respectively. However, the mean size and the median size of the DM was 93.30 and 71.79 μm, respectively.The mean diameter of the bulk sludge particle was larger than the DM layer, indicating that the DM efficiently intercepted biomass washed out with the effluent.

The SAA increased from 0.39 to 2.72 gN/g-VSS/d with an NLR increase from 0.1 to 0.2 kgN/m$^3$/d, but decreased to 1.55 gN/g-VSS/d when the NLR increased further to 0.25 kgN/m$^3$/d. However, SAOA and SNOA increased with NLR increases from 7.29 to 17.82 gN/g-VSS/d and from 0.7 to 2.77 gN/g-VSS/d, respectively. A specific SBR cycle was selected for the sampling measurements to monitor the variation in the N concentrations and DO at an NLR of 0.2 kgN/m$^3$/d. It can be concluded that the ratio of competitive ability of AOB and anammox to $NH_4^+$ was 1.54:1, anammox bacteria and NOB to $NO_2^-$ was 6.35:1 in aeration stage. And the specific ratio of anammox capacity in aeration and settlement phase was 2.15:1. The enriched anammox functional bacteria were *Candidatus Jettenia*, and the relative abundance was 11.06%. The AOB-like functional bacteria consisted primarily of *Nitrosomonas*, with a relative abundance of 2.76%, which ensured the nitrogen removal process reaction.

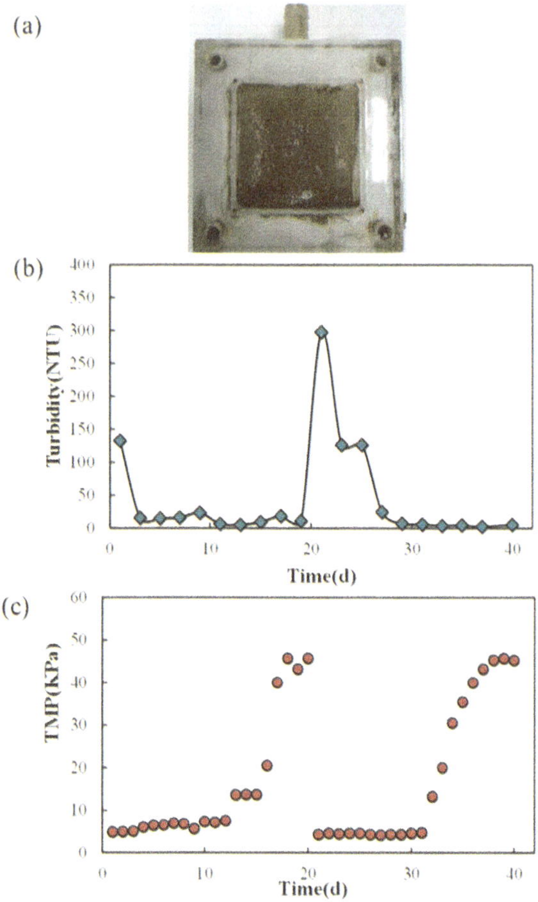


**Fig. 3** Analysis of the dynamic membrane: (**a**) diagram of the dynamic membrane; (**b**) the turbidity in the effluent; and (**c**) TMP of the dynamic membrane

In Fig. 4, nitrogen conversion in SBR cycle, SAA, SAOA and SNOA were investigated in S II to reveal the reaction kinetics of the microorganism. In SBR cycle, 60.5% of consumed $NH_4^+$-N was converted to $NO_2^-$-N by AOB and the SAOA was determined to be 16.8 gN/g-VSS/d achieved by *Nitrosomonas*-like AOB with the relative abundance of 2.76%. Apart from a small part of the transformed $NO_2^-$-N remained in the reactor, the rest $NH_4^+$-N and 86.5% of converted $NO_2^-$-N were changed to nitrogen gas by AnAOB which was 2.72gN/g-VSS/d dominated with *Candidatus jettenia* (11.06%) and *Candidatus brocadia* (1.59%) 13.5% of converted $NO_2^-$-N was oxidized to $NO_3^-$-N by NOB and the SNOA was measured 2.04 gN/g-VSS/d achieved by *Nitrospira* (1.24%), indicating the presence of NOB during the reactor operation. The relative abundance ratio of AOB to NOB was 2:1, while the ratio of SAOA and SNOA was 8:1, which can maintain the stable partial nitritation process. In addition, although SAA to SNOA was 1.33:1, *Candidatus Jettenia* and *Nitrospira*

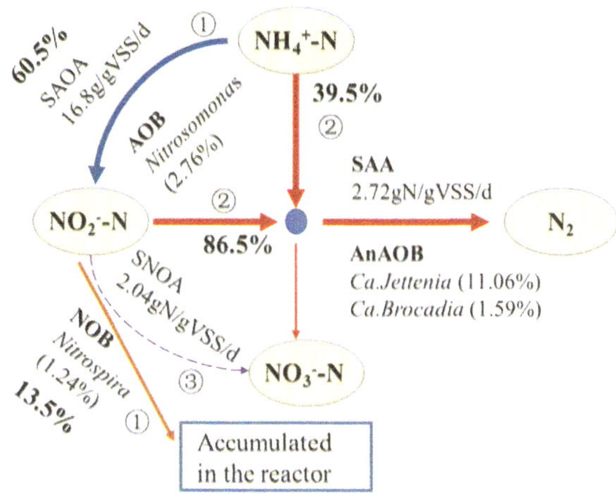


**Fig. 4** The analysis of nitrogen conversion in a specific SBR, microbial activity and microbial community characteristics during the operation of the reactor

were enriched 8.9:1, which conducive to the utilization of $NO_2^-$-N in the anammox process and stable, efficient nitrogen removal.

# Evaluating the Waste and Scrap Trade Risk in the Belt and Road Initiative Countries

**Xiaoqian Hu, Chao Wang, and Ming K. Lim**

**Abstract** China's Belt and Road Initiative (BRI) is a global development strategy, representing great potentials for multilateral trade cooperation and economic growth in Asia, Europe and Africa. The trade volume of commodities along the BRI countries booms in the last few years. The commodities trade attracts extensive attention worldwide, but the waste and scrap (WaS) trade received little attention. To fill this gap, this study reviews the dynamic evolution of the WaS trade network and evaluates the potential WaS trade risks in the BRI countries from three real world scenarios. First, this study constructs the WaS trade networks among the BRI countries from 1989 to 2018. Second, the synopsis of the WaS trade network is reviewed. Third, the shock models are built to analyze the impact of shocks in three scenarios on the stability of trade cooperation among the BRI countries. Finally, the policy implications are provided to promote the WaS trade collaboration among the BRI countries. This study is valuable because it is the first time to identify the WaS trade risk among the BRI countries. It will support policy-makers to build an effective collaboration mechanism to alleviate resource shortage and tackle the global WaS crisis.

**Keywords** Waste and scrap · Trade risk · Complex networks · Belt and road initiative

X. Hu
School of Management and Engineering, Capital University of Economics and Business, Beijing, China

C. Wang (✉)
Research Base of Beijing Modern Manufacturing Development, College of Economics and Management, Beijing University of Technology, Beijing, China
e-mail: chaowanghn@vip.163.com

M. K. Lim
College of Mechanical Engineering, Chongqing University, Chongqing, China

 
F. K. S. Chan et al. (eds.), *Proceedings of the 2020 International Conference on Resource Sustainability: Sustainable Urbanisation in the BRI Era (icRS Urbanisation 2020)*, Environmental Science and Engineering,
https://doi.org/10.1007/978-981-15-9605-6_10

# 1 Introduction

Since the outbreak of the global financial crisis in 2008, the globalization process has stalled. Trumpism and Brexitism are symptoms during this de-globalization wave [1]. The COVID-19 pandemic is accelerating this de-globalization trend, which may turn to regionalization. The global trade network is gradually giving way to regional economic integration, particularly in Asia. For example, in the first five months of 2020, China's imports and exports with Association of Southeast Asian Nations (ASEAN) reached 1.7 trillion RMB, an increase of 4.2% year-on-year. The growth rate of China's exports and imports to ASEAN were both higher than the overall growth rate of China's foreign trade imports and exports during the same period. ASEAN, China, Japan, and South Korea (10 + 3) held a special meeting of leaders to fight against the COVID-19, and stated that it would strengthen macroeconomic policy coordination and strive to reach a regional comprehensive economic partnership agreement (RCEP) within 2020.

The transition from globalization to regionalization is underway and provides the opportunities for the development of China's "Belt and Road Initiative" (BRI). BRI represents an ambitious programme to stimulate regional economic growth in Asia, Europe, and Africa, including policy coordination, facilities connectivity, unimpeded trade, financial integration and people-to-people connections [2]. According to statistics from the China's General Administration of Customs, despite the overall decline in foreign trade imports and exports in 2020, China's foreign trade imports and exports to the BRI countries have maintained a growth trend, with a total import and export volume of 2.07 trillion RMB, an increase of 3.2% year-on-year.

With the massive surge of the trade volume along the BRI countries in the last few years, the commodities trade attracts extensive attention. Complex network approach has been used to study several strategic commodities along the BRI countries, such as petroleum [3], agricultural products [4], and virtual water [5]. In the past decade, with the substantial growth in the international waste and scrap (WaS) trade and increasing stress on global waste management, trade network studies on various WaS have emerged, including e-waste [6, 7], scrap metals [8], plastic waste [9], and scrap copper [10]. These trade network studies explored the topological characteristics of trade system and enabled policymakers to adjust trade policies.

However, the WaS trade along BRI countries has received little attention. The rise of de-globalization trend and increasing concern of the environmental impacts for the BRI inspired this study to fill this research gap. Firstly, this study provides the panoramic view of the WaS trade network along the BRI countries. Then we evaluate the potential WaS trade risks in three real world scenarios, including the shocks via reduction of unilateral trade, the shocks via bilateral trade breakdown and the shocks via overall import reduction in a country. It is the first time to identify the WaS trade risk among the BRI countries. The investigation results highlight some policy implications for BRI countries to meet the challenges of the international trade friction and to maintain the smooth operation of the WaS flow in the BRI trade community.

The remainder of this paper is structured as follows. Sect. 2 presents the literature review. Section 3 introduces the data and methods. Section 4 discusses the dynamic evolution of the WaS trade networks. The simulation results of shocks in three scenarios are shown in Sect. 5. This section is followed by the discussion and policy implications in Sect. 6. Finally, conclusions are offered in Sect. 7.

## 2 Literature Review

In recent years, there have been a number of quantitative studies along BRI countries, as reviewed in [11–13]. Complex network is an effective method applied by researchers to explore the complicated relations among the BRI countries. This section classifies the most relevant studies into three categories.

First, some works sought to study the general international trade among BRI countries by complex network. Zhou and Liu [14] analyzed the characteristics and community evolution patterns of the BRI trade network. Song et al. [15] examined the topological relationship between the BRI and global trade networks. Liu et al. [16] focused on the top 2 trade relations of the BRI trade networks to illuminate the structure and evolution of BRI trade relations, the relative positions of different countries, and changes in the composition of trade communities (e.g., the community leaders) and the changing patterns of trade between them. Chong et al. [17] discussed the determinant factors of BRI trade relationships.

Second, the international trade networks based on specific commodities among BRI countries are also explored. Zhang et al. [3] investigated structural characteristics and evolution patterns with petroleum trade data along the BRI countries. Liu et al. [4] conducted an empirical analysis of the competitiveness and complementarity of agri-trade among the BRI countries based on the export similarity index and trade complementarity index from 2005 to 2016. Qian et al. [5] examined the evolution of virtual water trade in relation to agricultural products between China and the BRI countries during 2000–2016. Chen et al. [18] focused on the temporal cultural trade network between the 66 countries of this region between 1990 and 2016.

Third, apart from analyzing traditional international trade network, other complicated economic connections among the BRI countries are also studied. For example, Liao et al. [19] constructed correlation networks of exchange rates among the BRI countries and analyzed the risk contagion structure. He and Cao [20] studied the foreign direct investment network along the BRI countries from 2003 to 2017, revealing its structural and behavioral characteristics and evolution process. Huo et al. [21] built aviation e-services network along the BRI countries and studied the interconnections across different groups using block modeling. Zhang et al. [22] built the China railway express network under the BRI, and calculated the importance of the China railway express nodes including both inland nodes and seaport nodes.

In summary, complex network approach is an effective method to study commodities trade network along the BRI countries. This study will collect the WaS trade records along BRI countries from 1989 to 2018 and construct annual trade networks.

# 3 Data and Methods

## 3.1 Data Description and Network Construction

Based on the regional characteristics of the BRI region [23], 65 countries are grouped into six sub-regions, including Northeast Asia, Southeast Asia, South Asia, Central Asia, West Asia and North Africa, and Central and Eastern Europe in this study. The countries are tabulated in the Table 3 in Appendix 1. The international trade records among these countries are extracted from the United Nations Commodity Trade Statistics Database (UN Comtrade, https://comtrade.un.org/). To understand the WaS trade along the BRI countries, this study obtained all trade records of 12 categories of commodities during the period from 1989 to 2018, such as 7204 (Ferrous waste and scrap), 4707 (Waste and scrap of paper and paperboard), etc. The detailed list of the WaS commodities and data preprocessing details are shown in Table 4 in Appendix 1.

For a particular year $t$ ranging from 1989 to 2018, the WaS trade records among the BRI countries are used to construct a directed weighted network $G^{[t]}$. The nodes of network denote countries, which are represented as $V^{[t]}$. The trade relationships between countries are denoted by edges $E^{[t]} = \{(i, j)|i, j \in V^{[t]}\}$. The signal adjacency matrix of the network $G^{[t]}$ is $\mathbf{A}^{[t]} = \{a_{ij}^{[t]}|i, j \in V^{[t]}\}$, where $a_{ij}^{[t]} = 1$ if $(i, j) \in E^{[t]}$ or $a_{ij}^{[t]} = 0$ if $(i, j) \notin E^{[t]}$. The trade value between countries is represented by the matrix $\mathbf{W}^{[t]} = \{w_{ij}^{[t]}|(i, j) \in E^{[t]}\}$, where $w_{ij}^{[t]}$ is the total export trade value of 12 kinds of the WaS commodities from country $i$ to country $j$.

## 3.2 Measures of the Network Structure

For understanding the dynamic evolution of the WaS trade among the BRI countries, the following metrics are used to explore the structural characteristics of trade networks.

**Degree.** Node degree is one of the most important indicators to understand the characteristics of nodes, measuring the number of edges connected to a node [24]. In consideration of the directivity of edges, in-degree $k_i^{[t]}(in)$ and out-degree $k_i^{[t]}(out)$ are proposed and are expressed:

$$k_i^{[t]}(in) = \sum_{j \in V^{[t]}} a_{ji}^{[t]}, \quad k_i^{[t]}(out) = \sum_{j \in V^{[t]}} a_{ij}^{[t]} \tag{1}$$

The node degree $k_i^{[t]}$ is the sum of in-degree $k_i^{[t]}(in)$ and out-degree $k_i^{[t]}(out)$. In the WaS trade network, node degree reflects the importance of the BRI countries.

Degree distribution of the WaS trade network provides an intuitive understanding of trade patterns. To reflect the degree distribution of a network, the cumulative probability is defined as $P(nk^{[t]} \leq nk^*)$, showing the probability of nodes with normalized node degree less than $nk^*$. $nk^{[t]}$ is the node degree normalized by the maximum node degree in $G^{[t]}$ and $nk^*$ ranges between 0 and 1.

**Strength**. Based on the definition of node degree, node strength is proposed by considering the edge weights. Specifically, node strength of the country $i$ is calculated by aggregating the trade value with its all trade partners. To distinguish the imports and exports, the in-strength $s_i^{[t]}(in)$ and out-strength $s_i^{[t]}(out)$ is expressed as,

$$s_i^{[t]}(in) = \sum_{(j,i)\in E^{[t]}} w_{ji}^{[t]}, \quad s_i^{[t]}(out) = \sum_{(i,j)\in E^{[t]}} w_{ij}^{[t]} \tag{2}$$

The node strength is calculated by aggregating the in-strength and out-strength, which is $s_i^{[t]} = s_i^{[t]}(in) + s_i^{[t]}(out)$.

Similar to the node degree distribution, the distribution of node strength is defined as the probability of nodes with normalized node strength less than $ns^*$, which is $P(ns^{[t]} \leq ns^*)$. $ns^{[t]}$ is the node strength normalized by the maximum node strength in $G^{[t]}$ and $ns^*$ ranges between 0 and 1.

**Density and average clustering coefficient**. Density is an important indicator to reflect the tightness of nodes in the network [25]. It is calculated by the fraction of edges to the maximal number of possible edges. The specific definition is shown as follows,

$$d^{[t]} = \frac{m^{[t]}}{N^{[t]}(N^{[t]} - 1)} \tag{3}$$

where $m^{[t]}$ is the number of edges and $N^{[t]}$ represents the number of nodes in network $G^{[t]}$. A larger density in the WaS trade network reflects that the waste and scrap trade among the BRI countries maintains the smooth flow.

Although taking into consideration of the number of edges, the definition of density does not reflect the weight of edges. Therefore, the weighted clustering coefficient [26] is introduced to evaluate the tightness. For the directed weighted network $G^{[t]}$, the clustering coefficient of node $i$ is defined as the fraction of geometric average of the subgraph edge weights,

$$c_i^{[t]} = \frac{1}{k_i^{[t]}(k_i^{[t]} - 1) - 2r_i^{[t]}} \sum_{j,k} (\hat{w}_{ij}^{1/3} + \hat{w}_{ji}^{1/3})(\hat{w}_{ik}^{1/3} + \hat{w}_{ki}^{1/3})(\hat{w}_{jk}^{1/3} + \hat{w}_{kj}^{1/3}) \tag{4}$$

where $k_i^{[t]}$ is the degree of node $i$ and the edge weight $\hat{w}_{ij}^{[t]}$ is normalized by the maximum weight in the network $G^{[t]}$; that is, $\hat{w}_{ij}^{[t]} = w_{ij}^{[t]}/\max(w^{[t]})$. $r_i^{[t]}$ is the reciprocal degree of node $i$; namely, the number of nodes $l$ when $(i, l) \in E^{[t]}$ and $(l, i) \in E^{[t]}$. Therefore, the average clustering coefficient for the network $G^{[t]}$ is expressed as follows,

$$c^{[t]} = \frac{1}{N^{[t]}} \sum_{i \in V^{[t]}} c_i^{[t]} \tag{5}$$

where $N^{[t]}$ is the number of nodes in network $G^{[t]}$. The range of the clustering coefficient is from 0 to 1. And a larger average clustering coefficient of the WaS trade network denotes that countries share strong economic and trade ties.

**Heterogeneity and degree centrality**. Heterogeneity is proposed to provide a unique quantitative characterization about how heterogeneous the distribution of links in network is [27]. The detailed calculation is expressed as follows,

$$h^{[t]} = \frac{N^{[t]} - 2\sum_{(i,j)\in E^{[t]}} (k_i^{[t]} k_j^{[t]})^{-1/2}}{N^{[t]} - 2\sqrt{N^{[t]} - 1}} \tag{6}$$

where $N^{[t]}$ represents the number of nodes in network. $h^{[t]}$ ranges from 0 to 1 and a higher value reflects that the distribution of links is more uneven.

In terms of the direction of edges in the WaS trade network, degree centrality is introduced to measure the degree of monopoly in the export trade and the degree of competition in the import trade [24]. The specific definitions are as follows,

$$c^{[t]}(in) = \frac{\sum_{i\in V^{[t]}} (\max(k^{[t]}(in)) - k_i^{[t]}(in))}{(N^{[t]} - 1)^2} \tag{7}$$

$$c^{[t]}(out) = \frac{\sum_{i\in V^{[t]}} (\max(k^{[t]}(out)) - k_i^{[t]}(out))}{(N^{[t]} - 1)^2} \tag{8}$$

where $\max(k^{[t]}(in))$ and $\max(k^{[t]}(out))$ denote the maximum in-degree and out-degree of nodes respectively.

## 3.3 Shock Models

Facing the complicated and volatile international situations, it is crucial for policy makers to understand the consequences of shocks in the WaS trade along the BRI countries, and formulate timely and effective measures to response the trade crises. In consideration of actual economic and trade situations, this study proposes three types of shock models: reduction of unilateral trade, bilateral trade breakdown, and overall import reduction in a country. The specific definitions of shock models are expressed as follows.

**Shocks via the reduction of unilateral trade**. This scenario can be expressed as the reduction of weight in a single edge in the WaS trade network. For the edge $(i, j)$, the weight $w_{ij}^{[t]}$ is degraded by a fraction $d$ an $d$ ranges from 0 to 1. In this scenario, the export channel from the country $i$ to country $j$ subjects to limitations. We assume that countries in the WaS trade network have stable waste and scrap exports for the domestic waste disposal. Therefore, the reduction of exports from the country $i$ to $j$ will be redistributed to other importers of the country $i$ to remain total exports unchanged. Based on the trade weight preference, the reduction $dw_{ij}^{[t]}$ is proportional to the trade weight between the country $i$ and imports except for the country $j$. For the country $k$, a change in the import value indicates the absolute influence of the shock, which is calculated by

$$f_1^{[t]}(w_{ij}^{[t]}, d, k) = \frac{dw_{ij}^{[t]} w_{ik}^{[t]}}{\sum_{(i,h)\in E^{[t]}, h\neq j} w_{ih}^{[t]}} \tag{9}$$

where $k \neq i, j$. And the relative influence of the shock can be measured by the ratio of import increment to the total import value of the country $k$, which is expressed as,

$$g_1^{[t]}(w_{ij}^{[t]}, d, k) = \frac{f_1^{[t]}(w_{ij}^{[t]}, d, k)}{\sum_{(h,k)\in E^{[t]}} w_{hk}^{[t]}} \tag{10}$$

where $g_1^{[t]}(w_{ij}^{[t]}, d, k)$ ranges from 0 to 1. A larger $g_1^{[t]}(w_{ij}^{[t]}, d, k)$ indicates that the country $k$ has to improve the processing capacity of waste and scrap imports.

However, the increment of waste and scrap imports for importers is limited. When $g_1^{[t]}(w_{ij}^{[t]}, d, k)$ is larger than a threshold, the country $k$ could not meet the excess waste treatment capacity. To evaluate the impacts of the shock from the edge $(i, j)$ to the BRI trade community, we define an indicator as follows,

$$e_1^{[t]}(w_{ij}^{[t]}, d, \lambda) = \sum_{(i,k)\in E^{[t]}} I_1^{[t]}(k, \lambda) \tag{11}$$

where $I_1^{[t]}(k, \lambda)$ is an indicative function, namely, if $g_1^{[t]}(w_{ij}^{[t]}, d, k) >= \lambda$, then $I_1^{[t]}(k, \lambda) = 1$; otherwise, $I_1^{[t]}(k, \lambda) = 0$. And $\lambda$ is a threshold ranging from 0 to 1. Furthermore, $\max\limits_{(i,j)\in E^{[t]}} e_1^{[t]}(w_{ij}^{[t]}, d, \lambda)$ indicates the maximum impact of shocks via the reduction of unilateral trade. And the trade relationship between country pairs corresponding to a larger $e_1^{[t]}(w_{ij}^{[t]}, d, \lambda)$ need special attention for keep the stability of the WaS trade among the BRI countries. The simulation is conducted in the range of [10%, 100%] with increments of 10% for parameters $d$ and $\lambda$.

**Shocks via bilateral trade breakdown**. In the case of frequent bilateral trade disputes, we propose a model to describe the shocks from the termination of trade relations between two countries. In this case, all links between the country $i$ and $j$ will be removed in the network $G^{[t]}$.

In the case of trade relations breakdown, the impact of shocks on the WaS net exporting country is critical. Therefore, the net exporter between the country $i$ and

$j$ should be identified firstly. If $w_{ij}^{[t]} - w_{ji}^{[t]} > 0$, the country $i$ is the net exporter to country $j$; otherwise, the country $j$ is the net exporter to country $i$. Taking the net exporter $i$ as an example, we investigate the impact of shocks as follows. Due to the breakdown of trade relations between the country $i$ and $j$, country $i$ need proportionally redistribute the change value of trade $w_{ij}^{[t]} - w_{ji}^{[t]}$ to the importers except for country $j$. For the country $k$, a change in the import value indicates the absolute and relative influence of the shock, which are calculated by

$$f_2^{[t]}(w_{ij}^{[t]}, k) = \frac{(w_{ij}^{[t]} - w_{ji}^{[t]})w_{ik}^{[t]}}{\sum_{(i,h)\in E^{[t]}, h\neq j} w_{ih}^{[t]}} \tag{12}$$

$$g_2^{[t]}(w_{ij}^{[t]}, k) = \frac{f_2^{[t]}(w_{ij}^{[t]}, k)}{\sum_{(h,k)\in E^{[t]}} w_{hk}^{[t]}} \tag{13}$$

where $k \neq i, j$ and $g_2^{[t]}(w_{ij}^{[t]}, k)$ ranges from 0 to 1. Similar to the previous definition as Eq. (11), the indicator to evaluate the impacts of shocks is calculated by

$$e_2^{[t]}(w_{ij}^{[t]}, \lambda) = \sum_{(i,k)\in E^{[t]}} I_2^{[t]}(k, \lambda) \tag{14}$$

If $g_2^{[t]}(w_{ij}^{[t]}, k) >= \lambda$, then $I_2^{[t]}(k, \lambda) = 1$; otherwise, $I_2^{[t]}(k, \lambda) = 0$. And $\max_{(i,j)\in E^{[t]}} e_2^{[t]}(w_{ij}^{[t]}, \lambda)$ records the maximum number of countries bearing import pressure beyond their capacity.

**Shocks via overall import reduction in a country**. Different from the above two scenarios, the third shock affects all of a country's trade relations. For example, China imposed a series of import ban on solid waste in recent years [28]. In this scenario, the country $i$ reduces the value of imports by a certain proportion $d$, namely $ds_i^{[t]}(in)$. And the reduction of imports is proportional to the trade weight between the country $i$ and the exporting partners. Specifically, for the exporter $j$, the absolute and relative reduction in WaS exports are defined as follows,

$$f_3^{[t]}(i, j, d) = \frac{ds_i^{[t]}(in)w_{ji}^{[t]}}{\sum_{(k,i)\in E^{[t]}} w_{ki}^{[t]}} \tag{15}$$

$$g_3^{[t]}(i, j, d) = \frac{f_3^{[t]}(i, j, d)}{\sum_{(j,k)\in E^{[t]}} w_{jk}^{[t]}} \tag{16}$$

where $j \neq i$. Similar to the definition of $e_2^{[t]}(w_{ij}^{[t]}, \lambda)$ in Eq. (14), the indicator of integrative assessment for the impact of shocks $e_3^{[t]}(i, d, \lambda)$ is calculated by aggregating the indicative function $I_3^{[t]}(j, \lambda)$ for all edges $(j, i) \in E^{[t]}$. When $g_3^{[t]}(i, j, d)$ is larger than the threshold $\lambda$, $I_3^{[t]}(j, \lambda)$ is set as 1; otherwise, $I_3^{[t]}(j, \lambda)$ is 0. All nodes in $G^{[t]}$ are sorted in descending order by $e_3^{[t]}(i, d, \lambda)$. Countries at the top of the list are the primary focus because of significant impact on the WaS trade community in the BRI region.

# 4 Evolution of the WaS Trade Along the BRI Countries

## *4.1 Topological Structure*

To understand the dynamic evolution of the WaS trade along the BRI countries, we review the network topological structure from 1989 to 2018. As shown in Fig. 1a, the trade value of 12 kinds of WaS commodities in the trade network shows a continuous increase from 1989 to 2011 and then gradually decreased in the next 7 year. And the number of the involved countries in the WaS trade network increased dramatically from 1989 to 1993, as shown in Fig. 1b. In addition, Fig. 1c shows that the number of WaS trade relationships between countries in the BRI presented an upward trend since 1989 and remained stable from 2010. It indicates that countries paid more attention to the waste and scrap and increasingly joined in the international WaS trade.

The average node degree of the WaS trade network grew gradually and reached a peak in 2010 as shown in Fig. 1d. Specifically, it can be found that each country

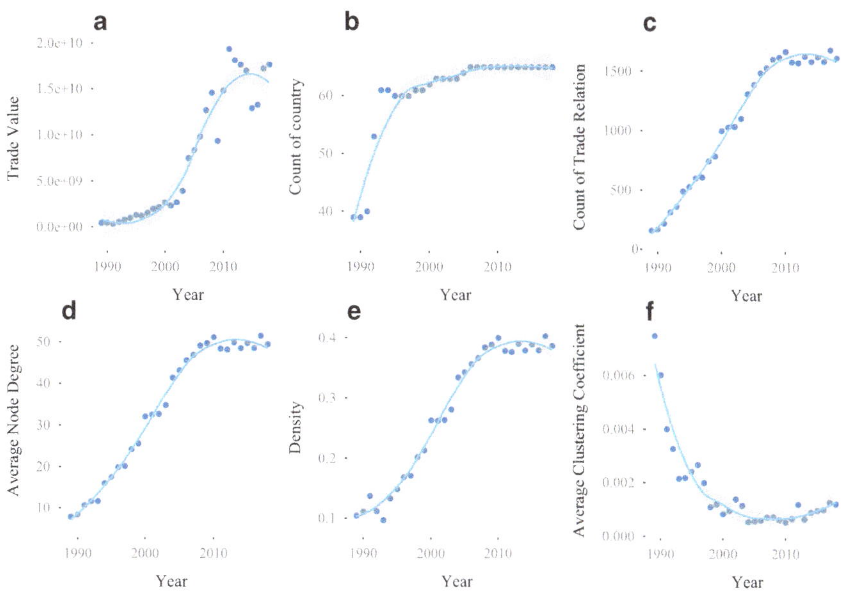


Fig. 1 Dynamic evolution of the network structure from 1989 to 2018

maintains trade relations with an average of 50 countries in 2018. And Fig. 1e indicates a similar upward trend of density to the average node degree. The results shown in Fig. 1d, e reflect the increasingly close trade connections among the BRI countries. However, Fig. 1f depicts a distinct trend of the average clustering coefficient. In other words, this indicator decreased sharply and kept stable at a low level since 2000. Combing the average node degree and density as well as average clustering coefficient, it is found that although the increasing trade relations were built among the BRI countries, the considerable disparity of trade value in edges led to a low level of tightness.

Furthermore, we investigate the distribution of node degree and node strength to understand the patterns of the WaS trade networks. Fig. 2 shows the cumulative probability distribution of in-degree, out-degree and degree in 1989, 2003 and 2018 as examples. Compare Fig. 2a, c, it is found that the cumulative probability distribution of in-degree in 1989 and 2018 are significantly different. In addition, the similar

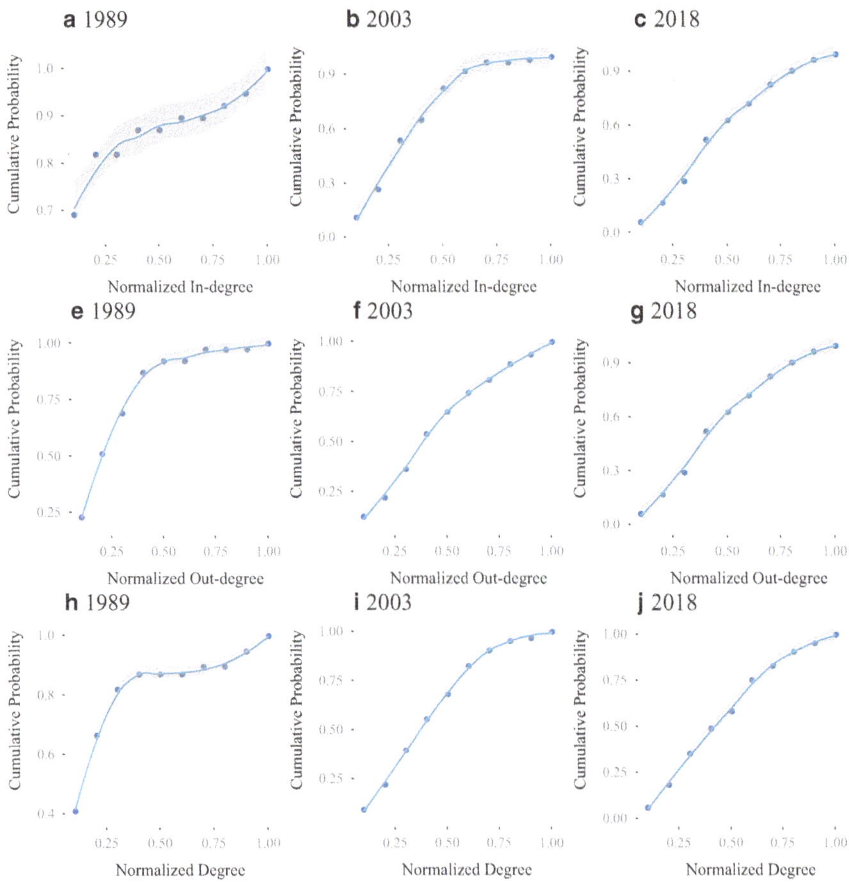


**Fig. 2** Distribution of in-degree, out-degree and degree in 1989, 2003 and 2018

situation happens in the cumulative probability distribution of out-degree and degree as shown in Fig. 2e–J. It indicates that most of the BRI countries in the WaS trade network only have a few trading partners in 1989. However, with the dynamic evolution of the WaS trade network structure, the cumulative probability distribution of in-degree/out-degree/degree shows a straight upward trend in 2018. It denotes that the ratio of countries having different number of trading partners is nearly equal.

Comparing to node degree, node strength takes weight of edges into consideration. Fig. 3 shows the cumulative probability distribution of in-strength, out-strength and strength in 1989, 2003 and 2018. It can be seen that the distribution patterns of the WaS trade network remained fairly static. Specifically, most of countries in the network have a small trade value and the WaS trade is dominated by a few countries. Combing the results shown in Figs. 2 and 3, it is found that the distribution of waste and scrap trade value is more heterogeneous than that of trade connections.

Although the cumulative probability distribution of degree and strength gives us an intuitive sense of the WaS trade network's patterns, it is necessary to provide immediate evaluation on the heterogeneity of edges in networks. Fig. 4a depicts a decreasing tendency of the heterogeneity from 1989 to 2018, indicating that the uneven distribution pattern of the WaS trade relations reached a relatively low level in 2018.

Figure 4b presents the changes of the in-degree centrality from 1989 to 2018, which indicates the level of the competition among waste and scrap importing countries in the BRI region. We find that the in-degree centrality decreased in the early years and fluctuated around 50% during the period of 2000 and 2018. As shown in Fig. 4c, the degree of monopolization measured by the out-degree centrality gradually increased from 1989 to 2018. It reflects that the WaS trade among the BRI countries shows an increasingly monopoly pattern.

## 4.2 Important Countries in the WaS Trade Network

In the WaS trade networks, some crucial countries make a significant difference with the whole waste and scrap trade community. In this sub-section, we identify the core countries based on in-degree, out-degree, in-strength and out-strength and explore the variation of the core countries.

As shown in Table 1, top 10 countries in the WaS trade network in 1989, 2003 and 2018 are listed. We find that countries with high in-degree are mainly geographically distributed in the Southeast Asia, like Indonesia, Thailand, and Singapore. As years passed, the countries in the Central and Eastern Europe, West Asia and North Africa, such as Turkey, United Arab Emirates (UAE), Poland and Czechia have obvious advantages on the number of importing channels. In terms of waste and scrap import value, the geographical location of the core countries had undergone a similar change from 1989 to 2018. Especially, the countries in the Central and Eastern Europe, including Poland, Russia, Belarus and Czechia were all in the top ten sorted by

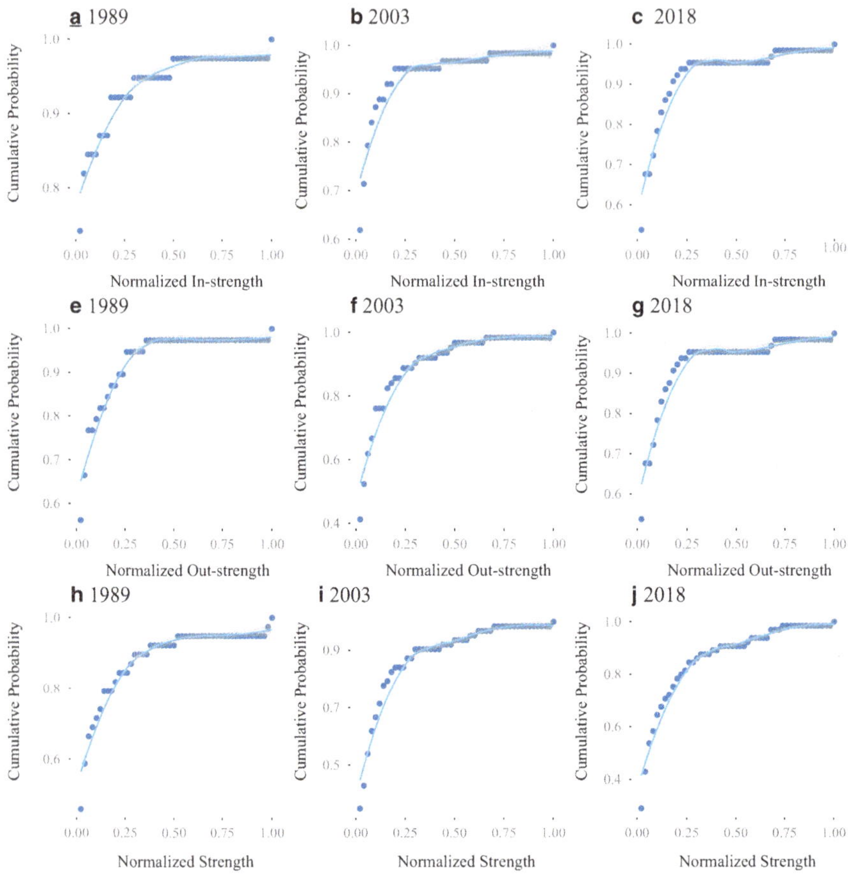


Fig. 3 Distribution of in-strength, out-strength and strength in 1989, 2003 and 2018

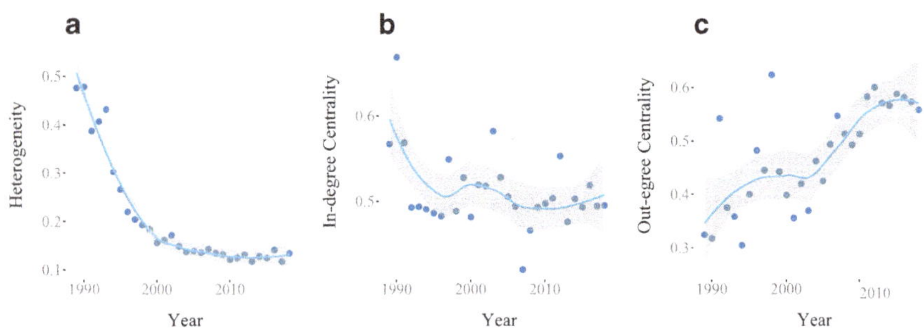


Fig. 4 Heterogeneity and degree centrality of the WaS trade network

import value. Notably, there is a great gap between the top three importing countries (India, China and Turkey) and other countries.

As for the waste and scrap exports, the core countries with a high out-degree is more geographically dispersed than that with a high in-degree in 2018. However, the number of core countries in the Central and Eastern Europe, West Asia and North Africa is larger than other regions. In addition, Table 1 indicates that China has the maximum exporting partners in 2003 and 2018, followed by Thailand and UAE. While, in terms of export value, it is surprise to find that China was not in the top ten countries in 2003 and 2018. By comparing the lists in 1989, 2003 and 2018, we find that the core countries changed drastically. And in 2018, the important exporting countries mainly geographical located in the Southeast Asia, Central and Eastern Europe, such as Singapore, Malaysia, Russia and Poland. Based on the above results, we find that the countries in the Southeast Asia, Central and Eastern Europe are located in the central positions measured by the waste and scrap imports and exports on the trade community in the BRI region.

## 5 Simulation Under Three Shock Scenarios

According to the shock models defined in Sect. 3.3, we analyze the impact from three types of shocks, including the reduction of unilateral trade, breakdown of the bilateral trade relationships and overall import reduction in a country. The simulation results will provide implications for policy-makers to understand the influence of different risks and formulate appropriate policies for countries in the BRI trade community.

### *5.1 Reduction of Unilateral Trade*

The reduction of unilateral trade is one of the most common shocks in the current world trading system. As defined in Sect. 3.3, this shock is expressed as the reduction of the weight of an edge $(i, j)$ and the redistribution of exports of the country $i$ to other exporters. And $e_1^{[t]}(w_{ij}^{[t]}, d, \lambda)$ is proposed to measure the impacts of the shock; that is, the number of countries with the pressure on import $g_1^{[t]}(w_{ij}^{[t]}, d, k)$ more than the threshold $\lambda$ when the unilateral trade $(i, j)$ with trade value $w_{ij}^{[t]}$ is degraded by a fraction $d$. For the given parameters $\lambda$ and $d$, the country pairs with a high value of $e_1^{[t]}(w_{ij}^{[t]}, d, \lambda)$ are the ballast stones for stable trade among the BRI countries. Therefore, we simulate the shocks based on the WaS trade network in 2018.

Table 2 shows the top six shocks in three types of scenarios. Based on the evaluation of $e_1^{[t]}(w_{ij}^{[t]}, 0.1, 1)$, we find that the reduction of unilateral trade from India to UAE and from UAE to India will have the worst impact. Specifically, 16 countries will be under the pressure to consume more waste and scrap, which is beyond their capacity. As mentioned in Table 1, India and UAE are the biggest waste and scrap

**Table 1** Top 10 countries in the WaS trade network in 1989, 2003, 2018

| Rank | In-degree | | | | | | Rank | Out-degree | | | | | |
|---|---|---|---|---|---|---|---|---|---|---|---|---|---|
| | 1989 | | 2003 | | 2018 | | | 1989 | | 2003 | | 2018 | |
| | Country | Value | Country | Value | Country | Value | | Country | Value | Country | Value | Country | Value |
| 1 | India | 25 | India | 53 | Thailand | 56 | 1 | Singapore | 16 | China | 40 | China | 60 |
| 2 | Indonesia | 25 | China | 45 | Turkey | 51 | 2 | Malaysia | 10 | Russia | 39 | Thailand | 57 |
| 3 | Thailand | 22 | Pakistan | 37 | China | 50 | 3 | India | 10 | Kuwait | 38 | UAE | 56 |
| 4 | Singapore | 18 | UAE | 36 | India | 50 | 4 | Thailand | 8 | UAE | 37 | Poland | 52 |
| 5 | Malaysia | 13 | Thailand | 34 | Pakistan | 46 | 5 | Saudi Arabia | 8 | Turkey | 36 | Turkey | 51 |
| 6 | Bangladesh | 10 | Turkey | 31 | Malaysia | 45 | 6 | China | 6 | India | 36 | Indonesia | 48 |
| 7 | Turkey | 8 | Russia | 31 | UAE | 43 | 7 | UAE | 6 | Thailand | 36 | Kuwait | 48 |
| 8 | Philippines | 4 | Malaysia | 29 | Indonesia | 42 | 8 | Oman | 6 | Poland | 32 | India | 47 |
| 9 | Pakistan | 4 | Sri Lanka | 29 | Poland | 42 | 9 | Turkey | 5 | Ukraine | 32 | Hungary | 44 |
| 10 | Myanmar | 3 | Singapore | 28 | Czechia/Qatar | 40 | 10 | Fmr USSR | 5 | Czechia | 29 | Russia | 43 |
| Rank | In-strength | | | | | | Rank | Out-strength | | | | | |
| | 1989 | | 2003 | | 2018 | | | 1989 | | 2003 | | 2018 | |
| | Country | Value | Country | Value | Country | Value | | Country | Value | Country | Value | Country | Value |
| 1 | India | 1.93 | Turkey | 10.25 | India | 35.03 | 1 | Singapore | 1.54 | Russia | 6.13 | UAE | 18.34 |
| 2 | Thailand | 0.94 | China | 6.96 | China | 24.26 | 2 | Fmr USSR | 0.54 | Singapore | 3.97 | Russia | 16.46 |
| 3 | Indonesia | 0.56 | India | 4.35 | Turkey | 23.56 | 3 | Czechoslovakia | 0.39 | Romania | 3.06 | Singapore | 12.26 |
| 4 | Malaysia | 0.34 | UAE | 1.99 | Singapore | 9.10 | 4 | Malaysia | 0.37 | Ukraine | 2.84 | Malaysia | 12.14 |
| 5 | Singapore | 0.33 | Singapore | 1.94 | UAE | 7.13 | 5 | UAE | 0.33 | UAE | 2.51 | Saudi Arabia | 8.97 |
| 6 | Turkey | 0.20 | Malaysia | 1.57 | Pakistan | 6.66 | 6 | Vietnam | 0.26 | Georgia | 1.90 | Poland | 8.20 |

(continued)

**Table 1** (continued)

| Rank | In-strength | | | | | | Rank | Out-strength | | | | | |
|---|---|---|---|---|---|---|---|---|---|---|---|---|---|
| | 1989 | | 2003 | | 2018 | | | 1989 | | 2003 | | 2018 | |
| | Country | Value | Country | Value | Country | Value | | Country | Value | Country | Value | Country | Value |
| 7 | Philippines | 0.09 | Belarus | 1.53 | Poland | 6.09 | 7 | China | 0.23 | Saudi Arabia | 1.76 | Indonesia | 7.74 |
| 8 | UAE | 0.06 | Indonesia | 1.21 | Russia | 5.88 | 8 | Lao PDR | 0.16 | Indonesia | 1.39 | Thailand | 7.26 |
| 9 | Bangladesh | 0.06 | Moldova | 0.94 | Belarus | 5.41 | 9 | Saudi Arabia | 0.15 | Kazakhstan | 1.35 | Lithuania | 6.54 |
| 10 | China | 0.04 | Slovakia | 0.88 | Czechia | 4.46 | 10 | Sri Lanka | 0.09 | Malaysia | 1.13 | Romania | 5.15 |

*Note* The unit of in-strength/out-strength is ten million US dollars

**Table 2** The lists of shocks with high impacts in three scenarios

| Rank | Scenario I | | Scenario II | | Scenario III | |
|---|---|---|---|---|---|---|
| | $(i, j)$ | $e_1^{[t]}$ | $i, j$ | $e_2^{[t]}$ | $i$ | $e_3^{[t]}$ |
| 1 | (India, UAE) | 16 | Russia, Turkey | 14 | India | 23 |
| 2 | (UAE, India) | 16 | Malaysia, China | 10 | Turkey | 19 |
| 3 | (Russia, Turkey) | 14 | UAE, India | 10 | China | 16 |
| 4 | (Malaysia, China) | 10 | Kazakhstan, Russia | 6 | Malaysia | 6 |
| 5 | (Russia, Belarus) | 7 | Thailand, China | 6 | Singapore | 6 |
| 6 | (Kazakhstan, Russia) | 7 | Russia, Belarus | 6 | Pakistan | 6 |

*Note* Scenario I, II and III represents the shocks via the reduction of unilateral trade, shocks via bilateral trade breakdown and the shocks via overall import reduction. The parameters of $d$ and $\lambda$ in Scenario I and III is set as 0.1 and 1. And the Scenario II sets the parameter $\lambda$ as 0.1

importer and exporter in 2018 respectively. Therefore, the effect of shocks from the trade between these two countries are significantly greater than the other shocks. And as shown in Fig. 5a, due to the reduction of trade value from India to UAE, the top

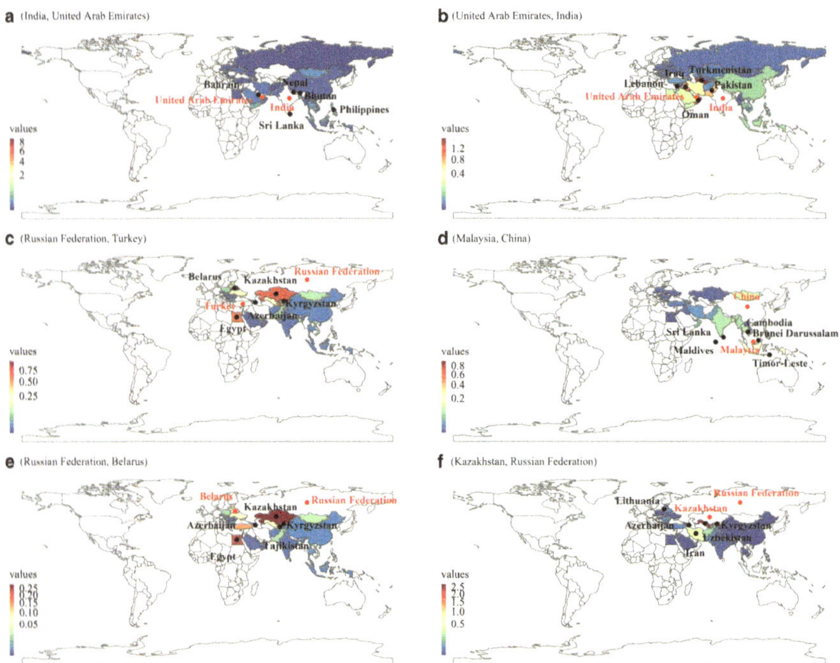


**Fig. 5** Impacts of the shocks via the reduction of unilateral trade. *Note* The black cycles denote the top 5 countries sorted by the indicator $g_1^{[t]}(w_{ij}^{[t]}, d, k)$ under the shock from $(i, j)$, and the red cycles represent the country $i$ and $j$. The warmer color indicates a high value

five affected countries are mainly located in South Asia, including Nepal, Bhutan and Sri Lanka. Comparing to the shock from (India, UAE), the core countries affected by the shock from (UAE, India) are closer to the UAE, as shown in Fig. 5b. Specifically, the countries with a high $g_1^{[t]}(w_{ij}^{[t]}, d, k)$ are centralized around the West Asia, such as Lebanon, Oman and Iraq. In addition, the shocks from (Russia, Turkey), (Russia, Belarus), (Kazakhstan, Russia) and (Malaysia, China) also have a tremendous influence on the trade community in the BRI region as shown in Fig. 5c–f. Notably, the great influence from all former three shocks occurs in the Central Asia, West Asia and North Africa. The countries with a high $g_1^{[t]}(w_{ij}^{[t]}, d, k)$ affected by the shock from the reduction of trade value from Malaysia to China mainly focus on the Southeast Asia and South Asia, like Brunei Darussalam, Cambodia, Timor-Leste and Sri Lanka.

## 5.2 Breakdown of the Bilateral Trade Relationships

With the escalation of economic and trade frictions, the risk of disruption of bilateral trade relationships exacerbates. Although it is the most extreme case, it is urgent to predict the impacts of shocks in this scenario and enhance the emergency response preparedness for the BRI countries.

Based on the shock models introduced in Sect. 3.3, we obtain the top six shocks in this scenario, as shown in Table 2. By comparing to the lists of Scenario I and Scenario II, we find that the countries with high impacts in shocks are similar, including Russia, Turkey, Malaysia, China, UAE, India, Belarus and Kazakhstan. Therefore, the change of trade relations between these countries should be given careful attention. However, there are still some differences in the rank of shock between countries in Scenario I and Scenario II. Specifically, the breakdown of the bilateral trade relationship between Russia and Turkey will lead to the maximum value of the indicator $e_2^{[t]}$; that is, 14 countries will have to face the pressure of disposing more than 10% original WaS imports under this shock. It will significantly impact the global circle of the waste and scrap, and threat the security of the WaS trade in the BRI region. In addition, Fig. 6a shows that the countries which are greatly affected by the shock from the breakdown of trade relation between Russia and Turkey are mainly distributed in the Central Asia and South Asia, like Kazakhstan, Kyrgyzstan and Azerbaijan. And similar results occur in the shock from (Kazakhstan, Russia) and (Russia, Belarus), as shown in Fig. 6d, f. The second biggest impact comes from the shock between Malaysia and China, which brings 15% countries in the BRI trade community exceeding the threshold. Besides, the countries with a high value of $g_2^{[t]}(w_{ij}^{[t]}, k)$ affected by the shocks from (Malaysia, China) and (Thailand, China) are mainly geographical distributed in South Asia and Southeast Asia, as shown in Fig. 6b, e. Therefore, countries in these regions, like Cambodia, Brunei Darussalam, Lao PDR and Philippines, should pay special attention to the WaS trade relationships between these country pairs.

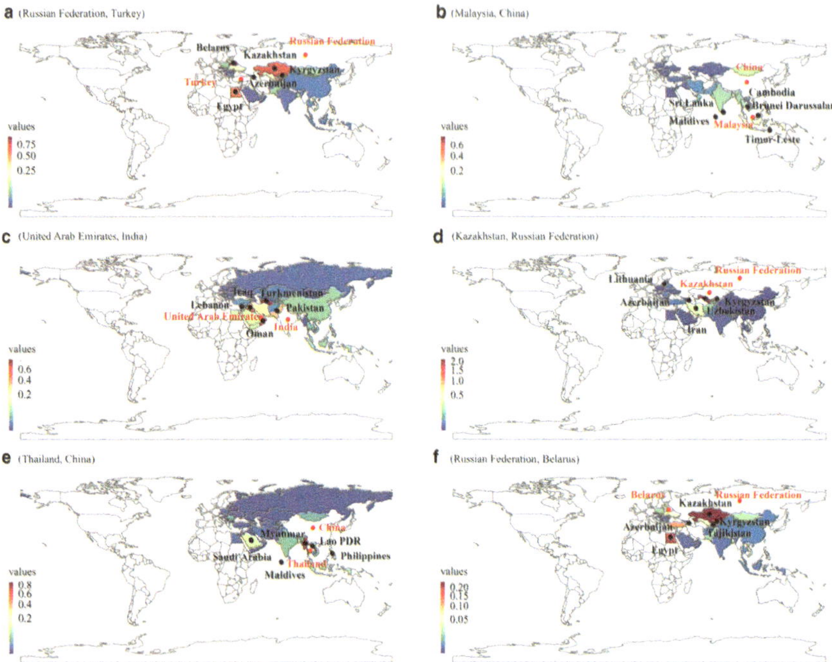


**Fig. 6** Impacts of shocks via bilateral trade breakdown. *Note* The black cycles denote the top 5 countries sorted by the indicator $g_2^{[t]}(w_{ij}^{[t]}, k)$ under the shock from the breakdown of bilateral trade relationship between country $i$ and $j$, and the red cycles represent the country $i$ and $j$

## 5.3 Overall Import Reduction in a Country

Overall import reduction in a country will also bring a large impact on the WaS trade community in the BRI region. For example, China enacted an import ban on the solid waste in 2018. Besides, the economic recession of a country will also lead to a significant drop in the imports [29]. Therefore, we model this type of shocks in Sect. 3.3 and analyze the simulation results in this sub-section.

Table 2 shows that the reduction of waste and scrap imports in India, Turkey, China, Malaysia, Singapore and Pakistan will have a considerable impact on the WaS trade community in the BRI region. Especially, around one third of countries in the trade community cannot export waste and scrap normally. And Turkey and China are second and third respectively, which having a great influence significantly more than other countries. As shown in Fig. 7a, the affected countries by the shock from India are mainly geographical distributed in the South Asia, including Nepal, Bangladesh, Sri Lanka and Maldives. In addition, we find that the shocks from Turkey and China have influence on the West Asia, Central and Eastern Europe, and Southeast Asia respectively, as shown in Fig. 7b, c. Similar to China, the reduction of waste and scrap imports in Malaysia and Singapore has a noticeable effect on the

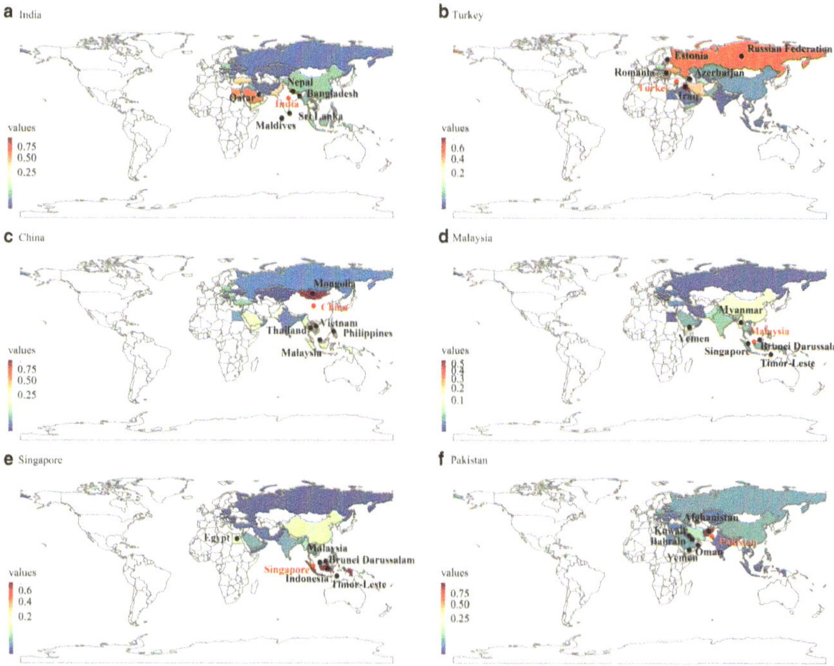


**Fig. 7** Impacts of shocks via overall import reduction in a country. *Note* The black cycles denote the top 5 countries sorted by the indicator $g_3^{[t]}(i, j, d)$ under the shock from country $i$, and the red cycle represents the country $i$

Southeast Asia in Fig. 7d, e. And Fig. 7f shows that the affected countries with a high $g_3^{[t]}(i, j, d)$ shocked by Pakistan are mainly located in the West Asia, such as Oman, Kuwait and Yemen. Based on above results, we find that the shocks via the reduction of scrap imports in a country have prominent geographical characteristics. In other words, the affected countries are mainly geographically distributed in the region around the country with waste and scrap import drops.

# 6 Discussion and Policy Implications

## *6.1 Dynamic Evolution of the Waste and Scrap Trade*

In Sect. 4, we review the dynamic evolution of WaS trade among the BRI countries. Although the trade value drops in recent years, an increasing number of countries were involved in the WaS trade and built trade connections with other countries. It is interesting that the number of trade relationships built by a country is distributed uniformly, which is different with many realistic networks. However, a distinctive characteristic is the heterogeneity of the trade values between countries. It indicates that a mass of the WaS trade is dominated by a few core countries. As shown in Table 1, India, China and Turkey, ranked as the top three countries, occupy huge waste and scrap imports, which is far more than any other countries. And the countries in the Central and Eastern Europe also have a great influence on the waste and scrap imports in 2018. In terms of waste and scrap exports, the important countries mainly geographical located in the Southeast Asia, Central and Eastern Europe. Besides, UAE and Saudi Arabia also have outstanding performance in the list of exporters. In addition, the level of the competition among waste and scrap importers fluctuated at historical lows, but the waste and scrap exporters show an increasingly monopoly pattern.

## 6.2 Policy Implications from Shocks in Three Scenarios

As defined in Sect. 3.3, three types of practical trade shocks are modeled. The simulation results will support policy-makers understanding the impacts of shocks with different scale of shock strength. And combing the threshold of acceptable changes for countries, the shocks with great influence are obtained. These analyses highlight some policy implications for countries to keep trade security and maintain the smooth operation of waste and scrap flow in the BRI trade community.

As shown in Table 2, the list shows the shocks with great effect in three scenarios, when the parameters $d$ and $\lambda$ is 0.1 and 1 respectively. In the first scenario, the shocks, like (India, UAE), (UAE, India), (Russia, Turkey) and (Malaysia, China), will put increased pressure on importing the waste and scrap. And the geographical distribution of affected countries under different shocks are really different. Therefore, for the BRI countries, it is suggested to focus on the reduction of unilateral trade between specific country pairs. For example, it is necessary for BRI countries in the South Asia to pay attention to the change of trade value in (India, UAE) and (UAE, India). And the countries in the Central Asia and West Asia should focus on the waste and scrap trade in (Russia, Turkey), (Russia, Belarus) and (Kazakhstan Russia). The reduction of the WaS trade from Malaysia to China has a big impact on the Southeast Asian countries.

Based on the analyses on the second scenario in Sect. 5.2, the top six shocks are from the breakdown of the bilateral trade relationships in (Russia, Turkey), (Malaysia, China), (UAE, India), (Kazakhstan, Russia), (Thailand, China) and (Russia, Belarus). Therefore, the stability of the bilateral trade relationship in these country pairs should be taken seriously. Particularly, the country pairs including Russia mentioned above have a great influence on the countries in the Central Asia and West Asia. And the countries in the Southeast Asia need to focus on the WaS trade between Malaysia and China, Thailand and China. Once there is deterioration in the WaS trade relations between the country pairs mentioned above, the countries in the corresponding regions need timely adjustment strategies on the WaS trade to keep the stable waste and scrap trade.

In the third scenario, the shocks from the overall import reduction in India, Turkey, China, Malaysia, Singapore and Pakistan have a considerable influence on the BRI trade community. And the affected countries under the shocks are mainly geographically distributed around the BRI countries with the reduction of imports. The shocks from India will have a great impact on the countries in the South Asia. The countries in the Southeast Asia will be strongly influenced by the reduction of waste and scrap imports in China, Malaysia and Singapore. And countries in the West Asia should follow the change of the WaS imports in Pakistan. In addition, the countries affected strongly under the shocks should strengthen other export relations and reduce the dependence on any one country.

## 7 Conclusion

Due to the different roles of countries in the international trade, the cooperation between countries can be conducive to regional stability and development. International WaS trade, as an important way for countries to handle and recycle waste and scrap, should be fully investigated. For the BRI countries, a comprehensive understanding of the WaS trade can provide insights of the WaS trade trend and highlight policy implications to deal with trade shocks.

Therefore, this study uses the trade records of BRI countries to construct the WaS trade networks. We review the development tendency of WaS trade from 1989 to 2018 and investigate the dynamic evolution of the network structure based on the complex network theory. In addition, for the circumstance of the growing stressful trade relations, we define shock models in three different scenarios, including the shocks via the reduction of unilateral trade, the shocks via the breakdown of the bilateral trade relationships and the shocks via the reduction of overall imports in a country. The simulation results indicate the lists of shocks with a great influence on the BRI countries in three scenarios and the corresponding countries affected strongly by these shocks. Therefore, for these shocks, some policy implications are provided to help countries to keep trade security.

In the further work, we will discuss the WaS trade of BRI countries in the context of the global waste and scrap trade. Specifically, the trade relations between the BRI trade community and other countries will be further investigated and thus we will provide a wider world view to understand the influence of trade shocks.

## Appendix 1: Data Preprocessing

See Tables 3 and 4.

The database of UN Comtrade records the trade flow, trade countries, the unique code of commodities (HS code from the interpretation of the Harmonized System) and the trade value (US dollar). Due to the difference of the statistical methods used

**Table 3** Countries in belt and road region

| Regions | Countries | Count |
|---|---|---|
| Northeast Asia | Mongolia, China | 2 |
| Southeast Asia | Indonesia, Thailand, Malaysia, Vietnam, Singapore, Philippines, Myanmar, Cambodia, Lao PDR, Brunei Darussalam, Timor-Leste | 11 |
| South Asia | India, Pakistan, Bangladesh, Sri Lanka, Afghanistan, Nepal, Maldives, Bhutan | 8 |
| Central Asia | Kazakhstan, Uzbekistan, Turkmenistan, Kyrgyzstan, Tajikistan | 5 |
| West Asia and North Africa | Saudi Arabia, United Arab Emirates, Oman, Iran, Turkey, Israel, Egypt, Kuwait, Iraq, Qatar, Jordan, Lebanon, Bahrain, Yemen, Georgia, Azerbaijan, Armenia, Syria, State of Palestine | 19 |
| Central and Eastern Europe | Russian Federation, Poland, Romania, Czechia, Slovakia, Bulgaria, Hungary, Latvia, Lithuania, Slovenia, Estonia, Croatia, Albania, Serbia, Ukraine, Belarus, Moldova, TFYR of Macedonia, Bosnia and Herzegovina, Montenegro | 20 |

**Table 4** HS code of the world scrap trade

| Commodity | HS code | Description |
|---|---|---|
| Ferrous | 7204 | Ferrous waste and scrap; remelting scrap ingots of iron or steel |
| Paper | 4707 | Waste and scrap of paper and paperboard |
| Nonferrous | | |
| Copper | 7404 | Copper; waste and scrap |
| Aluminum | 7602 | Aluminum; waste and scrap |
| Nickel | 7503 | Nickel; waste and scrap |
| Lead | 7802 | Lead; waste and scrap |
| Zinc | 7902 | Zinc; waste and scrap |
| Plastics | 3915 | Waste, parings and scrap, of plastics |
| Rubber | 4004 | Waste, parings and scrap of rubber (other than hard rubber) and powders and granules obtained therefrom |
| Precious metals | 7112 | Waste and scrap of precious metal or of metal clad with precious metal; other waste and scrap containing precious metal compounds, of a kind uses principally for the recovery of precious metal |
| Textiles | 6309 | Textiles; worn clothing and other worn articles |
| Glass | 7001 | Glass; cullet and other waste and scrap of glass, glass in the mass |

in reporting countries, there are some discrepancy on export and import (re-export and re-import) between two countries. For unifying the trade value, we aggregate the export and re-export value (import and re-import value) and treat the maximal value from different reporting countries as the trade value between countries.

## References


1. Van Bergeijk, P.A.: On the brink of deglobalisation… again. Cambridge J. Regions Econ. Soc. **11**(1), 59–72 (2018)
2. Wang, C., et al.: Railway and road infrastructure in the belt and road Initiative countries: estimating the impact of transport infrastructure on economic growth. Transport. Res. Part A Policy and Practice **134**, 288–307 (2020a)
3. Zhang, C., Fu, J., Pu, Z.: A study of the petroleum trade network of countries along "The Belt and Road Initiative." J. Cleaner Prod. **222**, 593–605 (2019)
4. Liu, C., Xu, J., Zhang, H.: Competitiveness or complementarity? A dynamic network analysis of international agri-trade along the belt and road. Appl. Spatial Anal. Policy 1–26 (2019)
5. Qian, Y., et al.: Driving factors of agricultural virtual water trade between China and the Belt and Road countries. Environ. Sci. Technol. **53**(10), 5877–5886 (2019)
6. Lepawsky, J.: The changing geography of global trade in electronic discards: time to rethink the e-waste problem. Geogr. J. **181**(2), 147–159 (2015)
7. Petridis, N.E., Petridis, K., Stiakakis, E.: Global e-waste trade network analysis. Resour. Conserv. Recycl. **158**, 104742 (2020)
8. Hu, X., et al.: Characteristics and community evolution patterns of the international scrap metal trade. J. Cleaner Prod. **243**, 118576 (2020)
9. Wang, C., et al.: Structure of the global plastic waste trade network and the impact of China's import Ban. Resour. Conserv. Recycl. **153**, 104591 (2020b)
10. Wang, C., et al.: Mapping the structural evolution in the global scrap copper trade network. J. Cleaner Prod. **275**, 122934 (2020c)
11. Lim, W.X.: China's One Belt One Road initiative: a literature review. China's One Belt One Road initiative 113–132 (2016)
12. Shahriar, S.: Literature survey on the "Belt and Road" initiative: a bibliometric analysis. In: Foreign Business in China and Opportunities for Technological Innovation and Sustainable Economics, pp. 79–115. IGI Global (2019)
13. Thürer, M., et al.: A systematic review of China's belt and road initiative: implications for global supply chain management. Int. J. Prod. Res. **58**(8), 2436–2453 (2020)
14. Zhou, J., Liu, W.: Trade network of China and countries along "Belt and Road Initiative" areas from 2001 to 2013. Scientia Geographica Sinica **36**(11), 1629–1636 (2016)
15. Song, Z., Che, S., Yang, Y.: The trade network of the Belt and Road Initiative and its topological relationship to the global trade network. J. Geog. Sci. **28**(9), 1249–1262 (2018)
16. Liu, Z., et al.: The structure and evolution of trade relations between countries along the Belt and Road. J. Geog. Sci. **28**(9), 1233–1248 (2018)
17. Chong, Z., Qin, C., Pan, S.: The evolution of the Belt and Road trade network and its determinant factors. Emerg. Markets Finance Trade **55**(14), 3166–3177 (2019)
18. Chen, Q., Cheng, J., Wu, Z.: Evolution of the cultural trade network in "the Belt and Road" region: implication for global cultural sustainability. Sustainability **11**(10), 2744 (2019)
19. Liao, Z., Wang, Z., Guo, K.: The dynamic evolution of the characteristics of exchange rate risks in countries along "The Belt and Road" based on network analysis. PLoS ONE **14**(9), e0221874 (2019)
20. He, Q., Cao, X.: Pattern and influencing factors of foreign direct investment networks between countries along the "Belt and Road" regions. Sustainability **11**(17), 4724 (2019)

21. Huo, D., et al.: Complex network of aviation E-services in the Belt and Road initiative: a heuristic study of small data based on block modeling. Emerg. Markets Finance Trade **55**(14), 3151–3165 (2019)
22. Zhang, X., Zhang, W., Lee, P.T.W.: Importance rankings of nodes in the China Railway Express network under the Belt and Road initiative. Transport. Res. Part A Policy Practice **139**, 134–147 (2020)
23. Yang, W., et al.: Network structure and proximity of the trade network in the Belt and Road region. Geograph. Res. **37**(11), 2218–2235 (2018)
24. Freeman, L.C.: Centrality in social networks conceptual clarification. Social Networks **1**(3), 215–239 (1978)
25. Fischer, C.S., Shavit, Y.: National differences in network density: Israel and the United States. Social Networks **17**(2), 129–145 (1995)
26. Fagiolo, G.: Clustering in complex directed networks. Phys. Rev. E **76**(2), 026107 (2007)
27. Estrada, E.: Quantifying network heterogeneity. Phys. Rev. E. **82**(6), p. 066102 (2010)
28. Wang, C., et al.: Structure of the global plastic waste trade network and the impact of China's import Ban. Resour. Conserv. Recycling **153** (2020)
29. Foti, N.J., Pauls, S., Rockmore, D.N.: Stability of the world trade web over time - an extinction analysis. J. Econ. Dyn. Control **37**(9), 1889–1910 (2013)

# Pyroligneous Acids as Herbicide: Three-Years Field Trials Against *Digitaria sanguinalis*, *Cyperus rotundus*, *Capsella bursapastoris* and *Amaranthus lividus*

**Huidong Maliang, Zhikun Li, Anliang Chen, Haiping Lin, and Jianyi Ma**

**Abstract** Food quality and food safety issues has drawn wide public concern, and thereby organic agriculture has experienced rapid growth worldwide. In the production of organic crops, weeds control is the biggest problem. The comprehensive utilization of agricultural and forestry residues is an attractive option. Biochar as potential agricultural application benefits, has attracted wide attention. However, this will lead to a large overproduction of the by-product pyroligneous acids (PAs) involved in bamboo/wood/straw vinegar. Therefore, research needs to be done to improve the vinegar weed control effect, reduce the use-cost and accordingly to expand the application scope of PAs. Three tests of new herbicide discovery were conducted to evaluate the weeding effect of 4 PAs and acetic acids (AA) using greenhouse tests and field trials using *Digitaria sanguinalis* (L.), *Cyperus rotundus* (L.), *Capsella bursa-pastoris* (Linn.) Medic. and *Amaranthus lividus* L. Field trial showed that a good herbicidal effect of PAs on four weeds. Broadleaf weeds were more sensitive than narrowleaf weeds. Biomass tar had strong herbicidal synergistic effect of AA. PAs with the sum content of the AA and tar (>6%) have the potential for development as a bioherbicide in organic agriculture and non-crop land areas.

**Keywords** Agricultural and forestry residues · Pyroligneous acids · Vinegar herbicide,Ecological agriculture · Organic food

## 1 Introduction

Agricultural and forestry (A&F) residues have an important influence on climate mitigation in the next few decades, as a kind of energy source with considerable potential alternative to fossil fuels, greenhouse gas to reduce emissions [1]. The integrated utilization of A&F residues could help boost the rural economy, bring job

H. Maliang · Z. Li · A. Chen · H. Lin · J. Ma (✉)
School of Forestry and Biotechnology, Zhejiang A and F University, 311300 Hangzhou, People's Republic of China
e-mail: 811506796@qq.com

 
F. K. S. Chan et al. (eds.), *Proceedings of the 2020 International Conference on Resource Sustainability: Sustainable Urbanisation in the BRI Era (icRS Urbanisation 2020)*, Environmental Science and Engineering,
https://doi.org/10.1007/978-981-15-9605-6_11

opportunities and benefit from environmental protect. The global biomass production output from A&F residues is estimated to be 146 billion tons per year, which is mostly discarded in landfills or burnt to ashes [1]. There are 280 million tons (MT) of agricultural straw that have not been fully utilized in China [2]. Biomass resources totaling 680 MT and agricultural residues for energy production 155 MT could be made available per year in the United States. The 11 European countries obtaining a total estimate of 365 MT per year potential A&F residues, and after the application of different technical and environmental restrictions, the available resources were calculated to be 205 MT per year [3]. Those biomass is considered to be one of the most promising renewable energy sources. Pyrolyzed biomass can be converted into solid (biochar), liquid (biooil), and gaseous carbonaceous products. Although the carbon efficiency of the thermal conversion of biomass into biochar is only 30–50% [4]. In recent years, biochar has attracted wide attention due to its potential application in agriculture, which as a promising soil conditioner and carbon sequestration that can improve soil quality and increase crop yield [5]. Biochar soil amendment can also be mitigated greenhouse gas emissions in agriculture and can adsorb heavy metals ( Cd, Cu, and Zn) from soil [6]. However, this will lead to a large by-product of pyroligneous acids (PAs), which have a long tradition as a fertilizer and growth promoterin many Asian countries [7]. However, no scientific evidence to prove that PAs were once used as herbicides [8]. For expanding the wide application of PAs, this work is to improve the efficacy of vinegar herbicides, particularly for narrowleaf weed control and to reduce the cost for their wider use.

Food quality and food safety have increasingly aroused people's concern. Current and future increase in food production must go along with production of food with better quality and with less toxic contaminants. Alternative paths to the intensive use of crop protection chemicals are open, such as organic farming and development of food technologies. In 2014, organic agriculture was conducted in 172 countries, covering an area of 1% of agricultural land, and the size of the organic market reached US$80 billion [9]. However, there has been little research to support such scaled-up production systems and the market offers few crop protection products, such as herbicides. Poor weed control is often blamed for the decline in yields of organic crops. Manual weeding and tillage are alternatives to herbicides, but the cost is greatly increased while the effect is decreased [10]. It is well known that the control effect of natural herbicides, such as natural acetic acid (AA) named vinegar, on stiltgrass has not been evaluated. As a post-emergence herbicide, AA can reduce some weed species, although its dose is higher than traditional herbicides [11]. Vinegar is oxidized by aerobic bacteria that ferment ethanol in grains, fruit juices, or almost any liquid containing alcohol. The concentration is determined by the percentage of AA in vinegar, such as 20% AA in 200 grains of vinegar. Clove oil is the essence of the herbaceous part of the clove tree. A clove oil-based herbicide contains 34% (v/v) clove oil. The active ingredient in clove oil is eugenol, a volatile phenol. The application of vinegar and clove oil herbicides is similar to that of traditional herbicides and requires the use of protective equipment [12]. Vinegar can cause skin burns and eye damage if proper precautions are not taken. There may be opportunities to improve the effectiveness and cost-effectiveness of these herbicides [13]. Organic growers have identified effective weed control as the most important production facilitator. Scientists at the U.S. Department of Agriculture have shown that vinegar can achieve an 80–100% weeding effect at a concentration of 10–20% AA, and is more effective than 5% AA, the concentration used in food [12]. However, because these concentrations of AA do not meet the minimum risk pesticides, all pesticides must be registered by the EPA before they are sold in the United States.

# 2 Materials and Methods

## *2.1 Chemicals*

NBV, HWV, CWV and HSV were provided by the Ningbo Fenghua bamboo charcoal factory (Zhejiang Province, China), Zhejiang Huashang Energy Co., LTD. (China), Cantimber Biotech Inc (Canada) and Haiquan Fenglei New Energy Power Generation Co., LTD. (Anhui Province, China), respectively. Analytical grade including AA was purchased from Aladdin Reagent Co. (Shanghai, China). The positive control chemicals 20% glufosinate-ammonium aqueous solutions (GAS), made in YongNong BioSciences CO., LTD. Zhejiang Province, China) were obtained buying in the market.

## *2.2 Gas Chromatography–Mass Spectrometry Analysis (GC-MS)*

PAs samples were prepared through distillation at a pressure of 0.1MPa and a temperature of $52 \pm 2$ °C. The distillates of these PAs were mainly composed of AA, and the content of AA therein was measured according to GBT5009.41-2003 in China for the analysis of hygienic standards of vinegar. Residual phenols were then dissolved by DMF solvent. Detailed GC-MS analysis conditions and methods can be found in Bilehal and Kim [14], and Wang et al. [15].

## *2.3 Greenhouse Tests*

Greenhouse tests was planned to establish the herbicidal activity of PAs and AA against 4-week-old the same model plants with Petri-dish tests. Tested seeds of two model plant were raised from locally collected seeds in 15cm diameter pots in a greenhouse at $25 \pm 3$ °C temperature, $70 \pm 5$% relative humidity, 12/12h light/dark photoperiod and light intensity of approximately 3000 lx. For this 1200g of garden soil (soil:sand, 3:1, w/w) was taken in each pot and 12 seeds were sown in each pot, respectively. A week after emergence, pots were thinned to 10 plant per pot. When the plants were 4-week-old, they were spray treated with 1.0, 2.0, 4.0, 6.0, 8.0% (percentage of AA) PA solution and 2.0, 4.0, 6.0, 8.0%, 12.0% water solution of AA to serve as treatments, spray water within the same dose of surfactant solution to serve as control. In such a manner that each pot received 20ml of treatment solution. For each treatment four replications were maintained. On the 7 day after treatment, the plant fresh weight were recorded and used for the estimation inhibition of percentage as weed control effect. Control efficacy (fresh weight) % = (control- treatment) / control × 100% [16].

## *2.4 Field Weed Control Efficacy Test*

To test the herbicidal activity of the PAs and AA under field conditions, experiments were conducted in the garden fields selected Zhejiang A&F University (Hangzhou city, Zhejiang Province, China. 30°14′ N, 119°42′ E) in 2017 and 2018. The soil belonged to the silty loam red soil class, and soil samples had an organic C content of 17.40 g $kg^{-1}$, total N content of 1.72 g $kg^{-1}$, and carbon to nitrogen ratio of 10.23 t [17]. A randomized complete block design was used for the experiment with four replicates per treatment. The plot size was 50 × 50 cm, and all plots were separated from the others by a 50 cm wide buffer zone. In the weeds early seedling stage, in the absence of planting crops, looks like the choice the same weeds. To determine the efficacy of the herbicide, weed species and weed quantity were assessed at 7 and 14 days after herbicide applications for all plots. In addition, the aboveground biomass of 0.25$m^2$ of each plot was collected and weighed in accordance with the same protocol in the previous survey. The weed species assessed included those of the *D. sanguinalis*, *C. rotundus*, *C. bursa-pastoris* and *A. lividus*. The concentrations of 4 PAs were set as 1:0 (1L PAs:0L water), 1:1(1L PAs:1L water), 1:3(1L PAs:3L water) and 1:7 (1L PAs:7L water) (HWV lack of 1:7 treatment), and AA set as 2, 4, 8 and 16% with 0.1% adjuvant alkyl polyglucoside. The positive control herbicides set as 0.2% active ingredients. All tested PAs, AA, positive control herbicides and control water with 0.1% adjuvant alkyl polyglucoside were applied using WH-16 Manual backpack Sprayer, Huangyan Farming Machinery Plant, Zhejiang, China) with flat-fan nozzles delivering 450 L/ha dilutions t [18]. Control effects were observed on day 1, 3, 7 and 14 day after spraying. Three samples (0.25 $m^2$ per sample) in each plot were randomly selected and the effects were calculated as follow: Fresh weight control effect (%) = 100% × (weed fresh weight in control plot—weed fresh weight in treatment plot)/weed fresh weight in control plot [16].

## 2.5 Statistical Analysis

The data of greenhouse test and field trials in this study were statistically analyzed using SPSS 16.0 and were is played as the means and SDs by LSD's multiple-range test ($p = 0.05$ and 0.01). Every time data in greenhouse test and every day data in field trials were analyzed separately for variance analysis, respectively.

# 3 Results

## 3.1 PAs Chemicals

The GC-MS analysis of the underivatised active fractions of the PAs provided high-resolution information on the 22, 14, 16 and 19 components of NBV, HWV, CWV and HSV, respectively. All of these PAs exhibited clear and distinct chromatograms in the GC-MS analysis. Because of the differences in their constituents and properties, these PAs likely also should have different potential effects on soil activity and health. These components mainly included AA and tar (NBV 9.2%, 5.0%; HWV1.8%, 4.3%; CWV 5.3%, 7.5%; HSV 8.7%, 7.0%).

## 3.2 Greenhouse Efficacy

Greenhouse tests showed that the efficacy of AA and AA to control 2 model plants increased with the increasing acids concentration (Table 1). NBV in diluted 0–15 times (9.2–0.58% AA), inhibition rate range of oilseed rape and sorghum were 62–98% and 58–90%, respectively. There was no significant difference between NBV1:3 (2.3% AA and 1.3% tar) and 8%AA, NBV1:3 phytotoxicity were significantly higher than 20%GAS. NBV synergistic effect nearly 2.5 times higher than AA according to the content of AA. HWV in diluted 0–7 times (1.8–0.2% AA), inhibition rate range were 26–92%, and 34–80%, respectively. There was no significant difference between HWV1:0 (1.8% AA and 4.3% tar) and 8%AA/20%GAS, HWV synergistic effect was 3.4 times higher than AA. HSV in diluted 0–15 times (8.7–0.5% AA), inhibition rate range of oilseed rape and sorghum were 52–99% and 51–91%, respectively. HSV showed the highest toxicity of two plants, HSV1:1 (4.4% AA and 3.6% tar) phytotoxicity were significantly higher than 8%AA/20%GAS and synergistic effect was 0.8 times higher than AA. CWV in diluted 0–7 times (5.3–0.7% AA), inhibition rate range of oilseed rape and sorghum were 54–89% and 54–83%, respectively. There was no significant difference between CWV1:0 (5.3% AA and 7.6% tar) and 8%AA/20%GAS. CWV synergistic effect was 0.5 times higher than AA.

**Table 1** Model plant control effect of 4 PAs and AA in greenhouse tests (mean ± SD%)

| Treatment | Oilseed rape | Oilseed rape (replicated) | Sorghum | Sorghum (replicated) |
|---|---|---|---|---|
| NBV1:0 | 97.54 ± 2.12a[a] | 95.46 ± 2.54a[a] | 89.57 ± 2.59a[a] | 87.54 ± 3.21a[a] |
| NBV1:1 | 95.56 ± 2.79bc | 92.47 ± 2.67ab | 82.45 ± 4.01b | 83.25 ± 2.25abc |
| NBV1:3 | 84.55 ± 3.08ef | 81.54 ± 3.65de | 78.54 ± 3.33bc | 76.45 ± 3.00de |
| NBV1:7 | 74.10 ± 2.05hij | 69.87 ± 4.21gh | 72.46 ± 3.67d | 71.05 ± 2.80ef |
| NBV1:15 | 65.42 ± 2.04k | 62.33 ± 3.05i | 62.15 ± 3.24f | 58.44 ± 3.21hi |
| HWV1:0 | 91.55 ± 2.47bcd | 88.89 ± 3.44bc | 79.48 ± 3.21b | 75.89 ± 2.56de |
| HWV1:1 | 74.55 ± 1.86hi | 72.56 ± 2.10fgh | 68.87 ± 4.03de | 65.79 ± 4.01fg |
| HWV1:3 | 45.22 ± 2.65m | 48.56 ± 3.58j | 52.30 ± 2.89g | 50.23 ± 3.22jk |
| HWV1:7 | 26.54 ± 2.12o | 29.50 ± 3.08k | 36.21 ± 3.04h | 34.58 ± 2.87l |
| HSV1:0 | 98.21 ± 1.12a | 95.23 ± 2.47a | 90.42 ± 2.58a | 86.25 ± 2.45ab |
| HSV1:1 | 94.20 ± 2.52ab | 90.12 ± 5.41abc | 82.65 ± 3.24b | 80.00 ± 1.98cd |
| HSV1:3 | 78.64 ± 3.52gh | 75.50 ± 4.06efg | 73.45 ± 2.70cd | 71.45 ± 3.27ef |
| HSV1:7 | 69.46 ± 3.27jk | 66.67 ± 4.12hi | 63.54 ± 3.57ef | 62.56 ± 2.58gh |
| HSV1:15 | 56.20 ± 2.00l | 52.87 ± 3.25j | 51.00 ± 4.05g | 48.79 ± 4.59k |
| CWV1:0 | 88.99 ± 2.89cde | 85.47 ± 1.89cd | 82.57 ± 3.21b | 80.52 ± 3.05bcd |
| CWV1:1 | 79.87 ± 3.05fg | 76.60 ± 4.21ef | 73.34 ± 4.20cdDE | 71.49 ± 4.56efEF |
| CWV1:3 | 69.87 ± 2.58ijkIJ | 67.89 ± 5.06hiGHI | 62.66 ± 4.44f | 63.25 ± 3.58gh |
| CWV1:7 | 56.37 ± 5.12l | 54.56 ± 4.21j | 54.21 ± 2.57g | 55.55 ± 4.64ij |
| AA1% | 34.64 ± 3.24n | 35.64 ± 3.45k | 21.33 ± 3.21i | 23.54 ± 4.56m |
| AA2% | 55.28 ± 3.26l | 52.64 ± 2.58j | 32.46 ± 2.98h | 33.56 ± 3.62l |
| AA4% | 71.24 ± 4.25ij | 72.56 ± 5.12fgh | 60.06 ± 4.56f | 62.54 ± 4.87gh |
| AA8% | 87.56 ± 3.84de | 84.59 ± 4.65cd | 72.55 ± 2.57d | 74.78 ± 4.08de |
| GAS20% | 81.25 ± 2.58fg | 84.56 ± 4.78cd | 70.45 ± 2.56d | 72.35 ± 4.00e |

[a]The different small letters in each column indicate significant difference at $P = 0.05$ level

## *3.3 Field Efficacy Against* D. sanguinalis

The control effects of PAs and AA against *D. sanguinalis* was enhanced with increasing concentrations and time after spraying 7 (Table 2), 14 day (Fig. 1a–e). NBV1:0 (9.2% AA and 5.0% tar) showed higher weed control effect reached over 83% after spraying 14 day, which was the same with 16%AA and 20%GAS. NBV synergistic effect was over 0.7 times higher than AA according to the content of AA. The control effects of HWV1:0 (1.8% AA and 4.3% tar) was 72–76% which was lower than NBV1:0, but was the same with 4%AA after spraying 14 day. HWV synergistic effect was 1.2 times higher than AA. HSV1:0 (8.7% AA and 7.0% tar) showed the highest weed control effect was up to 90% than other PAs after spraying 14 day, There is no significant difference between HSV1:0 and 16%AA/20%GAS. HSV synergistic effect was over 0.8 times higher than AA. The control effects of CWV1:0 (5.3% AA and 7.5% tar) was 81–86%, which was lower than 16%AA/20%GAS, but was the same with 8%AA after spraying 14 day. CWV synergistic effect was over 0.5 times higher than AA.

## 3.4 *Field Efficacy Against* C. rotundus

The control effects of PAs and AA against *C. rotundus* was enhanced with increasing concentrations and time after spraying 7 (Table 2), 14 day (Fig. 2a–e). NBV1:0 showed higher weed control effect reached over 80% after spraying 14 day, there is no significant difference between NBV1:0 and 16%AA/20%GAS, repeat test also shows the same result. NBV synergistic effect was 0.7 times higher than AA according to the content of AA. The control effects of HWV1:0 was 63–68% which was lower than NBV1:0, but was higher than 4%AA after spraying 14 day. HWV synergistic effect was 1.2 times higher than AA. HSV1:0 showed higher weed control effect was up to 83% after spraying 14 day, which was the highest than other PAs. There is no significant difference between NBV1:0 and 16%AA/20%GAS. HSV synergistic effect was over 0.8 times higher than AA. The control effects of CWV1:0 was 72–80%, which was lower than 16%AA/20%GAS, but was the same with 8%AA after spraying 14 day. CWV synergistic effect was over 0.5 times higher than AA.

## 3.5 *Field Efficacy Against* C. bursa-pastoris

The control effects of PAs and AA against *C. bursa-pastoris* was enhanced with increasing concentrations and time after spraying 7 (Table 2), 14 day (Fig. 3a–e). The control effects of NBV1:0–7 (9.2–0.6% AA) were 24–56%, 25–68%, 42–84% and 46–93% after spraying 3,7,14 day, respectively. NBV1:0 showed higher weed control effect reached over 77% and 86% after spraying 7 and 14 day, respectively. There is no significant difference between NBV1:0 and 16%AA/20%GAS after spraying 7 and 14 day. NBV synergistic effect was 0.7 times higher than AA according to the content of AA. HWV1:0 showed lower weed control effect reached over 64% and 75% after spraying 7 and 14 day, respectively. The control effects of HWV1:0 was lower than that of NBV1:0, but was the same with 8%AA after spraying 14 day. HWV synergistic effect was 3.2 times higher than AA. The weed control effects of CWV1:0 was 77–85% and 82–91%, after spraying 7 and 14 day, respectively. CWV1:0 was lower than 16%AA/20%GAS, but was higher than 8%AA after spraying 14 day, which synergistic effect was over 0.5 times higher than AA. HSV1:0 showed the highest weed control effect was up to 84% and 92% after spraying 7 and 14 day, respectively. There is no significant difference between NBV1:0 and 16%AA/20%GAS. HSV synergistic effect was over 0.8 times higher than AA.

**Table 2** Control effect of 4 PAs and AA on four weeds at 7 day in field trial (mean ± SD%)

| Treatments | *D. sanguinalis* | *C. rotundus* | *C. bursa-pastoris* | *A. lividus* |
|---|---|---|---|---|
| NBV1:0 | 75.21 ± 4.51bc (78.24 ± 4.26bc) [71.54 ± 2.80bc] | 70.12 ± 2.57cd (73.25 ± 3.56bc) [68.55 ± 2.65cd] | 83.56 ± 3.27c (80.12 ± 4.05bc) [77.45 ± 2.78c] | 85.00 ± 2.05ab (84.92 ± 2.21ab) [80.23 ± 3.24ab] |
| NBV1:1 | 68.54 ± 4.56d (70.12 ± 3.67def) [64.26 ± 3.05de] | 62.45 ± 3.15e (64.58 ± 3.47d) [60.47 ± 2.68e] | 72.33 ± 2.58e (73.56 ± 2.98de) [68.74 ± 3.04de] | 67.58 ± 2.72e (71.05 ± 3.41c) [61.78 ± 1.89f] |
| NBV1:3 | 60.23 ± 3.24e (63.25 ± 2.58fgh) [56.45 ± 2.15f] | 50.36 ± 1.89f (52.12 ± 3.45ef) [47.95 ± 2.67fg] | 64.54 ± 2.79f (60.29 ± 4.56hi) [55.42 ± 3.46g] | 56.23 ± 1.83h (60.56 ± 4.05d) [50.47 ± 2.77hi] |
| NBV1:7 | 51.23 ± 2.34gh (52.64 ± 3.24jk) [46.58 ± 1.86hij] | 43.48 ± 3.58gh (43.56 ± 4.65g) [41.26 ± 2.56hi] | 52.26 ± 3.41h (48.54 ± 2.59j) [42.57 ± 1.73i] | 50.02 ± 4.05jk (47.20 ± 3.01g) [42.78 ± 2.74kl] |
| HWV1:0 | 67.58 ± 4.30d (65.87 ± 5.05fg) [63.56 ± 3.45e] | 62.26 ± 3.47e (64.88 ± 4.02d) [60.32 ± 2.57e] | 74.65 ± 3.54de (71.64 ± 1.99def) [64.24 ± 2.46ef] | 71.22 ± 2.33e (74.51 ± 2.45c) [67.47 ± 3.45d] |
| HWV1:1 | 58.84 ± 4.61ef (57.89 ± 4.61hij) [55.23 ± 3.04fg] | 45.66 ± 4.56fgh (46.30 ± 3.06fg) [43.65 ± 2.68gh] | 57.99 ± 3.45g (61.23 ± 4.09h) [55.08 ± 2.97g] | 62.33 ± 3.25fg (58.97 ± 2.74d) [52.40 ± 3.07gh] |
| HWV1:3 | 47.89 ± 4.04hi (49.89 ± 3.54kl) [45.55 ± 3.33ij] | 40.12 ± 5.78h (41.25 ± 3.25g) [37.66 ± 2.87i] | 51.51 ± 3.00h (54.58 ± 2.74i) [47.46 ± 1.79h] | 46.78 ± 3.12kl (49.56 ± 3.22fg) [42.78 ± 2.78kl] |
| HSV1:0 | 79.55 ± 2.45ab (82.55 ± 3.80ab) [76.32 ± 2.74ab] | 75.98 ± 2.78abc (79.32 ± 2.18a) [72.46 ± 3.21bc] | 92.58 ± 2.02a (90.10 ± 1.98a) [84.60 ± 2.40ab] | 85.66 ± 3.44a (90.23 ± 2.82a) [81.30 ± 0.92a] |
| HSV1:1 | 70.45 ± 1.38cd (73.53 ± 3.30cde) [68.56 ± 2.55cd] | 69.44 ± 4.32d (71.26 ± 3.54c) [66.64 ± 2.78d] | 84.65 ± 3.65c (80.18 ± 4.78bc) [80.46 ± 3.04bc] | 80.25 ± 2.87bc (82.13 ± 1.85b) [76.48 ± 3.10bc] |
| HSV1:3 | 58.79 ± 4.30ef (60.32 ± 5.41ghi) [55.64 ± 3.04fg] | 60.39 ± 2.23e (63.48 ± 4.21d) [57.64 ± 3.35e] | 75.89 ± 1.78de (79.54 ± 4.03bc) [71.47 ± 2.28d] | 71.35 ± 2.78de (70.12 ± 5.03c) [66.64 ± 3.47de] |
| HSV1:7 | 51.23 ± 4.20gh (48.59 ± 5.60kl) [45.56 ± 3.24ij] | 49.57 ± 4.89fg (51.36 ± 4.08ef) [47.45 ± 2.56fg] | 64.31 ± 2.54f (67.47 ± 3.41fg) [60.27 ± 1.84f] | 55.32 ± 3.87hi (51.78 ± 4.20fg) [46.78 ± 2.79ijk] |
| CWV1:0 | 72.54 ± 3.26cd (77.54 ± 3.25bc) [70.00 ± 2.85c] | 70.33 ± 3.45bcd (72.48 ± 1.98c) [70.21 ± 2.44cd] | 84.54 ± 2.89c (82.45 ± 4.08b) [77.66 ± 3.47c] | 76.56 ± 3.41cd (81.23 ± 1.78b) [72.61 ± 2.90c] |
| CWV1:1 | 60.21 ± 3.50e (66.58 ± 6.02efg) [57.32 ± 2.07f] | 62.54 ± 5.04e (63.55 ± 2.28d) [60.23 ± 2.45e] | 75.48 ± 3.20de (71.45 ± 5.08def) [70.34 ± 4.21d] | 67.45 ± 3.43ef (70.05 ± 1.78c) [62.44 ± 2.22ef] |
| CWV1:3 | 54.22 ± 4.52efg (53.26 ± 6.00ijk) [51.23 ± 3.25gh] | 51.20 ± 2.39f (51.06 ± 3.47ef) [48.25 ± 2.49fg] | 66.54 ± 3.87f (68.55 ± 4.03efg) [62.33 ± 2.79f] | 59.42 ± 2.76gh (62.33 ± 3.66d) [55.55 ± 3.07g] |

(continued)

**Table 2** (continued)

| Treatments | *D. sanguinalis* | *C. rotundus* | *C. bursa-pastoris* | *A. lividus* |
|---|---|---|---|---|
| CWV1:7 | 47.58 ± 4.05hi<br>(51.45 ± 3.48jk)<br>[44.65 ± 2.56jk] | 45.23 ± 3.69fgh<br>(44.28 ± 4.59g)<br>[42.06 ± 3.08hi] | 52.33 ± 2.87h<br>(57.84 ± 4.65hi)<br>[46.72 ± 3.46hi] | 50.46 ± 2.00ijk<br>(53.46 ± 4.21ef)<br>[45.72 ± 3.05jk] |
| AA2% | 42.58 ± 4.52i<br>(43.68 ± 4.56l)<br>[40.05 ± 3.04k] | 40.89 ± 4.03h<br>(42.58 ± 3.21g)<br>[38.46 ± 2.78i] | 48.45 ± 3.28h<br>(46.48 ± 2.78j)<br>[42.44 ± 2.55i] | 43.55 ± 4.21l<br>(47.21 ± 2.33g)<br>[40.44 ± 3.06l] |
| AA4% | 53.26 ± 2.48fgh<br>(54.21 ± 4.05ijk)<br>[50.13 ± 3.02hi] | 51.01 ± 2.09f<br>(53.44 ± 2.97e)<br>[48.99 ± 3.27f] | 66.66 ± 4.03f<br>(63.54 ± 2.89gh)<br>[60.45 ± 3.25f] | 54.56 ± 2.55hij<br>(58.46 ± 3.47de)<br>[50.00 ± 1.82hij] |
| AA8% | 72.33 ± 2.21cd<br>(74.59 ± 6.03cd)<br>[69.46 ± 3.46c] | 69.54 ± 4.05d<br>(67.54 ± 2.80cd)<br>[66.56 ± 3.05d] | 77.89 ± 3.64d<br>(74.56 ± 2.77cd)<br>[70.46 ± 2.78d] | 67.46 ± 4.05ef<br>(72.56 ± 2.89c)<br>[62.40 ± 3.06ef] |
| AA16% | 81.54 ± 3.78a<br>(84.25 ± 4.68ab)<br>[77.75 ± 2.99a] | 76.39 ± 4.12ab<br>(78.67 ± 4.23ab)<br>[75.05 ± 2.80ab] | 90.25 ± 3.65ab<br>(90.79 ± 1.98a)<br>[86.45 ± 2.76a] | 86.25 ± 3.65a<br>(89.23 ± 2.38a)<br>[82.46 ± 1.89a] |
| 20%GAS | 82.52 ± 4.60a<br>(85.67 ± 2.97a)<br>[80.09 ± 3.26a] | 80.25 ± 4.23a<br>(79.87 ± 4.08a)<br>[77.77 ± 2.58a] | 85.45 ± 2.87bc<br>(88.65 ± 1.49a)<br>[80.45 ± 3.05bc] | 85.44 ± 2.77ab<br>(88.45 ± 2.67a)<br>[80.44 ± 3.34ab] |

The data outside brackets, in brackets and in square brackets are the 2018 summer, 2019 spring and 2017 autumn, respectively.The different small letters in each column indicate significant difference at $P = 0.05$ level

## *3.6 Field Efficacy Against* A. lividus

The weed control effects of PAs and AA against *A. lividus* was enhanced with increasing concentrations and time after spraying 7 (Table 2), 14 day (Fig. 4a–e). The weed control effects of NBV1:0–7 (9.2–0.6% AA) were 20–51%, 31–66%, 42–85% and 51–91% after spraying 7,14 day, respectively. NBV1:0 showed higher weed control effect reached over 80% and 85% after spraying 7 and 14 day, respectively. There is no significant difference between NBV1:0 and 16%AA/20%GAS after spraying 7 and 14 day. NBV synergistic effect was 0.7 times higher than AA according to the content of AA. HWV1:0 showed lower weed control effect reached over 67% and 74% after spraying 7 and 14 day, respectively. The control effects of HWV1:0 was lower than that of NBV1:0, but was the same with 8%AA after spraying 14 day. HWV synergistic effect was 3.2 times higher than AA. The weed control effects of CWV1:0 was 72–82% and 81–87%, after spraying 7 and 14 day, respectively. CWV1:0 was lower than 16%AA/20%GAS, but was the same with 8%AA after spraying 14 day, which synergistic effect was over 0.5 times higher than AA. HSV1:0 showed the highest weed control effect was 81–91% and 89–96% after spraying 7 and 14 day, respectively. There is no significant difference between NBV1:0 and 16%AA/20%GAS. HSV synergistic effect was over 0.8 times higher than AA.

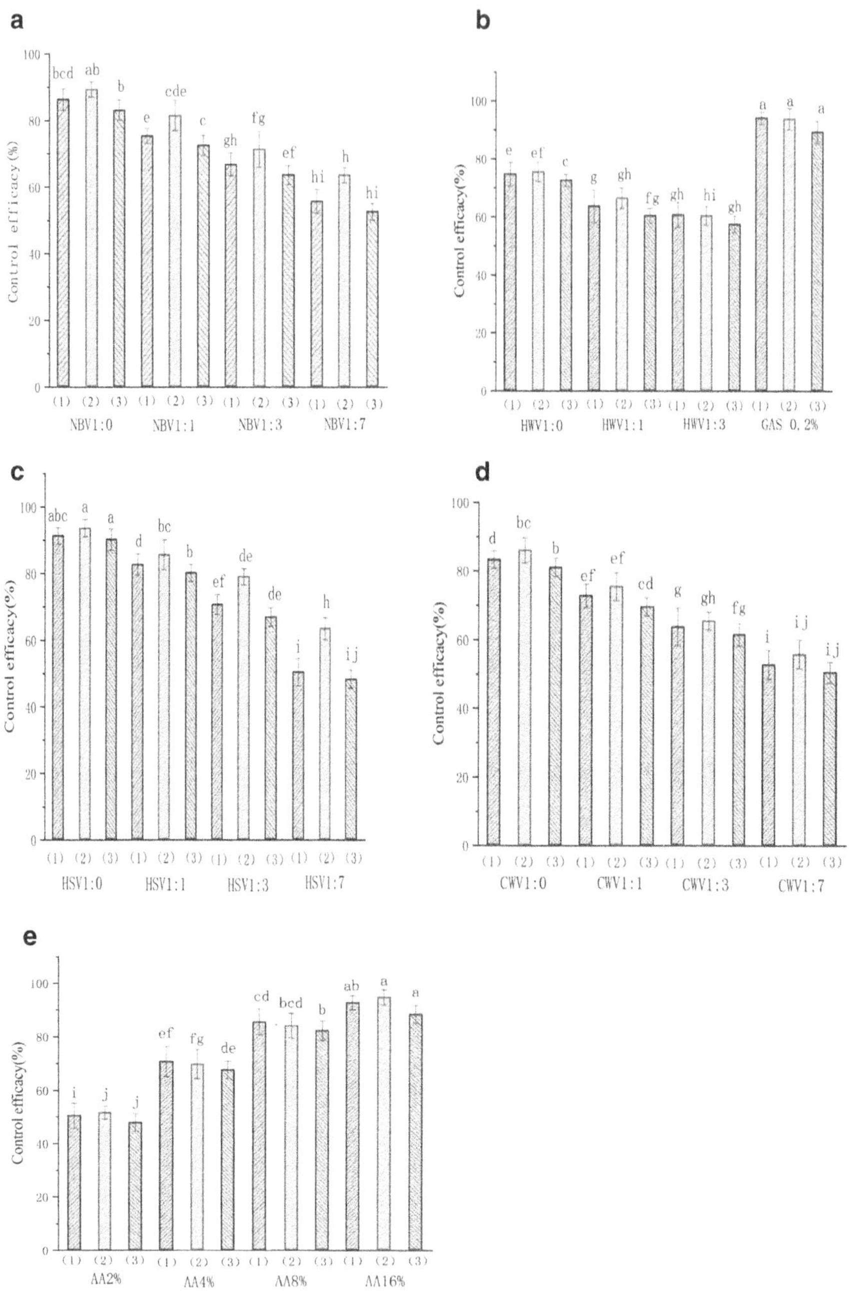


**Fig. 1** Control effect of 4 PAs and AA on *D. sanguinalis* at 14 day in field trial (Mean ± SD%). The (1), (2) and (3) are the 2018 summer, 2019 spring and 2017 autumn, respectively.The different small letters in each column indicate significant difference at $P = 0.05$ level

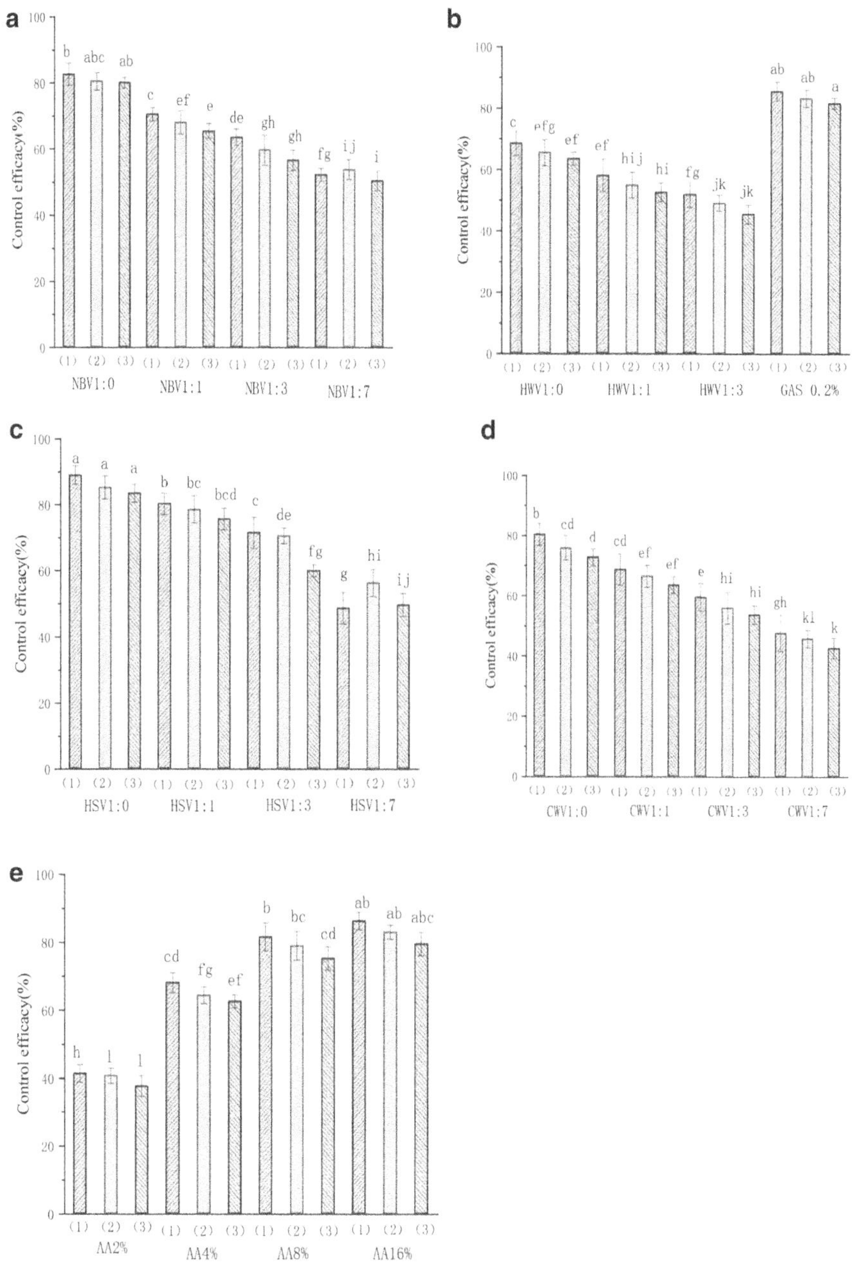


**Fig. 2** Control effect of 4 PAs and AA on *C. rotundus* at 14 day in field trial (mean ± SD%). The (1), (2) and (3) are the 2018 summer, 2019 spring and 2017 autumn, respectively.The different small letters in each column indicate significant difference at $P = 0.05$ level

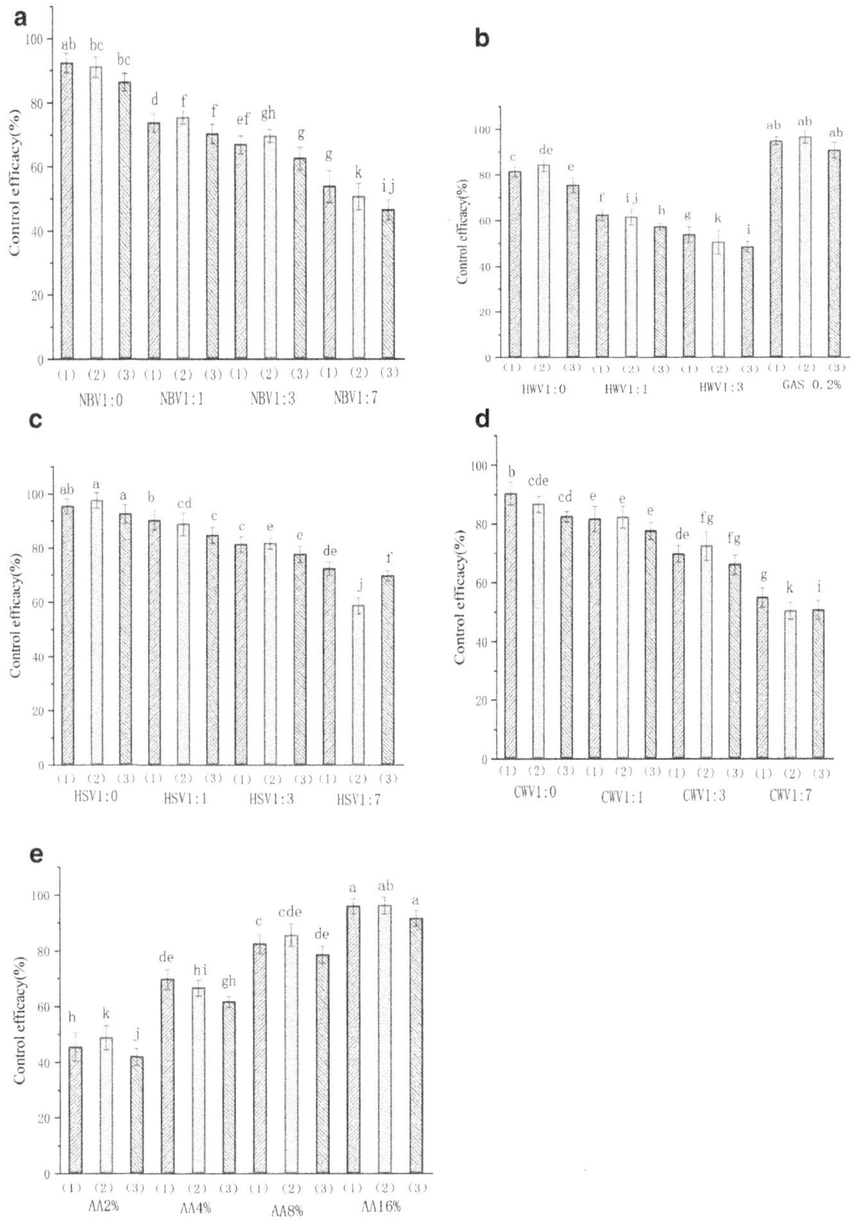


**Fig. 3** Control effect of 4 PAs and AA on *C. bursa-pastoris* at 14 day in field trial (mean ± SD%). The (1), (2) and (3) are the 2018 summer, 2019 spring and 2017 autumn, respectively. The different small letters in each column indicate significant difference at $P = 0.05$ level

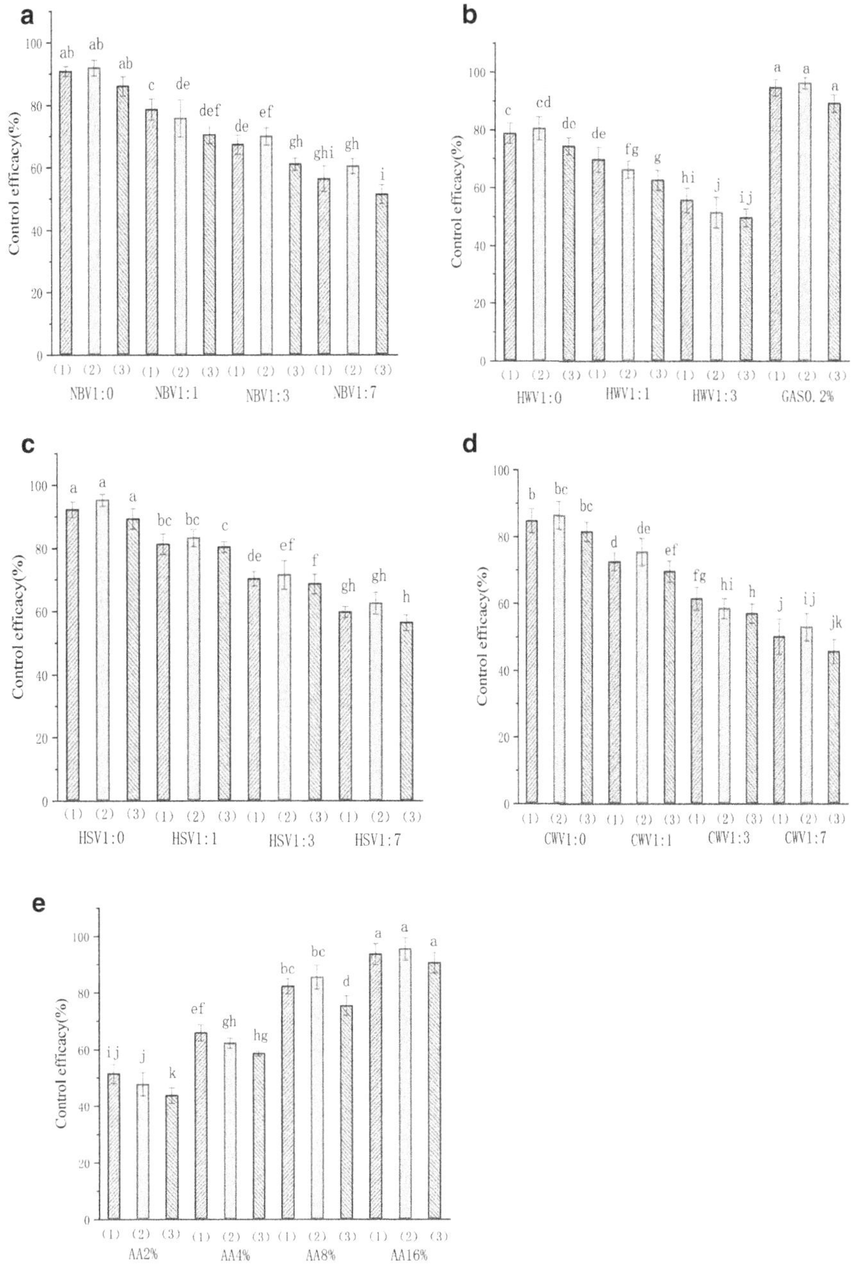


**Fig. 4** Control effect of 4 PAs and AA on *A. lividus* at 14 day in field trial (mean ± SD%). The (1), (2) and (3) are the 2018 summer, 2019 spring and 2017 autumn, respectively. The different small letters in each column indicate significant difference at $P = 0.05$ level

## 4 Discussion

Little is known about the optimal use of vinegar-based herbicides because few field effect had been reported. Research reported that 20% grain vinegar applied at 636 L/ha provided 100% control of two-leaf redroot pigweed but the same treatment on two-leaf velvetleaf provided only 73% control and 18% mortality after spraying 6 day [12]. Another field trials conducted with treatments with 5–20% AA vinegar concentrations at 107 L/ha, they showed that 15% vinegar were needed for adequate control of mustard [12]. The other work found that the effect of reduced stiltgrass on treatments PA and AA, which can reduced stiltgrass cover and seed production, reduced seed production by more than 90% [19]. In addition, greenhouse experiments were conducted to evaluate the effectiveness of 5% AA, 30% AA, 10% CA, against the broadleaf weeds stranglervine, wild mustard, black nightshade, sicklepod, velvetleaf, and redroot pigweed and to narrowleaf weeds crowfootgrass, Johnsongrass, annual ryegrass, goosegrass, green foxtail, and yellow nutsedge. Their results showed that wild mustard was most sensitive to these herbicides while redroot pigweed was the least sensitive. Herbicides did not control narrowleaf weeds except for 30%AA. Delayed application until the 4–6 leaf stage significantly reduced efficacy; AA was less sensitive to growth stage than other herbicides [10]. Our greenhouse tests showed that synergistic relationship with the content of AA kinds and concentration, also has something to do with the tar content, the higher the tar content, the synergistic effect the greater, this mean tar may have herbicidal activity or synergistic effect of AA.

Control effect of 4 PAs and AA on four weeds in field trial were increased with the increasing acids concentration. The weed control effects in descending order was: HSV (acids + tar = 15.7%) > NBV (14.2%) > CWV(12.8%) > HWV(6.1%), this effect order was accord with the sum content of the acids and tar, rather than a single content of acids or tar, this mean tar may have herbicidal activity or synergistic effect of AA, whether tar have herbicidal activity still need further test. Field trials showed that a good herbicidal effect of HSV 、NBV and CWV with diluted by water (AA: water = 1: 0–3, v/v) while herbicidal effect of HWV was low owing to low content of the acids and tar. There is less dead weeds in the treatment with low concentration, but the growth including the size and weight of weeds was observably inhibited and were lower than those of the control. Herbicidal effect of PAs and AA on two broadleaf weeds is good after spraying 1 day (over or nearly 50%). In addition, herbicidal effect on two broadleaf weeds were significantly higher than that on two narrowleaf weeds after spraying 14 day. These results may be due to the fact that narrowleaf weeds with less sticking herbicide dose than that broadleaf weeds. Herbicidal effect of PAs were much higher than that of AA with the same acids content, this reason may be that PAs is easier to stick to a narrowleaf weeds on the small leaf and reducing AA volatilizing speed.In order to further improve the control effect of weeds especially in narrowleaf weeds, Manufacturing formula need improve the viscosity of the herbicide formulations to increase sticking dose, or the spray method need spray wet through cauline leaf of narrowleaf weeds.

The registration of vinegar herbicide by Ecoval Corporation (Trade mark name: EcoClear) and by Pharm Solutions, Inc. (Weed Pharm) in America and Canada [20, 21]. The above two products, which are components of 20–30% AA, can be used to organic agriculture [22, 23]. Two herbicide were recommended for the control or suppression of herbaceous broadleaf and grassy weeds on non-crop land areas, such as: sidewalks, driveways and patios; railroad rights-of-way, power stations, and industrial sites; around farm buildings, storage areas, tank farms, in and around greenhouses, plant nurseries and golf courses; and in fencerows and vacant lots [21]. Their use concentration is 12–30% AA concentration. Two products will greatly increase the cost of special packaging, transportation and using owing to dangerous chemicals (more than 10% AA). However, AA content in PAs are less than 10%, with tar and other AA synergistic effect, PAs weed control effect will achieve the above 20–30% AA vinegar-herbicides weed control effect and solved the problem of poor effect on narrowleaf weeds. Not only save packaging, transport and use cost, but also greatly consumed the overproduction of PAs. PAs is used as a herbicide is lower than the cost of the use of chemical herbicides 20% GAS, completely changed the traditional concept of natural pesticide control effect is poor and using expensive. PAs is a natural, pyrolysis products useless artificial chemical process, which used as a herbicide, only need to add a small amount of natural surfactants (<0.1%), can achieve the organic farming standards around the world and may be widely used in organic agriculture, solve organic farming is the high cost of weeding technology difficult problem. PAs can also be used for organic fertilizer and plant growth regulator with using 500 times dilution. In addition, PAs as auxiliary effect such as insecticide, fungicides and increasing fertility are significant [24, 25], While weeding, whether on crop and soil in pests repellent effect, antifungal (or antibacterial) effect and fattening effect, more research is still needed.

Our previous research works showed that the use of PAs and AA at high doses (2000 kg $ha^{-1}$) which is the actual weeding dosage of 2–4 times (500–1000 kg $ha^{-1}$) and high frequencies (sprayed once every 6 days, for a total of 3 times) have no long-term effects on soil pH, soil water holding capacity, and three soil enzyme activity involving in urease, protease and sucrase [26]. Furthermore, The animal safety tests in our previous study showed that the acute oral toxicity of BV $LD_{50}$ was >5000 mg/kg for male and female mice, and the dermal sensitization rate was zero in the rabbits, and indicates that BV does not induce mutation on the basis of polychromatophilic erythrocytes with micronuclei in the mice [27]. This conclusion may provide a theoretical basis and practical support for those vinegar-based herbicides.

## 5 Conclusion

Four PAs and AA were phytotoxic to sorghum and oilseed rape in the greenhouse tests, in addition, sorghum and oilseed rape as model plants are suit to screen herbicidal active. Three years field trial showed that a good herbicidal effect of HSV, NBV and CWV with diluted by water (PAs:water = 1:0–3, v/v) on two narrowleaf weeds *D. sanguinalis*, *C. rotundus* and two broadleaf weeds *C. bursa-pastoris*, *A. lividus* while HWV was secondary herbicidal effect owing to low content of the acids and tar. Broadleaf weeds were more sensitive to narrowleaf weeds. Biomass tar had strong herbicidal synergistic effect of AA. Therefore, PAs with the sum content of the AA and tar (>6%) have the potential for development as a bioherbicide in organic agriculture and non-crop land areas.

**Acknowledgements** This study was supported by the Scientific and Technological Projects of Zhejiang Province (No. 2019C02024).

## References

1. Grewal, A., Abbey, L., Gunupuru, L.R.: Production, prospects and potential application of pyroligneous acid in agriculture. J. Anal. Appl. Pyrol. **135**, 152–159 (2018)
2. Chen, X.: Economic potential of biomass supply from crop residues in China. Appl. Energy **166**, 141–149 (2016)
3. Esteban, L.S., Carrasco, J.E.: Biomass resources and costs: assessment in different EU countries. Biomass Bioenerg. **35** (supp-S1) (2011).
4. Wu, Q., Yu, S.T., Hao, N.J., Wells, T., Meng, X.Z., Li, M., Pu, Y.Q., Liu, S.X., Ragauskas, A.J.: Characterization of products from hydrothermal carbonization of pine. Bioresour. Technol. **244**, 78–83 (2017)
5. Hagner, M., Tiilikkala, K., Lindqvist, I., Niemelä, K., Rasa, K.: Performance of liquids from slow pyrolysis and hydrothermal carbonization in plant protection. Waste Biomass Valoriz, 1–12 (2018)
6. Wang, R., Shafi, M., Ma, J., Zhong, B., Guo, J., Hu, X., Xu, W., Yang, Y., Ruan, Z., Wang, Y., Ye, Z., Liu, D.: Effect of amendments on contaminated soil of multiple heavy metals and accumulation of heavy metals in plants. Environ. Sci. Pollut. R **25**, 28695–28704 (2018)
7. Huang, S., Shan, M., Chen., J, Penttinen, P., Qin, H.: Contrasting dynamics of polychlorinated biphenyl dissipation and fungal community composition in low and high organic carbon soils with biochar amendment. Environ. Sci. Pollut. R **25**, 33432–33442 (2018)
8. Tiilikkala, K., Fagernäs, L., Tiilikkala, J.: History and use of wood pyrolysis liquids as biocide and plant protection product. Open Agr. J. **4**, 111–118 (2012)
9. Gomiero, T.: Food quality assessment in organic vs. conventional agricultural produce: findings and issues. Appl. Soil Ecol. **123**, 714–728 (2018)
10. Abouziena, H.F.H., Omar, A.A.M., Sharma, S.D., Singh, M.: Efficacy comparison of some new natural-product herbicides for weed control at two growth stages. Weed Technol. **23**, 431–437 (2009)
11. Stuart, W.J., Mervosh, T.L.: Nonchemical and herbicide treatments for management of japanese stiltgrass (microstegium vimineum). Invas. Plant Sci. Mana **5**, 9–19 (2012)
12. Glennj, E., Robinr, B., Martinc, G.: Herbicidal effects of vinegar and a clove oil product on redroot pigweed (*Amaranthus retroflexus*) and velvetleaf (*Abutilon theophrasti*). Weed Technol. **23**, 292–299 (2009)
13. Brainard, D.C., Curran, W.S., Bellinder, R.R., Ngouajio, M., Vangessel, M.J., Haar, M.J., Lanini, W.T., Masiunas, J.B.: Temperature and relative humidity affect weed response to vinegar and clove oil. Weed Technol. **27**, 156–164 (2013)
14. Bilehal, D., Kim, Y.H.: Gas Chromatography–mass spectrometry analysis and chemical composition of the bamboo-carbonized liquid. Food Anal. Method **5**, 109–112 (2012)
15. Wang, P., Maliang, H., Wang, C., Ma, J.: Bamboo charcoal by-products as sources of new insecticide and acaricide. Ind. Crop Prod. **77**, 575–581 (2015)

16. Zhou, B., Wang, H., Meng, B., Wei, R., Wang, L., An, C., Chen, S., Yang, C., Qiang, S.: An evaluation of tenuazonic acid (TeA) as a potential biobased herbicide in cottons. Pest Manag. Sci. **75**, 2482–2489 (2019)
17. Cai, X., Lin, Z., Penttinen, P., Li, Y., Li, Y., Luo, Y., Yue, T., Jiang, P., Fu, W.: Effects of conversion from a natural evergreen broadleaf forest to a moso bamboo plantation on the soil nutrient pools, microbial biomass and enzyme activities in a subtropical area. For. Ecol. Manag. **422**, 161–171 (2018)
18. Zhang, D.X., Zhang, X.P., Luo, J., Li, B.X., Wei, Y., Liu, F.: Causation analysis and improvement strategy for reduced pendimethalin herbicidal activity in the field after encapsulation in polyurea. ACS Omega **3**, 706–716 (2018)
19. Ward, J.S., Mervosh, T.L.: Nonchemical and herbicide treatments for management of Japanese stiltgrass (*Microstegium vimineum*). Invasive Plant Sci. Manage. **5**, 9–19 (2012)
20. Evans, G.J., Bellinder, R.R.: The potential use of vinegar and a clove oil herbicide for weed control in sweet corn, potato, and onion. Weed Technol. **23**, 120–128 (2009)
21. Ivany, J.A.: Acetic acid for weed control in potato (*Solanum tuberosum* L.). Can. J. Plant Sci. **90**, 537–542 (2010)
22. Evans, G.J., Bellinder, R.R., Hahn, R.R.: Integration of vinegar for in-row weed control in transplanted bell pepper and broccoli. Weed Technol. **25**, 459–465 (2011)
23. Moran, P., Greenberg, S.M.: Winter cover crops and vinegar for early-season weed control in sustainable cotton. J. Sustain. Agr. **32**, 483–506 (2008)
24. Wang, C., Liu, Q., Wang, P., Mei, Y., Cui, G., Ying, L., Chen, A., Ma, J.: Bactericidal and antiviral effects of bamboo vinegar with different concentrations on domestic waste. Fresenius Environ. Bull. **28**, 1631–1638 (2019)
25. Wang, H.F., Wang, J.L., Wang, C., Zhang, W.M., Liu, J.X., Dai, B.: Effect of bamboo vinegar as an antibiotic alternative on growth performance and fecal bacterial communities of weaned piglets. Livest. Sci. **144**, 173–180 (2012)
26. Maliang, H., Tang, L., Lin, H., Chen, A., Ma, J.: Influence of high-dose continuous applications of pyroligneous acids on soil health assessed based on pH, moisture content and three hydrolases. Environ. Sci. Pollut. R **27**, 15426–15439 (2020)
27. Maliang, H., Wang, P., Chen, A., Liu, H., Lin, H., Ma, J.: Bamboo tar as a novel fungicide: its chemical components, laboratory evaluation, and field efficacy against false smut and sheath blight of rice, and powdery mildew and Fusarium wilt of cucumber. Plant Disease (2020, in Press). https://doi.org/10.1094/PDIS-06-20-1157-RE

# Activation of Traditional Construction Techniques Used in Linpan Based on the Concept of Sustainability: Case Study of Bamboo Materials

**Ding Ding, Qianqian Xu, Chunlu Liu, and Dingxin Zhang**

**Abstract** Some traditional habitations, which are vanishing with urban expansion, are sustainable. Linpan contains abundant and excellent construction skills, especially that of using bamboo materials, reflecting the architectural wisdom of local village craftsmen. This paper focuses on the activation of traditional bamboo construction techniques under the background of sustainable urbanization in a logical sequence of "extraction–evolution–activation". Through a literature review and field investigations, this paper preliminarily extracts the typical bamboo techniques in Linpan, such as bamboo roofs, walls, and window lattices. Then, this paper uses SketchUp and BECS software to conduct qualitative and quantitative analyses on bamboo construction techniques and their corresponding building performances. Although the heat conductivity coefficients of bamboo materials are relatively low, the thermal performance of traditional bamboo roofs and walls are still poor, due to the absence of the heavy construction layer. To inherit sustainable architecture and preserve rural memory during rapid urbanization, this study introduces a modern bamboo material—bamboo steel, analyzing its properties and construction methods. Finally, this research practices the activated bamboo construction techniques in the design stage of BambooUp, namely the Chengdu Bamboo Culture Exhibition Center. This project is also a medium to further demonstrate and inherit excellent bamboo construction skills.

**Keywords** Bamboo material · Linpan · Sustainable urbanization

D. Ding (✉)
School of Civil Engineering, Architecture, and Environment, Xihua University, 610039 Chengdu, Sichuan, China
e-mail: ding7777ding@foxmail.com

Q. Xu
Epiphany Architects LLC., 610037 Chengdu, Sichuan, China

C. Liu
School of Architecture and Built Environment, Deakin University, Geelong, VIC 3220, Australia

D. Zhang
Shandong Sheyu Architectural Design Consulting Co. Ltd., 250101 Jinan, Shandong, China



F. K. S. Chan et al. (eds.), *Proceedings of the 2020 International Conference on Resource Sustainability: Sustainable Urbanisation in the BRI Era (icRS Urbanisation 2020)*, Environmental Science and Engineering,
https://doi.org/10.1007/978-981-15-9605-6_12

# 1 Introduction

The Belt and Road Initiative projects to further accelerate the already-rapid development of urbanization and to produce urban challenges on social and environmental issues in the upcoming decades [1]. Some traditional habitations, which are vanishing with urban expansion, are products of the regional environment, agricultural production, and lifestyle [2]. They conform to the concept of resource sustainability. Linpan (a rural home garden agroforestry system, Fig. 1) refers to the rural settlement just like a green plate in the field of western Sichuan Plain, where farmers dwell in the forests of trees and bamboos since ancient times [3, 4]. The architectural heritage of Linpan contains abundant and excellent construction skills, especially that of using bamboo materials, reflecting the architectural wisdom of local village craftsmen. However, with the wave of urbanization, Lipan is currently facing problems such as weakened function and structural degradation [5].

This paper focuses on the activation of traditional bamboo construction techniques under the background of sustainable urbanization in a logical sequence of "extraction–evolution–activation". First, through a literature review and field investigations, this paper preliminarily extracts the typical bamboo techniques in Linpan, such as bamboo roofs, walls, and window lattices. Then, this study uses modeling software such as SketchUp to carry out visual construction modeling [6]. Based on the qualitative study, this paper uses performance simulation software such as BECS to conduct quantitative analyses on the relationship between bamboo construction and the built environment [7]. Second, this study introduces a modern bamboo material—bamboo steel, analyzing its properties and construction methods. This stage is the key to evolving traditional construction techniques to meet the standard of contemporary buildings. Last, this research practices the activated bamboo construction techniques in the design stage of BambooUp, namely the Chengdu Bamboo Culture Exhibition Center. This project takes this building as a medium to further demonstrate and inherit excellent bamboo construction skills.

**Fig. 1** A typical image of Linpan

## 2 Extraction: Traditional Bamboo Construction Techniques in Linpan

The most widely used bamboo material in Linpan is *Neosinocalamus affinis*, which is appropriate for roofs, walls, and window lattices [8]. The original bamboo, the bamboo slice, and the bamboo strip are three basic application forms of the bamboo rod after primary processing [9]. The traditional connection modes of bamboo components mainly include the inserting joint, the riveting joint, and the binding joint (Fig. 2).

### 2.1 *Bamboo Roofs*

In the western Sichuan Plain, a rainy and humid area, traditional rural residences always have triangular roofs [10]. Bamboo materials are suitable for roof coverings, such as purlins, rafters, boards, and tiles, above the wood structural frame. First, purlins are components that transmit roof loads to beams. They are subjected to downward pressure only. The specifications of bamboo purlins are usually 100–120 mm in diameter, and 3.0–3.6 m in length [9]. Second, the connections of multiple short rafters form the pitched roof, which makes it convenient to fix tiles and to drain water. Third, boards, basketry, or mats made of bamboo slices and strips can cover rafters and bear tiles. Fourth, bamboo tiles, which include concave tiles and cover tiles, can be installed both on boards or directly on rafters. However, bamboo tiles are no longer commonly used in recent years, substituted by ceramic tiles.

Although the heat conductivity coefficients of bamboo materials are relatively low (approximately 0.11 W/m K for the 8 mm bamboo tube and 0.15 W/mK for the 5–8 mm bamboo slice [11]), the thermal performance of a typical all-bamboo roof shown in Fig. 3 is still poor. Because the roof layers are thin, the roof's heat transfer coefficient is as high as approximately 3.60 W/m$^2$ K, estimated by BECS.

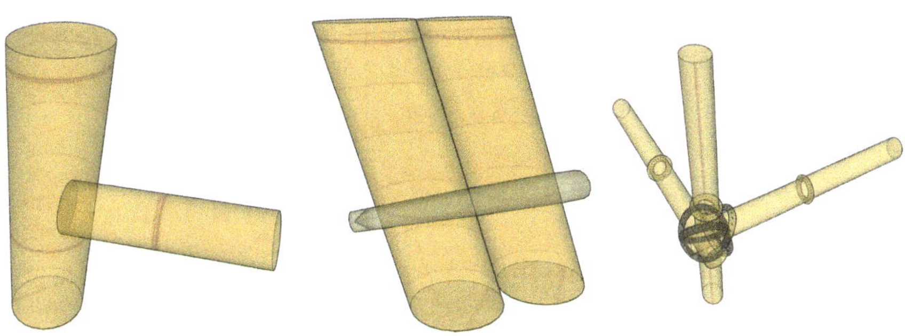

(a) Inserting joint (b) Riveting joint (c) Binding joint

**Fig. 2** 3D models of traditional connection modes of bamboo components

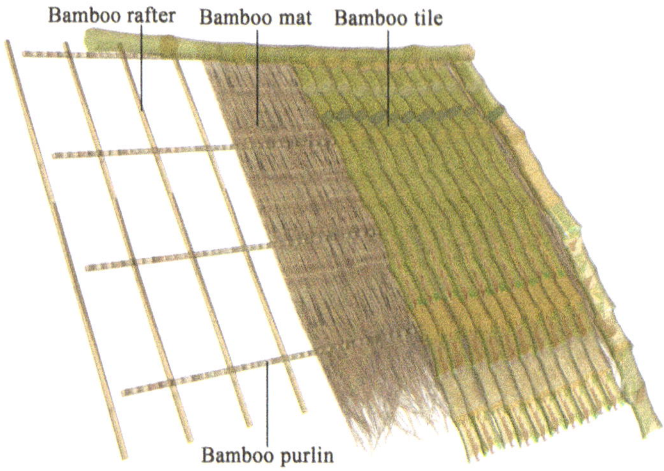


**Fig. 3** 3D model of a typical all-bamboo roof in Linpan

## 2.2 Bamboo Walls

The main types of bamboo walls mainly include bare bamboo walls, bamboo reinforced clay walls, bamboo finishes, and bamboo fences (Fig. 4). First, bare bamboo walls are patched, weaved, or pasted using original bamboo, bamboo pieces, or bamboo mats. These walls can not only stand as independent walls nailed on building frames but also as parapets attached to brick exterior walls. Second, bamboo reinforced clay walls take the original bamboos as their frameworks and bamboo basketry and clay as their finishes. These walls are usually used as partitions or triangular gable walls in Linpan. Third, bamboo materials can also decorate walls' surfaces, even with fine carvings. Fourth, bamboo fences are mostly log-cabin constructions that are used in outdoor spaces.

Similar to bamboo roofs, thermal performances of the bare bamboo wall and the bamboo reinforced clay wall are high, for the absence of the heavy construction layer (Table 1). For a 240 mm brick wall, the addition of a 5 mm bamboo finish decreases the heat transfer coefficient by approximately 0.15 $W/m^2$ K.

## 2.3 Bamboo Window Lattices

The traditional bamboo windows are the miniature of the bamboo walls, which take the original bamboo as structural frames and weave bamboo pieces or strips to be the enclosure. Different weave patterns lead to different transmittances (Fig. 5), which provide abundant textures and shadows for buildings.

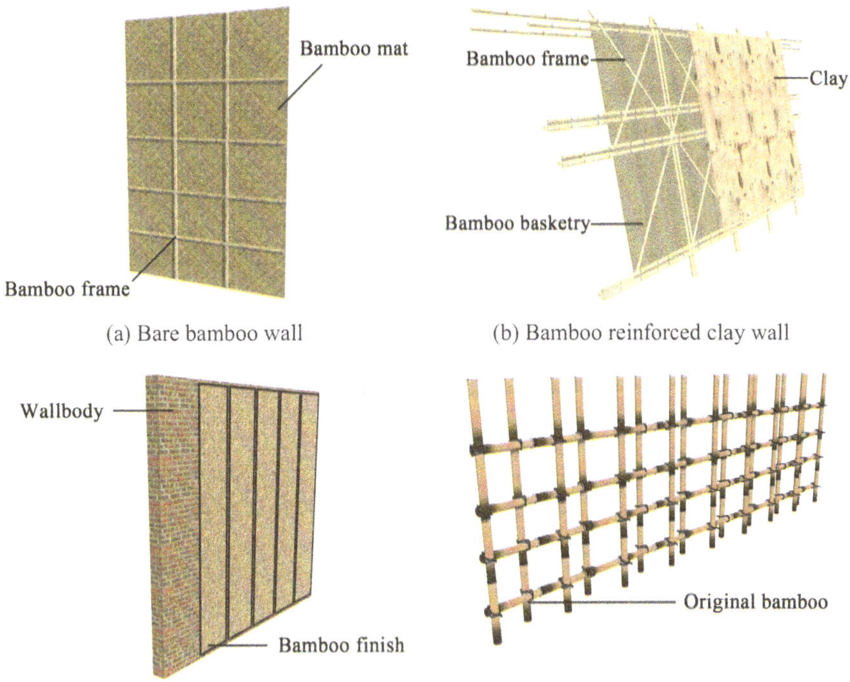


(a) Bare bamboo wall (b) Bamboo reinforced clay wall

(c) Bamboo finish (d) Bamboo fence

**Fig. 4** 3D models of typical bamboo wall types

**Table 1** Construction layers and heat transfer coefficients of typical bamboo solid walls

| Wall type | Construction layer | Heat transfer coefficient (W/m$^2$ K) |
|---|---|---|
| Bare bamboo wall | 5 mm bamboo mat | 5.44 |
| Bamboo reinforced clay wall | 10 mm clay<br>5 mm bamboo basketry<br>80 mm bamboo frame | 3.96 |
| Brick wall with a bamboo finish | 5 mm bamboo basketry<br>240 mm brick | 2.00 |

**Fig. 5** 3D models of typical bamboo window lattices of different weave patterns

## 3 Evolution: Bamboo Steel

To inherit sustainable architecture and preserve rural memory during rapid urbanization, it is significant to activate traditional construction techniques. That is, improving the characteristics of natural building materials according to contemporary and urban architecture requirements while retaining the materials' appropriate features and appearances. For example, in conventional Linpan residences, the woven textures of the bamboo materials are unique external images and the common aesthetic in western Sichuan. However, the mechanical and thermal properties of natural bamboo materials are not sufficient for modern buildings.

Bamboo steel is a new type of bamboo material, which is made of *Neosinocalamus affinis* and phenolic resin through a series of hot-pressing processes. This material has many good characteristics such as high strength, low carbon emission, and good weather resistance (Table 2).

The bamboo steel products are usually large boards (2500 mm × 1280 mm) with multiple thicknesses (e.g. 12, 25, and 35 mm) [13]. Although their specifications are not abundant, they can be easily planed, sawed, or joined, just like normal wood boards. Therefore, using these monotonous products, architects can form rich construction modules.

These modules mainly include decorative boards and structural rods. On the one hand, the use of bamboo steel boards is similar to that of ordinary wood boards, which can be attached to bearing walls with screws or adhesives. On the other hand, bamboo steel rods (straight or curve, two or more) are usually connected by steel sleeves (Fig. 6).

**Table 2** Properties of bamboo steel [12]

| Property | Index | Value |
|---|---|---|
| Density | | 1–1.4 g/cm$^3$ |
| Weather fastness | Thickness expansion rate after absorbing circulated water for 28 h | 0.6% |
| Environmental protection | Formaldehyde emission | E0 (0.1 mg/L) |
| Fire resistance | | B1 |
| Heat conductivity coefficient | | 0.173 W/m K |
| Extended stability | Compression strength along the grain | ≥180 MPa |
| | Extension strength along the grain | ≥300 MPa |
| | Elasticity modulus | ≥30 GPa |

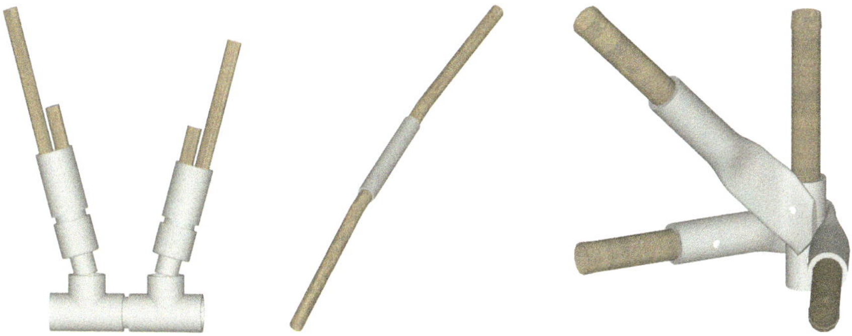

**Fig. 6** 3D models of connection modes of bamboo steel rods

# 4 Activation: BambooUp

Although the traditional construction techniques are extracted mostly from rural residences in Linpan, it is essential to activate the use of bamboo materials in urban public buildings, especially landmarks in Chengdu, the capital city of Sichuan. These representative constructions are not only carriers of building materials but also media of architectural cultures.

## *4.1 Project Overview*

BambooUp (Chengdu Bamboo Culture Exhibition Center) is one of the seven winners of the Jincheng Park International Architectural Competition in 2019. The lot of this project is located near the Qingshui River in the Wenjiang District of Chengdu. Containing rural residences, trees and bamboos, waters, and farmlands, this lot is a typical Linpan landscape (Fig. 7). However, the existing buildings are already deserted and cannot be further used. On the contrary, the other elements of the Linpan should be conserved as much as possible.

BambooUp is a creative exhibition space that includes exhibition halls, workshops, new media offices, and retails (Fig. 8). We arranged all these functions in clusters and distributed them among the vegetations. Then, we took linear service areas as links, drawing these clusters together. Visitors can walk or ride inside or on these links to view the architecture inside nature. This design highly respects the environment, trying to blur the boundary between the building and the landscape.

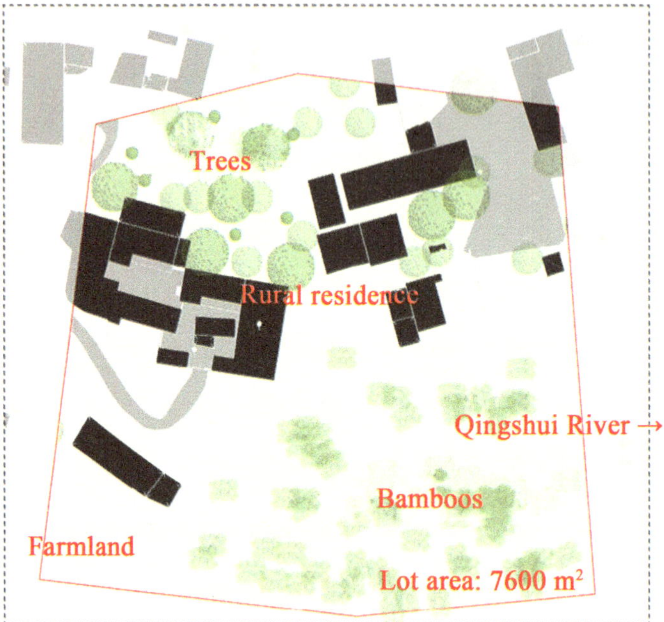


**Fig. 7** The current condition of the lot of BambooUp

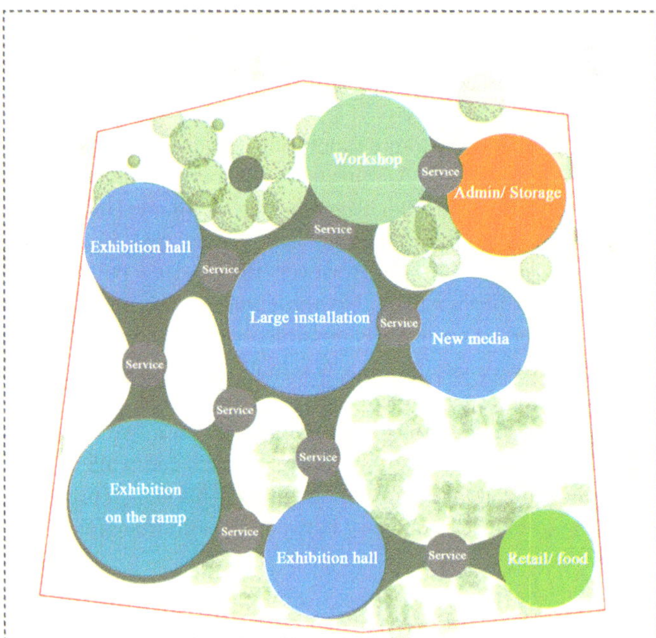


**Fig. 8** Function distribution and connection of BambooUp

## 4.2 Use of Bamboo Steel

The bamboo steel materials are used to form roofs, walls, and building textures, just similar to traditional bamboo residences in Linpan.

**Bamboo steel domes and skylights**. The traditional gable roof is not suitable for such a modern long-span exhibition architecture. However, bamboo has good resilience and is a proper material to form arches and domes. Although these forms are not used in traditional buildings in Linpan, they are commonly seen in household products (such as baskets, armchairs, and pot covers) since ancient times. BambooUp adopts these techniques and translates them into architectural languages (Fig. 9). Therefore, we cover function clusters using multiple bamboo steel domes with different lattices extracted from traditional Linpan houses. Besides, we add skylights to the indoor courtyard and make part of the rooftops accessible to visitors.

The bamboo steel dome in BambooUp is not the bearing structure but the skin of the roof. Supported by steel columns, the bamboo lattice layer is installed upon a glue-laminated timber dome and a tempered laminated glass layer (Fig. 10). This structure provides shelter and abundant shadows for the indoor environment.

**Bamboo steel walls**. Walls of a landmark are not only building envelopes but also cultural media for the public. Bamboo materials are a representative of traditional culture, as bamboo slips had been used for literal expressions for thousands of years. For some outer leaves of opaque walls, we add LED lights and frosted acrylic covers on to bamboo tubes, forming two types of bamboo screens. The screens can be

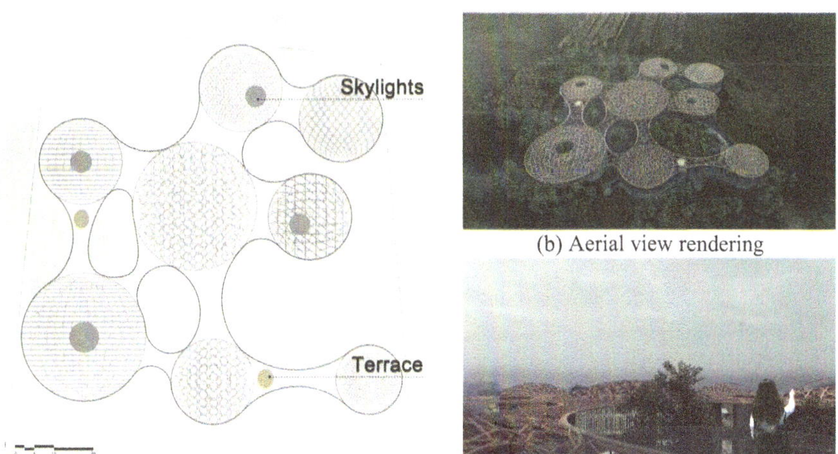


(a) Roof plan

(b) Aerial view rendering

(c) Rooftop view rendering

**Fig. 9** Roof plan and renderings of BambooUp

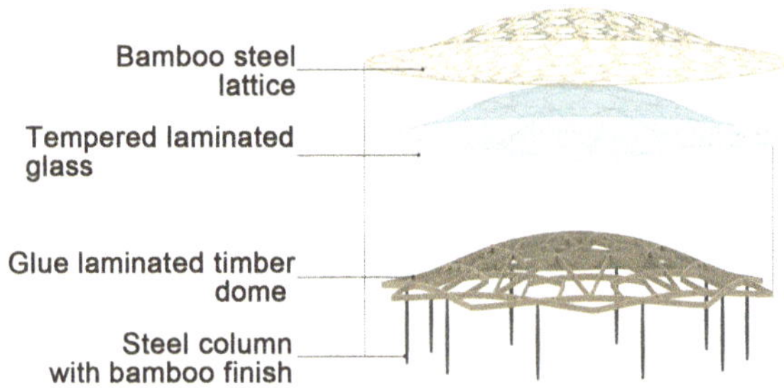


**Fig. 10** Dome structure and indoor view of an exhibition hall of BambooUp

installed on solid walls with light steel joists (Fig. 11). For some inner leaves of transparent walls, the bamboo walls can be closed or opened due to different exhibition requirements (Fig. 12).

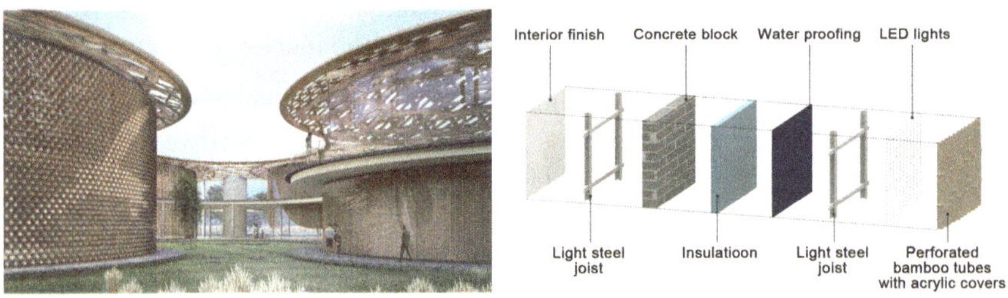


(a) Renderings of bamboo screens (b) Structure of an opaque wall

**Fig. 11** Bamboo screens renderings and the opaque wall structure of BambooUp

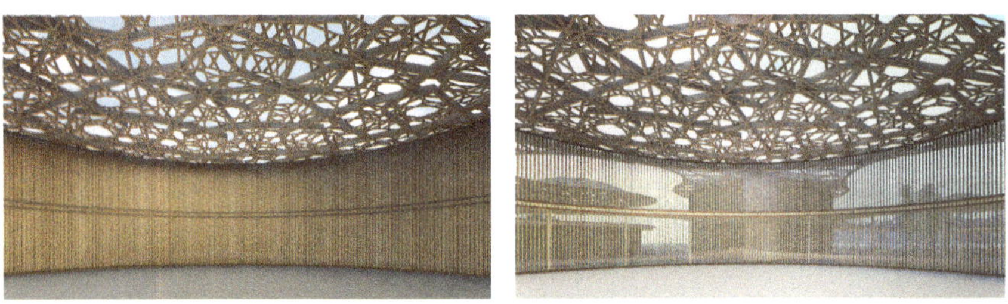

(a) Close mode (b) Open mode

**Fig. 12** Close mode and open mode of an exhibition hall of BambooUp

## 5 Conclusion

Some traditional habitations, which are vanishing with urban expansion, are sustainable. Linpan contains abundant and excellent construction skills, especially that of using bamboo materials, reflecting the architectural wisdom of local village craftsmen. The most widely used bamboo material in Linpan is *Neosinocalamus affinis*, which is appropriate for roofs, walls, and window lattices. The woven textures of the bamboo materials are unique external images and the common aesthetic in western Sichuan Plain. However, the mechanical and thermal properties of natural bamboo materials are not sufficient for modern buildings. Therefore, we introduced a new bamboo material—bamboo steel, which has many good characteristics such as high strength, low carbon emission, and good weather resistance. To activate the bamboo culture in urban landmarks, we designed BambooUp (Chengdu Bamboo Culture Exhibition Center), using the bamboo steel to form domes, skylights, and changeable walls.

Two probable future directions are: (1) more detailed applications of construction techniques of bamboo materials, and (2) public acceptability of the bamboo steel architecture concerning traditional memories.

**Acknowledgements** This work was supported by the Chengdu Key Research Base of Philosophy and Social Science (CCRC2020-4) and Xihua University (Z202040), Sichuan Science and Technology Department (2020YFS0308).

## References

1. Yu, S., Lu, H.: Relationship between urbanisation and pollutant emissions in transboundary river basins under the strategy of the Belt and Road initiative. Chemosphere **203**, 11–20 (2018)
2. Gao, X., Xu, A., Liu, L., et al.: Understanding rural housing abandonment in China's rapid urbanization. Habitat Int. **67**, 13–21 (2017)
3. Wang, X.: Accessibility and population density in the Linpan Landscape: a study of urbanization in the Chengdu Plain, Sichuan, China. University of Washington, Washington (2015)
4. Sun, D., Chen, Q., Hu, T., et al.: Community types and the biodiversity of farmhouse forest in western Sichuan province. J. Sichuan Agric. Univ. **29**, 22–28 (2011)
5. Tippins, J.L.: Planning for resilience: a proposed landscape evaluation for redevelopment planning in the Linpan Landscape. University of Washington, Washington (2014)
6. Li, Z., Chen, H., Lin, B., Zhu, Y.: Fast bidirectional building performance optimization at the early design stage. Build. Simul. **11**, 647–661 (2018)
7. THS: http://www.thsware.com/. Last accessed 15 April 2020
8. Yang, L., Huang, Z.: Mould proof treatment technology of bamboo and its effects on mechanical properties in *Neosinocalamus affinis*. J. Bamboo Res. **38**, 52–57 (2019)
9. Yang, J.: Applied research of plant materials in ancient Chinese architecture—Grass, bamboo, wood. Southeast University, Nanjing (2016)
10. Liu, Q, Peng, P.-H., Wang, Y.-K., et al.: Microclimate regulation efficiency of the rural home-garden agroforestry system in the Western Sichuan Plain, China. J. Mountain Sci. **16**, 516–528 (2019)

11. Huang, Z., Sun, Y., Musso, F.: Experimental study on hygric physical properties of bamboo based on HAM model. J. Hunan Univ. (Nat. Sci.) **45**, 150–161 (2018)
12. Hongya Bamboo Era Science and Technology Co., Ltd.: Bamboo steel (in Chinese) (2020)
13. Beijing Institute of Architectural Design (Group) Co., Ltd.: Wood and bamboo (in Chinese) (2020)

# A Multi-agent Platform to Inform Strategies for Briefing Age-Friendly Communities in Urban China

**Liqun Xiang, Geoffrey Shen, and Yongtao Tan**

**Abstract** The World Health Organisation has been working for more than a decade to guide global cities and communities to consider, plan and implement age-friendly places in response to the rapid ageing. In China, promoting age-friendly communities (AFCs) is critical due to the fact of both rapid demographic change and people's increasing awareness of making preparations to deal with their ageing issues. Although many efforts have been made to promote AFCs in urban China, challenges remain during the construction process. The engagement of stakeholders, budget and policy issues always challenge communities' pathways to become age-friendly.

Simulation as a tool of analysis and prediction can be applied to propose constructing strategies and explore solutions regarding the above concerns. For a multi-agent system that contains several agents, an agent itself may lack resources, information and capabilities in solving the whole problem; however, interactions between each identifiable agent provide aggregated attributes, which can facilitate decision-making procedures afterwards.

Therefore, this paper aims to design a multi-agent platform to simulate the briefing stage and explore strategies for promoting AFCs in urban China. The theory and process to develop the platform according to a case study are described; The simulation results and how the platform can help are discussed. This research will serve as a reference for researchers and practitioners to further explore the briefing stage and efficient strategies for promoting AFCs.

**Keywords** Multi-agent platform (MAP) · Age-friendly community (AFC) · Briefing

L. Xiang · G. Shen
The Hong Kong Polytechnic University, Hong Kong S.A.R, China
e-mail: 17903659r@connect.polyu.hk

G. Shen
e-mail: geoffrey.shen@polyu.edu.hk

Y. Tan (✉)
School of Engineering, RMIT University, Melbourne, VIC 3001, Australia
e-mail: yongtao.tan@rmit.edu.au



F. K. S. Chan et al. (eds.), *Proceedings of the 2020 International Conference on Resource Sustainability: Sustainable Urbanisation in the BRI Era (icRS Urbanisation 2020)*, Environmental Science and Engineering,
https://doi.org/10.1007/978-981-15-9605-6_13

# 1 Background

## *1.1 Challenges of Promoting AFCs in Urban China*

Challenges of promoting AFCs in urban China can be categorised into three aspects, which are the policy considerations, the environment and facility factors, and the market conditions [1, 2].

Developing AFCs has become a significant theme in China's public policy during the past decade. Although it is a breakthrough that only four years were taken for the 'liveable environment' to develop from a theoretical concept to several legal clauses, many issues remain to be tackled when promoting the AFCs. For example, requirements from senior citizens are likely to be overlooked during the infrastructure development process and the provision of social services [1, 3]; Connections between the legal causes and construction strategies are almost blank, which brings difficulties for stakeholders to promote AFCs.

The latest results of the national sampling survey in regard to the living conditions of China's urban and rural older persons indicated that 58.7% of the seniors considered their accommodations insufficiently age-friendly [4]. The lack of facilities limited senior citizens' participation in outdoor activities, caused safety problems to those with limited Activities of Daily Living (ADL) levels and brought pressures to stakeholders engaged in the construction process of AFC projects.

In recent years, real estate projects on housing for seniors in urban areas emphasised more on the wealthy groups' needs, therefore, numerous middle- and low-income senior citizens' needs are likely to be overlooked. Although the investors' investments for promoting such projects, and the customers' payment for purchasing such apartments are quite high, the sales conditions remain optimistic [2]. Such conditions inevitably hinder the process of promoting AFCs that may need longer time to pay back. However, the wealthy seniors account for only a small percentage of the entire ageing group. For the remainder of the senior citizens, the fact is that they become old before getting rich, thereby depriving them of the ability to purchase such apartments and enjoy the related care services.

## *1.2 Applications of the MAS in Construction Projects*

The rise of computation has generated a new field of knowledge named "complex systems". As complex systems show features such like emergence, nonlinearity, decentralization and adaptation [5], the multi-agent system (MAS) has developed rapidly in recent years due to its capacity of dealing with complicated problems regardless the subject area [6]. The most important and special feature of MAS is the agent perspective that takes to view any system as consisting of agents [7]. An agent itself may lack resources, information and capabilities in solving the whole problem, however, interactions between each identifiable agent provide aggregated attributes, which can facilitate decision-making procedures afterwards [8].

Almost all the construction projects contain complex systems since multiple stakeholders work together but generally concentrate on different prioritises. Besides, construction projects have co-evolution of developments and processes by self-modifications: The values of a certain project are set in the initial stages but will keep developing through the whole life cycle [5]. The application of MAS during the briefing stage of AFC projects will give researchers and practitioners a chance to stimulate, identify and analyse issues arising from stakeholders with different backgrounds or belonging to geographically separated teams and working in the dynamic environment.

# 2 Case Study: Modelling the Process of Stakeholders' Consensus Formation

## *2.1 Background and Configuration*

The target case study community was located in Shanghai and was built in 1994, with most of the original residents grew from their middle-age to old age after 2010. The elderly care facilities are in great need and the investor would like to renovate the original community centre into a small-scale care facility for the seniors. Although residents acknowledged the value of the elderly care facility, they did not want it to be built near their community, regardless changing their original community centre into one [9]. The objections from residents made the project hard to begin. Therefore, an agent-based model is built to simulate and investigate the consensus formation process of residents.

In this study, the networks and the opinions amongst residents are idealised. To apply the agent-based model, assumptions are made on the basis of several opinion dynamic models [10–12]. Considering the total number of households, ageing rate and structure of the owners' committee in the target community, three kinds of agents are set for simulation: (1) $A_m$: The household that contains a member from the owners' committee; (2) $A_s$: The household contains at least one member who aged 65 yr or over; (3) $A_o$: The other household. The implemented model can be treated as a multi-state opinion model where agents' initial attitudes are clarified as approval, neutral or disapproval ones, with $+\ 1$, 0, or $-1$ to imply the three kinds, respectively; The weight of approval and disapproval are considered more powerful than the neutral ones [13]. During the consensus formation process, agents would make interactions with others and change the initial attitudes according to their neighbours' opinions with a probability related to different influences. Therefore, the attitude of a certain agent is changing on the basis of its neighbours' opinions. The attitude of an agent would not change directly from approval (+1) to disapproval ($-1$) and vice versa, instead, it will pass a neutral (0) status which indicates a phase of undecided. The simulation results would depict how different initial approval rate and the connections amongst residents, would affect the final opinions of the residents.

## 2.2 Properties of Agents and Simulation Performed

Apart from holding an attitude, each agent is initialised with two properties: (1) A social impact factor, which is an integer number in the range [0,10], reflecting the agent's social importance. (2) An 'influenceability', which is a random real number in the range [0, 1], indicating the possibility that an agent directly changes its opinion to the opposite side without passing through the neutral stance. If the value of this parameter is no less than 0.5, an agent would have the possibility to directly change its opinion. Otherwise, it would pass the neutral status first.

The agent-based model will be run by *NetLogo 6.1.1*, a multi-agent programmable modelling environment that can be applied to explore interactions between agents and the phenomenon emerges as a result of such interactions [14, 15]. Figure 1 illustrates the interface of the simulation model. Two main stages, which are, setup of the initial condition (t = 0) and the opinion dynamics (t > 0), can describe the simulation performed.

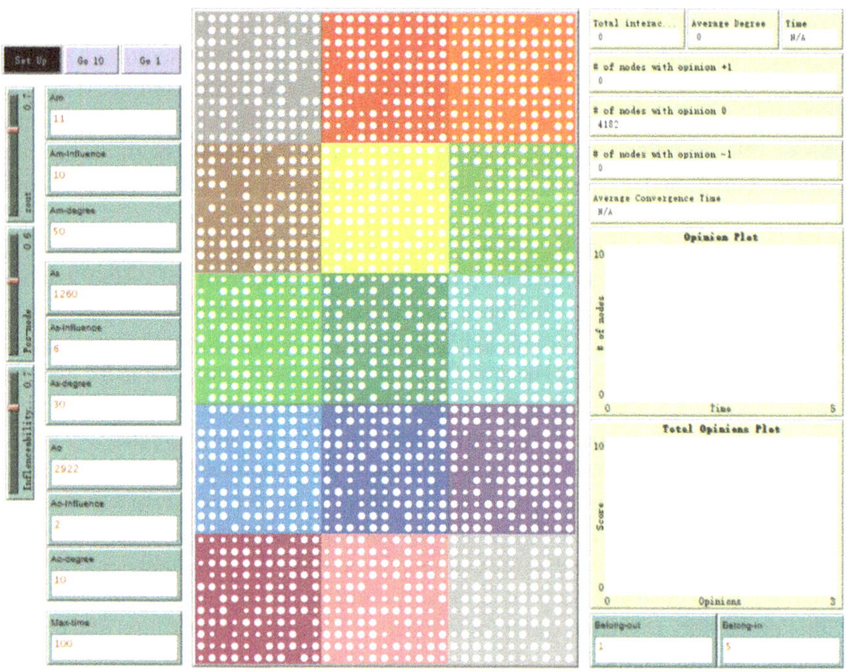


**Fig. 1** The interface of the simulation model

**Setup of the Initial Condition (t = 0).** The masterplan indicates that the target community consists 15 subregions. Therefore, the 4,193 households (agents) are divided into 15 groups, and each group contains approximately 280 agents. The connections amongst agents in each group is built to form the social network. The number of connections that an agent can make depends on the different influence factors. The agents outside the group will have lower impacts on a certain agent than those inside the group. Such assumptions are reasonable since members from the owners' committee are considered as the representatives of other residents and they would be able to contact with more residents, while the seniors spend most of their time in the community and may also get familiar with others. Finally, an opinion is assigned to all the agents when t = 0.

Details of the agents' properties are summarised in Table 1.

**The Opinion Dynamics (t > 0).** Assuming xi(t) is the opinion of agent i at time t, the opinion of this agent at time t + 1 will be a function of both xi(t) and vi(t), where vi(t) is the vector filled with the weighted opinions n(t) of all the neighbours that agent i has connections with [13].

The dynamics of opinions can be represented through the line chart with three curves, with each one represents the change of an opinion. Figure 2 illustrates a single event that after struggling amongst three opinions, all the agents reached consensus by the end of the simulation. Figure 3 depicts ten events with the same parametres but different random variables setting for simulation. The histogram shows at the

**Table 1** Properties of agents for the simulation

| Variable | Number of agents | Average connection | Social impact factor |
|---|---|---|---|
| $A_m$ | 11 | 50 | 10 |
| $A_s$ | 1260 | 30 | 6 |
| $A_o$ | 2922 | 10 | 2 |

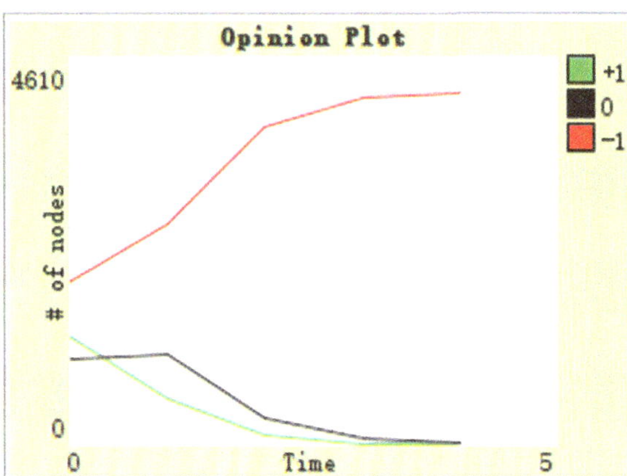


**Fig. 2** Opinion plot for a single event

Fig. 3 Distribution of opinions for ten events

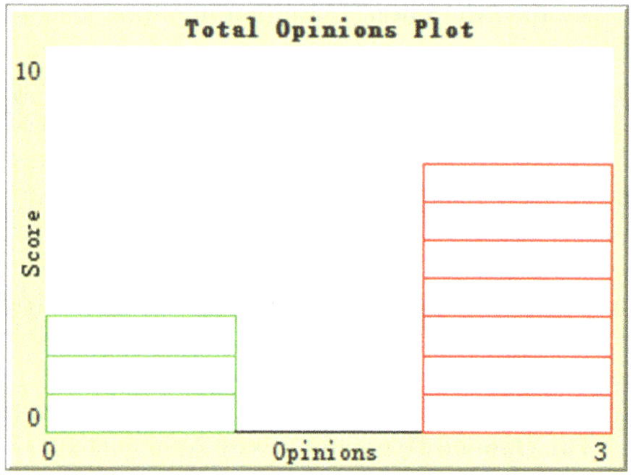


end of the given events, the predominant opinion is either consensus or dissent. Both figures indicate that the neutral opinion is only a transition phase.

Since the simulation aims to discuss how different initial approval rate and the connections amongst agents would affect residents' final opinions, two main parameters can be modified: (1) The initial approval rate, which is the percentage of agents who are in favour of the proposal at the beginning of the simulation; (2) The percentage of connections an agent makes with those outside the group (Outside connection rate). Both parametres can be changed from 0 to 100%.

It is assumed when the investor proposes a plan of renovating the community centre into a small-scale care facility for the seniors, the initial approval rate is 20%. Since it is realistic that one may have more connections with those who live nearby, the outside-connection rate is also set as 20%, meaning for $A_m$, $A_s$ and $A_o$. The connections they make outside the group are 10, 6 and 2, respectively. During the simulation, the two parametres are changed to explore how they would affect the final results.

## 2.3 Analysis of Results

The authors paid attention to the events when all of the agents held the same opinion in the end. Particularly, the complete consensus (all the opinions equal to + 1) or dissent (all the opinions equal to −1) is more meaningful than the neutral one (all the opinions equal to 0).

To clarify the final opinions, a unique parameter $C$ (Convergence achieved rate) was calculated to imply the result when the convergence was achieved [13]: $E_{+1}$ and $E_{-1}$ indicate the times when the simulation processes ended with complete consensuses or dissents, respectively, while $E$ stands for the total number of events. The simulation ends up with all the opinions equal to 0 is not taken into consideration, since all the residents hold a neutral opinion would not be helpful for the investors

**Table 2** Simulation results indicating by parameter $C$ $(t \leq 100)$

| Convergence achieved rate (C) | | Outside connection rate (%) | | | | | | |
|---|---|---|---|---|---|---|---|---|
| | | 20 | 30 | 40 | 50 | 60 | 70 | 80 |
| Initial approval rate (%) | 20 | -0.8 | -0.4 | -0.8 | -0.8 | -0.6 | -0.8 | -0.8 |
| | 30 | -0.1 | -0.2 | -0.3 | -0.7 | -0.2* | -0.1 | 0* |
| | 40 | 0.3 | 0.2 | 0.7 | 0.2 | 0* | + 1* | -0.2* |
| | 50 | 0.8 | 0.8 | 0.9 | 0.9 | + 1* | + 1* | 0.8* |
| | 60 | + 1* | + 1* | + 1* | + 1* | + 1* | + 1* | + 1* |
| | 70 | + 1* | + 1* | + 1* | + 1* | + 1* | + 1* | + 1* |

*All the 10 events ended up with either consensuses or dissents

to implement their proposal in practice. The value of $C$ ranges from $-1$ to 1, where $-1$ indicates that all the events were ended with complete dissents, while $+1$ means such events came to an end with complete consensuses.

$$C = \frac{E_{+1} - E_{-1}}{E} \tag{1}$$

During the simulation process, a limitation on the time $(t \leq 100)$ was set to exclude the cases when it took too long to reach a convergence. It should be noticed that $C$ does not stand for any of the step during the simulation process, it only informs the final status when consensuses or dissents are reached. Table 2 illustrates the simulation results when $E = 10$ with the parameter $C$ defined above.

It can be seen from the simulation results that both the initial positive rate and the outside connection rate would influence $C$. To ensure a convergence for each event, the initial positive rate should be no less than 50%, together with an outside connection rate which is no less than 60%. Besides, when the initial positive rate is lower than 40%, even convergences are achieved in the end, the residents are more likely to hold disapproval attitudes towards the investor's proposal. The outside connection rate can be considered as the "weak tie" [16], which is helpful to facilitate the information exchange from different groups. The simulation results indicate that when the outside connection rate is no less than 60%, convergence is more likely to be reached for all the events. With such value equals to 40% or more, chances are even the initial positive rate is lower than 50%, residents would hold complete approval attitudes towards the investor's proposal when the simulation ended.

The average convergence time $(T)$ listed in Table 3 is another parameter that the authors paid attention to. It stands for how many rounds of information exchange happened before the convergence is reached, rather than an exact duration. Since the simulation is conducted with 10 events as a group, when a certain event in the group did not reach convergence before $t = 100$, it would be neglected when calculating $T$.

According to Table 3, when all the 10 events in a group ended up with convergences, $T$ appears to be at least 3.000, with no less than 70% of the agents hold the positive opinion at $t = 0$. When the initial approval rate is 40% and the outside

**Table 3** Average convergence time *(T)* of the simulation

| Average convergence time *(T)* | | Outside connection rate (%) | | | | | | |
|---|---|---|---|---|---|---|---|---|
| | | 20 | 30 | 40 | 50 | 60 | 70 | 80 |
| Initial approval rate (%) | 20 | 3.125 | 3.667 | 3.125 | 3.125 | 4.000 | 3.750 | 4.375 |
| | 30 | 4.200 | 3.750 | 3.600 | 4.286 | 4.900* | 4.667 | 5.500* |
| | 40 | 4.286 | 3.500 | 3.571 | 3.750 | 7.200* | 4.400* | 5.400* |
| | 50 | 3.125 | 3.500 | 4.556 | 4.444 | 4.300* | 4.400* | 5.700* |
| | 60 | 3.200* | 3.100* | 3.300* | 3.900* | 3.900* | 3.600* | 3.900* |
| | 70 | 3.000* | 3.000* | 3.000* | 3.200* | 3.400* | 3.100* | 3.100* |

* All the 10 events ended up with either consensuses or dissents

connection rate is 60%, $T$ peaks at 7.200, meaning more than seven rounds of information exchange are averagely in need to achieve the convergence. Although a higher initial approval rate and a lower outsider connection rate may reduce $T$, three to five rounds of information exchange are still in need before the consensuses or dissents are formed.

# 3 Discussions

## *3.1 Implications for Stakeholders of Promoting AFCs*

The simulation results obtained from modelling residents' opinion dynamics give the investors implications on facilitating the renovation proposal, mainly by improving both the initial approval rate and the outside connection rate.

The initial approval rate is highly related to residents' concerns. As for the target case study community, the facility that investors intended to promote is an *Embedded Retirement Facility* (ERF) which was gradually emerged from 2014 in mainland China [17]. An ERF is treated as a small-scale, multifunctional community-based care facility with a total construction floor area of no more than 800m$^2$, service radius not longer than 450m and capacity of no more than 45 beds for senior citizens; An ERF can either offer day respite services, long-term residence or both; ERFs not only assist senior citizens in their daily lives but also help them maintain good health conditions by setting canteens, organising social and recreational activities and providing regular health examinations, amongst others [18, 19]. Therefore, the investors need to communicate with residents, let them know the properties of the facility, so as to eliminate their prejudice and alleviate their concerns with the renovating proposal. Besides, the investors should also consider leaving some places for younger residents to use. Although the community-based care facility for the seniors is in need, younger residents still have the rights to use the community centre, renovating the whole centre without considering their needs is unfair.

Increasing the outside connection rate means to facilitate interactions among residents live in different regions of the community. One potential method is organising activities for the whole community to provide opportunities for residents to exchange information and ideas. It may also be helpful if the investors convince and get support from Am or the As, as they may have more opportunities to interact with others and obtain more social impacts.

The practical experiences of the case confirmed the strategies proposed by the authors: The investors, designers and other constructing group members visited a great number of residents, explained the meaning of renovating the community centre to obtain their supports; Changes were also made regarding the initial proposal and indoor spaces are kept for other residents to conduct physical activities. The effective communication and collaboration amongst stakeholders at the briefing stage ensured the success of promoting the project and provided experiences for similar projects to follow [9].

## *3.2 Components of the Multi-agent Platform (MAP)*

An agent-based model is formed by three sets of elements in general: (1) the agents, (2) the interactions amongst agents, together with (3) the interactions between agents and the environment [15]. An agent stands for one cluster of stakeholders that engage in promoting AFC projects. According to the [2], seven stakeholders, which are senior citizens (S1), caregivers (S2), government and policymaking institutions (S3), research institutions (S4), project investors and real estate developers (S5), urban planners, architects and interior designers (S6), as well as NGOs (S7), are identified as key stakeholders that would have influences on the construction process of AFC projects in urban China.

The widely accepted and implementable approach to promote AFCs in urban China is inserting or renovating community-based elderly-care facilities in the built regions, during which process three clusters formed by five stakeholders play important roles [19–21]: (1) The end user cluster includes S1 and S2, whose needs should be clarified according to different ADL levels and income conditions. Residents who live in the same community with S1 and S2 also belong to this cluster; (2) The supervision cluster made up by S3, whose applications of the incentive policies would influence choices of S5 regarding promoting AFCs; and (3) The constructing cluster containing S5 and S6, who are responsible for the quality of the constructing work. The above clusters are defined as three agents when designing the MAP, namely the end user agent, the supervision agent, and the constructing agent.

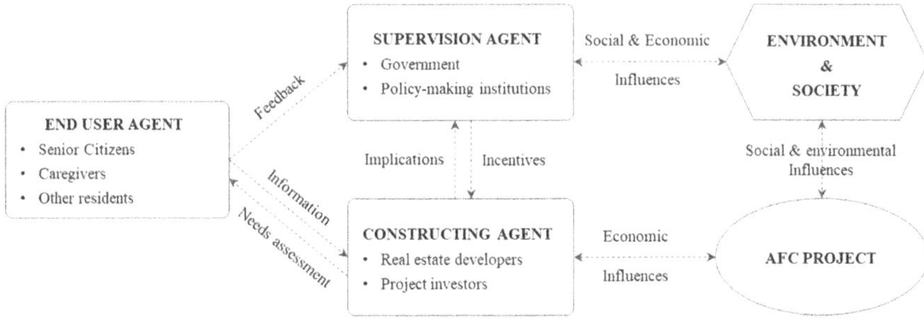


**Fig. 4** Relationship amongst agents in promoting AFCs

The relationship amongst three agents during the briefing stage of AFC projects is illustrated in Fig. 4. The supervision agent gives the constructing agent policy incentives such like public financial support, tax and land use benefits, and gains implications to adjust policies and regulations [20]. The constructing agent takes responsibilities of figuring out conditions of built communities, assessing needs from the end user agent, and carrying out the strategic and implemental plans. The end user agent is able to provide the constructing agent first-hand information pertaining to their current accommodations and communities, such like whether current service provisions can assist in the daily lives, or whether the barrier-free facilities are adequate to make participations of physical activities possible. In addition, feedback from the end user agent is collected by the supervision agent, particularly after the completion of AFC projects, to evaluate the quality of the work that the constructing agent has done.

Figure 4 also depicts the influences that the agents would have on the AFC projects, the environment and the society. The constructing agent typically invests in the AFC projects with subsidies from the supervision agent and looks forward to being paid back several years after the completion. Contributions that AFC projects would make can be clarified both in the environmental and social levels, for the aim of promoting AFCs is to ensure communities as inclusive and equitable places that even the most vulnerable seniors can live in, regardless of others [22]. Even though difficulties usually occur when promoting AFCs projects due to the burdens of resources [23], the government and policy-making institutions as the supervision agent, would gain social and economic benefits from the age-friendly society in the long term.

## 3.3 The Structure of the MAP

The MAP designed in this study is supposed to help the government and the constructing group to make better decisions during the briefing stage of AFC projects. The platform aims to support users both in the project and in the policy levels: For the project level, suggestions would mainly give to the constructing group regarding how to deal with conflicts between stakeholders; While for the policy level, suggestions are supposed to make for the government regarding the extent that incentives should be given to the constructing group.

The MAP is developed on the basis of the Proactive Construction Management System [24]. Figure 5 illustrates the three tiers of this platform: (1) the presentation tier, (2) the processing tier, and (3) the data tier. The presentation tier is designed for users to set specific scenarios for simulation. The processing layer is where the agent-based model runs within the environment. While the data tier contains general indicators and algorithms that would not change case by case. After the simulation process, results regarding specific scenarios will be displayed through the presentation layer for users to interpret and obtain strategies. Figure 6 depicts the potential interface of the MAP where the user logged in as a project investor.

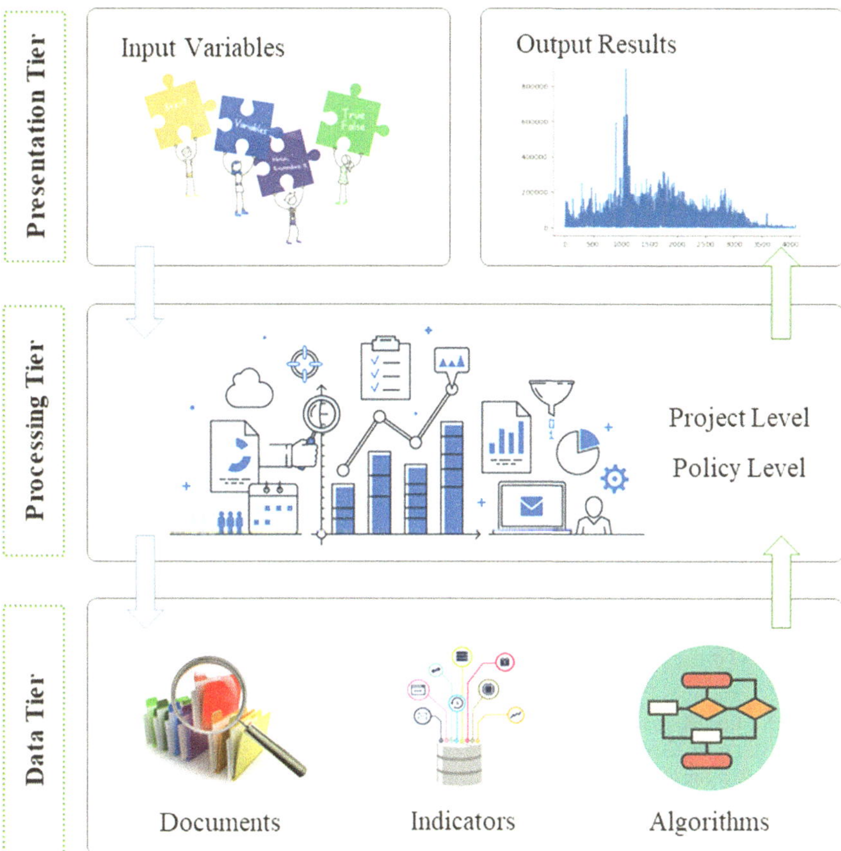


**Fig. 5** Three ties of the MAP

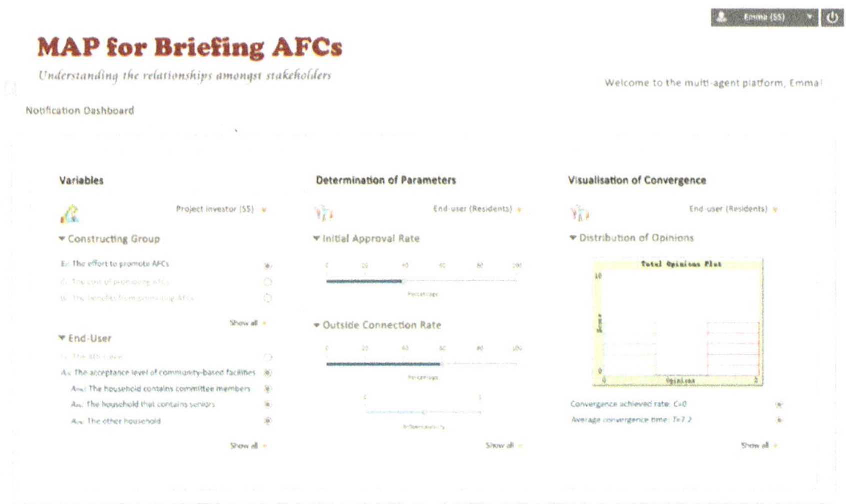


**Fig. 6** The interface of the MAP

# 4 Conclusions

The briefing stage is treated as an indispensable part to promote AFC projects in urban China, during with stage many aspects are needed to be taken into consideration by different stakeholders before setting out the strategic plan. The agent-based simulation of stakeholders' consensuses formation process conducted based on a real case provides the foundation to design a MAP, which would not only help to understand and manage relationships of multiple stakeholders affected by various factors, but to inform strategies during the briefing stage of promoting AFC projects. The components of the three-tier structured MAP are described, the potential interface of the MAP is also illustrated.

# References

1. Hu, X.: Livable space and building concept of elderly friendly city. Shanghai Urban Manage. **23**(03), 18–23 (2014)
2. Xiang, L., Tan, Y., Jin, X., Shen, G.: Understanding stakeholders' concerns of age-friendly communities at the briefing stage: a preliminary study in urban China. Eng. Constr. Architectural Manage. (ahead-of-print) (2020).https://doi.org/10.1108/ecam-01-2020-0070
3. Wu, X., Qu, J.: In: China Report of the Development on Livable Environment for the Elderly. Social Science Academic Press, Beijing (2015)
4. Dang, J.: In: Survey Report on the Living Conditions of China's Urban and Rural Older Persons. Social Sciences Academic Press, Beijing (2018)

5. Son, J., Rojas, E.M., Shin, S.W.: Application of agent-based modeling and simulation to understanding complex management problems in CEM research. J. Civil Eng. Manage. **21**(8), 998–1013 (2015)
6. Liang, X., Shen, G.Q., Bu, S.: Multiagent systems in construction: a ten-year review. J. Comput. Civil Eng. **30**(6), 04016016 (2016)
7. Macal, C.M.: Everything you need to know about agent-based modelling and simulation. J. Simul. **10**(2), 144–156 (2016)
8. Motieyan, H., Mesgari, M.S.: An agent-based modeling approach for sustainable urban planning from land use and public transit perspectives. Cities **81**, 91–100 (2018)
9. Vanke Weekly: (2020). https://www.vankeweekly.com/?p=81189. Last Accessed 16 Jul 2020
10. Hegselmann, R., Krause, U.: Opinion dynamics and bounded confidence models analysis and simulation. J. Artif. Soc. Soc. Simul. **5**(3), 1–32 (2002)
11. Galam, S.: Minority opinion spreading in random geometry. Eur. Phys. J. B **25**(4), 403–406 (2002)
12. Chen, P., Redner, S.: Consensus formation in multi-state majority and plurality models. J. Phys. A: Math. Gen. **38**(33), 7239–7252 (2005)
13. Le Pira, M., Ignaccolo, M., Inturri, G., Pluchino, A., Rapisarda, A.: Modelling stakeholder participation in transport planning. Case Stud. Transp. Policy **4**(3), 230–238 (2016)
14. Lu, M., Cheung, C.M., Li, H., Hsu, S.C.: Understanding the relationship between safety investment and safety performance of construction projects through agent-based modeling. Accid. Anal. Prev. **94**, 8–17 (2016)
15. Wilensky, U., Rand, W.: An Introduction to Agent-Based Modeling: Modeling Natural Social and Engineered Complex Systems with NetLogo. MIT Press, Cambridge (2015)
16. Granovetter, M.S.: The Strength of Weak Ties. Social Networks, pp. 347–367 (1977)
17. Hu, H., Wang, Y., Wang, X., Zhang, L.: Situation evaluation and improving path of embedded retirement pattern. Soc. Secur. Stud. **02**, 10–17 (2015)
18. Zhang, S., Zhao, Y.: A study on the design of small care facilities for the elderly embedded in urban communities. Architectural J. **10**, 18–22 (2017)
19. Xiang, L., Yu, A.T.W., Tan, Y., Shan, X., Shen, Q.: Senior citizens' requirements of services provided by community-based care facilities: a China study. Facilities **38**(1/2), 52–71 (2020)
20. Li, H., Zheng, H.: In: Elderly Livable Environment Construction. China Architecture Publishing and Media Co. Ltd, Beijing (2019)
21. Zhou, Y., Li, J.: China Report of the Development on Livable Environment for the Elderly. Social Science Academic Press, Beijing (2015)
22. WHO Homepage: (2020). https://www.who.int/ageing/publications/gnafcc-report-2018/en. Last Accessed 30 Aug 2020
23. Buffel, T., McGarry, P., Phillipson, C., De Donder, L., Dury, S., De Witte, N., Smetcoren, A., Verté, D.: Developing age-friendly cities: case studies from Brussels and Manchester and implications for policy and practice. J. Aging Soc. Policy **26**(1–2), 52–72 (2014)
24. Li, H., Lu, M., Chan, G., Skitmore, M.: Proactive training system for safe and efficient precast installation. Autom. Constr. **49**, 163–174 (2015)

# How to Make Elderly Mobility Safe: Voice of Residents

**Ryosuke Ando, Y. Mimura, S. Tsuboi, and M. Ishii**

**Abstract** Regarding traffic safety, Japan has been generally evaluated as a successful country in the world being with less traffic accident fatalities, but the traffic safety issue is still one of the most important issues. In Japan, elderly people share 28.5% of total population in 2020. How to make elderly's mobility sustainable and safe has been a very serious issue in Japan since several years ago. The project, to be reported here, focused on Aichi Prefecture which had been famous with the worst position of traffic accident fatality ranking in Japan for 16 yr from 2003 to 2018. However, the traffic accident fatalities in 2019 told us that Aichi succeeded exiting from the worst position. In this paper, the authors are mainly going to report on what we did in the 2018 project and mainly on what kinds of voice of residents had been taken into consideration.

**Keywords** Elderly mobility · Traffic safety · Public involvement

## 1 Introduction

In China, the fast urbanisation in the pasted years has brought the urban planning and urban economy to develop rapidly. Surely, this development may show some good models to the related countries along the Belt and Road. However, at the same time, the rapid urbanisation has caused the traffic accidents more and more. The traffic safety issue must be dealt with well-matched to the urbanisation. Meanwhile, the aging society is another important issue in China. Although the one-child policy was loosened several years ago, the situation of the aging had not been improved as expected and the aging has been as a new normal. Therefore, how to make to treat the elderly mobility safely in the aging/aged society has been a very urgent mission in the coming years, too.

R. Ando (✉) · Y. Mimura · S. Tsuboi · M. Ishii
TTRI (Toyota Transportation Research Institute), Toyota 471-0024, Japan
e-mail: ando@ttri.or.jp



F. K. S. Chan et al. (eds.), *Proceedings of the 2020 International Conference on Resource Sustainability: Sustainable Urbanisation in the BRI Era (icRS Urbanisation 2020)*, Environmental Science and Engineering,
https://doi.org/10.1007/978-981-15-9605-6_14

Regarding the aging/aged society, Japan is a good reference country as it is the most aged country in the world now. According to Japan Institute for Labour Policy and Training [1], elderly people in Japan shared 28.5% of total population in 2020. This is not only higher than 12.1% in China, but also higher than 18.4% in UK and 24.0% in Italy which is the second aged country in the world.

In Japan, as shown in Table 1 (data sources: [2] and [3]), the number of traffic accidents are continuously decreasing since 2015 and consequently that the traffic accidents fatalities and the fatalities per 100,000 people are also decreasing proportionally.

The share of elderly fatalities caused by traffic accidents is recorded more than 50% in the last eight years and has shown an increasing trend as illustrated in Fig. 1.

How to make elderly's mobility sustainable and safe has been a very serious issue since more than ten years ago. When comparing the elderly traffic fatalities per 100,000 population [2], 9.3 in Japan was lower than 13.6 in USA but higher than 4.2 in UK that is the lowest country. What we can know from these data is that traffic safety with elderly people's mobility is the most important issue in Japan.

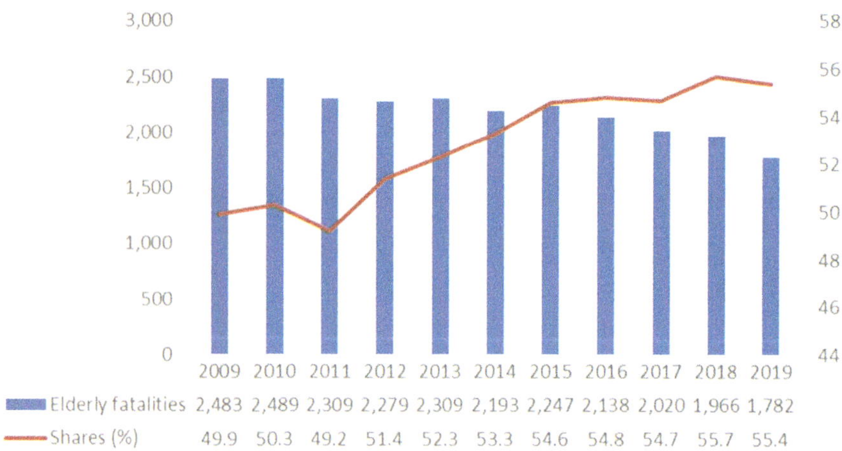


**Fig. 1** Numbers and shares of traffic accident elderly fatalities in Japan

**Table 1** Number of traffic accidents and fatalities in Japan.

| Year | No. of traffic accidents | Fatalities caused by accidents | Fatalities per 100,000 people |
|---|---|---|---|
| 2015 | 536,899 | 4117 | 3.24 |
| 2016 | 499,201 | 3904 | 3.07 |
| 2017 | 472,165 | 3694 | 2.91 |
| 2018 | 430,601 | 3532 | 2.79 |
| 2019 | 381,002 | 3215 | 2.54 |

Amongst all 47 prefectures of Japan, Aichi had been ranked the worst position by the traffic accident fatality for 16 yr from 2003 through 2018. Regarding the elderly people, in 2017, the traffic accident fatality in Japan is 3694 including 2020 elderly persons. Ten years before, the elderly fatalities because of traffic accidents were 2749 in 2007. These told us that the reduction rate is 26.5%. However, in Aichi, the total traffic accident fatalities were 200 in 2017. Of them, 110 were elderly people. Comparing with 137 elderly fatalities in 2007, the reduction rate is only 19.7%. These data also implies the main target should be elderly people when we discussed traffic safety countermeasures in Aichi. Conversely, the key issue to give effective countermeasures for traffic safety is elderly's mobility policy.

However, in 2019, the number of traffic accident fatalities in Aichi was reduced to 56, 12 less than Chiba Prefecture which became the worst alternatively. One of the important factors that resulted in the reduction of accidents was that the traffic policymakers of Aichi conducted a serious study named "the grand design of traffic safety countermeasures for elderly people in Aichi" in 2018. (See details in [4] and [5]) Many practical initiatives and steps were taken ever since in terms of the finding of the aforementioned study. In this paper, the authors are mainly going to report on what kinds of voice of residents had been taken into consideration.

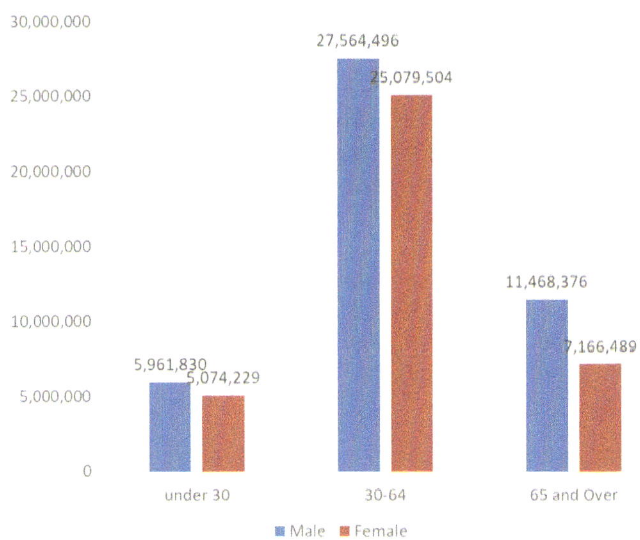


**Fig. 2** Number of people having driving licenses (March, 2019)

## 2 Drivers and Elderly Drivers

In Japan, as shown in Fig. 2, in all three age groups, male drivers are more than female drivers. Regarding the elderly people, the difference between the male and the female is the biggest comparing to other two age group. About 11.5 Mio. of elderly male and 7.2 Mio. of elderly female have the automobile driving licenses.

In Japan, there is no age limitation for elderly people to have the driving licences and their private cars. Therefore, many elderly drivers despite of being 80 yr old or even older are still driving cars in their daily lives. In Aichi, as shown in Fig. 3, most people make use of private cars in daily life. 59% of the elderly who are 70 yr old and older are driving their private cars and 1% of the 70 yr old and older are riding the motorcycles. Although these percentages are lower than the people who are older than 30 and under 70 yr old, we have to say that are very large percentages.

In order to reduce the traffic accident caused by the elder drivers because of aging. Japanese police authority have promoted campaigns for a long time. That is calling the elderly driver to return their driving licenses. As shown in Fig. 4, the number of elderly drivers who returned their automobile driving licenses is increasing in the recent years. However, the number per year is still very small comparing to the number of elderly drivers, being counted up to only one fortieth of total elderly drivers. Consequently, we can conclude that the solutions in consideration of driving cars are also necessary. Therefore, the alternative policies and traffic safety countermeasure have been discussed from the viewpoint of safer private cars promotions too.

In order to support the drivers to drive cars safely, the ADAS (Advanced Driver-Assistance Systems) equipped vehicles have been promoted in Japan. Figure 5 shows the diffusion of three representative systems: Advance Emergency Braking System (AEBS), Adaptive Cruise Control (ACC), and Wrong Pedal Operation protection System. All these three systems may help the drivers during driving cars. Especially, the ADAS have been evaluated improving the traffic safety of the elderly drivers very much [7]. Thus, when we discussed the countermeasures with the Grand Design for Elderly's Traffic Safety Countermeasures in Aichi Prefecture, promoting the ADAS vehicles was taken into consideration.

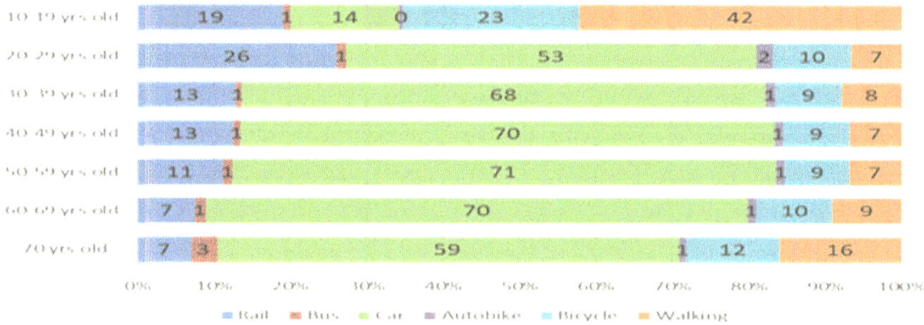


**Fig. 3** Travel modes by age group in Aichi (data: the 5th person trip survey in Chukyo metropolitan area [6])

## 3 Surveys to Hear from the Residents

In order to find feasible and effective solutions for realizing the safe and sustainable elderly mobility in Aichi, two surveys were conducted by targeting residents in Aichi. Survey A was conducted during mid of October through mid of November in 2018. The target people are elderly residents in Aichi. The survey sheets were distributed to the elderly people who came to the local police offices. 1978 answers had been collected in total. 177 were excluded as there had been some uncompleted answers. Finally, 1801 effective elderly responses were obtained. While Survey B is with 500 non-elderly participants. Survey B was implemented through Rakuten Insight (one of the most popular website survey companies in Japan) on October 15th and 16th, 2018. Both surveys let us summarize voice of residents on their mobility smoothly and understand response for the alternative policy and countermeasures.

According to Survey A, only 13% of elderly people do not have driving licenses in past. 1316 persons, 73% of 1801 elderly people, are still having driving licenses. Although 11% of 1316 answered that they planned to return their licenses within three years, 39% didn't plan to return their licenses. Again, we can know that solutions in consideration of driving cars are also necessary. Therefore, the alternative policies and countermeasure had been discussed from several viewpoints: 1. Public transport system improvement, 2. Pedestrian safety, 3. Safer private cars promotion, 4. Encouragement of returning driving licenses, and 5. New approaches for elderly's mobility.

At first, regarding public transport system improvement, five approaches were proposed: 1-A. taxi sharing, 1-B. making use of school bus or commuter bus of private companies, 1-C. shuttle service provided by the neighbourhood, 1-D. training of public transport system ride, and 1-E. information proving. All of them have not gotten more than 50% support ("very good" and "good"). Of them, regarding 1-C

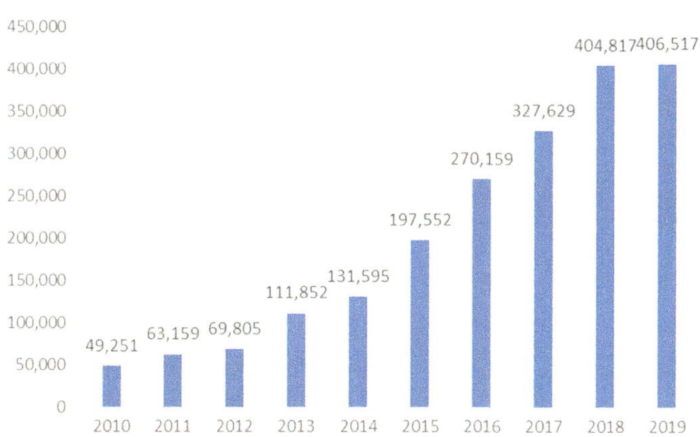


**Fig. 4** Number of driving licenses returned by elderly people (data: [3])

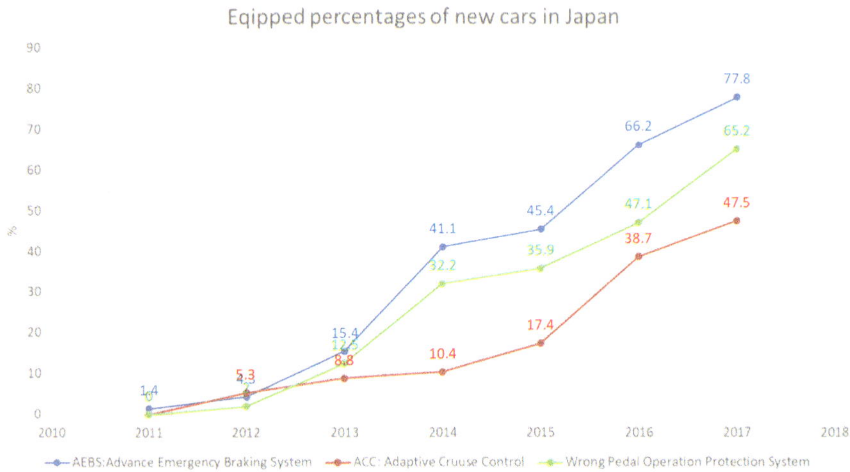


**Fig. 5** Diffusion of the ADAS systems in Japanese car market (data: [8])

and 1-D, all three age groups of elderly people show greater support percentages than the non-elderly people as given in Figs. 6 and 7.

Secondly, about pedestrian safety, in addition to the education when renewal of driving license which has been carried out long and long ago, four more approaches had been discussed: 2-A. stick to call on, 2-B. award for good behaviour, 2-C. education based on the survey and 2-D. campaign by the mass media. 2-A, 2-C and 2-D have been highly supported by the elderly people (higher than non-elderly and higher than 50%). Especially, as given in Fig. 8, more than 70% of "85 yr old and over" evaluated 2-C being "very good" and "good".

**Fig. 6** How do you think about shuttle service provided by the neighbourhood?

**Fig. 7** How do you think about training of public transport system ride?

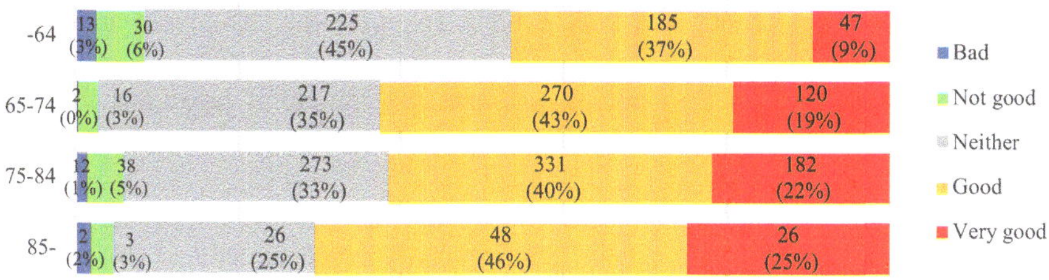


**Fig. 8** How do you think about pedestrian safety education based on the surveys?

Thirdly, on safer vehicles promotion, 3-A. pamphlet distribution, 3-B. experience workshop, 3-C. information providing when renewal of driving license, 3-D. subsidy to purchase an ADAS vehicle and 3-E. mandatory rule by law for specified elderly drivers were proposed as the countermeasures. Amongst them, 3-D needs financial budget by the government and 3-E needs a change of regulation.

As shown in Fig. 9, all age groups thought that subsidy are good as for 3-D. Especially, the 65–74 yr old group shows a highest percentage than the other age groups. By the way, Japanese central government has begun to promote the so-called SAPOKA vehicles which are ADAS (Advanced Driver-Assistance Systems) vehicles and defined with some specified functions being able to support the elderly drivers. In order to promote the SAPOKA vehicles, many local governments provide the subsidy for the elderly people when they purchase ADAS vehicles.

Regarding a new mandatory rule by law for specified elderly drivers to use the ADAS vehicles, the elderly people show lower percentages than that of the non-elderly people as shown in Fig. 10. This is a reasonable result. However, the important things are that the positive answers ("very good" and "good") are much higher than

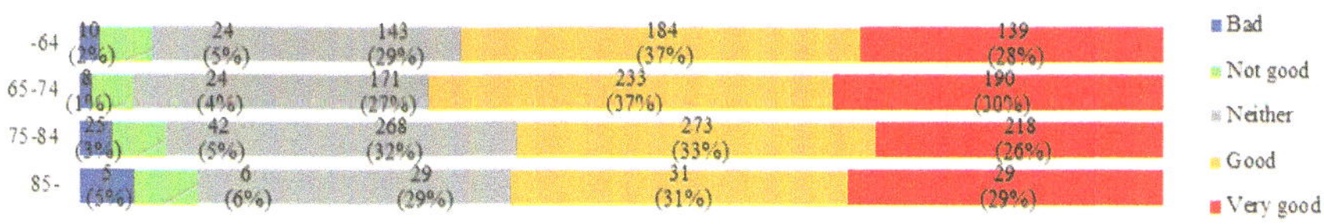


**Fig. 9** How do you think about subsidy to purchase ADAS vehicles?

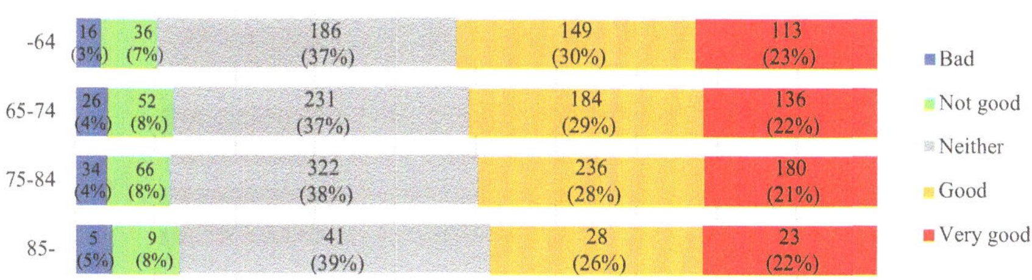


**Fig. 10** How do you think about mandatory rule by law for specified elderly drivers?

the negative answers ("bad" and "not good"). Even for the "85 yr old and over" group, the positive percentage is 48%, about 4 times of the negative 13%. Surely it was not based on this survey only, the central government of Japan has changed the law and will force the specified elderly drivers to drive the ADAS vehicles from 2022.

In order to encourage returning driver licenses, the following approaches have been discussed. Voice of the residents tell us that all are acceptable and more merits (4-B) are expected to be more effective.

- 4-A: publicity of the favourable treatment for elderly drivers to return licenses,
- 4-B: more merits for elderly drivers to return licenses,
- 4-C: publicity of the demerits for the elderly to drive cars,
- 4-D: more opportunities to test driving ability,
- 4-E: enlarging the good evaluation of returning licenses.

As the new methodology, some ideas on the basis of social psychology and behavioural economics have been discussed. Amongst the different approaches, the sign and marking on the road or roadside to call on safe drive obtained all people especially the elderly people supported as shown in Fig. 11. In addition, as given in Fig. 12, "joining traffic safety workshop with grandchildren" may nudge the elderly drivers more effectively.

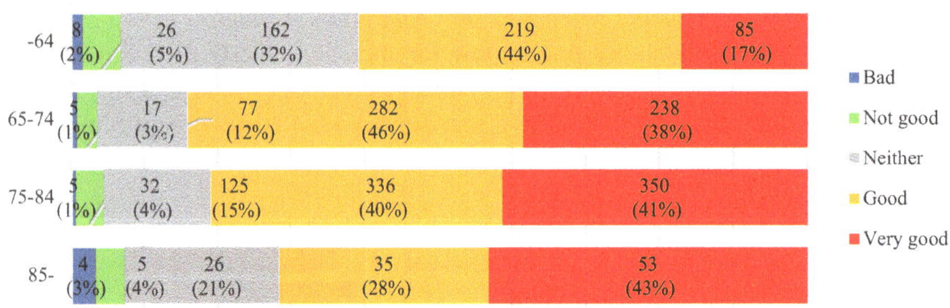


**Fig. 11** How do you think about using sign and marking to call on safe drive?

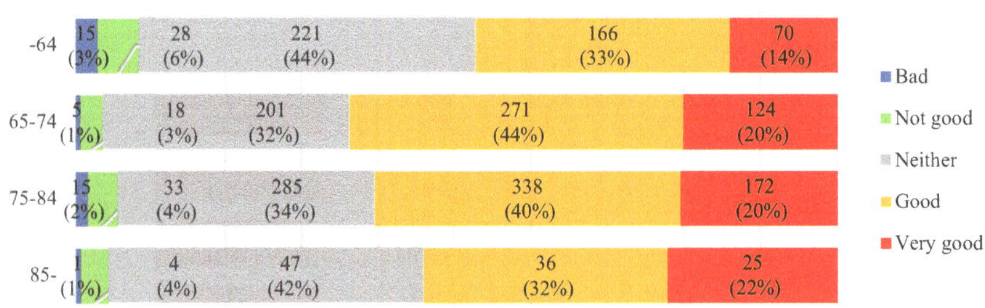


**Fig. 12** How do you think about joining traffic safety workshop with your grandchildren?

After summarizing the voice of the residents, the voice of enterprises was added. Finally, the integrated framework has been set up. There are four directions: A. to build a living environment where people may not need to drive cars; B. to build a living environment where people can safely drive cars; C. to build a living environment where people can safely and freely walk around; D. to build a living environment where people can safely and freely ride bicycles. In terms of the above four directions, seven domains have been finally decided as follows.

- To assure the mobility of elderly people being supported by public transport system and related services;
- To prepare a good environment for elderly drivers to return their driving license;
- To promote the ADAS cars for elderly drivers;
- To link CSR (corporate social responsibility) and CSV (creating shared values) of enterprises with traffic safety participation and activities;
- To foster awareness of car drivers for protection of pedestrians;
- To make use of new tools and methodology of behavioural science in traffic safety education;
- To do the publicity capably resonating with people.

## 4 Summary and Conclusion

Generally, the traffic safety countermeasures are based on the 3E (engineering, enforcement and education) in Japan. To propose and discuss the grand design for the mobility of elderly people in Aichi, the 3P (policy, people and participation) have been added into the consideration. The voice of residents including the elderly people and non-elderly people have been included into the policy making. Surely, the policies and countermeasures are not only based on the voice of the residents, but also discussed in terms of other data analysis and the opinions of all experts of the committee. The voice of the residents let us successfully get to seven domains from four directions. The actions made only one year in 2019 have let us exit from the worst position of the traffic fatalities ranking in Japan. The more active missions to be implemented by 2030 can be expected much more.

China and the Belt and Road related countries are not completely as same as Japan. Therefore, the voices of the residents may be different. However, the participation of people should be as important as that in Japan. This is because the traffic safety is the result of the people's behaviour. Hope this study may contribute to them in the future.

**Acknowledgements** The authors express our deep acknowledgements to Aichi Prefecture Police Headquarters, all members of the committee (chaired by Prof. Dr. Minoru KAMADA, present Emeritus Professor of Univ. of Tokyo, President of Japan Automobile Research Institute (JARI)) and all related organizations, companies and residents in Aichi for their kind supports during the study.

## References


1. Japan Institute for Labour Policy and Training: Data book of International Labour Statistics (2017). https://www.jil.go.jp/kokunai/statistics/databook/2017/documents/Databook2017.pdf. download on 18 Aug 2020
2. Institute for Traffic Accident Research and Data Analysis Homepage: https://www.itarda.or.jp/materials/traffic_accident/free. Last Accessed 03 Aug 2020
3. National Police Agency Homepage: https://www.npa.go.jp/publications/statistics/koutsuu/toukeihyo.html. Last Accessed 03 Aug 2020
4. Aichi Prefectural Police Headquarters: Report on Grand Design for Elderly's Traffic Safety Countermeasures in Aichi Prefecture, Nagoya (2019)
5. Aichi Prefectural Police Headquarters: References of Grand Design for Elderly's Traffic Safety Countermeasures in Aichi Prefecture, Nagoya (2019)
6. Council for Comprehensive Urban Transport Planning in Chukyo Metropolitan Area Homepage: https://www.cbr.mlit.go.jp/kikaku/chukyo-pt/index.html. Last Accessed 18 Aug 2020
7. Ando, R., Mimura, Y., Nishihori, Y.,Yang, J.: Effects of advanced driver assistance system for elderly's safe transportation. In: Proceedings of Smart Accessibility 2018, pp. 36–41. IARIA, Rome (2018)
8. Ministry of Land, Infrastructure, Transport and Tourism Homepage: https://www.mlit.go.jp/common/001213451.pdf#search=%27ADAS%E6%99%AE%E5%8F%8A%E7%8E%87%27. Last Accessed 18 Aug 2020

# Research on Determinants of Urban Residential Electricity Conservation Behaviors Based on Intervention Experiment -the Evidence from Household Survey of 4 megacities in China

**Chang Shu, Feng Xu, and Nan Xiang**

**Abstract** As a major composition of the total carbon emission in China, the carbon emissions from residential electricity consumption continue to grow. Residential electricity conservation has special significance for carbon emission reduction. Therefore, it is necessary to study the determinants of residential electricity conservation behaviors. Based on data from survey questions, this article analyzed the determinants of electricity saving behaviors by means of multivariate regression model, and assessed residential electricity conservation potential under four price intervention experiments. The results show that family income, geographic features show significant influences towards resident electricity conservation behaviors. The groups with higher income lack control over the electricity consumption. In this case, citizens in Beijing and Guiyang show higher awareness to save energy than those in Guangzhou and Hangzhou. Also, confronting electricity price increase, residents' energy-saving potential are higher along with price adjustments. When the electricity price rises by 0.05 CNY, 0.1 CNY, 0.2 CNY, and 0.5 CNY, residents are willing to reduce electricity consumption by 4.36, 8.33, 13.39 and 21.04%, respectively. Based on these findings, this paper indicates that it is effective to expand ladder price differences across different income groups which can better meet the overall demand for electricity.

**Keywords** Residential electricity behaviors · Intervention experiment · Electricity conservation potential

C. Shu · F. Xu
School of Economics and Management, Beijing University of Chemical Technology, Beijing 100029, China

N. Xiang (✉)
College of Economics and Management, Beijing University of Technology, Beijing 100124, China
e-mail: xiangnan@bjut.edu.cn



F. K. S. Chan et al. (eds.), *Proceedings of the 2020 International Conference on Resource Sustainability: Sustainable Urbanisation in the BRI Era (icRS Urbanisation 2020)*, Environmental Science and Engineering,
https://doi.org/10.1007/978-981-15-9605-6_15

## 1 Introduction

Being the top $CO_2$ emitter with huge economy amounts and high growth rates [5], China took up 28.66% of global $CO_2$ emissions in 2018 [6]. Residential sector is an important composition of the total carbon emission in China, and residential electricity demand continues to grow faster compared with other kinds of energy [18, 20]. According to Chinese energy department, residential consumption comprises 13.99% of the total electricity demand in China in 2017 [11]. During the period of 2000 to 2017, residential electricity use increased at an annual rate of 11%, which is around 6.25 times increase (907.1 billion kWh in 2017). However, both per capita residential electricity consumption amount and ratio will still continue to grow with income growth and urbanization process speeding up. China has established a plan for urbanization, with the urbanization rate will increase to 70% in 2020 from 56% in 2015, which means 2 billion people will move to urban area by 2020; this trend implicates that urban residential electricity control will be much more important in residential energy conservation and low carbon society development [10]. Therefore, it is necessary to prompt household energy conservation behavior in China to control residential carbon emission.

As one of the important influencing factors, there are some problems in electricity price at present in China. Liu and Lin [8] points out that with the rapid increase in residential electricity consumption, retail residential electricity price remained low. What's more, insufficient reflection in regional differences is another major problem [14]. Therefore, more relevant factors need to be taken into account in the formulation of electricity price, and there is a large room for improvement.

There are two main objectives of this paper. First, determinates of residential electricity consumption and conservation behavior will be estimated. This will be done by constructing an electricity demand model based on geography, prices, income, dwelling size, and energy-saving awareness. These determinates have evolved considerably for Chinese urban area, and could possibly effect electricity demand. This should be examined by variance analysis method and multivariate regression model. Secondly, we discuss the implications of these determinates on their influence on electricity demand, and analyze residential electricity conservation potential based on electricity price intervention experiments results. This paper provides answers to the above important questions that are valuable to the further energy pricing reform and residential energy conservation attempt in China. Also, this paper is based on micro household-level survey data in four megacities of China, that are more informative and time-efficient.

## 2 Literature Review

Existing literature generally recognizes the considerable significance of residential energy-saving behavior to the reduction of total energy consumption and the improvement of atmospheric environment [15]. Researchers concluded various determinates of residential energy consumption [4, 9, 16], such as economy, policy, administration, environmental awareness, living habit and information disclosure [3, 7, 13, 17, 19, 21]. This paper summarized these attributes impacting on residential electricity consumption intro three categories, electricity price, consumer factors and external factors, as shown in Fig. 1. These three categories of influence factors determine the quantity of commodity consumption. Research about residents' household electricity consumption and energy consumption emphasized more on the consumer attributes to do in-depth analysis, including residents' income, geographic features and environmental consciousness [1, 9]. It also can be seen that using economic measures, including the key measure for changing residents' propensity to save energy, which is proposed from the angle of commodity's price attribute and consumer's income attribute, has greater influence on residents' energy-saving behavior. The energy-conservation subsidy aimed at low-income group probably has better effect [12].

However, in the existing literature, research on electricity price analysis on energy saving behavior is really limited in this field. The analysis of economic measures in current studies is limited to qualitative analysis and lays special emphasis on the consumer's income attribute without conducting in-depth analysis of the policy effect of price attribute of the electricity, as a commodity, on changing residents' energy-saving behavior. And the studies do not conduct quantitative analysis of the influence of specific economic measures. As a result, they cannot provide foundation for measuring and calculating the conservation potential of electricity consumption under the circumstance that the price is adjusted and also have difficulty in providing specific suggestions about changing residents' propensity to save electricity.

## 3 Questionnaires and Data

### *3.1 Survey Samples*

The survey questionnaires comprised about 50 questions covering basic information on the profile of respondents as well as information on their knowledge, awareness, stated preferences and behaviors with respect to energy use and energy saving behaviors. The respondents were selected by using a combination of convenience sampling and judgment sampling methods at four megacities, from locations namely approaching the respondents in selected shopping centers, in residential areas, and in suburb and rural areas.

Considering the impact factors on residential energy behaviors, we selected Beijing, Guangzhou, Hangzhou, and Guiyang of China to conduct survey and intervention study. These four cities are significant different by regional economic level, city position and geological characteristics; can demonstrate different types of city in China. For instance, Beijing is the capital of China, which citizens' environment protection conscious is relatively high; Guangzhou is economic center of southern China with high economic level and hot weather; Hangzhou is in the central east of China, which has relatively high economic development and active low carbon society construction practices; Guiyang is in the west of China, in which the economic development is the slowest among the four cities.

Convenience sampling from four typical megacities of China should generate a relatively representative sample. Face-to-face delivery of structured questionnaires was used for the research. Therefore, it was planned to distribute 640 questionnaires (160 in each city) through face to face interview, and 635 questionnaires are effectively received. Some sample information is shown in the Fig. 2.

**Fig. 1.** Location of four megacities in China

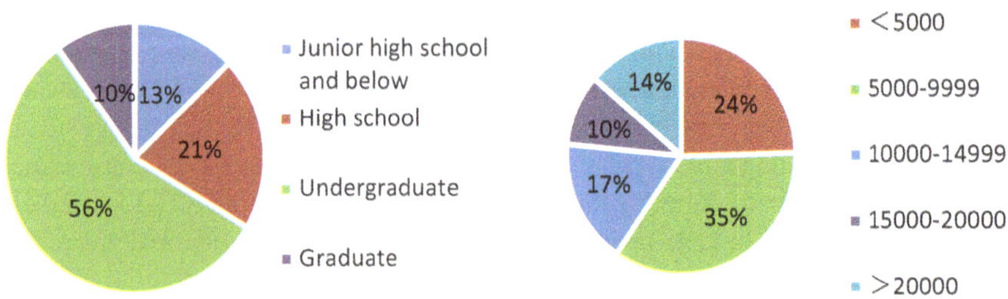


**Fig. 2.** Distribution of education and income levels of respondents.

## 3.2 Description of Questionnaires

The design of questionnaires is based on the pre-survey in Beijing city. Survey questions are divided into three categories: the status quo of household electricity consumption, resident attributes and impacts on residential electricity conservation, and residents' reaction towards electricity price changes. The main questions and their statistical description are reported below.

## 3.3 Survey of Statistics

(1) *the status quo of residents' electricity consumption in different regions.* According to face to face survey, residents' electricity consumption varied by region and (see Fig. 3). The ratio of person who don't aware about their electricity consumption are high in Beijing and Guiyang (northern China and western China, distinctly), while the electricity consumption amount below level 3 is about 85%.

(2) *the status quo of electricity consumption of different income groups.* For different income group, resident's energy consumption disparity is obvious: with income growth, people's awareness on electricity consumption declines (8-20%, see Fig. 4), and the amount of electricity increases. Richer residents are less sensitive to electricity cost and intend to use more electricity compared with middle-and lower-income people.

(3) *residents' energy saving awareness in different groups.* Residents' energy saving awareness is differed in four regions: citizens in Guiyang and Beijing have relatively high awareness to save energy in daily life. Guiyang have the lowest

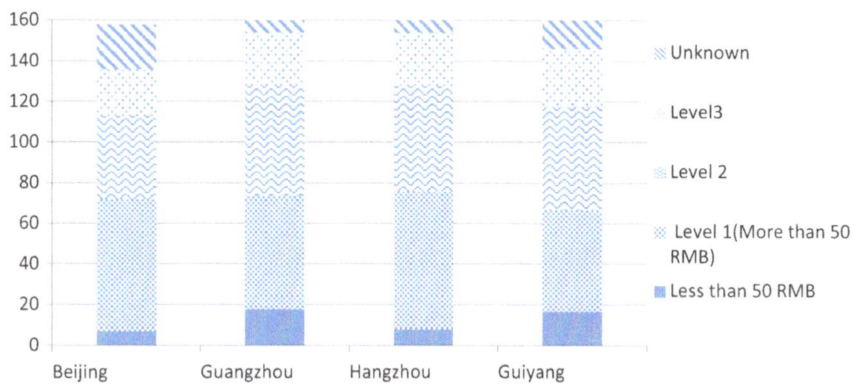


**Fig. 3.** Residents' electricity consumption amount situation in disparity groups

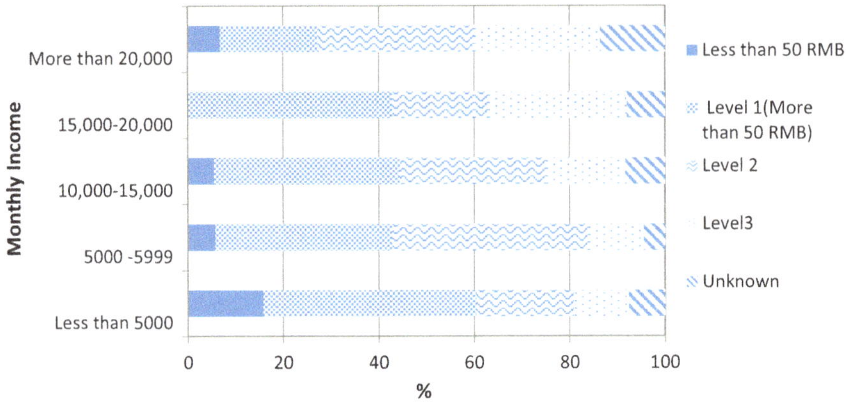


**Fig. 4.** Residents' electricity consumption amount situation in disparity groups

economic development situation, however the residents who have the highest conscious to conduct energy saving behavior, partially because their high need to save consumption. As for Beijing, with a high energy saving awareness city, low carbon education and promotion contributed for this phenomenon. While that in Guangzhou and Hangzhou is less, for their high economic development level and relatively high personal income (Table 1).

Residents' awareness of energy saving has significant variance between different electricity consumption groups. With higher electricity consumption amount utilization, residents' awareness of energy saving is lower. Therefore, it is necessary to improve households conscious on energy saving and low carbon society participation.

**Table 1** Awareness of energy saving and low carbon participation among different groups

| | | Awareness of energy saving | | | | | Total |
|---|---|---|---|---|---|---|---|
| | | High | Regular | Occasionally | Never | Disinterest | |
| Region | Beijing | 99 | 39 | 14 | 3 | 2 | 157 |
| | Guangzhou | 86 | 46 | 18 | 8 | 1 | 159 |
| | Hangzhou | 80 | 56 | 20 | 2 | 0 | 159 |
| | Guiyang | 141 | 14 | 4 | 1 | 0 | 160 |
| Total | | 406 | 155 | 57 | 14 | 3 | 635 |

**Table 2** Variance of residents' attribute

| | F | | | |
|---|---|---|---|---|
| | ΔP = 0.05 | ΔP = 0.1 | ΔP = 0.2 | ΔP = 0. 5 |
| Family income | 0.1429 | 0.0420 | 0.5831 | 0.720 |
| Geographic features | 0.007 | 0.0001 | 0.0002 | 0.0005 |
| Age | 0.0149 | 0.0452 | 0.0044 | 0.0172 |
| Environmental awareness | 0.1126 | 0.0438 | 0.0703 | 0.0056 |
| Education | 0.8219 | 0.3091 | 0.3138 | 0.6336 |
| Dwelling size | 0.1885 | 0.5477 | 0.5106 | 0.6728 |
| Total | 0 | 0 | 0 | 0 |
| R-squared | 0.0991 | 0.1265 | 0.1282 | 0.1131 |

Sample number: 635.

# 4 Empirical Results

## *4.1 Residential Electricity Determinates Analysis*

In order to analyze major determinates of residents' electricity consumption, and further to identify impact factors for household energy saving potential under different electricity price intervention experiments, this paper utilized variance analysis and regression models.

### 4.1.1 Analysis of Variance

Analysis of variance of residents attributes is utilized to demonstrate the correlation of consumers attributes with electricity price reaction (Table 2).

### 4.1.2 Electricity Consumption Changes and Its Determinants Analysis

Four regression models are adopted to demonstrate major determinants of residents' electricity conservation behavior under four price intervention experiments (Table 3).

$$\Delta Qi = \alpha_{i1}^{*}\text{income} + \alpha_{i2}^{*}\text{geographic} + \alpha_{i3}^{*}\text{aware} + \alpha_{i4}^{*}\text{age} + \alpha_{i5}^{*}\text{edu} + \alpha_{1i6}^{*}\text{dwellingsize} + \varepsilon$$

*t* statistics with "* $p < 0.10$ ** $p < 0.05$ *** $p < 0.01$".

**Table 3** Model estimation results

| Variables | ΔP = 0.01 | | ΔP = 0.1 | | ΔP = 0.2 | | ΔP = 0.5 | |
|---|---|---|---|---|---|---|---|---|
| *Income* | *-0.0588* | *** | *0.1297* | *** | *-0.1411* | *** | *-0.1565* | *** |
| *Geographic features* | *0.0169* | | *0.1213* | ** | *0.1639* | *** | *0.1740* | *** |
| Age | -0.0622 | ** | -0.0860 | *** | -0.1448 | *** | -0.1843 | *** |
| Environmental awareness | 0.1414 | | 0.1834 | | 0.1586 | ** | 0.1887 | ** |
| Education | -0.0486 | | 0.0159 | | 0.0434 | | 0.0163 | |
| Dwelling size | -0.0006 | | -0.0004 | | -0.0259 | | -0.0394 | |
| Constant | 1.1285 | *** | 1.4781 | *** | 2.0460 | *** | 3.1160 | *** |
| **$R^2$** | 0.1211 | | 0.1466 | | 0.1521 | | 0.1385 | |
| **Sample number** | 635 | | 635 | | 635 | | 635 | |

Estimation results proves that family income, geographic features have significant influences towards resident electricity conservation behaviors. Furthermore, family income's negative impacts on electricity conservation are strengthening along with electricity price increase. Education and environmental awareness are proved to be the positive factors on encouraging energy saving.

## 4.2 *Residents' Potential to Save Electricity*

Confronting electricity price increase, residents' energy saving potential are higher along with price adjustments. When price increased by 0.05CNY, residents in Guiyang are willing to reduce electricity consumption by 6.94%, compared with 3.08% in Beijing. However, when the price increase 0.5CNY, energy saving potential in Guiyang is 24.88%, and Beijing is 15.90%. From above analysis, we can calculate that when the electricity price rises by 0.05 CNY, 0.1 CNY, 0.2 CNY, and 0.5 CNY, residents are willing to reduce electricity consumption by 4.36, 8.33, 13.39 and 21.04%, respectively (Fig. 5).

# 5 Policy Implications

Based on intervention studies, this paper adopted and improved price intervention approach to explore residential electricity conservation determinates and analyze conservation potential. Family income, geographic features show significant influences towards resident electricity conservation behaviors. Economic measures are proved to have the greatest influence on middle-income residents. Raising the

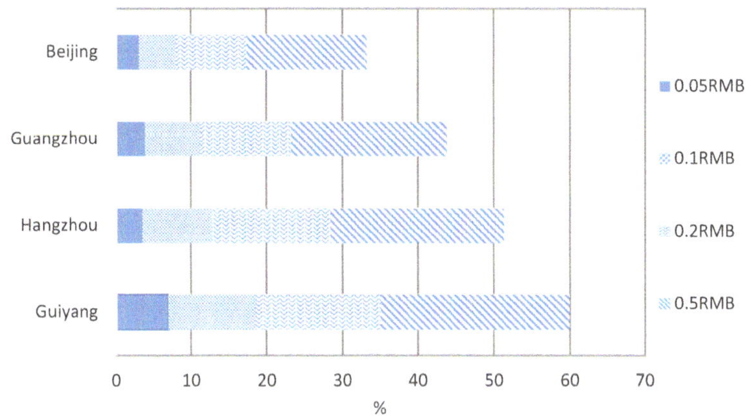


**Fig. 5.** Residents' electricity conservation potential under price interventions

electricity price has significant influence on the reduction in residential electricity consumption, particularly for the middle- and low-income groups. Meanwhile, based on empirical data, the adjustment of electricity prices in our cases will reduce household electricity consumption by 4.36–21.04 on average.

The research above comes up with suggestions about changing the preferences of residents in China's different regions and different income levels to save energy, policy suggestions are thus proposed. The adjustment of electricity price should be based on the regional differences and income level differences. In terms of the city differences, Guiyang is significantly different from other cities. Whereas for the differences in income level, with the increase in incomes, the residents' control over electricity consumption is digressive, those who have the lowest level of electricity consumption have significant depressive control as their incomes increase. In terms of income level differences, the increase in income makes the residents less sensitive to electricity consumption and their electricity-saving awareness constantly decrease as well. When the electricity price is changed, the bigger the increase in electricity price is, the more effective changing the propensity of high-income groups to save electricity can be. Therefore, multistep electricity price with bigger difference can be adopted to change high-income groups propensity to consume electricity and meanwhile protect economic benefits of low-income ones.

The conclusions above have rich policy implications. Economic intervention and social awareness promotion should be focused together to improve residential electrical consumption in long term urbanization process. As for further work, there is still much to be done. For instance, some further research can be done to estimate reasonable electricity price adjustment plan for different regions to reduce electricity consumption and improve electricity efficiency.

## References


1. Borgstede, C., Andersson, M., Johnsson, F.: Public attitudes to climate change and carbon mitigation-Implications for energy-associated behaviours. Energy Policy **57**, 182–193 (2013)
2. Craig, C.A., Feng, S.: Exploring utility organization electricity generation residential electricity consumption and energy efficiency: a climatic approach. Appl. Energy **185**((Part 1)), 779–790 (2017)
3. Ek, K., Soderholm, P.: The devil is in the details: Household electricity saving behavior and the role of information. Energy Policy **38**(3), 1578–1587 (2010)
4. Faruqui, A., Sergici, S., Sharif, A.: The impact of informational feedback on energy consumption-a survey of the experimental evidence. Energy **35**(4), 1598–1608 (2010)
5. International Energy Agency (IEA): $CO_2$ Emissions from Fuel Combustion 2016. OECD Publishing, Paris. (2016). https://doi.org/10.1787/co2_fuel-2016-en,lastaccessed2020/8/30
6. International Energy Agency (IEA): Global Energy and $CO_2$ Status Report 2019, IEA, Paris (2019). https://www.iea.org/reports/global-energy-and-co2-status-report-2019. Last Accessed 30 Aug 2020
7. Kavousian, A., Rajagopal, R., Fischer, M.: Determinants of residential electricity consumption: Using smart meter data to examine the effect of climate, building characteristics appliance stock and occupants' behavior. Energy **55**(15), 184–194 (2013)
8. Liu, C., Lin, B.Q.: Is increasing-block electricity pricing effectively carried out in China? a case study in Shanghai and Shenzhen. Energy Policy **138**, 111278 (2020)
9. Ma, G., Andrews, S.P., Zhang, J.D.: Chinese consumer attitudes towards energy saving: The case of household electrical appliances in Chongqing. Energy Policy **56**, 591–602 (2013)
10. National Bureau of Statistics (NBS): China Statistical Yearbook. Chinese Statistics Press, Beijing (2015). https://www.stats.gov.cn/tjsj/ndsj/2015/indexch.htm. Last Accessed 30 Aug 2020
11. National Bureau of Statistics (NBS). China Energy Statistical Yearbook. Chinese Statistics Press, Beijing (2018). https://www.stats.gov.cn/tjsj/ndsj/2018/indexch.htm. Last Accessed 30 Aug 2020
12. Niu, S.W., Jia, Y.Q., Ye, L.Q., Dai, R.Q., Li, N.: Does electricity consumption improve residential living status in less developed regions? an empirical analysis using the quantile regression approach. Energy **95**, 550–560 (2016)
13. Wang, B., Wang, X.M., Guo, D.X., Zhang, B., Wang, Z.H.: Analysis of factors influencing residents' habitual energy-saving behaviour based on NAM and TPB models: Egoism or altruism? Energy Policy **116**, 68–77 (2018)
14. Wang, C., Zhou, K., Yang, S.L.: A review of residential tiered electricity pricing in China. Renew. Sustain. Energy Rev. **79**, 533–543 (2017)
15. Wang, Z., Zhao, D.T., Yu, W.T.: Research on the driving forces of the increase in inbuilt emission of Carbon in the consumption of Chinese residents. Forum Sci. Technol. China **7**, 56–62 (2012)
16. Wang, Z.H., Zhang, B., Yin, J.H., Zhang, Y.X.: Determinants and policy implications for household electricity-saving behavior: evidence from Beijing China. Energy Policy **39**(6), 3550–3557 (2011)
17. Zhang, C.Y., Yu, B.Y., Wang, J.W., Wei, Y.M.: Impact factors of household energy-saving behavior: an empirical study of Shandong province in China. J. Clean. Prod. **185**, 285–298 (2018)
18. Zhang, Y.J., Peng, H.R.: Exploring the direct rebound effect of residential electricity consumption: an empirical study in China. Appl. Energy **196**, 132–141 (2017)
19. Zheng, X.Y., Wei, C., Qin, P., Guo, J., Yu, Y.H., Song. F., Chen, Z.M: Characteristics of residential energy consumption in China: findings from a household survey. Energy Policy 75, 126–135 (2014)

20. Zhou, S.J., Teng, F.: Estimation of urban residential electricity demand in China using household survey data. Energy Policy **61**, 394–402 (2013)
21. Zhu, Z., Shen, Y.Q., Huang, M.: Empirical analysis of residential low-carbon consuming behavior and driving forces of carbon emissions-investigation based on Hangzhou. Resour. Dev. Mark. **27**(09), 831–834 (2011)

# Comprehensive Model for Simulation of Economic Transformation and Air Pollutant Control for Steel Cities

**Fushang Cui, Feng Xu, and Nan Xiang**

**Abstract** The steel-manufacturing intensive cities (steel cities) in China are facing a difficult position of economic transformation and air pollutant control. In this case, taking Tangshan as the research object, this study intends to simulate the developing trends of steel cities and verify the effect of economic transformation by 2020 to explore the development rule of steel cities. The theories and methods of system dynamics, input-output theory and econometrics were adopted to establish a comprehensive evaluation model based on the detailed analysis of economy-energy-environment (3Es) coupling mechanism in Tangshan. The multi-objective dynamic simulation from 2016 to 2020 was carried out by the comprehensive evaluation model. The simulation results show the gross production of Tangshan would get slow in the period of prediction. Compared with the data in 2016, the gross production increases only 11% in 2020. It even shows a decline of 1.4% in 2019, which conforms to the reality. The steel industry decreased 34.6% during this period, showing that its development was heavily impacted with the target of both economic transformation and air quality improvement. The simulation results are consistent with the real data. The accuracy of the comprehensive evaluation model and reliability of the prediction can be confirmed. It is difficult for Tangshan to achieve the dual goals of economic development and air pollution reduction by 2020 if it maintains the current level of economic structure and pollution control level. Adjustment of industrial structure would be the key way for a more reasonable development.

**Keywords** Steel cities · 3Es model · Comprehensive evaluation model · Dynamic simulation

F. Cui · F. Xu
School of Economics and Management, Beijing University of Chemical Technology, Beijing, China

N. Xiang (✉)
College of Economics and Management, Beijing University of Technology, Beijing, China
e-mail: xiangnan@bjut.edu.cn



F. K. S. Chan et al. (eds.), *Proceedings of the 2020 International Conference on Resource Sustainability: Sustainable Urbanisation in the BRI Era (icRS Urbanisation 2020)*, Environmental Science and Engineering,
https://doi.org/10.1007/978-981-15-9605-6_16

# 1 Introduction

Steel-manufacturing industry has been a pillar industry in China for long time. However, due to the increasingly prominent environmental problems in recent years, the development of the steel industry is facing dilemma. As the largest steel-manufacturing intensive city (in short, steel city), Tangshan has also faced more and more serious environmental problem. Among the 74 cities monitored under the new standards in China, Tangshan's comprehensive air quality index was 8.27, ranking only 70th. It is the industrial structure dominated by heavy industry and the energy structure dominated by fossil energy that causes Tangshan city facing increasingly serious atmospheric environmental problems. The economic growth of Tangshan slowed down since 2011, and the economic growth rate in 2014 was only 5.1%, lower than the national average, indicating that environmental problems have become the bottleneck restricting economic development.

Facing severe economic and environmental pressure, the government of Tangshan proposed in the 13th five-year plan for economy and environment that it should strive to solve economic and environmental conflicts and strive to achieve the dual goals of economic growth and environmental improvement. However, whether a series of policies issued by the central and local governments have any effect on economic growth, industrial structure adjustment and atmospheric environment improvement, whether the dual goals of economic growth and environmental improvement can be taken into account, and whether sustainable development of Tangshan can become a reality need to be further analyzed and confirmed.

At present, improving the atmospheric environment has become a focus of scholars. In the process of environmental pollution from generation to concentrated outbreak, Some scholars [1, 2] started from the policies of several countries to analyze their successful experience in improving the atmospheric environment, so as to provide reference for the governance of atmospheric environment in China. In the context of China's endless air pollution prevention and control policies, some scholars compare the advantages and disadvantages of various environmental improvement policies horizontally and compare the improvement effects vertically, so as to study the environmental policies that are most suitable for China's current situation and have the best effect [3]. From about the document of atmospheric pollution control, it is shown that there are some scholars adopt all kinds of research methods to analyze future development direction under different environmental policies, so as to show how to refer to the appropriate methods for different questions to the proper solution [4–7].

The studies above focusing on qualitative analysis and lacking of empirical evidence are difficult to provide a well-founded basis. For China's specific environmental problems, specific analysis and innovative research methods are needed to obtain more detailed research data, so as to provide targeted countermeasures for China's environmental problems. Moreover, the quantitative analysis method can not only explain the severity of environmental pollution in detail, but also show the solutions of different policies and the actual impacts on the environment and economy. Therefore, more scholars adopt quantitative research methods at present.

Some scholars focus on the impact of economic growth [8], industrial structure [9, 10] and technological improvement [11] on environmental pollution problems. Some scholars also explore new research methods to evaluate current economic and environmental policies, or put forward corresponding improvement measures and Suggestions, or evaluate environmental and economic goals to formulate more reasonable economic and environmental policies. For example, Li [12] established the performance evaluation system of atmospheric environment governance based on the pressure-state-response model based on the environmental quality evaluation and atmospheric environment pressure load evaluation methods. Bollen et al. [13] used a multi-sectorial, multi-regional and international CGE model to analyze the relationship between air pollution and climate policy in the European Union. Zhang et al. [14] integrated the American option method and the two-factor learning curve model, established an evaluation model including non-renewable energy cost, carbon price, renewable energy cost and price subsidy, to evaluate the unit determination value of renewable energy development and calculate the interest balance of both sides. Zheng et al. [15] used the GAINS-China model to evaluate the effects of current policies on reducing atmospheric pollutants and improving climate in the industrial sector between 2005 and 2030. Amann et al. [16] used the GAINS model to study cost-effective policies for air pollution control and greenhouse gas reduction. Yang et al. [17] used the logarithmic production function model to study the win-win goals of China's economic development and environmental protection.

Based on the analysis of related literatures, it is found that the qualitative analysis method is not conducive to the solution of problems. At present, few scholars can study the current economic and environmental problems by comprehensively analyzing the current economic and environmental situation for steel cities. In this study, the three systems of economy, environment and energy were integrated to build a comprehensive policy evaluation model to solve the problems of economic transformation and air pollutant control for Tangshan.

# 2 Introduction

## *2.1 Model Framework*

Based on the principle of system dynamics, input-output law, this study constructed a 3Es model to link energy consumption, economic production and atmospheric environment through human activities. The input-output model is the basis for the economic model. It combines two sub-models (energy model and environment model) together through industrial production. Inter-industry development

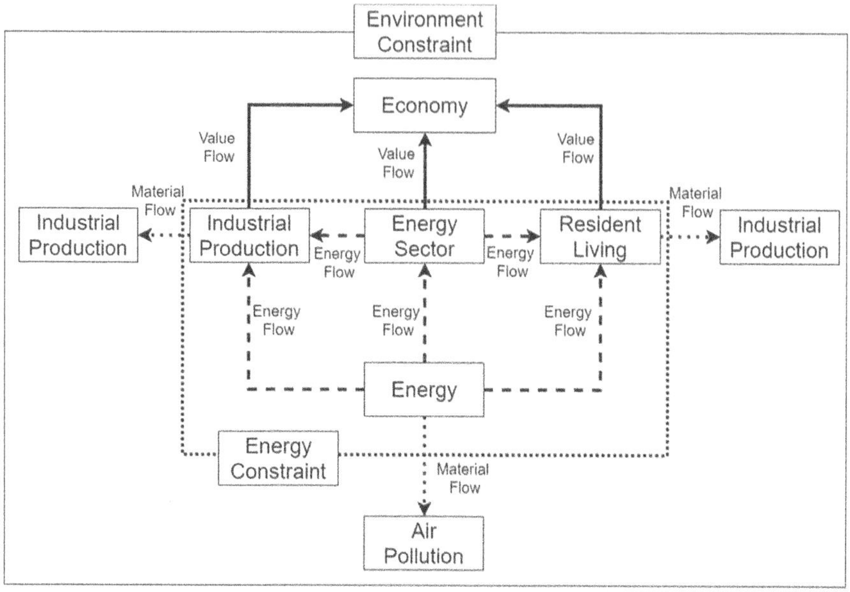


**Fig. 1** The model frame diagram

constraints are adopted as the basis of industrial and economic development. Furthermore, the sub-models shall reflect the material and energy flows within or between models. The model frame diagram is shown in Fig. 1.

## 2.2 *Objective and Constrain Function*

In the context of intensified economic and environmental conflicts, economic slowdown and pressure on environmental improvement, not only economy shall be developed, but also the emission of air pollutants and environmental conditions shall be focused. So the dual goal to be set is

$$\mathrm{MAX} \sum \frac{1}{(1-\rho)^{t-1}} GRP(t) \tag{1}$$

$$MIN\,TP(j,t) \tag{2}$$

$$(j = 1 : NO_X;\ j = 2 : SO_2)$$

where GRP(t) is gross regional production in year t (en), ρ is the social discount rate (0.05, average value of interest rates of loans from 2006 to 2015 in China; ex), (endogenous variable, referred to as "endogenous"). Here, "en" is the abbreviation for endogenous and "ex" is the abbreviation for exogenous. TP(j, t) is the total amount of j pollutants discharged in year t (en), j represents the atmospheric pollutants observed in this study.

In social-economy sub-model, the GRP is mainly contributed by industry, and the added value of each industry is equal to the product of its total output value multiplied by added value rate. In this study, the gross regional product was calculated based on 11 industrial sectors in Tangshan.

$$GRP(t) = \sum_{i=1}^{11} grp(i, t) \tag{3}$$

$$grp(i, t) = IVA(i) \times xn(i, t) \tag{4}$$

where grp(i, t) is the added value of i industry in year t (en), xn(i, t) is the output value in year t (en), IVA(i) is the added value rate of i industry (ex). Here, i represent the 11 major industries in Tangshan in this study. (i = 1: Agriculture, forestry, husbandry, fishery and their service industries; i = 2: Metal smelting and rolling industry; i = 3: Mining; i = 4: Equipment manufacturing; i = 5 Chemical industry; i = 6: Non-metallic mineral products; i = 7: Production and supply of electricity, heat, gas and water; i = 8: Other manufacturing; i = 9: Construction; i = 10: Transportation, warehousing and postal services; i = 11: Services).

According to the input-output theory and the law of economic operation, the production and consumption of social economy must meet the balance of input-output, namely:

$$xn(i, t) \geq A \times xn(i, t) + tc(i, t) + gcf(i, t) + ex(i, t) - im(i, t) \tag{5}$$

where xn(i,t) is the total output value of i industry in Tangshan in t year (en), A is the input coefficient (ex), tc(i,t) is the total consumption in t year (en), gcf(i, t) is the gross capital formation of i industry in t year (en), ex(i,t) is the total exports of i industry in t year (en), im(i,t) is the total imports of i industry in t year (en).

In environment sub-model, the production of air pollutants is mainly caused by industrial production and residents living. This study calculated the total pollutant emission from industrial production and living sector by:

$$tp(j, t) = \sum_{i=1}^{11} ef(i, j) \times xn(i, t) + R(t) \times ref(j) \tag{6}$$

where ef (i, j) is Pollutant emission coefficient of i industry (ex), ref(j) is the resident emission factor of j pollutant (ex), R(t) is The number of permanent residents in year t (en).

The average population growth rate is calculated based on the permanent population in 2016 and the natural population growth rate of Tangshan in the past 5 years:

$$R(t) = (1 + ez) \times R(t - 1) \tag{7}$$

## 2.3 Parameter and Scenario Setting

In the social-economic sub-model, the development of leading industries in Tangshan, the industrial policies in the planning period, the correlation of various economic industries and the feasibility of research are comprehensively considered, and all the economic industries are integrated into 11 economic sectors, and the input and output table is compiled based on the input and output theory. The pollutant emission data of various industries mainly come from the survey statistics and Tangshan statistical yearbook. Based on the total $NO_X$ and $SO_2$ emissions of each industry in 2016 divided by the total output value of the year, the pollutant emission coefficients of each industry are calculated and sorted out.

On the basis of these data above, this study established the Comprehensive Evaluation Model of Air Pollution Prevention and Control Policy.

Accordingly, two scenarios were set up in this study to simulate the economic development trend and environmental improvement situation of Tangshan under different paths, for the in-depth exploration of the above problems (Table 1).

**Table 1** Scenario setting

| Setting | Scenario | |
|---|---|---|
| | Scenario-1 Business as usual scenario | Scenario-2 Dual goal realization scenario |
| Industrial structure | Production adjust in range of −3 ~ 8% | Free adjustment |
| Economic development | Annual GRP growth rate less than 5% | No limitation |
| Pollutant emissions | No limitation | ≤141274.38 ton NOX ≤ 126179.04 ton SO2 |
| Purpose | Sensitive analysis | Development direction simulation |

# 3 Dynamic Simulation Results Analysis

## 3.1 Scenario-1 Analysis and Sensitive Analysis

In this scenario, the current economic development level, industrial structure and pollutant treatment level are kept as constraints, and the maximum economic aggregate is taken as the objective function to conduct dynamic simulation.

The simulation results are shown in Figs. 2 and 3. Under the condition of maintaining the existing industrial structure and pollutant control means, the economy of Tangshan will grow very slow at first and turn to decline since 2019. Comparison shows that the simulation result shows good degree of fitting within 3% with the real data, so the accuracy of comprehensive model and reliability of the prediction value of scenario simulation can be confirmed.

By observing the simulation results of pollutant discharge, it can be seen that the two pollutants do not show a downward trend in the forecast period. By 2020, emissions of $NO_X$ and $SO_2$ would be 192,322 tons and 170,346 tons respectively, failing to meet the emission reduction target of pollutants. This means that under the current situation, the dual goals of will not been achieved.

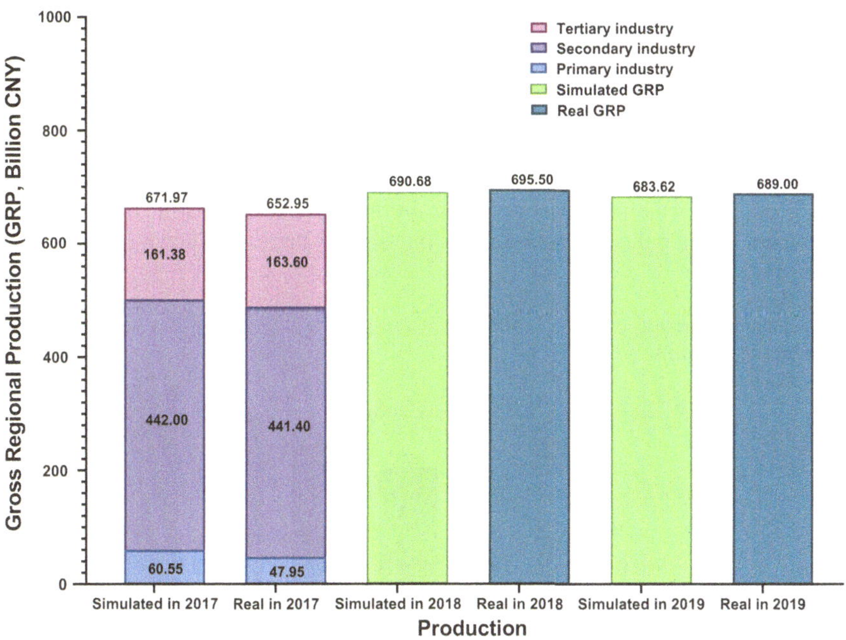


**Fig. 2** Simulation results of economic development under scenario-1 and contrast with real GRP

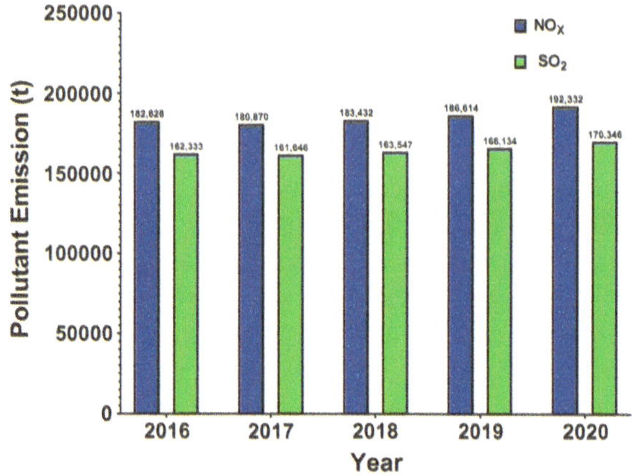


**Fig. 3** Simulation results of emission of air pollutants under scenario-1

## 3.2 *Scenario-2 Analysis and Development Simulation*

In this scenario, economic development and environmental improvement are set the dual objectives, and economic maximization is sought.

The simulation results show that, the GRP in 2020 would be 932.8 billion CNY, reaching the planning target of 900 billion CNY. Total $NO_X$ and $SO_2$ emissions would be 131,068 tons and 126,179 tons respectively by 2020, both meeting the planning target as shown in Figs. 4 and 5. Therefore, under the development path of scenario-2, the goals of economic development and environmental improvement would be achieved.

However, in this scenario, the realization of the dual goals is achieved through the adjustment of the industrial structure. In this context, the result shows that the adjustment of the industrial structure of Tangshan is large. As shown in Fig. 6, the

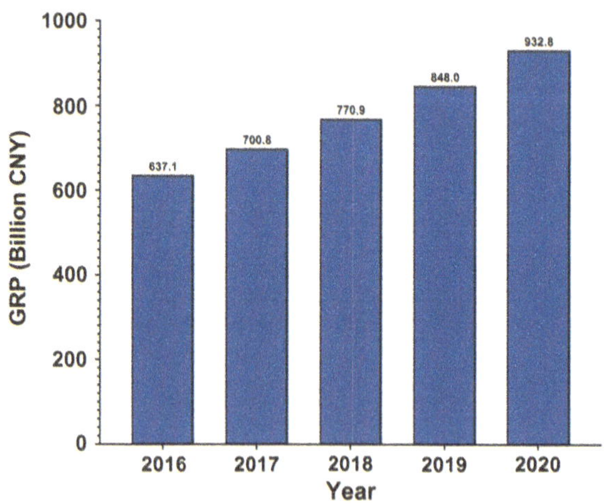


**Fig. 4** Simulation results of economic development under scenario-2

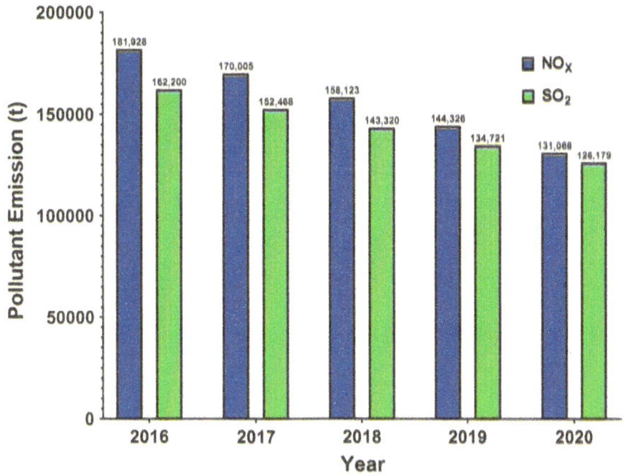


**Fig. 5** Simulation results of emission of air pollutants under scenario-2

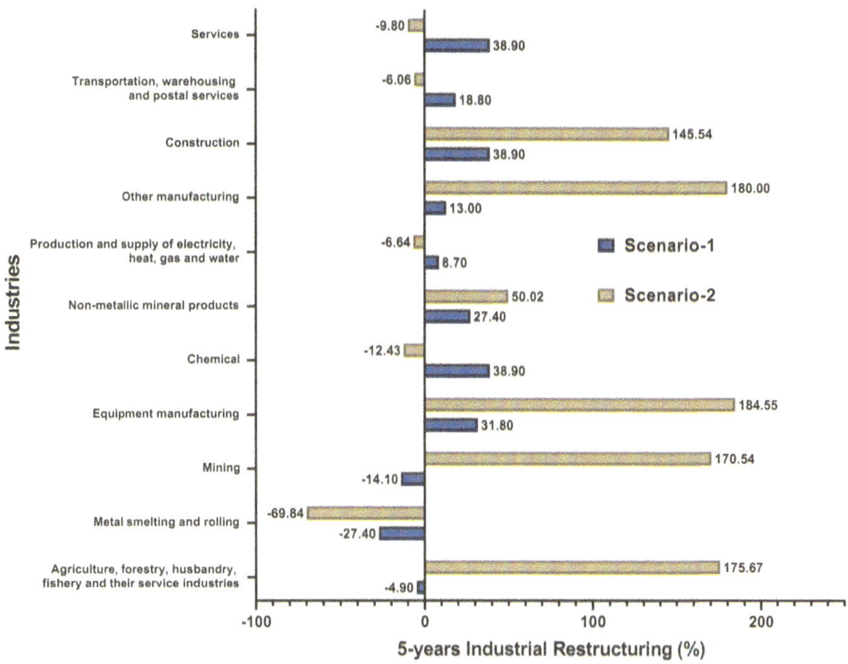


**Fig. 6** Simulation results of Industrial restructuring

realization of the dual goals will lead to a huge increase in agricultural, forestry, animal husbandry and fishery products and services, mining, equipment manufacturing, other manufacturing and construction industries. However, the development of these industries has been basically stable in recent years, and its changes are obviously inconsistent with the reality. In addition, due to the large base and high pollutant emission coefficient, the simulation result shows the total output value of the metal smelting and rolling industry would decrease by 69.84%, which would take a huge impact on the economy of Tangshan. Therefore, although the dual goals have been achieved, the industrial structure adjustment under this scenario is completely out of line with reality.

## 4 Discussion and Conclusion

Based on the analysis of the coupling mechanism of economy and environment in Tangshan city, this study constructed a comprehensive evaluation model of social economy and environment on the basis of system dynamics, input-output theory and econometrics for simulation of economic transformation and air pollutant control for steel cities. Taking Tangshan as the case, the economic development trend and environmental improvement status were predicted and analyzed, so as to explore whether the development goals of Tangshan city could be realized and how the industrial structure will be adjusted under the condition of realizing economic and environmental goals.

Through dynamic scenario simulation, this study shows that if the current level of economic development and economic growth rate are maintained, the planned goals of economic growth and pollution emission reduction cannot be achieved. If the dual goals of economy and environment are achieved simultaneously, the adjustment range of industrial structure would be too large and does not conform to the economic reality.

To sum up, under the current policy, it is difficult for the steel cities to achieve the dual goals of economic growth and environmental improvement without introducing new production and pollution material treatment technology. Therefore, to actively develop new production and pollutant treatment technologies and improve the level of pollutant treatment is the key direction that steel cities need to consider in order to achieve the dual goals of economic and environmental protection.

The current model constructed in this study provide a foundation for better evaluation of air pollution prevention and control policies in steel cities, and provide a good foundation for introduction of relevant technical data, upgrading optimization model, and improvement of research methods in the next step to achieve more breakthrough.

**Acknowledgements** We would like to specially thank reviewers for their helpful comments in this paper. This research was funded by National Natural Science Foundation of China [Grant Number. 41701635], Beijing Young Scientists' Supporting Grant [Grant Number. 2017000020124G187], Fundamental Research Funds for the Central Universities [Grant Number. BUCTRC201804], Beijing University of Chemical Technology Fund for Disciplines Construction and Development [Grant Number. XK1802-5].

## References


1. Dong, Z.F., Wang, J.F., Qu, A.Y.: Practical experience and enlightenment of cost-benefit analysis of environmental policies in OECD countries. Environ. Prot. China **45**(Z1), 93–98 (2017)
2. Goyal, P., Gulia, S., Goyal, SK., Kumar, R.: Assessment of the effectiveness of policy interventions for air quality control regions in Delhi city. Environ. Sci. Pollut. Res. **26**(30), 30967–30979 (2019). https://doi.org/10.1007/s11356-019-06236-1
3. Shi, C.C., Guo, F., Shi, Q.L.: Ranking effect in air pollution governance: evidence from Chinese cities. J. Environ. Manage. **251**, (2019)
4. Xu, F., Xiang, N., Tian, J.P., Chen, L.J.: 3Es-based optimization simulation approach to support the development of an eco-industrial park with planning towards sustainability: a case study in Wuhu China. J. Cleaner Prod. **164**, 476–484 (2017)
5. Zhang, Q.M., Wang, H.Y., Ouyang, J.: Study on environmental input-output efficiency measurement of urban agglomeration based on DEA. Chinese Population Res. Environ. China **21**(02), 18–23 (2011)
6. Hacatoglu, K., Dincer, I., Rosen, M.A.: A new model to assess the environmental impact and sustainability of energy systems. J. Cleaner Prod. **103**, 211–218 (2015)
7. Gerst, M.D., Wang, P., Roventini, A., Fagiolo, G., Dosi, G., Howarth, R.B., Borsuk, M.E.: Agent-based modeling of climate policy: an introduction to the ENGAGE multi-level model framework Environ. Environ. Modell. Softw. **44**, 62–75 (2013)
8. Ma, S.H., Wen, Z.G., Chen, J.N., Wen, Z.C.: Mode of circular economy in China's iron and steel industry: a case study in Wu'an city. J. Cleaner Prod. **64**, 505–512 (2014)
9. Guan, H., Liu, W., Zhang, P., Lo, K., Li, J., Li, L.: Analyzing Industrial Structure Evolution of Old Industrial Cities Using Evolutionary Resilience Theory: A Case Study in Shenyang of China. Chin. Geogra. Sci. **28**(3), 516–528 (2018). https://doi.org/10.1007/s11769-018-0963-5
10. Zhang, J., Jiang, H., Liu, G., Zeng, W.: A study on the contribution of industrial restructuring to reduction of carbon emissions in China during the five five-year plan periods. J. Cleaner Prod. **176**, 629–635 (2018)
11. Wei, W.X., Ma, X.L., Li, P.: The role of technological progress and tax revenue in regional air pollution control Chinese population. Res. Environ. China **26**(05), 1–11 (2016)
12. Li, C.Y.: Empirical Analysis of atmospheric environmental governance performance–principal component analysis based on PSR model. J. Central Univer. Finance Econ. China **03**, 104–112 (2016)
13. Bollen, J., Brink, C.: Air pollution policy in Europe: quantifying the interaction with greenhouse gases and climate change policies. Energy Econ. **46**, 202–215 (2014)
14. Zhang, M.M., Zhou, D.Q., Zhou, P.: A real option model for renewable energy policy evaluation with application to solar PV power generation in China. Renew. Sustain. Energy Rev. **40**, 944–955 (2014)
15. Zheng, J.J., Jiang, P., Qiao, W., Zhu, Y., Kennedy, E.: Analysis of air pollution reduction and climate change mitigation in the industry sector of Yangtze river delta in China. J. Cleaner Prod. **114**, 314–322 (2016)
16. Amann, M., Bertok, I., Borken-Kleefeld, J., Cofala, J., Heyes, C., Höglund-Isaksson, L., Klimont, Z., Nguyen, B., Posch, M., Rafaj, P., Sandler, R., Schöpp, W., Wagner, F., Winiwarter, W.: Cost-effective control of air quality and greenhouse gases in Europe: modeling and policy applications. Environ. Modell. Softw. **26**, 1489–1501 (2011)
17. Yang, L., Wang, J.M., Shi, J.: Can China meet its 2020 economic growth and carbon emissions reduction targets? J. Cleaner Prod. **142**, 993–1001 (2017)

# Exploring the Feasibility of Constructing Recyclable Prefabricated Buildings to Expedite Sustainable Urbanisation in Developing Countries

**Qiaopeng Xie and Hung-lin Chi**

**Abstract** Prefabricated buildings have been intensively studied for its multiple virtues, probably to become the main architectural form in the future. However, though prefabricated buildings have attracted enormous interest in the energy-saving field for its pollution reduction action, its contribution is paltry when compared to the great amount of waste produced in construction, demolition or reconstruction stage. The idea of constructing buildings with recyclable materials may provide a new train of thought. Current application of using recyclable materials to construct prefabricated buildings is still in the exploratory stage. The corresponding researches are also rare, especially in developing countries. Literature review and qualitative methods were adopted in this study to discuss whether the idea is conducive to facilitate the implementation of sustainable urbanisation, and consider another alternative—Additive manufacturing (AM) technology. This paper is expected to contribute to the data transformation of the construction industry.

**Keywords** Prefabricated buildings · Modular buildings · Recyclable materials · Urbanisation · Sustainable development

## 1 Introduction

Developing countries account for a large proportion at the second Belt and Road Forum [1]. These countries are still in the early stages of industrialization and urbanization, despite their abundant natural resources. They are the biggest beneficiaries of the cooperation under the Belt and Road Initiative (BRI) framework. Although urbanisation has been a driver for global economic development, it also leads to a lot

---

Q. Xie (✉) · H. Chi
Department of Building and Real Estate, The Hong Kong Polytechnic University, Hong Kong, China
e-mail: 17083861D@connect.polyu.hk



F. K. S. Chan et al. (eds.), *Proceedings of the 2020 International Conference on Resource Sustainability: Sustainable Urbanisation in the BRI Era (icRS Urbanisation 2020)*, Environmental Science and Engineering,
https://doi.org/10.1007/978-981-15-9605-6_17

of urban challenges on social and environmental issues, especially in those developing countries (icRS 2020). Thus, realizing sustainable development has become an urgent topic.

With the development of the city, the demand for building space and resource utilization is increasing, and the problem of the housing shortage is becoming more and more serious. Prefabricated and modular systems have been adopted to solve these problems in construction projects for reducing the impact of construction on the environment, including natural resource loss, cost and energy consumption caused by landfills. However, at the end of the structure's life cycle, there is a lack of structure that can be disassembled and reused. In the process of construction, demolition and renovation, only 20% of the construction waste could be recycled and reused [2]. Based on the lower material recovery rate, the idea that applying recyclable materials to construct prefabricated buildings may be a feasible way to reduce the waste of materials and energy consumption.

Current researches are mature in Prefabrication Technology, while in terms of recyclable materials, it is still in the experimental stage, mainly the trials of the low-rise temporary communal facilities. Literature review and qualitative research would be conducted to explore the feasibility of constructing prefabricated buildings with recyclable materials and whether it could contribute to expediting sustainable urbanisation in developing countries. The finding is that it would be helpful to some extent, but due to the technical conditions and other factors, further researches are needed.

## 2 Background

### *2.1 Urbanisation: Environmental Issues*

The rapid development of urbanization has promoted China's economic transformation in the past 30 yr, with a substantial increase in GDP, which has helped more than 500 Mio. People out of poverty. By 2030, according to the growing trend of per capita income, China's urbanization rate is expected to reach about 70%, and about one billion people will live in cities [3]. Economic growth and rapid urbanization have benefited China greatly in reducing poverty and improving people's living standards but has also brought China environmental pressure. Low efficiency of land development, pollution, increasingly scarce resources and other issues have emerged. As Sri Mulyani Indrawati stated at the launching meeting, urbanization creates opportunities, but cities also account for nearly 70% of world energy consumption and 80% of global greenhouse gas emissions [3]. Thus, a more coordinated urbanization process is needed.

### 2.2 *Solution: Sustainable Development*

The concept of sustainable development was proposed as a solution to the negative impact of the urbanization process. In *National New Urbanization Plan 2014–2020*, measures to improve the sustainable development ability of cities have been raised, including optimizing urban industrial structure, spatial structure and management pattern, conforming to the new concept and trend of modern urban development, promoting urban green development and enhancing urban innovation ability, improving intelligent level [4]. This concept could be traced back to the book published by UN-Habitat in 2012, and earlier *The State of Asian Cities 2010/11* and Training courses on Sustainable Urbanization launched by the International Urban Training Centre in 2007. The UN-Habitat note that the green economy is committed to leading the economy to a higher and fairer development direction based on low-carbon and energy-intensive utilization. It is considered to be a scientific and modern means to apply the principle of sustainable development to the present city, which integrates all aspects and protects the environment while promoting economic growth and employment [5].

### 2.3 *Building: Essential Element*

According to Malik, A., & Maheshwari, A. (2018), "the global construction industry is the world's largest consumer of raw materials, and constructed entities account for between 25 and 40 percent of total carbon emissions in the world," but less than a third of construction and demolition waste is recycled or reused [6]. It proves that construction has been an essential part of the progress of sustainable urbanisation. Urbanization drives the demand for sustainable development of the construction industry and transitioning. The United Nations Environment Programme estimates that improvements in the construction industry could help countries to reduce emissions economically and effectively and achieve energy savings of more than 30% [7]. Recycling waste materials reduces global emissions for the industry and makes its businesses more environmentally sustainable. There are also considerable commercial benefits in the use of circular economy in the construction industry.

## 3 Literature Review

### 3.1 *Prefabricated Buildings*

Prefabrication has environmental advantages in reducing components, reusability, adaptability and recyclability [8]. Prefabricated buildings almost eliminate the waste generated by demolition. After prefabrication, waste can be greatly reduced in plastering, wood formwork, concrete pouring, reinforcement and other on-site production activities, and the waste of plastering can be reduced by 100% [9]. The construction cycle of a prefabricated building is short, and it is easy to recycle, re-purpose and reuse. By adopting certain strategies for prefabricated buildings, a circular economy can be implemented in the construction industry.

Although more and more modular buildings are oriented to permanent buildings (modular structures are not intended to be relocated, just like traditional buildings), many enterprises still rely on the temporary use of modular buildings. When the rapid expansion of space is required, resettable buildings are brought to the site or used as swing space to accommodate students, patients, or employees in case of renovation or emergency. When additional space is no longer needed, the modular building will be removed from the ground and reused.

Unlike traditional building strategies, these modules will not be removed and the materials will not be thrown into the local landfill after one use. Instead, used modules are updated so they can be reprocessed and used in future projects, further reducing the need for additional raw materials and energy to create new things from scratch [10].

Although modular housing has its advantages, it is difficult to resell because of its low quality. Modern modular structures are trying to change the general perception of these houses, but it may take some time for the public to see them as the same as traditional brick and concrete structures. In addition to quality concerns, other problems such as excessive energy consumption during the construction process and the high cost of special design still exist, so that modular building can not give full play to their original value.

## 3.2 Recyclable Materials

As a way to promote waste reduction or prevention, 3R rules are created: reduction, reuse and recycling [11]. The first rule of 3R is the reduction, which is quite extensive. It can be understood as the reduction of materials, energy consumption and waste, especially the reduction of ecological footprint. Reduce the pollution of building materials, reduce the use of building materials, reduce the use of building materials, reduce the use of building materials. Reuse is a concept based on the multiple uses of materials and does not require processing and, therefore, does not require energy. In architecture, reuse can change from reusing materials and construction elements to reusing structure. This process is called adaptive reuse. Finally, recycling requires the transformation of a material before it can be reused. This means that manufacturing or manual processes that require energy are necessary to enable materials to take on new forms and uses. Recycling is a substitute for waste materials that do not meet the reuse conditions. Compared with the use of traditional materials, recycling is usually cost-effective.

The ideal recyclable materials should be reused, i.e., without energy consumption, among which the recycled plastic, glass, timber, recycled steel, mycelium, ashcrete, repurposed tile and aluminum have been adopted in a real case.

### *3.3 The Combination—Recyclable Prefabricated Building*

The sustainability characteristics of advanced off-site construction solutions have been well documented, greatly reducing waste and improving thermal efficiency and life cycle cost-effectiveness. Whereas recycled and renovated modular buildings take sustainable construction to a whole new level.

### *3.4 Case Study—The NEST "Urban Mining and Recycling" Unit*

In Switzerland, a residential module is built entirely of reusable, recyclable and compostable materials. It is a modular research and innovation building operated by EMPA and EAWAG in Dubendorf.

The growing scarcity of resources and the resulting desire to move away from today's one-off thinking means that the construction sector must give more consideration to the multiple uses and recyclability of materials, as well as other construction methods. The latest nest unit called "urban mining and recycling" implements these ideas; the result is a residential module whose structure and materials can be completely reused, reused, recycled or composted, and then decomposed.

"The materials we use will not just be used and processed; instead, they will be extracted from the loop and returned to it," Dirk E. Heber explained the concept [12]. As a result, a variety of continuously processed components are used in the "urban mining and recycling" unit; the various materials can be separated, classified and returned to their respective material cycles without any residue. In addition, the unit also used a new type of heat insulation board made of mushroom mycelium, innovative recycled stone, recycled insulation materials and rented carpets.

Besides Switzerland, U.K., U.S., Indonesia and even rural Chinese village has established some modular buildings utilizing local unique recyclable materials [12].

# 4 Limitations and Further Studies

## *4.1 Limitations*

**Cost** Prefabricated integrated building needs an integrated manufacturing base with a modern chemical plant. The construction of the base has a huge one-time investment, which raises the threshold for enterprises to enter and bring greater investment risk. Although the application of prefabricated buildings and recyclable materials can bring economic benefits, it also requires a large amount of R & D and design costs.

**Strength** The integrity and stiffness of prefabricated buildings are weak, so the seismic impact capacity of prefabricated buildings is also poor. In addition, the strength, durability and accessibility of the prefabricated buildings and recyclable materials remain to study further.

**Restrictions** There are some restrictions on the height and the number of floors of prefabricated buildings. The cases mentioned above are all low rise prefabricated buildings instead of high-rise example. There is also a lack of skilled workers, so the after-sales and maintenance will be relatively troublesome.

## *4.2 Further Studies*

In addition to the analysis weight, recovery, cost, durability, renewability, greenhouse gas emissions, energy consumption and other material characteristics, economic benefits and environmental impact, digital planning and management may become the next evolution. Wang et al. [13] present the potentially viable alternative for modular buildings, the Additive Manufacturing (AM) technology (also known as 3D-printing). The AM technology could counter the deficiency of prefabricated buildings in terms of cost, size, quality, safety, and environment. The whole process of AM technology is controlled by computers. Thus it could prevent fragile structures and structure risks, extend the life of the architectures. Moreover, it could avoid high costs for long-distance transportation, reduced carbon footprint and environmental impact, which simplifies the process of recycling and allow more freedom in design. Whether it could make the construction industry keep pace with the times requires more feasibility studies.

# 5 Conclusion

Employing recyclable materials in the construction of prefabrication buildings does improve the original cons of prefabrication buildings, the life cycle, availability and reusability of materials, and avoid demolition, thus reducing the emission of construction waste and carbon dioxide, which benefits the acceleration of the sustainable urbanisation. Nevertheless, new issues are also emerging. For instance, the durability, strength, accessibility and specific characteristics remain further research and attempt.

An interesting concept "build for unbuild" is noted during the progress of the literature review. Dirk E. Hebel [14] explains that "The materials that we utilize will not just be used and then disposed of; instead, they will be extracted from their cycle and later returned to it." That should be the goal when construction to reach sustainable development, producing the minimum waste and minimal impact on the environment.

Despite the fact that prefabricated buildings are widely used and praised, the Additive Manufacturing (AM) technology could be the new trend to replace prefabrication. It may even be more energy-efficient and cost-effective than the combination of recycled materials and prefabricated buildings. Technological progress and further research on M.A. technology are also required. Developing countries should seize the opportunity of BRI cooperation to complete the digital transformation and upgrading of the construction industry to realize sustainable urbanization.

## References


1. China.org.cn.: Developing Countries become BRI's Biggest Beneficiaries, Cision (2019). https://www.prnewswire.com/news-releases/developing-countries-become-bris-biggest-beneficiaries-300838067.html
2. Silva, F., Marielle, J., Bhagya, L., Waldmann, D., Hertweck, F.: Recyclable architecture: prefabricated and recyclable typologies. Sustainable (Basel, Switzerland) **12**(4), 1342 (2020)
3. Guo wu yuan fa zhan yan jiu zhong xin: In: Urban China: Toward Efficient Inclusive and Sustainable Urbanization. Development Research Center of the State Council, World Bank Publications (2014)
4. National new urbanization plan 2014–2020 (2014). Gov.cn. https://www.gov.cn/gongbao/content/2014/content_2644805.htm (in Chinese)
5. UN-Habitat: Sustainable Urbanization in Asia: A Sourcebook for Local Governments (2012)
6. Malik, A., Maheshwari, A.: Construction industry value chain: how companies are using carbon pricing to address climate risk and find new opportunities No. 132770, pp. 1–48. The World Bank (2018)
7. Doe, U.: An assessment of energy technologies and research opportunities. quadrennial technology review. United States Department of Energy (2015)
8. Hardcastle, J.L.: The Building Sector's 'Trillion-Dollar' Circular Economy Opportunity. Environment + Energy Leader (2016). https://www.environmentalleader.com/2016/05/the-building-sectors-trillion-dollar-circular-economy-opportunity/
9. Tam, V.W.Y., Hao, J.J.L.: Prefabrication as a mean of minimizing construction waste on site. Int. J. Constr. Manage. **14**(2), 113–121 (2014)
10. How Modular Construction is Keeping Waste Out of U.S. Landfills: Triple Pundit (2016). https://www.triplepundit.com/story/2016/how-modular-construction-keeping-waste-out-us-landfills/28726
11. Moreira, S.: Reduce Reuse and Recycle: the Three R's Rule Applied to Architecture. ArchDaily (2020). https://www.archdaily.com/945040/reduce-reuse-and-recycle-the-three-rs-rule-applied-to-architecture
12. KÄLin, S.: Building with waste and recycled material. Phys.Org. (2018). https://phys.org/news/2018-02-recycled-material.html

13. Wang, Q., Zhang, S., Wei, D., Ding, Z.: Additive manufacturing: a revolutionized power for construction industrialization. In: ICCREM 2018: Innovative Technology and Intelligent Construction, pp. 85–94. American Society of Civil Engineers, Reston, VA (2018)
14. Ruiz, A.: Building with Waste and Recycled Material, Waste Advantage Magazine (2018). https://wasteadvantagemag.com/building-with-waste-and-recycled-material/

# Optimal Rectangle Packing to Minimize Wastage

**Shijun Chen, Jiying Xu, Aiying Rong, and Weigang Zhou**

**Abstract** To mitigate the contradiction between resource shortage and population growth, it is important to promote resource conservation and environment protection across the manufacturing sectors because the manufacturing is the pillar industry for any industrialized countries. Conservation (saving) type economy is one of options for transition into future sustainable economy. This chapter addresses the problem of placing rectangles of a set of small rectangular pieces of different sizes into a larger rectangular sheet called rectangle packing problem to minimize sheet wastage (or maximize sheet utilization). The problem is one of widely studied problems in manufacturing. Here a new two-stage algorithm is developed to improve the sheet utilization of the old counterpart. The new version modified the old version in two ways. At the first stage, a new blank rectangle filling algorithm was applied instead of the improved bottom left filling (BLF). At the second stage, a more advanced neighborhood search algorithm was applied to optimize the sequence of the rectangle pieces. Numerical experiments with 21 classical instances in the literature showed that the new two-stage packing algorithm can achieve sheet utilization of 100% for 4 instances, over 99% for most instances and over 98% for a few instances. On the average, the new two-stage algorithm can improve sheet utilization by 4% over the old counterpart and new first-stage algorithm can achieve the comparable sheet utilization to the old two-stage algorithm. Overall, the new two-stage algorithm showed 19% improvement over the classical BLF.

S. Chen · J. Xu · A. Rong (✉) · W. Zhou
Group for Operational Research and Control, School of Mathematics and Statistics, Hubei University of Arts and Science, 441053 Xiangyang, P.R. China
e-mail: aiying.rong@hotmail.com

S. Chen
e-mail: csj@hbuas.edu.cn

J. Xu
e-mail: xjy@hbuas.edu.cn

W. Zhou
e-mail: zhouwgang@126.com

 
F. K. S. Chan et al. (eds.), *Proceedings of the 2020 International Conference on Resource Sustainability: Sustainable Urbanisation in the BRI Era (icRS Urbanisation 2020)*, Environmental Science and Engineering,
https://doi.org/10.1007/978-981-15-9605-6_18

**Keywords** Manufacturing sustainability · Wastage Reduction · Improved Two-stage Algorithm

## Abbreviations

| | |
|---|---|
| AC | Ant colony |
| BL | Bottom left |
| BLF | Bottom left filling |
| BRF | Blank rectangle filling |
| COP | Combinatorial optimization problem |
| CPLEX | Commercial mathematical programming optimization software (IBM) |
| GA | Genetic algorithm |
| HC | Hill climbing |
| NS | Neighborhood search |
| PSO | Particle swarm optimization |
| SA | Simulated annealing |
| TS | Tabu search |
| TSP | Traveling salesman problem |

## 1 Introduction

Manufacturing is a pillar industry for the economy of any industrialized country. However, the rapid development of manufacturing provides the material foundation for improving the living standard of people at the same time aggravates the resource shortage and environmental burden. Resource conservation (saving) economy is one option for transition into the future manufacturing sustainability. It emphasizes the high utilization of resource (raw material) and reduction of wastage [1].

The cutting and packing problems have found wide applications within manufacturing industries such as metal, cloth, paper and glass cutting. Generally, the problem tries to place a set of small pieces into a large sheet to maximize the utilization of the sheet (raw material). Optimization is one of the effective ways to reduce manufacturing cost and increase the utilization of the raw material. The traditional manual arrangement of pieces on the sheet not only consumes a large quantity of human and financial resources but also results in low utilization of raw materials. In addition, it is even prone to produce the error in the arrangement. This situation has negative effect on the profit of the enterprise. Therefore, it is important to increase the efficiency of making arrangements for the cutting and packing activities so that the utilization of the raw material can be improved.

This paper addresses the rectangle packing problem, which is a type of packing problem where a set of rectangular pieces must be arranged on a predefined larger rectangular sheet with no overlap to maximize the sheet utilization. The rectangle problem belongs to NP-hard combinatorial optimization problem (COP). Up to now, there has been no effective polynomial algorithm to find the optimal solution to the problem efficiently [2]. The classical exact algorithm is hard to deal with large-scale problems because of computational complexity. Some researchers proposed quick heuristics such as bottom-left (BL) algorithm [3]. However, the BL is prone to produce large wastage areas and result in low sheet utilization. Reference [4] developed the bottom-left filling (BLF) heuristic for filling the wastage area and improving the sheet utilization. Consequently, the BLF becomes the classical heuristic for the rectangle packing problem. Later, different heuristics like the best-fit [5], layer-based heuristic [6], corner incremental-based algorithm [7], anthropomorphic algorithm [8] were proposed to improve the sheet utilization. The above heuristics are sensitive to the sequence of the rectangle pieces though they are fast and easy to implement. Usually it is difficult to produce a good arrangement by intuition (e.g. the rectangular piece with the largest area is selected first) or by random selection.

With the swift development of intelligent techniques, some intelligent algorithms [9] such as swarm intelligent algorithms and neighborhood search (NS) algorithms have been applied to solve various complicated COP. Swarm intelligent algorithms include genetic algorithm (GA), ant colony (AC), and particle swarm optimization (PSO) as well as other types of intelligent algorithms mimicking mechanisms of biological behaviors while NS algorithms include hill climbing (HC), simulated annealing (SA) and tabu search (TS). As a result, there were numerous algorithms [10–15] by combining intelligent techniques and classical heuristics such as BL [3], BLF [4]. The lowest horizontal line method [15] as well as modified BL and BLF to solve the rectangle packing problem. The intelligent techniques were mainly used to optimize the sequence of rectangle pieces and improve the sheet utilization.

However, swarm intelligent algorithms suffered from high number of solution evaluations, large computational times, premature of convergence. In these years, NS algorithms have been widely applied to solve COPs such as rectangle packing problem [13], traveling salesman problem (TSP) [16] and vehicle scheduling problem [17]. Also, NS is easy to implement with lower computation complexity.

Based on our computational experience with the old two-stage algorithm by combining modified BLF and NS [13], it was found that BLF suffered disadvantages of larger wastage areas and lower sheet utilization. Therefore, we applied a modified BLF at the first stage [13]. In this paper, we developed a new two-stage algorithm by combining a new blank rectangle filling (BRF) algorithm at the first stage and a more advanced NS algorithm at the second stage. Numerical experiments with the 21 classical instances showed the new two-stage algorithm can achieve sheet utilization of 100% for 4 instances, over 99% for most instances and over 98% for a few instances. On the average, the new two-stage algorithm showed 4% improvement over the old counterpart and the new first-stage algorithm had the comparable sheet utilization to the old two-stage algorithm. By cross comparison of the old and new two-stage algorithms, it was found that the new two-stage algorithm showed 19% improvement over the BLF.

The paper is organized as follows. Section 2 gives problem description and formulation. Section 3 presents the new two-stage algorithm. Section 4 reports the results of numerical experiments. Section 5 gives a conclusion and points out that wastage reduction in manufacturing supports manufacturing sustainability.

## 2 Problem Description and Formulation

The rectangle packing problem considered in this paper is described as follows. Given a large rectangular sheet $R$ with width $W$ and height $H$, as well as $n$ small rectangular pieces of a given size (hereinafter referred to as 'mall rectangles'). Let denote the $i$th small rectangle $R_i$ with width $w_i$ and height $h_i$ and assume that $\sum_{i=1}^{n} w_i h_i \geq WH$ to avoid the trivial solutions. The rectangle packing problem is to fill the large rectangular sheet by selecting some small rectangles so that the total areas of the selected small rectangles is maximized satisfying the following constraints: (1) the sides of the small rectangles should be parallel to the width or height of the sheet, and (2) the packed small rectangles shall not exceed the boundary of the sheet with no overlap.

Take the lower left corner of the large rectangular sheet as the origin $O(0, 0)$, and set horizontal axis and vertical axis parallel to the width and height of the sheet respectively. Let $(x_{i1}, y_{i2})$ denote the coordinate of the lower left corner of the small rectangle $R_i$ and $(x_{j2}, y_{j2})$ the coordinate of the upper right corner of the small rectangle $R_j$. $z_i = 1$ indicates that $R_i$ is selected, and $z_i = 0$, Otherwise. The rectangle packing problem can be formulated as follows.

$$\max\ f(x, y, z) = \frac{\left(\sum_{i=1}^{n} w_i h_i z_i\right)}{W \cdot H} \times 100 \tag{1}$$

$$x_{j2} \leq x_{i1} \text{ or } x_{i2} \leq x_{j1} \text{ or } y_{j2} \leq y_{i1} \text{ or } y_{i2} \leq y_{j1},\ \forall i, j = 1, 2, \ldots, n. \tag{2}$$

$$0 \leq x_{i1} \leq W - w_i \text{ and } 0 \leq y_{i1} \leq L - h_i,\ \forall i = 1, 2, \ldots, n. \tag{3}$$

$$x_{i1}, x_{i2}, y_{i1}, y_{i2} \in N \text{ and } z_i \in \{0, 1\},\ \forall i = 1, 2, \ldots, n. \tag{4}$$

In the above model, the objective (1) is to maximize the sheet utilization. Constraints (2) indicate that any two small rectangles $R_i$ and $R_j$ do not overlap on the sheet. Constraints (3) ensure that the small rectangles to be packed does not exceed the boundary of the sheet. Constraints (4) indicate the domain of variables (integer packing position of the small rectangles and binary decision variables).

The model (1)–(4) means that the rectangle packing problem can be formulated as a combinatorial optimization problem (COP). With the increase of the number of small rectangles, the spatial scale of feasible solutions increases exponentially, which is difficult to be solved directly by commercial mathematical programming software such as CPLEX. In the following, we present a more effective new two-stage algorithm to improve the sheet utilization of the old counterpart [13].

# 3 New Two-Stage Algorithm

In the new two-stage algorithm, a more effective blank rectangle filling (BRF) algorithm was applied at the first stage and a more advanced NS algorithm was designed at the second stage.

## 3.1 Blank Rectangle Filling Algorithm

The general idea of the BRF algorithm is described as follows: the area that can be packed in the large rectangular sheet is regarded as the set of the blank rectangles. The small rectangles are arranged in order according to certain rules (such as the size of the area), and then the small rectangles are packed and the set of blank rectangles is updated one by one until all the small rectangles are packed or no blank rectangles are available. When a small rectangle needs to be packed, the most suitable blank rectangle is selected based on the evaluation rules. Specifically, the sorted small rectangles are denoted as $R_1$, $R_2$, ..., $R_{n,}$ and the set of available blank rectangles as $S = \{S_1, S_2,\ldots,S_m\}$. Initially, the large rectangular sheet is the only blank rectangle that can be packed, i.e. $S = \{R\}$.

**Algorithm 1**. Procedures of the BRF algorithm.

Step 1 All the small rectangles are sorted by some rules (e.g. according to the area) denoted by $R_1 \ldots R_n$, set $i = 1$.
Step 2 If there are no available blank rectangles or small rectangles, stop.
Step 3 For the $R_i$, select the most suitable blank rectangle based on the following conditions:

(1) If there is a blank rectangle can be packed by $R_i$;
(2) If there are multiple blank rectangles that satisfy the condition (1), select the lowest blank rectangle;
(3) If there are multiple blank rectangles that satisfy condition (2), the leftmost blank rectangle is selected;

Step 4 If a suitable blank rectangle exists, $R_i$ will be packed to the selected blank rectangle.
Step 5 Updates the set of blank rectangles; $i \leftarrow i + 1$, go to **Step 2**.

In **Step 1**, the order of small rectangular pieces to be packed directly affects the final sheet utilization. Therefore, by intuition and experience, five methods are used to sort the small rectangles based on five attributes: area, height, width, longest side and maximum aspect ratio, in descending order. In **Step 5**, it needs to update the set of blank rectangles. When a small rectangle packed overlap a blank rectangle, the blank rectangle should be deleted and some newly generated blank rectangles will be added to the set of blank rectangles.

### *3.2 Eliminating Redundant Blank Rectangles*

In the filling algorithm, the small rectangles just packed may overlap with some blank rectangles. In the worst case, the computational complexity is $O(NM)$, where $N$ is the number of small rectangles and $M$ is the number of blank rectangles. Overlapping usually generates at least two blank rectangles. Assuming that 2 new blank rectangles are generated on average, $2M$ blank rectangles will be generated for each packing operations. Therefore, the number of blank rectangles will be increased exponentially when implementing the above filling algorithm. In the process of the filling algorithm, the following facts hold: if no contain relationship happens between two blank rectangles in current packing state, when a new small rectangle is packed and two new blank rectangles will be generated, then the two blank rectangles will not contain other blank rectangles, but may be contained by other blank rectangles.

Therefore, the following method to eliminate abundant blank rectangles is adopted: when a new blank rectangle is generated, it needs to determine whether the new blank rectangle is included in an existing blank rectangle. If this is the case, the newly generated blank rectangle is eliminated. A lot of experiments show that the method of eliminating blank rectangles can significantly reduce the number of blank rectangles and improve the computational efficiency.

### *3.3 Iterative Neighborhood Search Algorithm*

The above BRF algorithm is a heuristic algorithm, and the performance of the algorithm depends on the sequence of the small rectangles. For this reason, an iterative NS algorithm is proposed to search the optimal sequence of small rectangles. The sequence of small rectangles is coded by the permutation of the serial number of rectangular pieces, and the BRF algorithm is used to decode and obtain the layout and sheet utilization.

Let $a_1 a_2 \ldots a_n$ be the permutation of the serial number of rectangular pieces (i.e. $1, 2 \ldots n$), and $a_i$ is the serial number of the $i$th small rectangle $R_{ai}$. In the iterative NS algorithm, two neighborhood operators are used, i.e., swap operator Swap $(i, j)$ and insert operator Insert $(i, j)$ based on random two points $(i, j)$. The operator Swap $(i, j)$ swaps $a_i$ and $a_j$, and the elements in the other positions do not change in the permutation, i.e. $a_1 a_2 \ldots a_i \ldots a_j \ldots a_n$ will change into $a_1 a_2 \ldots a_j \ldots a_i \ldots a_n$. The operator Insert $(i, j)$ inserts $a_i$ before or after the position of $a_j$ in the original permutation depending on the value of $i$ and $j$. If $i < j$, then $a_1 a_2 \ldots a_i \ldots a_j \ldots a_n$ will change into $a_1 a_2 \ldots a_{i-1} a_{i+1} \ldots a_j a_i a_{j+1} \ldots a_n$. If $i > j$, then $a_1 a_2 \ldots a_j \ldots a_i \ldots a_n$ will change into $a_1 a_2 \ldots a_i a_j a_{j+1} \ldots a_{i-1} a_{i+1} \ldots a_n$.

The process of iterative NS algorithm uses the above two neighborhood operators iteratively and alternately to search the solution space, i.e., one neighborhood operator is implemented on a certain number of continuous search (the maximum number of local search $lsNum$), another neighborhood search operator is used to search the optimal solution. At the same time, in order to reduce invalid search and prevent falling into the local optimal solution, when a continuous search by an operator fails to improve the current optimal solution to reach the specified upper limit number $nimprvNum$, another neighborhood operator will be used to continue the search for the solution. In the iterative neighborhood search algorithm, a global maximum search number $sNum$ is set to control the search time of the algorithm. When the total number of iterations exceeds $sNum$, the algorithm terminates and the solution is reported.

Let $x_0$ and $x_b$ be the initial and the current best solution respectively. Let SwapOpr = True mean that the operator Swap$(i, j)$ is used, and SwapOpr = False mean that the operator Insert $(i, j)$ is used. Then, the flow of the iterative NS algorithm is shown in Fig. 1.

In Fig. 1, 'Random select points' select two points $i$ and $j$ that are used by operator Swap$(i, j)$ and operator Insert$(i, j)$. When the total neighborhood search times larger

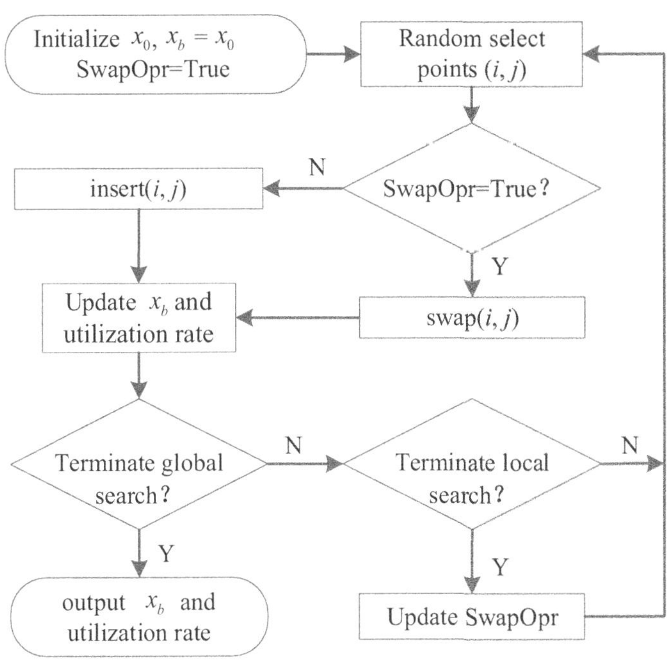


**Fig. 1** Iterative neighborhood search algorithm.

than $gsNum$ or the current sheet utilization has reached 100%, then terminate global search. When the number of times for the local search larger than $lsNum$, it means that the current solution cannot be improved by one operator for $nimprvNum$ times, terminate the current local research.

## 3.4 Algorithm Strategies

**Initializes the Order of the Small Rectangles.**

In the beginning, it needs to set the order of small rectangles to form the initial solution. A good initial solution can improve the efficiency of the NS algorithm. By intuition and experience, as compared with random order, the order considering large rectangles first usually produces better solutions. Therefore, 5 attributes of small rectangles are used to design 5 ordering methods: a. AreaBLF, b. WidthOrderBLF, c. HeightOrderBLF, d. TallestBLF, e. WHRationBLF. For $R_i$, its 5 attributes are as follows: a. $w_i h_i$; b.$w_i$; c. $h_i$; d. $\max\{w_i, h_i\}$; e. $\max\{w_i/h_i, h_i/w_i\}$ .

**A Two-Point Selection Strategy Based on Problem Characteristics.**

(a) Restricted two-point selection strategies. When Swap($i$, $j$) or Insert($i$, $j$) is executed, and if the distance between $i$ and $j$ is large, the rectangles with small attribute value would be packed earlier. Then a low sheet utilization rate is produced in this case. Therefore, a restricted two-point selection strategy is proposed: when two points are randomly selected, their distance should not exceed a given threshold.
(b) Specific point selection strategy. The sheet utilization is determined by the total area of blank rectangles in the final layout. In order to reduce unused areas created by blank rectangles, the serial number of the small rectangles that produce the blank rectangle is recorded. These small rectangles directly affect the overall sheet utilization. When the neighborhood operator is executed in next step, the positions corresponding to these small rectangles are selected first.

# 4 Numerical Experiments

To test the performance of the new two-stage algorithm for solving the rectangle packing problem, we used the old two-stage algorithm [13] as a benchmark. In addition, we tested the performance of the new first-stage algorithm (BRF algorithm). The evaluation of the new NS algorithm needs additional numerical test and the detailed report was delayed to the later. All of algorithms were implemented in C++ under Visual studio 2012 environment. All test runs were performed on a laptop (2.2 GHZ Intel Core (TM) i5-5200 CPU with 8 GB RAM) under Windows 10 operating systems.

**Table 1** Test instances

| Instance No. | 1 | 2 | 3 | 4 | 5 | 6 | 7 | 8 | 9 | 10 | 11 |
|---|---|---|---|---|---|---|---|---|---|---|---|
| No. of rectangles | 16 | 17 | 16 | 25 | 25 | 25 | 28 | 29 | 28 | 49 | 49 |
| Sheet model | 20, 20 | 20, 20 | 20, 20 | 40, 15 | 40, 15 | 40, 15 | 60, 30 | 60, 30 | 60, 30 | 60, 60 | 60, 60 |

**Table 2** Test instances (continued)

| Instance No. | 12 | 13 | 14 | 15 | 16 | 17 | 18 | 19 | 20 | 21 |
|---|---|---|---|---|---|---|---|---|---|---|
| No. of rectangles | 49 | 73 | 73 | 73 | 97 | 97 | 97 | 196 | 197 | 196 |
| Sheet model | 60, 60 | 60, 90 | 60, 90 | 60, 90 | 80, 120 | 80, 120 | 80, 120 | 160, 240 | 160, 240 | 160, 240 |

## 4.1 Test Instances

A total of 21 test instances were used in the test. These instances came from the literature as shown in Tables 1 and 2.

## 4.2 Numerical Results

Based on our numerical experience, usually small instances (1–14) can generate solution in a few seconds, medium instances (16–18) in a few minutes and large instances (19–21) in 30–40 min. In the computational experiment, each instance was run 10 times and the best result was chosen as the final result. In the following we present the sheet utilization of the instances for different algorithms in Table 3.

According to Table 3, Among 21 test instances, the new two-stage algorithm can achieve sheet utilization of 100% for 4 instances, over 99% for 13 instances and over 98% for 4 instances. On the average, the sheet utilization 99.23%, which shows 4% improvement over the old counterpart [13]. In addition, the new first-stage algorithm (BRF algorithm) can obtain the comparable sheet utilization to the old two-stage algorithm, especially for large instances (19–21), the sheet utilization shows significantly better result than old two-stage algorithm. This should be attributed to blank redundant rectangle removal procedure in the BRF algorithm as shown in Table 4.

According to Table 4, introducing procedure for removing redundant blank rectangle has positive effect on improving the computational efficiency and sheet utilization with the number of redundant blank rectangles reduced.

**Table 3** Sheet utilization of different algorithms (in percentage)

| Instance No. | Old two-stage algorithm | | New two-stage algorithm | |
|---|---|---|---|---|
| | First-stage | Second-stage | First-stage | Second-stage |
| 1 | 90.00 | 100.00 | 86.25 | *100* |
| 2 | 87.00 | 94.00 | 91.50 | *99* |
| 3 | 82.00 | 93.75 | 94.00 | *100* |
| 4 | 87.67 | 94.50 | 90.67 | *100* |
| 5 | 83.83 | 96.67 | 98.5 | *99.5* |
| 6 | 95.50 | 95.50 | 97.83 | *100* |
| 7 | 94.33 | 97.39 | 93.89 | *99.33* |
| 8 | 85.5 | 93.61 | 90.89 | 98.33 |
| 9 | 81.33 | 93.17 | 96.83 | *99.01* |
| 10 | 91.47 | 96.22 | 95.42 | *99* |
| 11 | 89.64 | 93.14 | 93.11 | 98.58 |
| 12 | 94.89 | 96.33 | 95.28 | *99* |
| 13 | 89.59 | 94.50 | 95.46 | *99.04* |
| 14 | 91.37 | 91.37 | 96.93 | *99.22* |
| 15 | 90.00 | 92.83 | 96.33 | *99.22* |
| 16 | 95.29 | 97.60 | 96.92 | *99* |
| 17 | 89.87 | 96.77 | 97.96 | *99.5* |
| 18 | 90.16 | 98.33 | 97.05 | *99.21* |
| 19 | 91.08 | 93.5 | 97.16 | 98.63 |
| 20 | 93.33 | 94.84 | 98.32 | *99.43* |
| 21 | 88.66 | 93.03 | 97.98 | 98.86 |
| Average | 89.64 | 95.10 | 95.16 | *99.23* |

**Table 4** Effect of blank redundant rectangle removal procedure

| Instance No. | Maximum number of blank rectangles | |
|---|---|---|
| | Without adopting blanking rectangle removing strategy | Adopting blank rectangle removing strategy |
| 19 | 1546 | 74 |
| 20 | 853 | 67 |
| 21 | 1334 | 80 |

According to the reports in [13], the old first-stage algorithm (a modified BLF) showed 9% improvement over the classical BLF [4]. Consequently, the new two-stage algorithm shows 19% improvement over the classical BLF. This justifies the effectiveness of the new algorithm presented in this chapter.

## 5 Conclusions

This paper developed a new two-stage algorithm for rectangle packing by combining a new blank-rectangle algorithm and a more advanced neighborhood search algorithm. Numerical experiments with the classical instances showed the superiority of the new algorithm. On the average, the new algorithm can obtain the sheet utilization of over 99% with minimum utilization exceeding 98%. In addition, the new two-stage algorithm showed 4% improvement over the old counterpart [4] and the new first-stage algorithm had the comparable sheet utilization to the old two-stage algorithm [4]. By crosschecking the old and new two-stage algorithms, it was found that new two-stage algorithm showed 19% improvement over the classical BLF [4]. High utilization of the raw materials helps the enterprises increase the production efficiency, reduce wastage and save valuable raw materials. Since wastage recycling needs to consume energy, which in turn increase the impact on the environment [1]. Also, many raw materials such as iron and steel used by manufacturing were produced by non-renewable minerals such as iron ores. Therefore, saving raw material facilitates the enterprises to achieve the target of manufacturing sustainability.

**Acknowledgments** The research was partially supported by National Natural Science Foundation of China (No. 71501064) and talent program from the university in China (No. kyqdf2020001).

## References

1. Ouyang, Z.: Revisit ‘circular economy’ and ‘conservation economy’ pilosophy and social science edition. Huaiyin Normal College J. **27**, 429–431 (2005) (in Chinese)
2. Oliveira, J.F., Alvaro, N.J., Silva, E.M., et al.: A survey on heuristics for the two-dimensional rectangular strip packing problem. Pesquisa Operacional **36**(2), 197–226 (2016)
3. Baker, B.S., Coffman, E.G., Rivest, R.L.: Orthogonal packings in two dimensions. SIAM J. Comput. **9**(4), 846–855 (1980)
4. Hopper, E., Turton, B.C.H.: A genetic algorithm for a 2D industrial packing problem. Comput. Ind. Eng. **37**(1–2), 375–378 (1999)
5. Burke, E.K., Kendall, G., Whitwell, G.: A new placement heuristic for the orthogonal stock-cutting problem. Oper. Res. **52**(4), 655–671 (2004)
6. Leung, S.C.H., Zhang, D.F.: A fast layer-based heuristic for non-guillotine strip packing. Expert Syst. Appl. **38**(10), 13032–13042 (2011)
7. Chen, Z., Chen, J.: An effective corner increment-based algorithm for the two-dimensional strip packing problem. IEEE Access **6**, 72906–72924 (2018)
8. Deng, J., Wang, L., Yin, A.: A global anthropomorphic algorithm for solving a two-dimensional rectangle packing problem. Comput. Eng. Sci. **40**(2), 331–340 (2018) (in Chinese)
9. Wang, L.: Intelligent Algorithms and their Applications. Tsinghua University Press, Beijing, China (2001) (in Chinese)
10. Hopper, E., Turton, B.C.H.: An empirical investigation of meta-heuristic and heuristic algorithms for a 2D packing problem. Eur. J. Oper. Res. **128**(1), 34–57 (2001)
11. Babalik, A.: Implementation of bat algorithm on 2D strip packing problem. In: Lavangnananda, K., Phon-Amnuaisuk, S., Engchuan, W., Chan, J. (eds.), Intelligent and Evolutionary Systems. Proceedings in Adaptation, Learning and Optimization, vol. 5. Springer, Cham (2016)

12. Babaoglu, I.: Solving 2D strip packing problem using fruit fly optimization algorithm. Proc. Comput. Sci. **111**, 52–57 (2017)
13. Xu, J., Chen, S., Zheng, Q.: A two-stage algorithm for the rectangle packing problem. Comput. Times **5**, 13–15 (2020) (in Chinese)
14. Anand, K.V., Babu, A.R.: Heuristic and genetic approach for nesting of two-dimensional rectangular shaped parts with common cutting-edge concept for laser cutting and profile blanking processes. Comput. Ind. Eng. **80**(2), 111–124 (2015)
15. Liu, H., Zhou, J., Wu, X.: Improved combined lowest horizontal line method and genetic algorithm for the rectangle packing problem. Graph. J. **36**(4), 526–531 (2015) (In Chinese)
16. Fan, Z., Liang, G., Lin, W.: Self-adaptive neighborhood search method and its extension for solving TSP. Comput. Eng. Appl. **12**, 75–78 (2008) (In Chinese)
17. Li, Y., Li, J., Gao, Z.: Large-scale neighborhood search algorithms for solving the time-varying vehicle scheduling problem. Manage. Sci. J. **15**(1), 22–32 (2012) (In Chinese)

# Potential Reduction of $CO_2$ Emissions Under Rebalancing Process in China

**Ran Wu, Xiaoying Chang, and Ping Ma**

**Abstract** This chapter analyses how economic growth structure impacts on the growth of $CO_2$ emissions from an overall perspective, supply-side perspective and demand-side perspective. Growth contribution of secondary industry is the most important factor on increasing emissions growth in supply-side. Growth contribution of consumption and net outflow are the most significant ones in demand-side, but due to the tiny share of net outflow and inventory change, consumption actually becomes the most important factor to induce emissions as a result. Supply-side structural change can be more helpful to reduce $CO_2$ emissions while increasing share of consumption in demand-side structural change will hinder emission reduction. Potential reduction is discussed under future rebalancing process in China with reference to the forecasts from official organisations. Comparing the most radical situation with the most conservative one under the same GDP growth level, an approximate 0.2% decline of $CO_2$ emissions under different supply-side growth structures can be observed. A possible potential 0.21%, 0.13% and 0.13% fluctuation of $CO_2$ emissions can be observed corresponding to 6%, 5.5% and 5% GDP growth rate under different demand-side growth structures. Additionally, lots of information like energy structure change, emission reduction policies or technological progress are hided into year effects which need to be discussed carefully in further studies.

**Keywords** China · $CO_2$ emissions · Rebalancing · Structural change · Potential reduction

R. Wu · X. Chang · P. Ma (✉)
School of Management, Harbin Institute of Technology, Harbin 150000, China
e-mail: 93834334@qq.com

R. Wu
e-mail: raynawu@163.com

X. Chang
e-mail: 984419060@qq.com

 
F. K. S. Chan et al. (eds.), *Proceedings of the 2020 International Conference on Resource Sustainability: Sustainable Urbanisation in the BRI Era (icRS Urbanisation 2020)*, Environmental Science and Engineering,
https://doi.org/10.1007/978-981-15-9605-6_19

After years of rapid growth, China has entered the stage of late industrialization, facing the transition from high speed growth to medium-high speed growth. Average annual growth rate of GDP between 2000 and 2010 was over 10% while economic growth has slowed and levelled off at around 7% since 2010. Rapid economic growth resulted in a large amount of carbon emissions and China has become the largest carbon emitter in the world. At the end of 2017, China has cut $CO_2$ emissions caused by per unit of GDP by 46% compared with that in 2005, fulfilling its commitment to reduce carbon emissions by 40–45% from the 2005 level by 2020. Although China has reached its 2020 carbon emissions target mentioned in 2009 Copenhagen Climate Conference three years ahead of schedule, according to BP statistical review of world energy [1], China's $CO_2$ emission is still the highest in the world, up to 9232.6 Mio. Tonnes in 2017, accounting for almost 30% of global $CO_2$ emissions. As the key contributor of global $CO_2$ emissions growth [1], this is a great challenge for global environment.

After China became of a member of WTO and participated more and more into global economic system, increasing external demand pulled China's export sharply and simulated the development of secondary industry, especially manufacturing industry. From the perspective of final demands, there is no doubt that export was a significant factor for China's growth in the past years, while from the perspective of industry structure, growth of secondary industry contributed most. Secondary industry is most closely to energy consumption, leading to the largest amount of $CO_2$ emissions among all the industries, and China's main source of energy is still coal till now. The rise of $CO_2$ emissions in recent years is due to rapid industrialization [3, 4], and China's overall coal consumption is likely to increase slightly in the following years, which will raise the pressure to reduce emissions [5]. Export was a key engine of China's economic growth before global financial crisis but it was weakened due to the collapse of external demand, and then economic growth turned to rely on investment most in recent years. The sharp rise in the share of investment expenditure in China's GDP has been mirrored by a fall in the share of final consumption expenditure [6]. Previous extensive mode of growth relied heavily on investment, external demand and resource consumption, which is not sustainable and green development path will be the inevitable choice for China. Government has also adopted policies to generate a sustainable growth path which can meet country's future demand.

China's economy has stepped into a "new normal" phase nowadays and is experiencing shift from high-speed growth to high-quality growth. According to IMF report [7], China's rebalancing process is consisted of four key elements: external, internal, environmental and distributional rebalancing. All the elements are interlinked closely and reinforcing each other. Figure 1 presents the contents of rebalancing and basic relationship among different elements. Upgrading economic structure will be carried on from the following aspects. Firstly, composition of demand will change as future growth will rely less on external demand and investment and turn to domestic consumption-led mode [8]. Currently, the share of household consumption in China's GDP is still much more lower than that of developed countries like the United States and the United Kingdom [9]. Secondly, although tertiary industry has become the main growth contributor since 2014, industry structure will furtherly shift

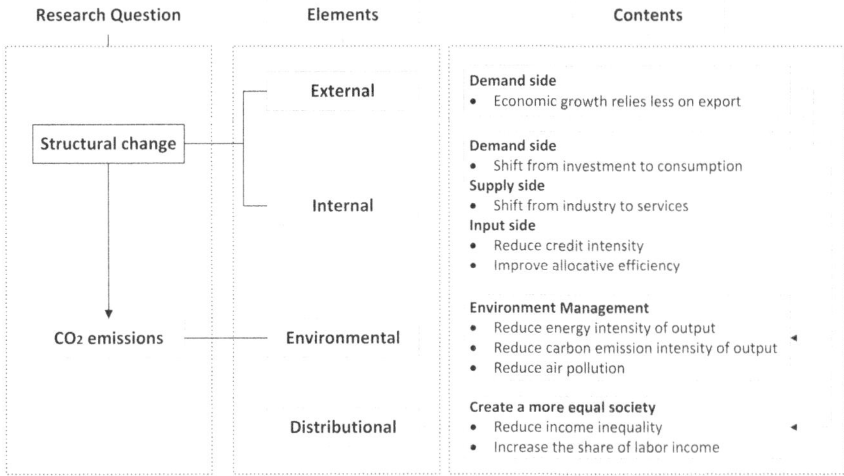


**Fig. 1** Rebalancing contents and research question

from industry to services. Thirdly, reducing credit intensity and improving allocative efficiency are the goals from the perspective of input side. Both external element and internal element will impact on improving environment including reducing energy intensity, carbon emissions and air pollution. Additionally, a healthy growth mode will help to create a more equal society as the distributional element shows.

Above all, it is of significance to explore the potential reduction of $CO_2$ emissions during China's future rebalancing process. In this paper, we will mainly focus on how external and internal rebalancing impact on environmental rebalancing in terms of reducing $CO_2$ emissions. More precisely, how will the change of supply-side growth structure and demand-side growth structure impact on carbon emissions growth? How much will the potential reduction of carbon emissions be during China's future rebalancing process? Our contributions include: (1) analyzing how the change of economic growth and growth structure influence $CO_2$ emissions growth from overall perspective, supply-side perspective and demand-side perspective; (2) predicting the reduction potential of $CO_2$ emissions from both supply-side and demand-side under different growth levels, which is a new perspective on the questions related to environmental problems; (3) using $CO_2$ emissions data at provincial level for a long period from 1997 to 2015, which provides a more persuasive evidence. Additionally, outliers are excluded from the samples in order that the estimation results can be more accurate.

# 1 Literature Review

Problems related to China's peak $CO_2$ emissions has been brought into focus for a long time. Applying a Monte Carlo approach and using Kuznets function, a research projects that China's $CO_2$ emissions will be likely to peak between 2021 and 2025 [10]. Studies using other methods including multi-objective optimization model [11] and decomposition method based on Kaya identity [12] also get the same results. Additionally, China's export emissions have already peaked, which has made great contribution to global emission reduction [13]. As China becomes richer, there will be a long-term trend decoupling of emissions and GDP, and there may be intermittent peak emission due to economic cycle [14]. Carbon intensity in most provinces is expected to fall furtherly based on multi-regional computable general equilibrium model, therefore reducing $CO_2$ emissions [15]. However, using updated and harmonized energy consumption and clinker production data and new sets of measured emission factors for Chinese coal, China Emission Accounts and Datasets team holds a different idea that previous $CO_2$ emissions in China might have been overrated [16]. Research based on LMDI method from the same institution also shows that China's $CO_2$ emissions has already peaked in 2014 [17], and China's carbon emissions have plateaued due to energy efficiency gains and structural upgrading [18].

Researchers also pay highly attention to the driving forces of $CO_2$ emissions in China. Large number of researches have proved that economic growth is the main driver of $CO_2$ emissions, as rapid growth is always related to energy consumption. The decrease of energy intensity is the main indicator to reduce emissions. Decomposition method has been applied in many studies when analyzing driving factors of emissions, and researches based on LMDI method have shown that economic growth is the main driver of emission growth, while emission intensity and efficiency are two offsetting factors [19–21], but the reduction effect of inhibiting factors is smaller than the driving effect of economic growth [22]. Industrial structure adjustment has the biggest potential to reduce emissions [23], while another study based on dynamic spatial panel models propose that technical progress plays the most important role in reducing carbon intensity, and industrial structure adjustment plays a secondary role [24]. Apart from elements related to emissions mentioned above, there are many other factors which should be responsible for the increase of $CO_2$ emissions, including population scale, energy consumption structure [25] and urbanization [26].

More precisely, from the view of industry structure, most studies support the idea that industry structure is closely related to $CO_2$ emissions. China's industrial restructuring exerts a positive impact on reduction of $CO_2$ emissions and the impact varies with sectors' share within the economic structure [27]. Industrial sectors have accounted for more than 50% of China's final energy consumption in the past 30 year and remained heavily reliant on coal, inducing the largest percentage of $CO_2$ emissions. It is expected that industrial energy demand and $CO_2$ emissions will approach a plateau between 2030 and 2040, and then decrease gradually [28]. Correspondingly, the adjustment of industry structure has made great contribution to reduce $CO_2$ emissions in recent years [29], and energy-intensive heavy industries including cement, iron and steel, and electricity sectors, took the lead in capping $CO_2$ emissions [30]. Emissions related to agriculture mainly come from fertilizer, pesticide, agricultural film, agricultural diesel, irrigation and plowing, accounting for a small share in total $CO_2$ emissions, and agricultural emissions will increase with the development of agricultural economy [31]. Additionally, increasing the share of tertiary industry is also helpful to emission reduction [32]. Conversely, few studies suggest that industry structure has limited impact on emissions [33, 34].

Analyzing carbon emissions from the perspective of final demand structure, current structure of final demands is inhibiting the decline of carbon intensity, as a research based on input-output SDA shows [35]. Another study proposes that Chinese export production has contributed most to emission increase, followed by capital formation, and carbon emissions related to consumption of services by urban households and governmental institutions are growing components for emissions [36], and especially emissions related to urban households consumption should be paid more attention [37, 38]. The effect of external demand also drives carbon emissions as China is a developing country with big economic scale and its increasing important role in global production system significantly affects $CO_2$ emissions [39]. Net export effects in developed countries are greater than that in developing countries, which means developing countries like China are becoming pollution haven, and China is indeed a carbon emissions exporter [40]. However, there still exists opposite idea that the impact of trade openness on China's $CO_2$ emissions is negligible [41].

Above all, rapid economic growth is mainly responsible for the increase of China's $CO_2$ emissions. Decomposition methods including IDA, SDA and LMDI are popular in most studies, combining with input-output analysis. Econometric analysis like co-integration model and Granger causality analysis are also applied in some studies. As to the research contents, researches have paid more attention to the driving factors of $CO_2$ emissions, and relationship between different elements and emissions has been studied deeply. More precisely, previous studies also focused on how industry structure and final demand impact on $CO_2$ emissions, but there are few studies tracking how growth structure change influence emissions growth. Thus, we will discuss to what extent the environment can benefit from China's future rebalancing process from the perspective of $CO_2$ emissions. Specifically, this study will exploit a new research angle and fill the research gap by decomposing growth structure from the perspective of supply-side and demand-side, and then analyzing the relationship between structural change and emissions growth under China's future rebalancing process. Furthermore, potential reduction of $CO_2$ emissions under different growth circumstances will be predicted and discussed.

## 2 Methodology and Data

### 2.1 Basic Estimation

To evaluate the overall emission elasticity of GDP, we focus on the average effect of economic growth and $CO_2$ emissions growth. The basic estimation equation is:

$$\Delta \ln C_{i,t} = \alpha \Delta \ln Y_{i,t} + \beta_i + \beta_t + \varepsilon_{i,t} \tag{1}$$

where $C_{i,t}$ is the $CO_2$ emissions of province $i$ in year $t$, $Y_{i,t}$ is gross regional domestic product of province $i$ in year $t$. $\Delta$ is the first difference operator, and $\Delta \ln C_{i,t}$ and $\Delta \ln Y_{i,t}$ represents the growth rate of $CO_2$ emissions and the growth rate of GDP respectively. Also, the first order difference makes the data stationary. $\beta_i$ is a province-specific effect and $\beta_t$ is a time-specific effect, and $\varepsilon_{i,t}$ is an error item. $\alpha$ is the overall emission elasticity of GDP. Subtracting $\Delta \ln Y_{i,t}$ from both sides of Eq. (1) shows the impact of a 1% GDP growth on the carbon intensity of GDP which is given by $(\alpha - 1)\%$.

### 2.2 Supply-Side Estimation

To test how sectoral composition of economic growth impacts on overall $CO_2$ emissions, total value added is divided into value added of (1) primary industry (agriculture, forestry, animal husbandry and fishery and services); (2) secondary industry (industry and construction industry); (3) tertiary industry (service industry). Then we have the equation $O = P + S + T$, where $O$, $P$, $S$ and $T$ represent total value added by output approach, value added of primary industry, value added of secondary industry and value added of tertiary industry respectively.

More precisely, differentiating the equation with respect to time $t$, diving by $O$, and denoting *dx/dt* by a dot over the variable, with reference to Burke's method [42], the following equation can be got:

$$\frac{\dot{O}}{O} = \frac{P}{O}\frac{\dot{P}}{P} + \frac{S}{O}\frac{\dot{S}}{S} + \frac{T}{O}\frac{\dot{T}}{S} \tag{2}$$

Equation 2 states that the growth rate of total value added equals the sum of weighted average of sectoral value-added growth rates. To put it in another formation, we have Eq. 3 as following:

$$\Delta \ln O_{i,t} = \frac{P}{O}\Delta \ln P_{i,t} + \frac{S}{O}\Delta \ln S_{i,t} + \frac{T}{O}\Delta \ln T_{i,t} \tag{3}$$

Furtherly, the estimation model is built as Eq. 4. Each independent variable means how much the growth of specified item contributes to overall GDP growth.

$$\begin{aligned}\Delta\ln C_{i,t} &= \alpha_1\Delta\ln P_{i,t(weighted)} + \alpha_2\Delta\ln S_{i,t(weighted)} \\ &\quad + \alpha_3\Delta\ln T_{i,t(weighted)} + \beta_i + \beta_t + \varepsilon_{i,t}\end{aligned} \tag{4}$$

## 2.3 Demand-Side Estimation

Expenditure-based GDP (*E*) is also divided into several items: (1) rural consumption (*RC*); (2) urban consumption (*UC*); (3) government consumption (*GC*); (4) completed fixed assets (*FA*); (5) changes in inventories (*IN*) and (6) net outflow of goods and service (*NX*). Same with the equation above, $E = RC + UC + GC + FA + IN + NX$, and the estimation model is shown as Eq. 5:

$$\frac{\dot{E}}{E} = \frac{RC}{E}\frac{\dot{RC}}{RC} + \frac{UC}{E}\frac{\dot{UC}}{UC} + \frac{GC}{E}\frac{\dot{GC}}{GC} + \frac{FA}{E}\frac{\dot{FA}}{FA} + \frac{IN}{E}\frac{\dot{IN}}{IN} + \frac{NX}{E}\frac{\dot{NX}}{NX} \tag{5}$$

Consistent with the supply-side estimation model, Eqs. 6 and 7 can be got as below.

$$\begin{aligned}\Delta\ln E_{i,t} &= \frac{RC}{E}\Delta\ln RC_{i,t} + \frac{UC}{E}\Delta\ln UC_{i,t} + \frac{GC}{E}\Delta\ln GC_{i,t} \\ &\quad + \frac{FA}{E}\Delta\ln FA_{i,t} + \frac{IN}{E}\Delta\ln IN_{i,t} + \frac{NX}{E}\Delta\ln NX_{i,t}n\end{aligned} \tag{6}$$

$$\begin{aligned}\Delta\ln C_{i,t} &= \alpha_1\Delta\ln RC_{i,t(weighted)} + \alpha_2\Delta\ln UC_{i,t(weighted)} + \alpha_3\Delta\ln GC_{i,t(weighted)} \\ &\quad + \alpha_4\Delta\ln FA_{i,t(weighted)} + \alpha_5\Delta\ln IN_{i,t(weighted)} + \alpha_6\Delta\ln NX_{i,t(weighted)} \\ &\quad + \beta_i + \beta_t + \varepsilon_{i,t}\end{aligned} \tag{7}$$

In the database of expenditure-based GDP, detailed export data and import data cannot be accessed separately. Thus, when net outflow is negative, logarithmic transformation is based on the opposite number and the weight is negative. Net-export provinces are distinguished from net-import provinces in terms of the outflow item. More precisely, the extension of the model is shown as Eq. 8. Each independent variable represents the growth contribution of each item to overall GDP growth.

$$\begin{aligned}\Delta\ln C_{i,t} &= \alpha_1\Delta\ln RC_{i,t(weighted)} + \alpha_2\Delta\ln UC_{i,t(weighted)} + \alpha_3\Delta\ln GC_{i,t(weighted)} \\ &\quad + \alpha_4\Delta\ln FA_{i,t(weighted)} + \alpha_5\Delta\ln IN_{i,t(weighted)} + \alpha_6\Delta\ln EX_{i,t(weighted)} \\ &\quad + \alpha_7\Delta\ln IM_{i,t(weighted)} + \beta_i + \beta_t + \varepsilon_{i,t}\end{aligned} \tag{8}$$

## 2.4 Data Resources

Output-based GDP and expenditure-based GDP for each provinces from 1997 to 2015 are derived from China Statistical Yearbook and Provincial Statistical Yearbook. GDP measured by output approach covers the value added of primary industry (agriculture, forestry, animal husbandry and fishery and services), secondary industry (industry and construction industry), and tertiary industry (transportation, storage, postal and telecommunications services, wholesale, retail trade and catering services and other services). Indexes of value added for three industries (preceding year = 100) in output-based GDP are collected from Provincial Statistical Yearbook and therefore growth contribution of three industries can be calculated directly. GDP measured by expenditure approach covers final consumption expenditure (rural households consumption, urban households consumption and government consumption), total investment (completed fixed assets and change in inventories), and net outflow of goods and services. However, growth rate of different items in expenditure-based GDP are not available and price indexes for different items are not provided completely. Therefore, we estimate the real value added of these items in expenditure-based GDP using real GDP multiplied by the proportion of corresponding item in nominal GDP, in order to get the growth rate of different items. Different items of GDP data in Statistical Yearbook are counted as current price, and the earliest year 1997 is used as the base year when adjusting nominal GDP to real GDP.

$CO_2$ emissions data at provincial level from 1997 to 2015 are collected from CEADS (China Emission Accounts and Datasets), in which emissions data for province Tibet, Hong Kong, Macau and Taiwan are not given [43]. Specifically, the database covers 44 sectors, rural emissions and urban emissions, also specific emissions derived by different types of fuels. Based on Chinese National Economical Industry Classification (GB/T4754-2017), we merge sectoral emission data to calculate emissions for primary industry, secondary industry and tertiary industry. The database uses unified approach to calculate emission data for all the provinces in China, and it is the first provincial-level emission database for China. $CO_2$ emissions data for Hainan province in 2002 and Ningxia province from 2000 to 2002 are not available. Finally, the main samples cover 566 observations for 30 provinces over 19 year from 1997 to 2015.

As the existence of outliers may negatively affect the accuracy and significance of the models, BACON algorism [44, 45] is used to exclude outliers in the observations, and 0.05 is chosen as default percentile in order to keep most observations and improve the accuracy of estimations as much as possible in all the following estimations.

## 2.5 Descriptive Statistics

Table 1 presents the mean of some main indicators related to economic growth and

**Table 1** Mean of some main indicators

| Year | GDP (100 Mio. Yuan) | $CO_2$ emissions (million tons) | Emissions of secondary industry (%) | Emissions generated by raw coal (%) | Emissions of urban households (%) | Number of observations |
|---|---|---|---|---|---|---|
| 1997 | 2,563 | 97.86 | 81.0 | 65.4 | 4.2 | 30 |
| 2000 | 3,457 | 105.3 | 80.5 | 62.8 | 3.7 | 29 |
| 2003 | 4,548 | 138.5 | 82.6 | 62.2 | 3.1 | 30 |
| 2006 | 6,634 | 206.6 | 83.0 | 59.8 | 2.7 | 30 |
| 2009 | 9,478 | 255.2 | 83.3 | 57.2 | 2.4 | 30 |
| 2012 | 13,232 | 316.7 | 82.9 | 55.9 | 2.8 | 30 |
| 2015 | 16,912 | 318.6 | 81.4 | 52.1 | 3.0 | 30 |

$CO_2$ emissions in some years. They are GDP, $CO_2$ emissions, the proportion of $CO_2$ emissions from secondary industry, the proportion of $CO_2$ emissions generated by raw coal and the proportion of emissions from urban households. $CO_2$ emissions increased over years as economic grew fast. But during the period between 2012 and 2015, GDP grew by nearly 30% while $CO_2$ emissions didn't grow obviously. The proportion of emissions caused by secondary industry decreased in recent years. Meantime, from the perspective of energy structure, the proportion of emissions generated by raw coal in total emissions also fell down in recent years. Decrease of the proportion of both secondary industry emissions and raw coal emissions helped to reduce emission effectively. However, the proportion of urban household emissions was increasing in recent years, which was directly related to urban household consumption.

# 3 Results

## *3.1 Basic Estimation*

Table 2 shows the emission-income elasticity respectively using two differenced-log specifications and log specifications. Standard errors in parentheses are robust to heteroscedasticity and clustered by province. In the first column, both emission-income elasticity and constant are significant ($p < 0.1$) but the significance is not so satisfactory. After two outliers are excluded, result in column 2 is got. Figure 2 presents the scatter plot of all the observations. It can be seen that two outliers deviate a lot from the whole. Both coefficients and R-square improved remarkably compared with column 1 and not so many observations are missed. More precisely, 1% increase of economic growth will lead to 0.991% increase of $CO_2$ emissions, associated with a decline in the carbon intensity of GDP of 0.009%. Column 3 presents the result

**Table 2** Basic estimation of emission-income elasticity

| | (1) $\triangle \ln CO_2emissions_{i,t}$ | (2)* $\triangle \ln CO_2emissions_{i,t}$ | (3) $\triangle \ln CO_2emissions_{i,t}$ |
|---|---|---|---|
| $\triangle \ln GDP_{i,t}$ | 0.807** | 0.991*** | |
| | (0.331) | (0.202) | |
| $\ln GDP_{i,t}$ | | | 0.907** |
| | | | (0.390) |
| Constant | −0.057* | −0.073*** | −2.487 |
| | (0.033) | (0.024) | (2.940) |
| Observations | 534 | 532 | 566 |
| R-squared | 0.276 | 0.332 | 0.917 |
| Number of province | 30 | 30 | 30 |
| Province fixed effects | YES | YES | YES |
| Year fixed effects | YES | YES | YES |

Robust standard errors in parentheses, ***p < 0.01, **p < 0.05, *p < 0.1

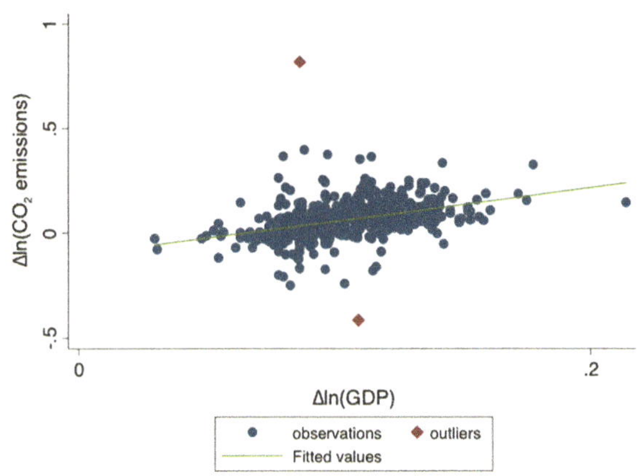


**Fig. 2** Scatter plot of $CO_2$ emissions growth rate and GDP growth rate

without using first-difference estimator as a kind of reference, and the coefficient also denotes emission-income elasticity, which is quite close to the previous estimations, but it doesn't have a strong reference value as there may exist unit root for both variables. Thus, estimation 2 is set as our basic model.

## 3.2 Supply-Side Estimation

Table 3 presents how sectoral composition of GDP growth impacts on overall

**Table 3** Supply-side estimation

| | (1) $\triangle \ln CO_2 emissions_{i,t}$ (weighted) | (2)* $\triangle \ln CO_2 emissions_{i,t}$ (weighted) | (3) $\triangle \ln CO_2 emissions_{i,t}$ (without weight) |
|---|---|---|---|
| $\triangle \ln P$ (value added) | 0.315 | 0.529 | 0.268 |
| | (0.885) | (0.896) | (0.399) |
| $\triangle \ln S$ (value added) | 0.664** | 0.898*** | 0.986** |
| | (0.303) | (0.181) | (0.433) |
| $\triangle \ln T$ (value added) | 0.554 | 0.627* | 0.749 |
| | (0.385) | (0.345) | (0.968) |
| Constant | −0.040 | −0.055** | 4.136*** |
| | (0.030) | (0.025) | (0.141) |
| Observations | 534 | 533 | 566 |
| R-squared | 0.273 | 0.300 | 0.909 |
| Number of province | 30 | 30 | 30 |
| Province fixed effects | YES | YES | YES |
| Year fixed effects | YES | YES | YES |

Robust standard errors in parentheses, ***$p < 0.01$, **$p < 0.05$, *$p < 0.1$

GDP growth. Output-base GDP growth are divided into primary industry growth, secondary industry growth and tertiary industry growth. Each sector has contributed differently to the overall growth. Independent variables mean the weighted growth rate for each sector, and the sum of all the variables equals to overall GDP growth. In the estimation using all 534 observations, coefficients of most variables are not significant, which is not consistent with the assumption, as sectoral growth would have impacts on the growth of emissions. Then one outlier is excluded, which extremely deviate the estimation result, and the result in column 2 is got. Significance of most coefficients and R-squared improve obviously. More precisely, 1% increase of the contribution of industry on GDP growth will lead to 0.898% increase of emissions while 1% increase of the contribution of service industry on GDP growth will lead to 0.627% $CO_2$ emissions growth. The coefficient of agriculture's contribution is not significant which is consistent with the assumption as the proportion of emissions caused by agriculture is tiny, and agriculture is not a carbon-intensive industry. If the weights are excluded from the independent variables, result in column 3 can be given. It's obvious that increase in secondary industry growth will lead to more emissions growth than the other sectors and $CO_2$ emission is not sensitive to primary industry growth and tertiary industry growth. In all, estimation 2 is set as the basic model from the perspective of supply-side.

### *3.3 Demand-Side Estimation*

Then we discuss how demand-side growth structure influence growth of $CO_2$ emissions as shown in Table 4. Column 1 shows the result using all the observations. Then one outlier is excluded from the observations and result in column 2 is got. A slightly improvement can be seen in terms of the significance of the coefficients and R-squared. As shown in column 2, 1% increase in the growth contribution of government consumption, urban consumption and inventory change in overall GDP growth can lead to 0.913%, 0.810% and 0.816% carbon emissions growth, respectively. Also, 1% increase of the growth contribution of fixed assets also increase 0.429% emissions growth positively. Consumption is related to production directly and causing emissions. Investment also plays an important role in China's growth and large number of investment have been put into secondary industry which is the largest emission sector in the past years. The growth contribution of net export also has positive effect on $CO_2$ emissions, as the increasing external demands can stimulate domestic production, producing more emissions as a result. Correspondingly, growth contribution of net import will reduce $CO_2$ emissions growth, but the degree is not that obvious as net export. Then the weight is removed from the model and result in column 3 is given. Obviously, the growth of fixed assets has the largest effect on emission growth, while the impact of urban consumption growth and government consumption growth play secondary roles. The coefficient of net export growth is also significant, which is consistent with the results in column 1 and 2. In all, estimation 2 is set as the basic model from the perspective of demand-side.

**Table 4** Demand-side estimation

| | (1) $\Delta \ln CO_2 emissions_{i,t}$ (weighted) | (2)* $\Delta \ln CO_2 emissions_{i,t}$ (weighted) | (3) $\Delta \ln CO_2 emissions_{i,t}$ (without weight) |
|---|---|---|---|
| $\Delta \ln RC$ | 0.275 | 0.335 | −0.006 |
| | (0.647) | (0.649) | (0.067) |
| $\Delta \ln UC$ | 0.771*** | 0.810*** | 0.105* |
| | (0.251) | (0.250) | (0.060) |
| $\Delta \ln GC$ | 0.868*** | 0.913*** | 0.081** |
| | (0.227) | (0.230) | (0.031) |
| $\Delta \ln FA$ | 0.400*** | 0.429*** | 0.108** |
| | (0.098) | (0.099) | (0.043) |
| $\Delta \ln IN$ | 0.769** | 0.816** | 0.009 |
| | (0.338) | (0.330) | (0.009) |
| $\Delta \ln EX$ | 0.689*** | 1.032*** | 0.019*** |
| | (0.237) | (0.221) | (0.006) |
| $\Delta \ln IM$ | −0.210** | −0.235*** | 0.003 |
| | (0.082) | (0.084) | (0.013) |
| Constant | −0.040** | −0.043** | −0.021 |
| | (0.019) | (0.018) | (0.020) |
| Observations | 479 | 478 | 479 |
| R-squared | 0.293 | 0.297 | 0.277 |
| Number of province | 30 | 30 | 30 |
| Province fixed effects | YES | YES | YES |
| Year fixed effects | YES | YES | YES |

Robust standard errors in parentheses, ***$p < 0.01$, **$p < 0.05$, *$p < 0.1$

# 4 Prediction

In this section, several growth scenarios are proposed to predict emissions using basic estimation model, basic supply-side model and basic demand-side model, respectively. The indicators in different scenarios come from official international organisations and academic institutions including World Bank, International Monetary Fund (IMF), United Nations (UN), Organization for Economic Cooperation and Development (OECD), Oxford Economics and Chinese Academy of Social Sciences (CASS). Estimations above have considered time trend although year effect hasn't been presented in the results. Coefficients of most years are significant and there doesn't exist a regular trend, while year effects in recent years fluctuate around 0. In this way, we suppose that long-term year effects of all the estimations will be stable at around 0.

## *4.1 Prediction of $CO_2$ Emissions Growth Based on Basic Estimation*

Firstly, we estimate the growth of $CO_2$ emissions from 1997 to 2018 using China's historical GDP growth data published by Chinese government. World Bank, BP statistical review and CEADS all published $CO_2$ emissions data at country level which can be used to test our estimation. As shown in Fig. 3, points of different patterns are the $CO_2$ emissions growth rate in different years and the solid line also shows the historical estimation. It can be seen that the historical estimation is consistent with the real trend of $CO_2$ emissions growth. According to the calculation, an annual GDP growth rate of 6%, 5.5% and 5% will correspond to annual $CO_2$ emissions growth rate of −0.78%, −1.43% and −1.8% respectively. When GDP growth rate falls from 6.0% to 5.5% and from 5.5% to 5%, approximate 37.04% and 27.03% decline of emissions growth rate can be observed respectively on condition that year effect keeps unchanged.

## *4.2 Supply-Side Prediction*

First of all, historical industrial growth contribution data is used to test our supply-side model. Based on supply-side estimation model in the last section, historical estimation result is shown in Fig. 4. The estimation is consistent with the overall trend of $CO_2$ emissions growth. Taking rebalancing process into consideration, tertiary industry will play a more important role in economic growth while secondary industry will become less important in China's future growth path. In this part, we will focus on the impact on $CO_2$ emissions growth from growth structural change, in terms of growth contribution of different industries. Figure 5 presents historical supply-side

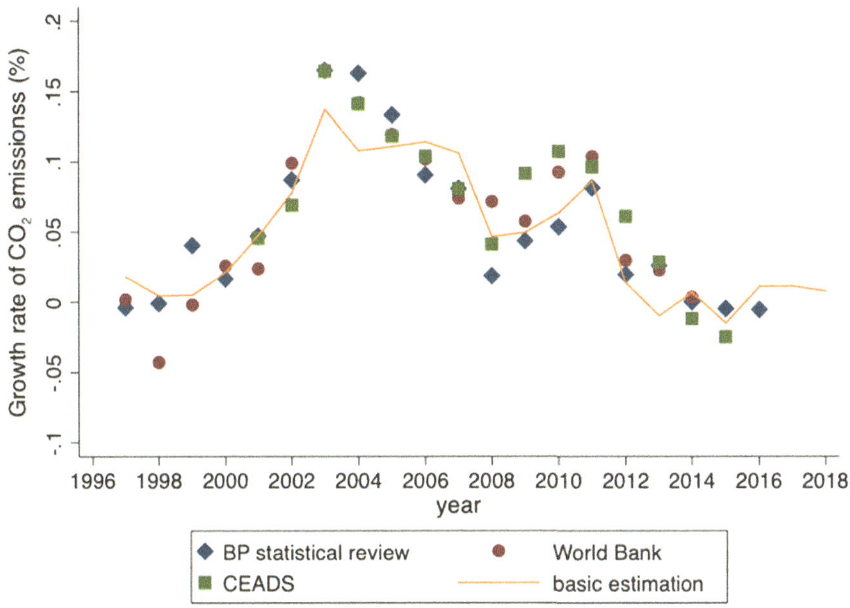


**Fig. 3** $CO_2$ emissions growth under basic estimation

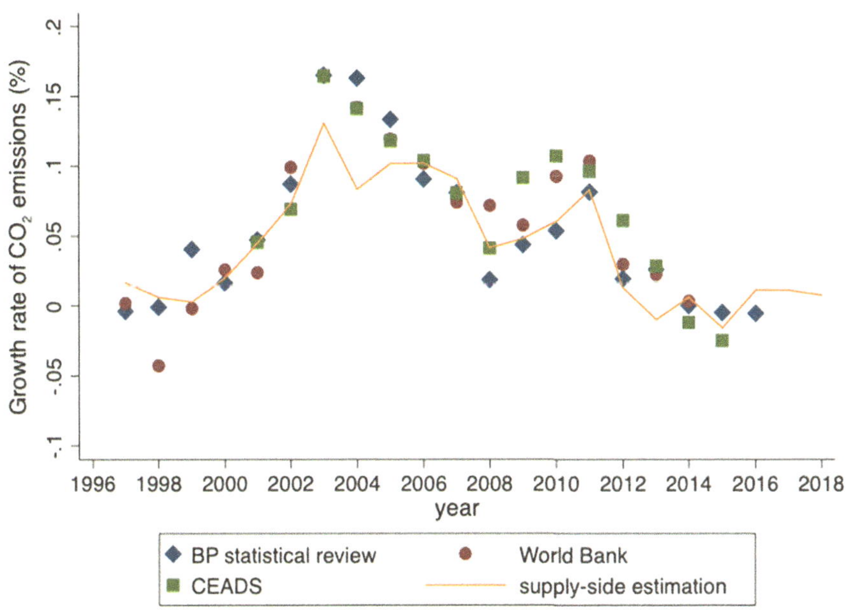


**Fig. 4** Supply-side historical estimation

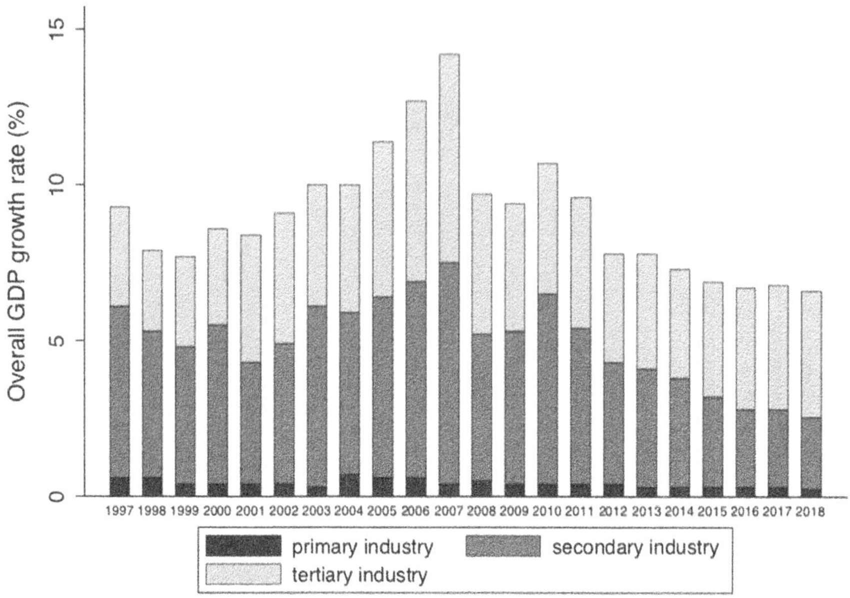


**Fig. 5** Historical supply-side growth structure

growth structure. It can be seen that the role of secondary industry growth played a less important role over years while the importance of tertiary industry was growing. The effect of primary industry was quite stable, accounting for a really small part in overall growth.

There is no accurate and unified prediction for the proportion and the growth rate of three industries, and only few scattered prediction can be collected. IMF uses the growth contribution of tertiary industry vs secondary industry as a main indicator to test the improvement of China's rebalancing process, and the first column of Table 5 presents this indicator. Thus, this indicator is taken as the assumption criteria when GDP growth rate is 6, 5.5 and 5%. According to IMF, growth rate of secondary industry and tertiary industry will be similar and the share of tertiary sector in GDP will be around 55% when GDP growth rate is 6%. The share of secondary industry was quite stable since 2015, at around 39.7–41.0%. If this indicator can follow the historical trend, it is reasonable to assume that the share of secondary industry will be around 37–40%, which is also consistent with the prediction from CASS [46]. Therefore, growth contribution of tertiary industry will be 1.4–1.5 times growth contribution of secondary industry. Actually, the growth contribution of tertiary industry versus secondary industry has reached 1.5 since 2016, and this number was up to 1.7 in 2018.

Based on the discussion above, different levels of growth contribution of tertiary industry vs secondary industry are considered, holding overall GDP growth rate unchanged. Five scenarios are presented for each economic growth level. Growth contribution of tertiary industry vs secondary industry is set as 1–1.8. 1 is a conservative assumption as IMF predicted, but actually tertiary industry has already dominated economic growth since 2013, and 1.8 is a radical assumption if tertiary industry growth rate will grow faster than expectation. Additionally, 1.8 scenario under 6% GDP growth rate is also consistent with the prediction from World Bank, so 1.8 is set as the default baseline. As to the growth contribution of primary industry, the indicator kept stable at 0.4% from 2007 to 2012 and 0.3% from 2013 to 2018, and it is unlikely to fall down sharply in the following years due to its small share. Thus, only a 0.05% decrease is suggested in growth contribution of primary industry corresponding to a 0.5% decrease in GDP growth rate.

Considering three GDP growth level 6, 5.5 and 5% in the following years, the prediction results are shown in Table 5. Taking the baseline 1.8 as reference, potential reduction under different development levels are compared. $CO_2$ emissions will fall around 0.35–0.36% when GDP growth rate falls 0.5%, assuming year effects in the following years are constant. $CO_2$ emissions growth rates under different growth structures are also compared at the same GDP growth level. There is no doubt that higher growth contribution of tertiary industry will lead to less $CO_2$ emissions, and $CO_2$ emissions decrease most when the growth contribution of tertiary industry versus secondary industry is 1.8. Vice versa, $CO_2$ emissions decrease least when the growth

**Table 5** Supply-side potential reduction under different growth structures

| Growth contribution of tertiary industry vs secondary industry | GDP growth rate (%) | Growth contribution of primary industry (%) | Growth contribution of secondary industry (%) | Growth contribution of tertiary industry (%) | $CO_2$ emissions growth rate (without year effect) (%) | Compared with baseline (%) |
|---|---|---|---|---|---|---|
| 1.8 | 6 | 0.3 | 2.04 | 3.66 | −1.26 | |
| 1.6 | 6 | 0.3 | 2.19 | 3.51 | −1.22 | 0.04 |
| 1.4 | 6 | 0.3 | 2.38 | 3.33 | −1.17 | 0.09 |
| 1.2 | 6 | 0.3 | 2.59 | 3.11 | −1.11 | 0.15 |
| 1 | 6 | 0.3 | 2.85 | 2.85 | −1.04 | 0.22 |
| 1.8 | 5.5 | 0.25 | 1.88 | 3.38 | −1.62 | |
| 1.6 | 5.5 | 0.25 | 2.02 | 3.23 | −1.58 | 0.04 |
| 1.4 | 5.5 | 0.25 | 2.19 | 3.06 | −1.53 | 0.09 |
| 1.2 | 5.5 | 0.25 | 2.39 | 2.86 | −1.48 | 0.14 |
| 1 | 5.5 | 0.25 | 2.63 | 2.63 | −1.41 | 0.21 |
| 1.8 | 5 | 0.2 | 1.71 | 3.09 | −1.97 | |
| 1.6 | 5 | 0.2 | 1.85 | 2.95 | −1.93 | 0.04 |
| 1.4 | 5 | 0.2 | 2.00 | 2.80 | −1.89 | 0.08 |
| 1.2 | 5 | 0.2 | 2.18 | 2.62 | −1.84 | 0.09 |
| 1 | 5 | 0.2 | 2.40 | 2.40 | −1.78 | 0.19 |

contribution of tertiary industry vs secondary industry is 1. When GDP growth rate is 6% and the growth contribution of tertiary industry vs secondary industry is 1.6, a 0.04% increase can be observed compared with baseline, while a 0.09%, 0.15%, 0.22% increase correspond to 1.4%, 1.2% and 1 scenario respectively. When GDP growth rate is 5.5 and 5%, similar increase of $CO_2$ emissions can be observed under different scenarios compared with baseline. Then the radical situation 1.8 scenario is compared with the conservative situation 1 scenario, and an approximate 0.2% decline of $CO_2$ emissions can be got under different GDP growth levels, and this is a very substantial result. Above all, if the growth structure is inclined to rely on more on tertiary industry, considerable potential reduction of $CO_2$ emissions can be observed. Either the rise of growth rate or the percentage of tertiary industry will help to reduce emissions effectively.

### *4.3 Demand-Side Prediction*

Firstly, the accuracy of the estimation is tested using historical country-level data. As shown in Fig. 6, demand-side estimation is also consistent with the real trend of $CO_2$ emissions growth. Considering rebalancing process, the percentage of consumption contribution to GDP growth will increase while investment and export will play less important roles. Figure 7 shows the demand-side growth structure in the past

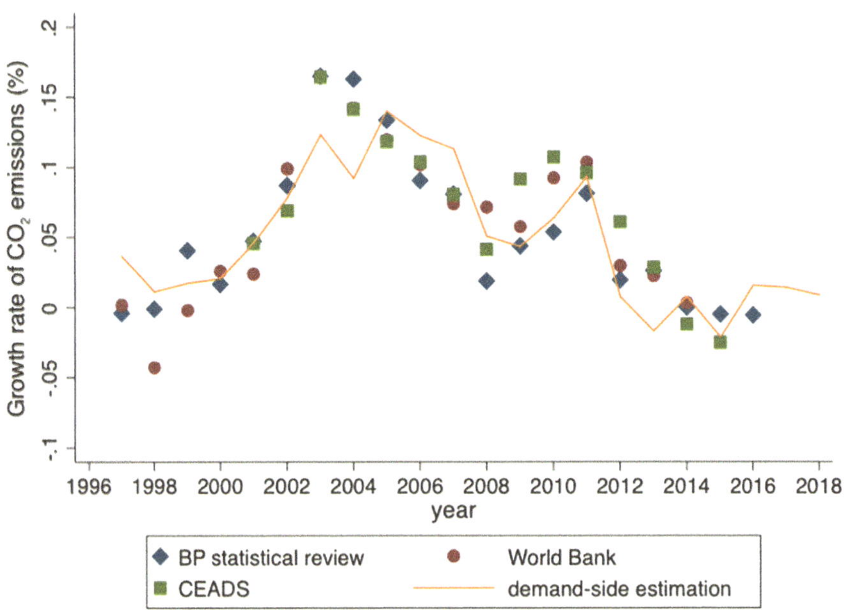


**Fig. 6** Demand-side historical estimation

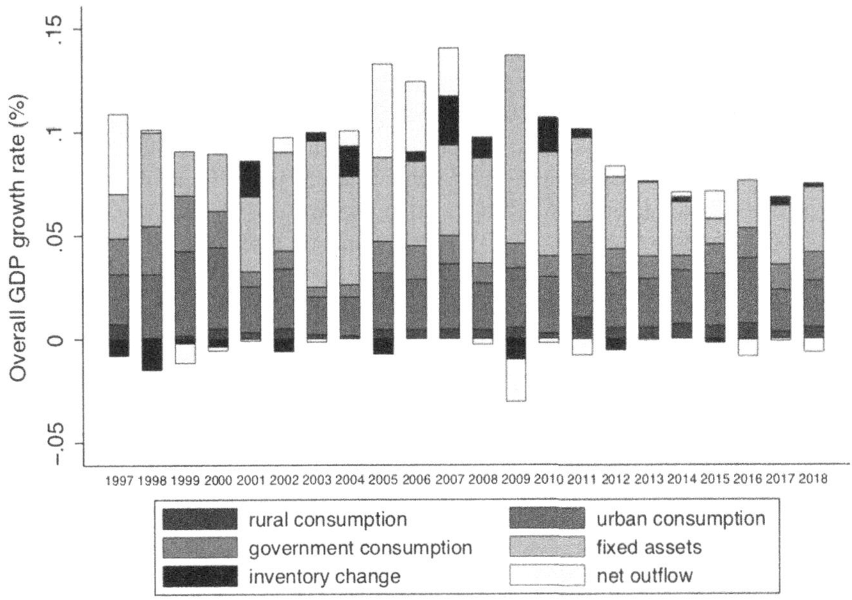


**Fig. 7** Historical demand-side growth structure

years. Growth contribution of consumption, especially urban consumption and rural consumption became dominant forces in GDP growth, and growth contribution of fixed assets also played an important role, while growth contribution of inventory change in recent years could be ignored. In addition, growth contribution of net export played a negative role in GDP growth in most years after 2007 due to global financial crisis, which helped to reduce China's $CO_2$ emission to a large extent.

With reference to the prediction from official organisations in terms of demand-side, different demand-side growth structures are discussed. Prediction from Oxford Economics, IMF and World Bank are mainly referred in this part. Oxford economics has predicted that the share of inventory will be stable at 1.7% in the following years, which is consistent with historical share of inventory (1.75% in 2017 and 1.79% in 2018). The growth contribution of inventory change will be also stable as IMF predicted. Thus, the percentage of inventory change is set as 1.7% and its growth rate is assumed to be similar to GDP growth rate, therefore the weighted growth rate of inventory change can be calculated, which is around 0.1% under different GDP growth levels. Then different scenarios are proposed based on the prediction of the change of fixed assets. When GDP growth rate is 6.0%, IMF predicted that the share of fixed assets investment would be around 39.9–40.5%, and the growth rate of fixed assets investment would be 6.0% and 5.2% correspondingly. However, Oxford Economics predicted that the growth rate of fixed assets investment would be 4.4% under the same economic growth level. Thus, three scenarios are considered under 6% growth level. In the third scenario, we set the share of fixed assets investment as 40.5% as the corresponding growth rate of fixed assets is closer to

Oxford Economics prediction. When GDP growth rate is 5.5%, IMF predicted the share of fixed assets investment would be 38.1%, and this prediction is applied in three scenarios. The first scenario is that when growth rate of fixed assets investment is 5.4%, predicted by IMF. However, Oxford Economics proposed a quite different growth rate 4.4%, and this forecast is taken as the third scenario. Since the difference between predictions from different organisations is too large, the average of two predictions is calculated, and the growth rate of fixed assets investment is 4.9%, which is set as the second scenario. When GDP growth rate is 5%, some indicators are assumed by ourselves with reference to the existed predictions. With the slow stable development of economic growth, we assume that share of fixed assets would be stable, and the share of fixed assets would be 38.1%, which is equal to 5.5% growth level. IMF shows that the growth rate of fixed assets will fall from 6.0% to 5.4% corresponding to GDP growth falling from 6.0% to 5.5%. Thus, when GDP growth falls to 5.0%, the growth rate of fixed assets will fall to 4.8% accordingly, and this is the first scenario. Oxford predicted the growth rate of fixed assets would be 3.8% and this is our third scenario. Taking the average of the two predictions above, we set the growth rate of fixed assets as 4.3% in the second scenario.

World Bank [47] has predicted that the growth contribution of net outflow would be −0.3% when GDP growth rate is 6% and we apply this prediction to the 6% growth level. According to IMF report, the growth contribution of net outflow will be −0.1% during the period when GDP growth rate falls down from 6 to 5.5%, and we apply this forecast to 5.5% growth level. IMF also pointed out that the contribution of net export would no longer impact economic growth in the long run, therefore we assume that net outflow will not impact economic significantly growth when GDP growth rate is 5%, and set the growth contribution of net outflow as zero.

Within the structure of growth contribution of final consumption, the share of three items including rural consumption, urban consumption and government consumption was stable during the past 5 year and followed the stabilized ratio at around 1:4.2:2.1. Therefore this ratio is used to predict internal growth contribution of consumption-related items.

Based on the discussions and assumptions above, Table 6 presents demand-side potential reduction under different growth structures. Assuming year effects in the future are all the same or equal to 0, a 0.22 and 0.28% reduction of $CO_2$ emissions growth can be observed when GDP growth falls from 6 to 5.5% and from 5.5 to 5%. This number is much more lower than supply-side estimation, as demand-side structural change tend to increase more consumption, which will actually increase emission growth. For each growth level, the first row is the basic scenario in which the growth rate of fixed assets is the highest. It can be seen that in all baseline scenarios will $CO_2$ emissions reduce more, as growth contribution of consumption tend to induce more $CO_2$ emissions in demand-side estimation model. When GDP growth rate is 6%, $CO_2$ emissions growth can be lower through adjusting final-demand structure such as decreasing the growth contribution of consumption or increasing the growth contribution of fixed assets. Compared with baseline, a possible increase of $CO_2$ emissions at 0.21% can be observed under the third scenario. 0.13% increase of emissions can be also got under different demand-side growth structures when GDP

**Table 6** Demand-side potential reduction under different growth structures (%)

| GDP growth rate | Rural consumption | Urban consumption | Government consumption | Fixed assets | Inventory change | Net outflow | Carbon emissions growth rate (without year effect) | Compared with baseline |
|---|---|---|---|---|---|---|---|---|
| 6 | 0.51 | 2.13 | 1.06 | 2.50 | 0.10 | −0.30 | −0.60 | |
| 6 | 0.55 | 2.30 | 1.15 | 2.19 | 0.10 | −0.30 | −0.50 | 0.10 |
| 6 | 0.59 | 2.50 | 1.25 | 1.86 | 0.10 | −0.30 | −0.39 | 0.21 |
| 5.5 | 0.46 | 1.93 | 0.97 | 2.15 | 0.09 | −0.10 | −0.82 | |
| 5.5 | 0.49 | 2.05 | 1.02 | 1.95 | 0.09 | −0.10 | −0.75 | 0.07 |
| 5.5 | 0.51 | 2.16 | 1.08 | 1.75 | 0.09 | −0.10 | −0.69 | 0.13 |
| 5 | 0.42 | 1.78 | 0.89 | 1.83 | 0.09 | 0.00 | −1.10 | |
| 5 | 0.45 | 1.89 | 0.94 | 1.64 | 0.09 | 0.00 | −1.04 | 0.07 |
| 5 | 0.47 | 1.99 | 1.00 | 1.45 | 0.09 | 0.00 | −0.97 | 0.13 |

growth rate is 5.5 or 5.0%. Above all, increase of growth contribution of consumption tend to hinder emission reduction. Conversely, decrease of growth contribution of net export will help to reduce emission effectively. It's a good new for emission reduction that net outflow will be no longer important in China's future growth.

## 5 Conclusion and Discussion

The chapter analyses the effect of economic growth structure on $CO_2$ emissions growth from an overall perspective, supply-side perspective and demand-side perspective using econometric method. We predict the potential reduction under China's future rebalancing process and test how structural change will impact on emissions. Supply-side structural change will be more helpful for emission reduction as tertiary industry will play a more important role. Demand-side structural change will hinder emission reduction with the increasing share of consumption. Detailed conclusions are as follows.

In terms of the overall relationship between economic growth and emission growth, 1% increase of economic growth will lead to 0.991% increase of $CO_2$ emissions, associated with a decline in the carbon intensity of GDP of 0.009%. An annual GDP growth rate of 6%, 5.5% and 5% will correspond to annual CO2 emissions growth rate of −0.78%, −1.43% and −1.8% respectively.

From the perspective of supply-side growth structural change, growth contribution of secondary industry is the most important influence factor on $CO_2$ emissions growth, and the role of growth contribution of tertiary industry is secondary. As future growth structure is inclined to rely more on tertiary industry, a considerable potential reduction of $CO_2$ emissions can be observed. $CO_2$ emissions will fall around 0.35 to 0.36% when GDP growth rate falls 0.5%. Taking the growth contribution of tertiary industry vs secondary industry as reference index, a possible approximate 0.2% potential reduction can be observed under different development levels when the index changes from 1 to 1.8. Either the increase of growth rate or the proportion of tertiary industry will help to reduce emission effectively.

From the perspective of demand-side growth structural change, consumption actually becomes the most important factor to induce emissions, and in the future rebalancing process, the increase of growth contribution of consumption will hinder emission reduction. Several scenarios are presented under different GDP growth levels by adjusting the growth rate and the percentage of fixed assets, combining with official possible prediction published by international organisations. $CO_2$ emissions will fall 0.22 and 0.28% when GDP growth rate falls from 6.0 to 5.5% and from 5.5 to 5.0%. And a possible potential fluctuation of 0.21%, 0.13% and 0.13% can be observed corresponding to 6%, 5.5% and 5% GDP growth rate.

In this paper, the estimation of year effect is assumed to be stable around zero and this may not be so accurate. As year effects include lots of information such as technological progress, publish of emission reduction policy and the change of energy structure. There exists so much uncertainty in the year effect and we are not able to take all the information into consideration. Additionally, supply-side analysis and demand-side analysis are separated in this paper and these two angles might be combined for further analysis in the next step.

## References


1. Dudley, B., et al.: Bp statistical review of world energy. In: BP Statistical Review, London, UK (2018)
2. Liu, D., Guo, X., Xiao, B.: What causes growth of global greenhouse gas emissions? Evidence from 40 countries. Sci. Total Environ. **661**, 750–766 (2019)
3. Long, X., Naminse, E.Y., Du, J., Zhuang, J.: Nonrenewable energy, renewable energy, carbon dioxide emissions and economic growth in China from 1952 to 2012. Renew. Sustain. Energy Rev. **52**, 680–688 (2015)
4. Shan, Y., Liu, J., Liu, Z., Xu, X., Shao, S., Wang, P., Guan, D.: New provincial $CO_2$ emission inventories in china based on apparent energy consumption data and updated emission factors. Appl. Energy **184**, 742–750 (2016)
5. Qiao, H., Chen, S., Dong, X., Dong, K.: Has China's coal consumption actually reached its peak? National and regional analysis considering cross-sectional dependence and heterogeneity. Energy Econ. **84**, 104509 (2019)
6. McKay, H., Song, L., et al.: Rebalancing the Chinese economy to sustain long-term growth. In: Rebalancing and Sustaining Growth in China, pp. 1–18. ANUE Press, Canberra (2012)
7. Zhang, M.L.: Rebalancing in China-progress and prospects. International Monetary Fund (2016)
8. Dorrucci, E., Pula, G., Santab´arbara, D.: China's economic growth and rebalancing. In: Banco de Espana Occasional Paper (1301) (2013)
9. Lardy, N.R.: China: rebalancing economic growth. In: The China Balance Sheet, pp. 1–24 (2007)
10. Wang, H., Lu, X., Deng, Y., Sun, Y., Nielsen, C.P., Liu, Y., Zhu, G., Bu, M., Bi, J., McElroy, M.B.: China's $CO_2$ peak before 2030 implied from characteristics and growth of cities. Nat. Sustain. **2**(8), 748–754 (2019)
11. Yu, S., Zheng, S., Li, X., Li, L.: China can peak its energy-related carbon emissions before 2025: evidence from industry restructuring. Energy Econ. **73**, 91–107 (2018)
12. Green, F., Stern, N.: China's changing economy: Implications for its carbon dioxide emissions. Climate policy **17**(4), 423–442 (2017)
13. Mi, Z., Meng, J., Green, F., Coffman, D., Guan, D.: China's "exported carbon" peak: patterns, drivers, and implications. Geophys. Res. Lett. **45**(9), 4309–4318 (2018)
14. Cohen, G., Jalles, J.T., Loungani, P., Marto, R., Wang, G.: Decoupling of emissions and gdp: evidence from aggregate and provincial Chinese data. Energy Econ. **77**, 105–118 (2019)
15. Li, J.F., Gu, A.L., Ma, Z.Y., Zhang, C.L., Sun, Z.Q.: Economic development, energy demand, and carbon emission prospects of China's provinces during the 14th five-year plan period: application of CMRCGE model. Adv. Clim. Change Res. (2019)
16. Liu, Z., Guan, D., Wei, W., Davis, S.J., Ciais, P., Bai, J., Peng, S., Zhang, Q., Hubacek, K., Marland, G., et al.: Reduced carbon emission estimates from fossil fuel combustion and cement production in china. Nature **524**(7565), 335 (2015)
17. Guan, D., Meng, J., Reiner, D.M., Zhang, N., Shan, Y., Mi, Z., Shao, S., Liu, Z., Zhang, Q., Davis, S.J.: Structural decline in China's $CO_2$ emissions through transitions in industry and energy systems. Nat. Geosci. **11**(8), 551 (2018)
18. Zheng, J., Mi, Z., Coffman, D., Milcheva, S., Shan, Y., Guan, D., Wang, S.: Regional development and carbon emissions in China. Energy Econ. **81**, 25–36 (2019)
19. Shan, Y., Zhou, Y., Meng, J., Mi, Z., Liu, J., Guan, D.: Peak cement-related $CO_2$ emissions and the changes in drivers in China. J. Ind. Ecol. (2019)

20. Du, K., Xie, C., Ouyang, X.: A comparison of carbon dioxide ($CO_2$) emission trends among provinces in China. Renew. Sustain. Energy Rev. **73**, 19–25 (2017)
21. Wang, C.: Differential output growth across regions and carbon dioxide emissions: evidence from us and China. Energy **53**, 230–236 (2013)
22. Zhang, Y.J., Da, Y.B.: The decomposition of energy-related carbon emission and its decoupling with economic growth in China. Renew. Sustain. Energy Rev. **41**, 1255–1266 (2015)
23. Zhang, C., Su, B., Zhou, K., Yang, S.: Decomposition analysis of China's $CO_2$ emissions (2000–2016) and scenario analysis of its carbon intensity targets in 2020 and 2030. Sci. Total Environ. **668**, 432–442 (2019)
24. Cheng, Z., Li, L., Liu, J.: Industrial structure, technical progress and carbon intensity in China's provinces. Renew. Sustain. Energy Rev. **81**, 2935–2946 (2018)
25. Xu, S.C., He, Z.X., Long, R.Y.: Factors that influence carbon emissions due to energy consumption in China: decomposition analysis using LMDI. Appl. Energy **127**, 182–193 (2014)
26. Ouyang, X., Lin, B.: Carbon dioxide ($CO_2$) emissions during urbanization: a comparative study between China and Japan. J. Clean. Prod. **143**, 356–368 (2017)
27. Zhang, J., Jiang, H., Liu, G., Zeng, W.: A study on the contribution of industrial restructuring to reduction of carbon emissions in China during the five five-year plan periods. J. Clean. Prod. **176**, 629–635 (2018)
28. Zhou, S., Kyle, G.P., Yu, S., Clarke, L.E., Eom, J., Luckow, P., Chaturvedi, V., Zhang, X., Edmonds, J.A.: Energy use and $CO_2$ emissions of China's industrial sector from a global perspective. Energy Policy **58**, 284–294 (2013)
29. Wu, R., Geng, Y., Cui, X., Gao, Z., Liu, Z.: Reasons for recent stagnancy of carbon emissions in China's industrial sectors. Energy **172**, 457–466 (2019)
30. Jiang, J., Ye, B., Liu, J.: Peak of $CO_2$ emissions in various sectors and provinces of China: recent progress and avenues for further research. Renew. Sustain. Energy Rev. **112**, 813–833 (2019)
31. Li, B., Zhang, J., Li, H.: Research on spatial-temporal characteristics and affecting factors decomposition of agricultural carbon emission in China. China Popul. Res. Environ. **21**(8), 80–86 (2011)
32. Friedl, B., Getzner, M.: Determinants of $CO_2$ emissions in a small open economy. Ecol. Econ. **45**(1), 133–148 (2003)
33. Zhang, M., Mu, H., Ning, Y.: Accounting for energy-related $CO_2$ emission in China, 1991–2006. Energy Policy **37**(3), 767–773 (2009)
34. Chen, C., Zhao, T., Yuan, R., Kong, Y.: A spatial-temporal decomposition analysis of China's carbon intensity from the economic perspective. J. Clean. Prod. **215**, 557–569 (2019)
35. Dong, F., Yu, B., Hadachin, T., Dai, Y., Wang, Y., Zhang, S., Long, R.: Drivers of carbon emission intensity change in China. Resour. Conserv. Recycl. **129**, 187–201 (2018)
36. Guan, D., Peters, G.P., Weber, C.L., Hubacek, K.: Journey to world top emitter: an analysis of the driving forces of China's recent $CO_2$ emissions surge. Geophys. Res. Lett. **36**(4) (2009)
37. Ye, B., Jiang, J., Li, C., Miao, L., Tang, J.: Quantification and driving force analysis of provincial-level carbon emissions in China. Appl. Energy **198**, 223–238 (2017)
38. Zhang, Y.J., Bian, X.J., Tan, W., Song, J.: The indirect energy consumption and co2 emission caused by household consumption in China: an analysis based on the input–output method. J. Clean. Prod. **163**, 69–83 (2017)
39. Andreoni, V., Galmarini, S.: Drivers in $CO_2$ emissions variation: a decomposition analysis for 33 world countries. Energy **103**, 27–37 (2016)
40. Jakob, M., Marschinski, R.: Interpreting trade-related $CO_2$ emission transfers. Nat. Clim. Change **3**(1), 19 (2013)
41. Du, L., Wei, C., Cai, S.: Economic development and carbon dioxide emissions in China: provincial panel data analysis. China Econ. Rev. **23**(2), 371–384 (2012)
42. Burke, P.J., Shahiduzzaman, M., Stern, D.I.: Carbon dioxide emissions in the short run: the rate and sources of economic growth matter. Global Environ. Change **33**, 109–121 (2015)
43. Shan, Y., Guan, D., Zheng, H., Ou, J., Li, Y., Meng, J., Mi, Z., Liu, Z., Zhang, Q.: China $CO_2$ emission accounts 1997–2015. Sci. Data **5**, 170201 (2018)

44. Weber, S.: Bacon: an effective way to detect outliers in multivariate data using Stata (and Mata). Stata J. **10**(3), 331–338 (2010)
45. Billor, N., Hadi, A.S., Velleman, P.F.: Bacon: blocked adaptive computationally efficient outlier nominators. Comput. Stat. Data Anal. **34**(3), 279–298 (2000)
46. Ping, L., Feng, L., Hongwei, W.: Analysis and forecast of China's total economy and its structure from 2016–2035. Strat. Study Chin. Acad. Eng. **19**(1), 13–20 (2017)
47. Mileva, E., Litwack, J., Zhao, L., Vashakmadze, E.: China Economic Update (2019)

# Dynamic Spatial Analysis of Economic Performance on Comprehensive Carrying Capacity in the Greater Bay Area Considering Mediating Effects

**Qinglong Shao**

**Abstract** The Guangdong-Hong Kong-Macao Greater Bay Area (GBA) is one of the regions with the strongest economic vitality in China. This study explores the nonlinear effect of economic performance on urban comprehensive carrying capacity (UCC) in the 11 GBA cities for the period 2004–2016. In doing so, we employ static and dynamic spatial panel econometric models to investigate the nonlinear relationship of the economic-UCC nexus and explore the mediating effect of technology innovation and energy consumption. Results confirm the inverted U-shaped relationship of GDP per capita and UCC in the contexts of geographic, economic, and economic-geographic matrices, respectively. External capital shows a promoting effect on UCC while the opposite holds for share of manufacturing sector to the economy. This study verifies that technology innovation and energy consumption are important intermediate variables. Innovation shows a positive spatial spillover effect that helps to improve the UCC level of neighboring cities, while energy consumption has a negative spillover effect. Therefore, this study explores the nonlinear effect of economic growth on environmental pressure in the GBA. The significant roles of technology innovation and energy consumption are illustrated. The spatial spillover effect in GBA cities is confirmed. The final section outlines proposals for policy recommendations to improve UCC, to ensure continuous growth of the economy, and to introduce environmentally friendly technologies.

**Keywords** Urban comprehensive carrying capacity (UCC) · Economic performance · Technology innovation · Energy consumption · Static and dynamic spatial econometric models · Guangdong-Hong Kong-Macao greater bay area (GBA)

Q. Shao (✉)
College of Management, Shenzhen University, 3688 Nanhai Ave., Shenzhen 518060, China
e-mail: shao.qinglong@foxmail.com

 
F. K. S. Chan et al. (eds.), *Proceedings of the 2020 International Conference on Resource Sustainability: Sustainable Urbanisation in the BRI Era (icRS Urbanisation 2020)*, Environmental Science and Engineering,
https://doi.org/10.1007/978-981-15-9605-6_20

## 1 Introduction

Urban comprehensive carrying capacity (UCC), that involves social, economic, and environmental systems, is the key to coordinating the relation between population, resources, and environment [1]. The evaluation index of UCC includes multiple systems such as resources, environment, ecology, society, and economy. With the development of research on UCC, the research object has shifted from a single element to a composite system, and the application of the system has also expanded. In particular, the Environmental Kuznets Curve (EKC) is reflected in the analysis of UCC. Usually, EKC is described as follows: "environmental degradation would initially increase as income per capita increases, but reduces subsequently as income per capita continues to increase above a certain threshold," which is reflected as an inverted U-shaped curve [2, p. 4]. Since the environmental subsystem is an important part of UCC, we can expect an inverted U-shaped curve to exist between UCC and economic performance. This will be confirmed in the empirical analysis in Sect. 3.

UCC can be positively promoted by economic development. The logic is that industrial structure upgrading and spatial optimization increase economic scale. Recently, urban agglomeration (UA) attracted widespread attention due to its role as the driver of coordinated economic development [3], and UA has become the driving force for high-quality economic growth in China. For this reason, scholars use the agglomeration of cities as an example to explore the relationship between economic development and carrying capacity. Focusing on the Yangtze River Economic Belt (YREB), one of the fastest-growing UAs in the world, Tian and Sun [4] used advanced spatial techniques to reveal the spatial differentiation of UCC. By dividing the comprehensive UCC into four hierarchies and 39 indicators, the authors found a downward trend for comprehensive UCC over time, implying that sustainability levels in the YREB have been reduced, and subsystems of the ecological environment made a negative contribution to UCC. A significant imbalance in the development of the YREB region had been confirmed. Another sustainability indicator, urban state carrying capacity, also revealed the unbalanced development between the eastern and the western cities of the YREB [5].

As an important indicator measuring carrying capacity in the UAs, UCC is rarely used to detect the relation of carrying capacity to economic performance, and research on the influencing mechanism of growth on UCC is rare in the existing literature. In this study, we will assess the city-level comprehensive carrying capacity in the Guangdong-Hong Kong-Macao Greater Bay Area (GBA) and study its relationship with respect to economic growth, among other factors.[1] Not only will economic development affect UCC, but UCC will become an important influencing indicator restricting potential economic development, as confirmed by, among others [4, 6]. To resolve the questions of this study, we explore the relationship between UCC and the

[1]The GBA includes the Special Administrative Regions of Hong Kong and Macao and the nine municipalities of Guangzhou, Shenzhen, Zhuhai, Foshan, Huizhou, Dongguan, Zhongshan, Jiangmen, and Zhaoqing in Guangdong Province.

economic performance of GBA cities during the research period 2004–2016 by first using static and dynamic spatial panel models and then verify the mediating effect.

The present paper attempts to contribute to the literature in four aspects. First, we confirm the inverted U-shaped relation between UCC and economic output. Second, static and dynamic spatial panel models are employed to confirm the robustness of the results. Third, we illustrate the significant roles of technological innovation and energy consumption as mediating indicators in the economic-UCC nexus. Fourth, this study provides policy recommendations on improving UCC and introducing environmentally friendly technologies to realize the sustainable development of the newly constructed GBA.

# 2 Materials and Methods

## 2.1 Data Selection

**Dependent Variable: Urban Comprehensive Carrying Capacity**

To obtain the values of UCC for the 11 cities in the GBA during the research period 2004–2016, we follow Wei et al. [7], Tian and Sun [4], and Shao et al. [8]. We construct an index system based on social, economic, environmental, and transportation perspectives by using the improved entropy method [9].

This study was based on the use of the four dimensions of social, economic, environmental and transportation aspects with 14 indicators. To estimate the UCC from the supply and demand aspects, the 14 indicators were divided equally into two parts, seven of which were attributed to the supply aspect (positive indicators), which was beneficial for improvement (higher value of UCC), while the other seven indicators (reverse indicators) were attributed to the demand aspect, which was detrimental to the UCC (lower value). The four evaluation systems and the grading standards for the GBA are presented in Table 2.

Data were extracted mainly from the China Statistical Yearbook (2001–2017), the China Energy Statistical Yearbook (2001–2017) and the Guangdong Statistical Yearbook (2001–2017). Data for Hong Kong and Macao were sourced from the Hong Kong Annual Digest of Statistics (2001–2017) and the Macao Yearbook of Statistics (2001–2017), respectively.

**Independent Variables: Economic Growth Level (AGDP) and Its Square Term (AGDPSQ)**

Square terms are always used in econometric literature to examine the existence of nonlinear curves. This study employs per capita GDP and its square term to investigate the nonlinear relationship between UCC and economic development. Following Zhang and Chen [10], we take the log-form for these two indicators while other variables are not.

**Control Variables**

(1) Population density (dens). Since the area of each city is quite different, population density is different from population size and can directly reflect its impact on UCC from the perspective of aggregation [11].
(2) Foreign direct investment (fdi). We use the proportion of municipal-level foreign direct investment to GDP to measure the openness of the cities in the GBA. Usually, a higher level of foreign direct investment share implies a higher degree of internationalization [8, 12].
(3) Industrial structure (manu). This study selects the proportion of the output value of the secondary industry in GDP to examine the influence direction and magnitude of the manufacturing sectors on UCC.
(4) Economic vitality (growth). The economic growth rate of each GBA city is used to represent the economic vitality. A high growth rate means a more dynamic economy.

**Intermediate Variables: Technology Innovation and Resource Consumption**

(1) Technology innovation (patent). The more the economy develops, the higher the requirements and demands for technology become, thus promoting the improvement of innovation. Technological innovation, especially environmentally friendly technologies, has a significant impact on UCC. This study uses the number of patents granted to represent the innovation level of the city.
(2) Energy consumption (energy). The initial economic growth is more likely to be the result of increased energy input. With the growth of the economy, dependence on energy gradually declines, which has an impact on UCC. We use the energy consumption per unit of GDP to represent energy consumption.

Table 1 presents descriptive statistics for the GBA variables for the period 2004–2016.

## *2.2 Spatial Panel Models*

This study uses the Spatial Autoregression Model (SAR), Spatial Error Model (SEM), and Spatial Durbin Model (SDM) to comprehensively investigate the following aspects: the mutual influence of UCC among cities, the spillover effects among different cities, and the impact of economic development on the UCC of neighboring cities. The direct and indirect effects will be discussed. Referring to Elhorst [13], Ning et al. [14], Chen and Zhou [15], and Yang [16], we make the following spatial econometric models:

$$\begin{aligned} UCC_{i,t} =& \tau \mathrm{UCC}_{i,t-1} + \delta \mathrm{W}_N \mathrm{UCC}_{i,t} + \eta \mathrm{W}_N \mathrm{UCC}_{i,t-1} + \alpha_1 \mathrm{lnagdp}_{i,t} + \alpha_2 \mathrm{lnagdpsq}_{i,t} \\ &+ \alpha_3 \mathrm{dens}_{i,t} + \alpha_4 \mathrm{fdi}_{i,t} + \alpha_5 \mathrm{manu}_{i,t} + \alpha_6 \mathrm{growth}_{i,t} + \alpha_7 \mathrm{patent}_{i,t} + \alpha_8 \mathrm{energy}_{i,t} + \nu_t \end{aligned} \quad (1)$$

**Table 1** Descriptive statistics for the variables of GBA during the period 2004–2016

| Varibles | Description | Unit | Mean | SD | Min | Max |
|---|---|---|---|---|---|---|
| *UCC* | Comprehensive carrying capacity | – | −0.1675 | 6.4115 | −15.1411 | 15.4547 |
| *lnagdp* | Per capita GDP | CNY | 11.2903 | 0.8269 | 9.4186 | 13.2692 |
| *lnagdpsq* | Per capita GDP square | – | 128.1498 | 18.8603 | 88.7095 | 176.0709 |
| *dens* | Population density | Person/km$^2$ | 6.9678 | 1.2275 | 5.5713 | 9.9711 |
| *fdi* | Foreign direct investment/GDP | % | 0.0595 | 0.0573 | 0.0088 | 0.2842 |
| *manu* | Manufacturing industry/GDP | % | 43.1545 | 18.5332 | 3.7000 | 65.6000 |
| *growth* | Economic growth rate | % | 11.2617 | 5.5920 | −21.6000 | 26.8000 |
| *patent* | Patent counts/10,000 persons | – | 34.9309 | 43.1400 | 0.4003 | 195.6450 |
| *energy* | Energy consumption per unit of GDP | – | 0.0789 | 0.0527 | 0.0008 | 0.3145 |

*Note* SD denotes standard deviation

$$\mu = \rho W_N v_t + \xi_t \tag{2}$$

Where $UCC_{i,t}$ denotes UCC in the GBA; $\text{lnagdp}_{i,t}$ represents economic development, $\text{lnagdpsq}_{i,t}$ means the square term of economic development; $\text{dens}_{i,t}$, $\text{fdi}_{i,t}$, $\text{manu}_{i,t}$, and $\text{growth}_{i,t}$ represent population density, proportion of foreign direct investment to GDP, share of manufacturing sector, and economic growth rate; $\text{patent}_{i,t}$ (technology innovation) and $\text{energy}_{i,t}$ (energy consumption) are intermediate variables; the subscripts it stands for city $i$ in year $t$, respectively. In addition, $\tau$ is the space delay factor for dependent variables capturing the spillover effects from neighboring cities, $N$ is the number of cities, and $W_N$ is the $N$-order spatial weight matrix. $\mu$ is the space error term, $\rho$ is the spatial error coefficient, and $v$ is the vector data set containing the independent and identically distributed errors. Furthermore, $\tau$, $\delta$, and $\eta$ represent the time lag factor, the space lag factor, and the time-space lag factor, respectively (Table 2).

**Table 2** Social, economic, environmental, and transportation evaluation index systems and grading standards for the GBA

| System layer | Indicator layer (Units) | Serial number | Attributes |
|---|---|---|---|
| Social subsystem | Population (10,000 persons) | X1 | – |
| | Population density (persons/km$^2$) | X2 | – |
| | University/College enrollment (persons) | X3 | – |
| | Health care institutions (unit) | X4 | + |
| | Number of beds in health institutions (unit) | X5 | + |
| Economic subsystem | GDP per capita (yuan) | X6 | + |
| | GDP growth rate (%) | X7 | + |
| | Share of tertiary industry to GDP (%) | X8 | + |
| Environmental subsystem | Electricity consumption per capita (kWh) | X9 | – |
| | Volume of tap water supplied (tons) | X10 | – |
| | Volume of industrial solid waste produced (10,000 tons) | X11 | – |
| | Public green areas per capita (m$^3$) | X12 | + |
| Transportation subsystem | Passenger traffic (10,000 persons) | X13 | – |
| | Turnover volume of passenger traffic (10,000 person/km) | X14 | + |

# 3 Empirical Results and Analysis

## *3.1 Static Spatial Panel Regression Results Using Geographic, Economic, and Economic-Geographic Matrix*

This section uses static spatial panel models to examine the effect of economic performance and other driving factors on UCC. All the models are controlled for the city- and year-fixed effects. Table 3 shows the results of the impact of economic growth on UCC, in which Model (1) is the baseline test that presents the ordinary least square (OLS) regression result; Models (2)–(4), Models (5)–(7), and Models (8)–(10) present SAR, SEM, and SDM regression results with a geographic matrix (W1), an economic matrix (W2), and an economic-geographic matrix (W3), respectively.

**Table 3** Impact of economic development on UCC in the GBA using static spatial geographic, economic, and economic-geographic matrices

| Variables | Dependent variable: comprehensive carrying capacity | | | | | | | | | |
|---|---|---|---|---|---|---|---|---|---|---|
| | | W1 | | | W2 | | | W3 | | |
| | Model (1) OLS | Model (2) SAR | Model (3) SEM | Model (4) SDM | Model (5) SAR | Model (6) SEM | Model (7) SDM | Model (8) SAR | Model (9) SEM | Model (10) SDM |
| *lnagdp* | 29.8424*** (2.938) | 28.9763*** (3.301) | 30.7875*** (3.329) | 27.9587** (2.551) | 27.2372*** (3.037) | 27.0736*** (3.098) | 39.7154*** (3.818) | 28.9555*** (3.300) | 30.7643*** (3.327) | 27.8586** (2.549) |
| *lnagdpsq* | −1.3448*** (−3.208) | −1.3022*** (−3.593) | −1.3794*** (−3.618) | −1.2747*** (−2.852) | −1.2291*** (−3.314) | −1.2150*** (−3.357) | −1.7814*** (−4.105) | −1.3013*** (−3.591) | −1.3783*** (−3.617) | −1.2697*** (−2.849) |
| *dens* | 3.7801 (0.970) | 3.3171 (0.984) | 2.0367 (0.563) | −4.9575 (−0.865) | 3.7365 (1.102) | 3.2615 (0.973) | 11.4621*** (3.286) | 3.3096 (0.982) | 2.0210 (0.558) | -5.0831 (-0.889) |
| *fdi* | 23.0568*** (2.656) | 21.9533*** (2.922) | 22.7803*** (2.878) | 45.5243*** (3.257) | 22.5609*** (2.984) | 22.8118*** (2.952) | −6.5632 (−0.843) | 21.9403*** (2.921) | 22.7772*** (2.877) | 45.7210*** (3.273) |
| *manu* | −0.2504*** (−3.833) | −0.2325*** (−4.070) | −0.2373*** (−3.906) | −0.1929** (−2.520) | −0.2443*** (−4.288) | −0.2641*** (−4.465) | −0.3427*** (−6.070) | −0.2324*** (−4.069) | −0.2373*** (−3.904) | −0.1928** (−2.517) |
| *growth* | 0.0579 (1.289) | 0.0559 (1.441) | 0.0599 (1.466) | 0.0671 (1.526) | 0.0602 (1.538) | 0.0630 (1.616) | 0.0962*** (2.959) | 0.0559 (1.442) | 0.0599 (1.467) | 0.0673 (1.532) |
| *patent* | −0.0210 (−1.369) | −0.0197 (−1.487) | −0.0204 (−1.432) | −0.0058 (−0.252) | −0.0171 (−1.260) | −0.0148 (−1.137) | −0.0503*** (−3.888) | −0.0197 (−1.484) | −0.0203 (−1.428) | −0.0052 (−0.225) |
| *energy* | −34.7888** (−2.051) | −36.7941** (−2.508) | −42.4489*** (−2.639) | −70.1813*** (−3.255) | −32.9205** (−2.224) | −30.1993** (−2.005) | 5.7784 (0.410) | −36.7720** (−2.507) | −42.4240*** (−2.638) | −70.4766*** (−3.275) |
| *spatial rho* | | −0.4669* (−1.953) | −0.5454* (−1.936) | −0.7411*** (−2.865) | −0.2646* (−1.727) | −0.3269** (−1.985) | −0.1562 (−1.007) | −0.4711** (−1.963) | −0.5494* (−1.944) | −0.7431*** (−2.865) |
| *cons* | −179.6979** (−2.444) | | | | | | | | | |
| *City-fixed effect* | YES | YES | YES | YES | YES | YES | YES | YES | YES | YES |

(continued)

**Table 3** (continued)

| Variables | Dependent variable: comprehensive carrying capacity | | | | | | | | | |
|---|---|---|---|---|---|---|---|---|---|---|
| | | W1 | | | W2 | | | W3 | | |
| | Model (1) OLS | Model (2) SAR | Model (3) SEM | Model (4) SDM | Model (5) SAR | Model (6) SEM | Model (7) SDM | Model (8) SAR | Model (9) SEM | Model (10) SDM |
| *Year-fixed effect* | YES | YES | YES | YES | YES | YES | YES | YES | YES | YES |
| *Hausman* | 21.86 (0.005) | 414.62 (0.000) | 16.75 (0.033) | 56.21 (0.000) | 175.27 (0.000) | 43.21 (0.000) | 163.90 (0.000) | 381.88 (0.000) | 16.55 (0.035) | 57.61 (0.000) |
| *Obs* | 143 | 143 | 143 | 143 | 143 | 143 | 143 | | | |
| *R-squared* | 0.3634 | 0.0474 | 0.0335 | 0.0054 | 0.0569 | 0.0645 | 0.1321 | 0.0472 | 0.0332 | 0.0051 |

*Note* t-statistics in parentheses; ***$p < 0.01$, **$p < 0.05$, *$p < 0.1$. Same apply to the following

According to Model (1), within our expectation, per capita GDP (lnagdp) is positively correlated with UCC at the 1% significance level, while the coefficients of its square term (lnagdpsq) are negative and significant, illustrating an inverted U-shaped relationship between economic performance and UCC in the GBA. This result corresponds to the study by Tian and Sun [6], who proposed an inverted U-shaped curve in the YREB. Similar results in the spatial panel models in Models (2)–(10) further verified the existence of an inverted U-shaped curve for the economics-UCC nexus. Moreover, across the models, we find that a 1% increase in GDP per capita leads to an average UCC increase of 30%. Regarding the control and intermediate variables, fdi positively and significantly correlated with UCC at a 1% level regardless of Model (7). The coefficients of manu are negative and significant at the 5 and 1% levels across the models, and energy is negatively correlated with UCC. Also, across all the models except for Model (7), the coefficients of spatial rho in the three matrices are significantly negative.

To further investigate the spatial effect of the influencing factors on UCC, Table 4 shows the spatial direct, indirect, and total effects under W1, W2, and W3. Direct effect refers to the direct effect of local influencing factors on local UCC, while indirect effect refers to the effect of local factors on the UCC of neighboring cities. Results show that lnagdp and UCC have an inverted U-shaped relationship but only show a negative influence on economically neighboring cities, Model (15). Furthermore, growth and fdi promote local UCC while manu negatively affects local UCC. In short, growth improves the UCC levels of economically neighboring cities, while fdi and manu show negative effects. Also, for Models (14)–(16), we find patent lower at local UCC levels but it promotes the UCC levels of economically neighboring cities. Considering the total positive effect displayed in Model (16), we find that a 1% increase in innovation leads to a 0.06% increase in overall UCC level, illustrating the positive effect of technology innovation on UCC levels. As for the geographically-adjacent cities, the increase of dens and energy will reduce the UCC levels of neighboring cities.

## *3.2 Dynamic Spatial Panel Regression Results*

This section presents the dynamic spatial panel regression results using SAR and SDM under W1, W2, and W3 (Table 5). The lagged UCC indicator shows a significant negative effect on UCC. Consistent with the results discussed above, GDP per capita and UCC show an inverted U-shaped curve. The proportion of FDI to GDP promotes UCC while the expansion of the secondary sector deteriorates UCC levels. Corresponding to Model (14), technological progress lowers the UCC level. Energy consumption shows a detrimental effect on UCC. The results of dynamic spatial models are in line with the static results in Sect. 3.1, confirming the robustness of the regression results.

**Table 4** Effect decomposition using spatial geographic, economic, and economic-geographic matrices

| Variables | Dependent variable: Comprehensive carrying capacity | | | | | | | | |
|---|---|---|---|---|---|---|---|---|---|
| | W1 | | | W2 | | | W3 | | |
| | Model (11) direct effect | Model (12) indirect effect | Model (13) total effect | Model (14) direct effect | Model (15) indirect effect | Model (16) total effect | Model (17) direct effect | Model (18) indirect effect | Model (19) total effect |
| *lnagdp* | 25.3533*** (2.711) | 28.0279 (0.684) | 53.3813 (1.185) | 42.0254*** (3.814) | −50.1601* (−1.712) | −8.1347 (−0.285) | 25.2662*** (2.707) | 27.9275 (0.681) | 53.1938 (1.181) |
| *lnagdpsq* | −1.1626*** (−3.037) | −1.2180 (−0.726) | −2.3806 (−1.293) | −1.8901*** (−4.099) | 2.3924* (1.905) | 0.5024 (0.410) | −1.1580*** (−3.031) | −1.2151 (−0.724) | −2.3730 (−1.289) |
| *dens* | 0.9919 (0.196) | −55.1264*** (−2.825) | −54.1345** (−2.474) | 11.2063*** (3.303) | 20.4365 (1.356) | −31.6428* (1.915) | 0.9705 (0.192) | −56.2319*** (−2.863) | −55.2614** (−2.510) |
| *fdi* | 44.4586*** (4.432) | 4.7389 (0.091) | 49.1975 (0.852) | −3.8250 (−0.479) | −81.5155*** (−2.589) | −85.3405** (−2.399) | 44.6963*** (4.452) | 4.4152 (0.085) | 49.1115 (0.852) |
| *manu* | −0.1961*** (−3.159) | 0.0460 (0.163) | −0.1501 (−0.486) | −0.3279*** (−5.913) | −0.4187** (−2.327) | −0.7466*** (−3.632) | −0.1951*** (−3.144) | 0.0367 (0.130) | −0.1584 (−0.513) |
| *growth* | 0.0661* (1.685) | 0.0326 (0.225) | 0.0987 (0.631) | 0.0907*** (2.670) | 0.2361* (1.746) | 0.3268** (2.326) | 0.0663* (1.694) | 0.0323 (0.222) | 0.0987 (0.629) |
| *patent* | −0.0117 (−0.681) | 0.0583 (0.708) | 0.0466 (0.494) | −0.0550*** (−4.259) | 0.1171*** (3.339) | 0.0621* (1.795) | −0.0114 (−0.669) | 0.0623 (0.755) | 0.0509 (0.539) |
| *energy* | −57.1550*** (−3.121) | −131.3598* (−1.912) | −188.5148** (−2.389) | 3.3970 (0.230) | 61.0979* (1.847) | 64.4949* (1.677) | −57.4015*** (−3.138) | −132.0134* (−1.921) | −189.4149** (−2.401) |

**Table 5** Impact of economic development on UCC in the GBA using dynamic spatial geographic, economic, and economic-geographic matrices

| Variables | Dependent variable: Comprehensive carrying capacity | | | | | |
|---|---|---|---|---|---|---|
| | W1 | | W2 | | W3 | |
| | Model (20) SAR | Model (21) SDM | Model (22) SAR | Model (23) SDM | Model (24) SAR | Model (25) SDM |
| *L.UCC* | −1.1622*** (−2.633) | −1.6838*** (−3.826) | −0.7212** (−2.011) | −0.2736 (−0.893) | −1.1743*** (−2.651) | −1.6909*** (−3.831) |
| *lnagdp* | 25.7808*** (2.610) | 37.5961*** (2.888) | 20.3133* (1.949) | 52.8466*** (3.916) | 25.7437*** (2.608) | 37.2531*** (2.876) |
| *lnagdpsq* | −1.1771*** (−2.900) | −1.6912*** (−3.185) | −0.9210** (−2.131) | −2.2677*** (−4.062) | −1.1753*** (−2.897) | −1.6765*** (−3.173) |
| *dens* | 4.8617 (1.287) | −3.8266 (−0.548) | 4.8815 (1.273) | 18.2887*** (4.599) | 4.8477 (1.284) | −4.0003 (−0.574) |
| *fdi* | 19.7361** (2.470) | 40.1795*** (2.744) | 23.9998*** (2.980) | −3.3894 (−0.414) | 19.7103** (2.468) | 40.3257*** (2.757) |
| *manu* | −0.1936*** (−3.000) | −0.2267** (−2.536) | −0.2188*** (−3.365) | −0.4719*** (−6.968) | −0.1934*** (−2.999) | −0.2256** (−2.523) |
| *growth* | 0.0370 (0.939) | 0.1066** (2.340) | 0.0324 (0.793) | 0.0812** (2.439) | 0.0369 (0.936) | 0.1069** (2.347) |
| *patent* | −0.0250* (−1.749) | 0.0038 (0.159) | −0.0055 (−0.355) | −0.0594*** (−3.919) | −0.0249* (−1.742) | 0.0045 (0.187) |
| *energy* | −62.6785** (−2.299) | −57.0255* (−1.658) | −21.5208 (−0.771) | −51.2534* (1.943) | −62.5819** (−2.297) | −57.4233* (−1.671) |
| *spatial rho* | −0.2950 (−1.113) | 0.4505 (1.630) | −0.2465 (−1.515) | 0.1825 (1.128) | −0.2983 (−1.122) | 0.4532 (1.636) |
| *City-fixed effect* | YES | YES | YES | YES | YES | YES |
| *Year-fixed effect* | YES | YES | YES | YES | YES | YES |
| *Hausman* | 47.97 (0.000) | 6.51 (0.590) | 57.50 (0.000) | 96.80 (0.000) | 46.01 (0.000) | 5.51 (0.702) |
| *Obs* | 132 | 132 | 132 | 132 | 132 | 132 |
| *R-squared* | 0.0414 | 0.0019 | 0.0747 | 0.0034 | 0.0684 | 0.1670 |

## 3.3 Mediating Effect Analysis

This section explores the mediating effect of technology innovation and energy consumption in the economic-UCC nexus (Table 6). As for Model (26) and Model (27), economic output promotes technology innovation, and innovation promotes UCC level, which is consistent with the result of Model (16). Regarding Model (28) and Model (29), economic output promotes energy consumption, and energy consumption lowers UCC level, which is consistent with the result of Model (13) and Model (19). Therefore, we confirm the significant mediating role of technology innovation and energy consumption in the effect of economic performance on UCC.

**Table 6** Mediating effect analysis for technology innovation and energy consumption using SAR with an economic matrix

| Variables | *patent* Model (26) | *UCC* Model (27) | *energy* Model (28) | *UCC* Model (29) |
|---|---|---|---|---|
| *lnagdp* | 172.9256*** (3.524) | | 0.1350*** (2.812) | |
| *lnagdpsq* | −7.5916*** (−3.700) | | −0.0047** (−2.352) | |
| *patent* | | 0.0111* (1.028) | | |
| *energy* | | | | −23.1673** (−2.032) |
| *dens* | 133.9015*** (7.367) | −3.5263 (−1.098) | −0.0000 (−0.002) | −1.9542 (−0.712) |
| *fdi* | 14.8526 (0.347) | 10.8140* (1.861) | 0.2671*** (6.571) | 13.8338** (2.307) |
| *manu* | −2.0496*** (−6.304) | −0.1174** (−2.557) | 0.0003 (1.035) | −0.1104** (−2.520) |
| *growth* | −0.1320 (−0.515) | 0.0746* (1.832) | −0.0000 (−0.028) | 0.0817** (2.025) |
| *spatial rho* | −1.4421*** (−5.510) | −0.4931** (−1.987) | −1.2089*** (−4.663) | −0.5380** (−2.150) |
| *City-fixed effect* | YES | YES | YES | YES |
| *Year-fixed effect* | YES | YES | YES | YES |
| *Obs* | 143 | 143 | 143 | 143 |
| *R-squared* | 0.0069 | 0.3278 | 0.0577 | 0.1268 |

# 4 Discussion

## *4.1 The Inverted U-shaped Relation of Economic Performance on UCC and Other Driving Forces of UCC*

Results illustrate an inverted U-shaped relationship between economic performance and UCC in the GBA. That is, with the increase of per capita GDP, UCC first increases and then declines. This is because traffic congestion and environmental pollution become increasingly serious as urban development progresses [4], in line with the EKC curve. During the historical evolution of international megacities, the UCC level rose along with the growth of the economy. This study confirmed the existence of an EKC relation between economic performance and UCC in the GBA.

As for the control and intermediate variables, a higher proportion of FDI to GDP can greatly promote UCC; the more open a city is, the better its capacity will be. This result refutes the view that external capital would deteriorate the local environment by introducing new enterprises, the so-called "Pollution Haven Hypothesis"

[8]. Our findings verify that this point of view does not stand in GBA cities, and that, introducing green intellectual capital could improve sustainable performance in business, environmental, and social-related areas [17, 18]. The expansion of the manufacturing sector in the economy will inevitably lower the carrying capacity. A better way to improve UCC would be to put more emphasis on the service economy. In addition, more consumption of energy and resources does not only increase carbon emissions [19, 20], and other harmful gases such as sulfur dioxide [21], but also the local carrying capacity.

As for the geographically-adjacent cities, the increase in population density and energy consumption will reduce the UCC levels of neighboring cities. The underlying reasons could be that, due to the short spatial distance, the population growth of the local city will cause people to migrate to neighboring cities. The increasing demand for energy will promote resource exploration and will increase the garbage disposal needs of neighboring cities, leading to a downscaling of carrying capacity. Shenzhen, for example, has a high population density. People choose to live in Huizhou or Dongguan but to work in Shenzhen. Also, most of the waste generated in Shenzhen is transported to Huizhou for landfill.

## *4.2 Mediating Effect of Technology Innovation and Energy Consumption*

This study suggests two channels that may link economic performance and UCC: (1) Technological innovation: The more the economy develops, the higher the requirement and demand for technology will be, thereby promoting technological innovation development. Innovation will change the carrying capacity. (2) Energy consumption: To a certain extent, economic development is built upon energy use, especially in the early stage. When the economy develops, dependence on energy will be gradually reduced. The process has an impact on UCC.

From the perspective of spatial spillover effect, inter-city innovation will enhance the innovation of neighboring cities, which indicates that innovation activities have strong vitality and bring significant positive externalities, whereas UCC in the GBA is competitive. Therefore, improving UCC can not only effectively tap the potential of innovation but also contribute significantly to solving environmental problems. Through the spatial spillover effect, the UCC of the 11 GBA cities still displays the apparent competitive relation of beggar-thy-neighbor. Geographically-adjacent cities and economically-adjacent cities compete for resources in the process of economic growth. Moreover, these phenomena correspond to the reality that economic development always comes at the expense of varying degrees of energy consumption.

## 5 Concluding Remarks

The Guangdong-Hong Kong-Macao GBA will not only be built into a vibrant world-class UA and an important support for the construction of the "Belt and Road" initiative but will also become a high-quality living area suitable for living and working, a model of high-quality sustainable development in the blueprint for China [22]. This paper uses panel data of 11 cities in the GBA from 2004 to 2016 to investigate the nonlinear relationship between UCC and economic performance. Three main conclusions are presented here.

First, we find a competitive relationship between the geographically and economically neighboring GBA cities, which will impede regional integration. By comparison, technology innovation has a significant positive spatial spillover effect in GBA cities.

Second, by using the SEM, SAR, and SDM models, an inverted U-shaped relationship between economic performance and UCC was found in GBA. With the development of per capita GDP, UCC first increases and then declines beyond the peak point.

Third, technological innovation and energy consumption are shown to be important intermediate variables for the economic-UCC nexus. On the one hand, economic development promotes innovation, and innovation promotes the improvement of UCC. On the other hand, economic growth consumes energy, and energy consumption deteriorates the carrying capacity.

**Acknowledgments** The author is grateful for the support of National Natural Science Foundation of China (71903131).

## References

1. Li, K., Jin, X., Ma, D., Jiang, P.: Evaluation of resource and environmental carrying capacity of China's rapid-urbanization areas—a case study of Xinbei District, Changzhou. Land **8** (2019). https://doi.org/10.3390/land8040069
2. Ike, G.N., Usman, O., Asumadu-sarkodie, S., Ike, G.N.: Testing the role of oil production in the environmental Kuznets curve of oil producing countries : new insights from method of moments quantile regression. Sci. Total Environ. 135208 (2019). https://doi.org/10.1016/j.scitotenv.2019.135208
3. Fang, C., Yu, D.: Urban agglomeration: an evolving concept of an emerging phenomenon. Landsc. Urban Plan **162**, 126–136 (2017). https://doi.org/10.1016/j.landurbplan.2017.02.014
4. Tian, Y., Sun, C.: Comprehensive carrying capacity, economic growth and the sustainable development of urban areas: a case study of the Yangtze river economic belt. J. Clean. Prod. **195**, 486–496 (2018a). https://doi.org/10.1016/j.jclepro.2018.05.262
5. Sun, C., Chen, L., Tian, Y.: Study on the urban state carrying capacity for unbalanced sustainable development regions: evidence from the Yangtze river economic belt. Ecol. Indic. **89**, 150–158 (2018). https://doi.org/10.1016/j.ecolind.2018.02.011
6. Tian, Y., Sun, C.: A spatial differentiation study on comprehensive carrying capacity of the urban agglomeration in the Yangtze river economic belt. Reg. Sci. Urban Econ. **68**, 11–22 (2018b). https://doi.org/10.1016/j.regsciurbeco.2017.10.014

7. Wei, Y., Huang, C., Lam, P.T.I., Yuan, Z.: Sustainable urban development: a review on urban carrying capacity assessment. Habitat Int. **46**, 64–71 (2015). https://doi.org/10.1016/j.habitatint.2014.10.015
8. Shao, Q., Wang, X., Zhou, Q., Balogh, L.: Pollution haven hypothesis revisited: a comparison of the BRICS and MINT countries based on VECM approach. J. Clean Prod. **227**, 724–738 (2019). https://doi.org/10.1016/j.jclepro.2019.04.206
9. Wang, S., Tan, F.: Environment regulation and haze decoupling effect: based on the perspective of heavy pollution industrial transformation. J. Beijing Inst. Technol. (Social Sci Ed) **19**, 1–7 (2017)
10. Zhang, C., Chen, C.: A dynamic study on the impact of local government competition on environmental regulation: based on Chinese decentralization perspective (in Chinese). Nankai Econ. Stud. **4**, 137–157 (2018). https://doi.org/10.14116/j.nkes.2018.04.009
11. Kareem, O.I.: The determinants of large-scale land investments in Africa. Land Use Policy **75**, 180–190 (2018). https://doi.org/10.1016/j.landusepol.2018.03.039
12. Jiang, L., Feng, Z.H., Bai, L., Zhou, P.: Does foreign direct investment drive environmental degradation in China? An empirical study based on air quality index from a spatial perspective. J. Clean Prod. **176**, 864–872 (2018). https://doi.org/10.1016/j.jclepro.2017.12.048
13. Elhorst, J.P.P.: Dynamic spatial panels: models, methods, and inferences. J. Geogr. Syst. **14**, 5–28 (2012). https://doi.org/10.1007/s10109-011-0158-4
14. Ning, L., Wang, F., Li, J.: Urban innovation, regional externalities of foreign direct investment and industrial agglomeration: evidence from Chinese cities. Res. Policy **45**, 830–843 (2016). https://doi.org/10.1016/j.respol.2016.01.014
15. Chen, J., Zhou, Q.: City size and urban labor productivity in China: new evidence from spatial city-level panel data analysis. Econ. Syst. **41**, 165–178 (2017). https://doi.org/10.1016/j.ecosys.2016.07.002
16. Yang, Z.: Unified M-estimation of fixed-effects spatial dynamic models with short panels. J. Econ. **205**, 423–447 (2018). https://doi.org/10.1016/j.jeconom.2017.08.019
17. Yusoff, Y.M., Omar, M.K., Kamarul Zaman, M.D., Samad, S.: Do all elements of green intellectual capital contribute toward business sustainability? Evidence from the Malaysian context using the partial least squares method. J. Clean Prod. **234**, 626–637 (2019). https://doi.org/10.1016/j.jclepro.2019.06.153
18. Yusliza, M.Y., Yong, J.Y., Tanveer, M.I., et al.: A structural model of the impact of green intellectual capital on sustainable performance. J. Clean Prod. **249**, 119334 (2020). https://doi.org/10.1016/j.jclepro.2019.119334
19. Lamb, W.F., Rao, N.D.: Human development in a climate-constrained world: what the past says about the future. Global Environ. Change **33**, 14–22 (2015). https://doi.org/10.1016/j.gloenvcha.2015.03.010
20. Yao, X., Guo, C., Shao, S., Jiang, Z.: Total-factor $CO_2$ emission performance of China's provincial industrial sector: a meta-frontier non-radial Malmquist index approach. Appl. Energy **184**, 1142–1153 (2016). https://doi.org/10.1016/j.apenergy.2016.08.064
21. Hu, C.Q., Zhang, C.X., Han, X.W., Yin, R.Y.: Sulfur flow analysis for new generation steel manufacturing process. J. Iron Steel Res. Int. **15**, 12–15 (2008). https://doi.org/10.1016/S1006-706X(08)60136-3
22. SC: Outline Development Plan for the Guangdong-Hong Kong-Macao Greater Bay Area. Beijing (2019)

# Understanding Environmental Justice Capital in China—A New Framework to Study Environmental Justice in Contexts

**Mengqi Shao, May Tan-Mullins, and Faith Ka Shun Chan**

**Abstract** Environmental justice has drew worldwide attention since the 1982 protest in USA to against duping toxic waste. In the globalization era, worldwide scholars and environmental activists are actively engaged in related studies and social movements. However, these researches and movements usually ignore the influences of local contexts on local environmental justice configurations, including related researches in China. Whilst, evidences have been provided that different forms of capital from contexts, such as economic, social, political, natural capital and cultural capital, will affect the local concept of environmental justice. That is to say environmental justice should have different discourse from what has been researched in western countries in different contexts. Thus, this research will discuss the common ground of environmental justice study framework and promote the new conceptual framework "environmental justice capital" for having a better understanding environmental justice in contexts. Additionally, the framework of "environmental justice capital" will be put in Chinese contexts as a preliminary discussion to get an initial image of environmental justice in China.

**Keywords** Environmental justice · Environmental justice capital · China

## 1 Theorizing Environmental Justice

First coined in 1982, environmental justice, a conceptual framework inherited from Rawls' justice theory [78] has been developed as an interdisciplinary concept [76]. To

M. Shao (✉) · M. Tan-Mullins · F. K. S. Chan
University of Nottingham, Ningbo Campus, Ningbo, China
e-mail: mengqi.shao@nottingham.edu.cn

M. Tan-Mullins
e-mail: may.tan-mullins@nottingham.edu.cn

F. K. S. Chan
e-mail: faith.chan@nottingham.edu.cn



F. K. S. Chan et al. (eds.), *Proceedings of the 2020 International Conference on Resource Sustainability: Sustainable Urbanisation in the BRI Era (icRS Urbanisation 2020)*, Environmental Science and Engineering,
https://doi.org/10.1007/978-981-15-9605-6_21

cope with conflicts within the social structure, Rawls [72] put forward "justice" and an important notion of "justice as fairness" in view of contemporary liberal theories. Here, the idea of "justice as fairness" emphasizes just distribution of social, political and economic goods and burdens [78], which has become a dominant academic concept to research what and how to distribute justly over past years (ibid). However, numerous debates towards Rawls' theory have been generated that justice should not solely focus on fair processes of distribution [29, 75, 104]. Young and Fraser have challenged Rawls' justice theories that distributive justice ignores the causes for unjust distribution [29, 104]. They argued that lacking recognition in social, economic, political and cultural realms will result in justice damages or unjust distribution to individuals and communities (ibid). Therefore, "justice as recognition" is as much important as distribution to achieve justice [29]. Besides that, Schlosberg [78] further argues there should also be another dimension of justice—procedural justice defined as democratic public participation in the political process, which is also often viewed as the way to realize distributive and recognitive justice.

Drawing on the consensus of "justice" as fairness, recognition and participation, various definitions of environmental justice have emerged (ibid). Attention to environmental justice originates in protests against unjust environmental risk distribution including siting hazardous waste sites and dumping PCB-laden dirt in predominately black neighborhoods in the USA (see [12, 59, 69]. Subsequently, in the light of the distributive justice study of Rawls [72], environmental justice is defined as the academic concept focusing on the injustice in the distribution of environmental risks and burdens [9, 11, 47, 82]. However, sharing the critics of "justice as fairness", scholars have argued that the environmental justice concept should also underline the significance of recognition and participation [26, 47, 82].

Thus, the three-pronged justice schema proposed by Fraser [29, 30] is widely adopted in the environmental justice research, which includes socio-economic distributional dimension, cultural dimension of recognition, and political dimension of representation and participation [30, 41, 75, 109]. The socio-economic distributional dimension provides fair opportunities to accesses the environmental resources and services between stakeholders [94]. The political dimension of representation and participation includes procedural justice, which concentrates on involving more people vulnerable to the decision-making process, such as allocation and management of environmental resources [86]. Here, the cultural dimension of recognition remains to be the cornerstone principle of achieving environmental justice, referring to embracing the value of plurality and prompting recognition and validation of local communities [56]. As environmental injustice problems have not only caused impacts to individuals but also to social groups and communities, scholars argue that environmental justice should address individuals, social groups and communities at the same time [24, 78, 76, 77, 87].

With the development of globalization and increasingly serious worldwide climate change, environmental justice concept has been continuously developed and expanded as an integrative and interdisciplinary concept which could be used both in academic research and environmental movements [18, 76, 88]. Nevertheless, just as Parker Follett [27] called for "unity without uniformity" a century ago, scholars hold the notion that a comprehensive, integrative and inclusive justice concept is not enough, and it is important to have different principles and understandings of environmental justice in various contexts [53, 78, 95].

## 2 Environmental Justice Capital

As stated by Schlosberg [78], "groups emphasize different notions of justice, on different issues, in various contexts". Supporting this idea, findings of various researches prove that different forms of capital including natural capital, economic capital, political capital, and social capital will affect the notion formation and acquisition of environmental justice. The following will be the review of various forms of capital, along with their connections with environmental justice.

### *2.1 Various Forms of Capital and Environmental Justice*

Natural capital here is defined as "the natural resource stocks from which resource flows and services useful for livelihoods are derived" [23], the main value of which to human beings is its contribution to transforming its stock to services, or in another word, other forms of capital [14]. Scholars found that differences between the quality and quantity of accessible environmental resources, as well as difference in the proximity to environmental resources or burdens would result in environmental injustice problem [33, 35, 68]. In this regard, natural capital plays a significant role especially to those people who heavily depend on natural resources on obtaining environmental justice [84]. Stucker's (ibid) environmental justice research in rural Tajikistan also echoes the discourse the discourse that the arrangement and access of natural resources is influenced by local culture and power [8, 17, 25].

Cultural capital is "an asset embodying cultural value" and contains both tangible and intangible forms. The intangible culture capital including beliefs, traditions and values and others will be shared socially and inter-generationally [28, 91], as well as influence the recognition of the society towards justice [35]. Such shared value of cultural capital could contribute to the creation of "regionness" [38] and "regional identity" [108], which can be interpreted as a collective consciousness, identification and belonging of a region [37, 108]. Insufficient respect and consideration of regional identity will arouse violent conflicts when utilizing and managing the natural resources [37]. Therefore, the cultural capital has a close bond with environmental justice, and helps to form the notion of environmental justice as it brings local perspective of culture into conversation at the very beginning [76]. Hensengerth's [37] research in dams of the Greater Mekong Subregion (GMS) again provides evidence that hydropower projects are prone to local conflicts as the government only focuses on energy production without fully considering the region identity including local water culture and belief system.

Economic capital refers to capital that is "immediately and directly convertible into money and may be institutionalized in the form of property rights", briefly means money and property [6]. Fraser [29] emphasizes the economic inequality deeply ties to justice. Quan [71] further discussed that economic status such as income influences environmental justice, specifically the patterns of environmental benefits. Morello-Frosch's [63] finding on environmental distributive justice also illustrates that the counties with the most serious economic inequalities are experiencing the highest environmental hazards. Segura and Boyce's (1994) research has shown that poverty of communities result in weak resistance to those who come to exploit the forests, such as timber and mining firms. Moreover, with the priority of market value uses of ecosystems, the Capitalist market economy is in an increasing demand for exploitation and extraction of materials and energy from nature where the indigenous, poor and minority often live. That will also lead to continuous negative ecological and social impacts [79, 80]. Meanwhile, the environmental burdens will reduce the property values, which may hence attract more poor people to move in and exacerbate environmental injustice problems [59–62]. As argued by Pastor [67], environmental deterioration and injustice always come with economic weakness.

Political capital is defined by Birner and Witttmer [5] as "the resources used by an actor to influence policy formation and realize outcomes that serve the actor's perceived interests", Drawing on the Hicks and Misra's [39] political capital research with an emphasis of political resources, as well as McCarthy and Zald's [55] resource mobilization theory. Environmental justice researches offer the findings that political capital, such as the ability to express their ideas and participation during the decision-making process has impacts on the achievement of environmental justice plan [59]. For instance, industry and government seek the path of least resistance to arrange pollution and polluting industries [10, 61, 62]. Knowing that the protest and opposition may happen in some communities where such facilities are placed, industry and government try to avoid controversy or delays to get a smoother and effective plan [59]. Thus, the communities where the poor, racial, and ethnic minorities live easily become the target locations (ibid).

Social capital is regarded as the set of relationships that have developed around shared values, norms and trust [22, 70], which is formed via the free and long-term collaboration between individuals and among groups [2, 14]. As a series of reciprocal norms, social networks, and trust, social capital can effectively solve collective action problems, moreover, improve the trust among social group and efficiency of social cooperation in a community [49, 50, 81]. Lin [51] also links social capital and justice by explaining that the location in a social network, determining the types and amount of resources available, influences injustice formation [15, 31]. In such circumstances, social capital strongly restricts people's capability to access and mobilize the resources including environmental resources [90]. Pastor [67] also addresses the central role of social capital in environmental justice with her findings that unequal exposure of some social groups to environmental burdens results from lacking social capital.

In light of above, different forms of capital in contexts will influence the way of understanding and interpreting of environmental justice. However, The researches discussed above usually focus on a single form of capital and its influence on environmental justice. Besides, the concept of each form of capital is too broad and has been used in the researches of various fields. Taking social capital as an example, it has been used in sociology, politics, economy, and organizational theory studies by an increasing number of scholars of these fields [2]. Moreover, the interconnection and the convertibility between these forms of capital are usually neglected. Therefore, an integrative conceptual framework to bring these forms of capital together to involve different socioeconomic structures and conditions, sociocultural values as well as and political systems is needed.

The World Social Science Report released by the United Nations Educational, Scientific and Cultural Organization [92] recognizes there are different but interconnecting dimensions of inequality covering economic, social, cultural, political, environmental inequality, etc. Moreover, early in 1994, Viederman [93] proposed the capitals of sustainability aiming to provide a clear and demystified definition of "sustainability" within a community to the public. As such, the broad capital could also be used to define what are important in contexts for understanding, interpreting and obtaining environmental justice. Sharing the thinking patterns and drawing on the characteristics of five forms of capitals above along with their relations with environmental justice, I propose the "environmental justice capital" framework.

## 2.2 *Environmental Justice Capital*

Environmental justice capital is an asset set that an individual, group or community control and prudent use to obtain environmental justice, which has five interconnecting dimensions including natural, cultural, economic, political and social dimensions (see Fig. 1). These five dimensions of environmental justice capital have

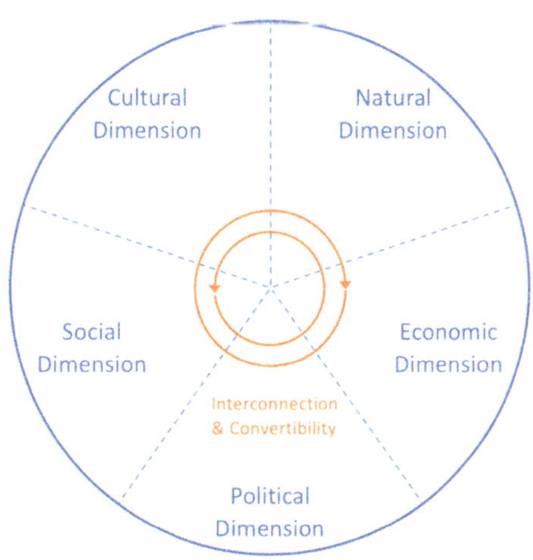


**Fig. 1** Environmental justice capital. *Source* the author

interconnection and convertibility with each other and determine the maximum value of environmental justice can be obtained. Specifically, the various notions of environmental justice could lie into the differences in the ways of controlling and using the environmental justice capital, the differences in quantity and/or quality of capitals accessible, and the differences in capability (opportunity) to access, mobilize or use through the capitals.

Sharing the characteristics of usefulness and durability of "capital" [16] as well as other forms of capital, environmental justice capital may have the following potential characteristics, some of which need to be further explored in future research. Firstly, environmental justice capital is not suitable for quantified measurement, as conventional economic capital, due to it contains intangible and unmeasurable elements, such as cultural value as "collective goods" and social networks. Secondly, like other forms of capital, environmental justice capital can be costed to trade for services and benefits, as well as invested and cumulated with the expectation of benefits in the future. Bourdieu [6] argued, capital accumulation contributes to analyzing the social world as an accumulated history, which takes historical development into consideration [5]. Features of other forms of capital can support this point, for example, cultural capital requires researchers to understand the relation between past, present, and future [42].

While, another main characteristic has already been proved in previous study—interrelation. There are interconnections and convertibility within the five dimensions of environmental justice capital. Such complementarities between various forms of has been largely accepted by economic researchers [67]: economic capital could easily convert to cultural and social capital for its liquidity [3, 83], the strength of social capital could largely determines the strength of economic, cultural and political capital held by the actors [6, 14, 67]; there is a mutually reinforcing character of natural and economic capital that natural capital is usually to be converted into economic capital for the minority and the poor [67]. However, such convertibility is not equally possible in all directions [7].

Thus, environmental justice capital should pay more attention on the balance between the benefits and costs of environmental justice capital, especially this balance referring to the disproportionate efficiency among the various dimensions of environmental justice. Simply put, the costs could link to the criteria of efforts and sacrifices—'individuals who make a greater effort or incur a greater sacrifice relative to their innate capacity should be rewarded more than those who make little effort and incur few sacrifices' and the benefits could link to the criteria of need—'individuals have rights to equal levels of benefit which means that there is an unequal allocation according to need' [72, 73]. The balance between costs and benefits, in another word, rational efficiency, is of great importance when study environmental justice capital for it is highly related both to the environment "where we live, work, and play" for human [66], and also environment regarding nature [76]. For example, Adams [1] and Pellow [68] studies have shown that one cause for environmental injustice is unbalanced costs and benefits of environmental justice (natural capital converts to economic capital).

The environmental justice capital could provide an exploratory and comprehensive framework to understand the local configurations and maximum values of environmental justice can be obtained in various contexts. However, the current framework is yet to be completed and needs to be examined and complemented through more theoretical and empirical studies. To conduct a preliminary examination of the feasibility of the framework, I will use newly proposed framework to discuss how the environmental justice capital manifests itself on Chinese soil.

## 3 Environmental Justice in Chinese Contexts

With China's active involvement in economic globalization and increasingly important role in global governance since Reform and Opening, environmental justice, a global research and environmental movement trend, has also received the attention of Chinese scholars with a focus on distributive justice. These studies emphasize environmental justice study model should be occupation-based and gender-based with the consideration of the disparities between east and west regions along with urban and rural, rather than race and income-based research models in America [52, 71]. However, Quan and Liu's researches have not successfully provided how Chinese contexts influencing the environmental justice and what are the reasons behind these environmental injustice issues. In the absence of sufficient literature references and field investigations of the environmental justice in China, this section will provide the very first exploration of Chinese environmental justice capital by understanding China social structure first, and diving in to environmental justice capital in Chinese contexts to understand environmental justice in Chinese contexts, as well as policies and measures in environmental protection over the last 40 years.

Heretofore, leaning about Chinese unique and resilient social structure is inevitable precondition for unfolding any subjects, phenomenon, achievements, or problems in China [107]. As stated by Wang [110], in the 21st century, China is the only society in the world that maintains the size, population and political culture of the former 19th century within the scope of a sovereign state and nation, where the changes brought about by the national movement and state building in modern China have been integrated into the internal structure. So to begin, the ultrastable structure in China since 1840 proposed by Jin and Liu [43] will be introduced in turn, which could also provide insights of what are the environmental justice capital in today's China.

Contrary to the "shock-response" argument shaping modern China under the Western Centralism [21, 43] argue that the modern China could be observed through two lenses including both internal structure and the impact of Western industrial civilization. They further argue the key that Chinese civilization could continue to today lies into the ultrastable system in China, where there are three sub-systems in Chinese ultrastable system, including ideological identity, political structure and economic structure working in a scenario of mutual adjustment and maintenance (see Fig. 2). Its evolutionary mode could be expressed as the disintegration of the

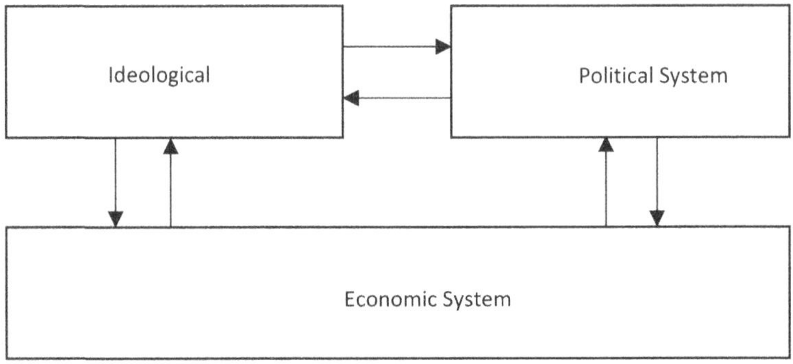


**Fig. 2** Ultrastable system in China. *Source* [43]

traditional integrated structure—the replacement of ideology—the establishment of a new integrated structure (ibid). Following this structure, we could try to understand the evolving environmental justice capital in China.

## *3.1 Cultural Dimension*

In a traditional society characterized by small peasant economy, Confucian Ideology provides a legitimate source of authority for the upper, middle and lower levels of the society, which could integrate and coordinate the three levels of the society [43, 107]. However, during the period of New Culture Movement (1925–1924), this traditional ideological identity was criticized and substituted as People realized they are far behind the world. Since the end of the Qing Dynasty, at the point when the country's very existence is at stake, individuals could sacrifice themselves as long as the ultimate goal of national salvation can be realized [96, 105]. At that time, the moral idealism of the group has replaced the moral idealism of the individual, and the utopia of the future has replaced the ancient great harmony ("da tong"). With anti-imperialist and anti-feudal aims, this new ideology differs from Confucianism but combines the deep structure of the traditional ideology, thus possesses greater organizational power to build an independent and powerful modern country and stand on its own among the nations of the world [43, 99]. During the Socialist revolution, the determination of Leninism and the strategy of encircling the city from the countryside meant the localization of the Socialist revolution. The pursuit of individual liberation and social equality in the socialist ideals once again gave way to national worries, which could be viewed as an new ideological identity at that time to maintain the ultrastable system (ibid). Therefore, modernity in China often means a powerful country, a prosperous nation, and a wealthy individual—these must be in a strict sequence [102]. Moreover, individual rights and freedom have to submit to national survival and material satisfaction, which is rather different in the Western society focusing on individual freedom, democracy and justice [102, 105].

Entering into the new era of Reform and Opening, Chinese people began again to recognize the outside world. In 1989, with the influences of "end of history" on ideological identity. Chinese people began to have central worship and believe in International and domestic politics are nothing but power competition [99]. That again subtly affect Chinese thinking patterns and development modes. "Lagging behind leaves one vulnerable to attacks" ("luo hou jiu yao ai da") has been summarized as non-forgotten lesson for China's development. Moreover, Western thinking patterns tend to be a binary thinking mode, which usually separates theory and practice. For instance, Rawls' justice theory is a hypothesis theory and the "original position" is an imaginary utopia [106]. However, Chinese thought has a very strong tendency to practice. A traditional view "zhi xing he yi" (knowledge as action) supports this tendency (ibid). In recent years, such an idea has become more and more utilitarian, which reflects people are pursuing immediate outcomes in reality [105]. The idea can be found both in the Chinese leader's speaking and Chinese policies. Deng Xiaoping's [111] famous economic-centric slogan in 1960s "it doesn't matter whether the cat is back or white, as long as it catches mice" has influenced China's economic development during the last decades. The main indicator to evaluate the social progress goal today is Materialistic, such as GDP [102]. The logic behind the GDP-oriented development can be also interpreted as the intention to strengthen a shared ideological identity to maintain the ultrastable structure for it could improve people's faith in the central government [107]. However, the economic-centric development discourse in China has resulted in different serious issues such as social inequality, environmental degradation, as well as environmental injustice problems [19, 97]. China has released the official Environmental Protection Law in 1989, while, it was not until 2003 that China began to initiate Environmental Impact Assessment (EIA) and publish environmental statistics [64]. Before 2005, environmental protection has been neglected, and then in the Eleventh Five-year Plan, the indicators of sulfur disulfide and sewage treatment finally appeared [107]. Meanwhile, The GDP increased from 1717.97 billion RMB of 1989 to 187318.9 billion RMB of 2005 [64].

## 3.2 Political Dimension

Political democracy has been a tradition with a history of four hundred years in western countries [105], which means the individual entitlement and freedom is protected by law, the material needs are no longer the main goal of social progress there [102]. In these western countries, democracy has been already well accepted by society and used as a set of principles for daily life and social relations [4, 102]. In such a context, drawing on Kant's theory, Rawls formed his justice theory that guides the development of environmental justice based on Liberalism [74] which emphasizes fair treatments and rights to individuals [72]. However, the corresponding political background is that China is a bureaucratic-authoritarian, one-party and centralized governance [46, 102].

The contemporary China has the enormous bureaucracy system including huge coverage of state organizations and the unmatched depth of governance levels compared to any political system in Chinese history (see Fig. 3). By the end of June of 2020, there are 91.914 Mio. CCP members and 4.681 Mio. Grass-roots party organizations [101]. That results in the missing of civil society and their voices. As stated in Zhou's [107] arguments, the political system in China could be understood as a 3-level model—Client, Contractor and Agent (see Fig. 4). He further states that in this upwardly responsible organizational structure, the core task is to efficiently complete top-down tasks, which is contrary to the bottom-up function of conveying public opinion (ibid).

Moreover, given that the market economy starts from scratch and China is undergoing the period of social transformation, centralized governance aims to promote and maintain a market economy and effectively allocate resources to all levels in

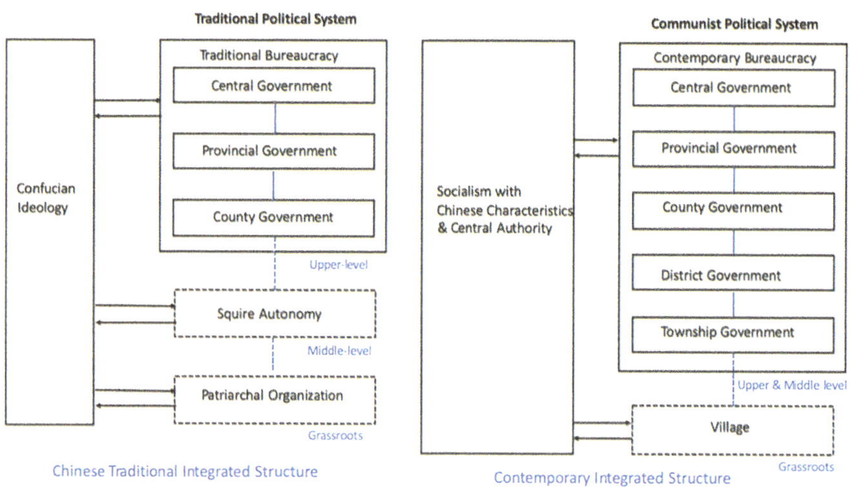


**Fig. 3** Traditional and contemporary. *Source* [43]

**Fig. 4** Chinese political system. *Source* [107]

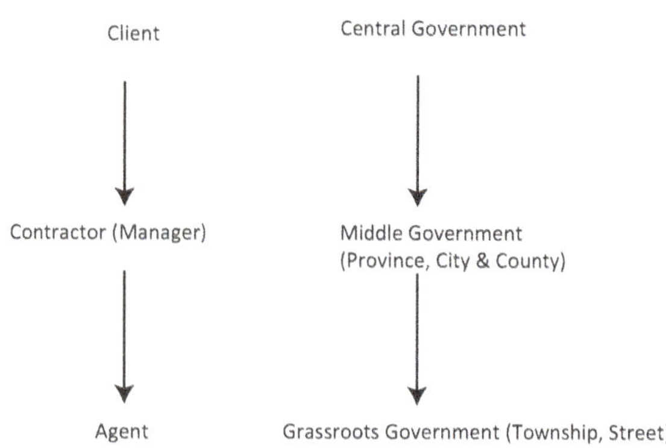

the society [54]. The market economy has magnified people's profit-making motivations and behaviors, which leads to that people in China have relatively indifferent consciousness of rules and laws (ibid).

Besides that, the deeply rooted tradition of rule by man is another reason causing such a phenomenon, which could be understood as an informal form of social capital. Unlike western society tends to regard justice as the core to realize the rule of law as well as strict compliance with rules and laws, Chinese tend to understand justice within the framework of ethics under the influence of Confucianism with "ren" ("virtue") as the core [54, 103, 105]. To be specific, the Chinese usually adopt the principle that "zun zun" ("hierarchical relationship") first, "qin qin" ("kinship relationship") second then others to decide the rules and resource allocation [40].

Drawing on above, the political dimension of environmental justice capital is rather different. All these above highly restrict the motivations and behaviors of citizens' political participation, which leads to that people are lacking attention of public affair and public interest [54]. That also provides the interpretation that why public participation has no solid foundation for in China. The government would like to turn to specialists for pieces of advice instead of the public, which results in that the public could seldom participate in any decision-making process [89]. Additionally, EIA, an important approach to achieve environmental justice, is only a planning tool mainly for a smooth project process to gain rapid benefit return. In addition, the result of assessment has limited influence on policymaking in China, rather than its significant role in policymaking and environmental governance in western countries [89].

## 3.3 *Natural and Economic Dimensions*

However, this phenomenon above is common in developing countries [85, 89], and it directs to another two dimensions shaping different notions of "justice" and environmental justice in China—natural capital endowment and economic development. Today, most of the discussions of justice are based on discussions of morality [72, 74]. Yin [103] states that the pursuit of morality must first meet the basic needs, that is, food, clothing, and housing. If these are not satisfied, the lowest morality will not be guaranteed, in another word, a society that has real morals must have a considerable economic foundation (ibid). As Kong Feili [40] states in 2014, the time differences between different development levels of various countries should be given enough consideration. This can provide a reasonable explanation of why environmental justice is important but it is still a "luxury good" for many developing countries or less developed countries [71]. The main goal of these countries is still poverty alleviation at the expense of the environment (ibid) (convertibility from natural dimension to economic dimension). As illustrated in the Fig. 5, the environmental performance is closely related to the economic development of the country. This supports that nations care little about pollution when they are poor and then grow markedly cleaner as they get richer [46].

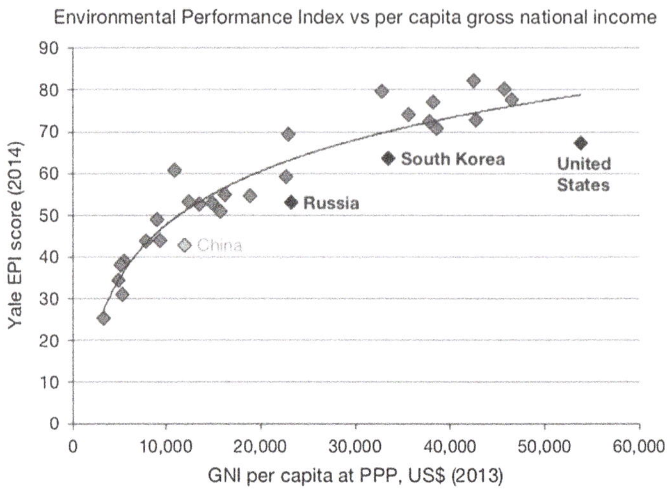


**Fig. 5** Environmental index versus GDP. *Source* [45]

Likewise, distributive environmental injustice is also serious in China due to these two dimensions of environmental justice capital. Western China usually suffers more environmental injustice due to the lack of natural resource endowment and economic weakness [52]. For example, the amount of water supply in the coastal areas with higher GDP is bigger than that in western China with lower GDP (see Fig. 6).

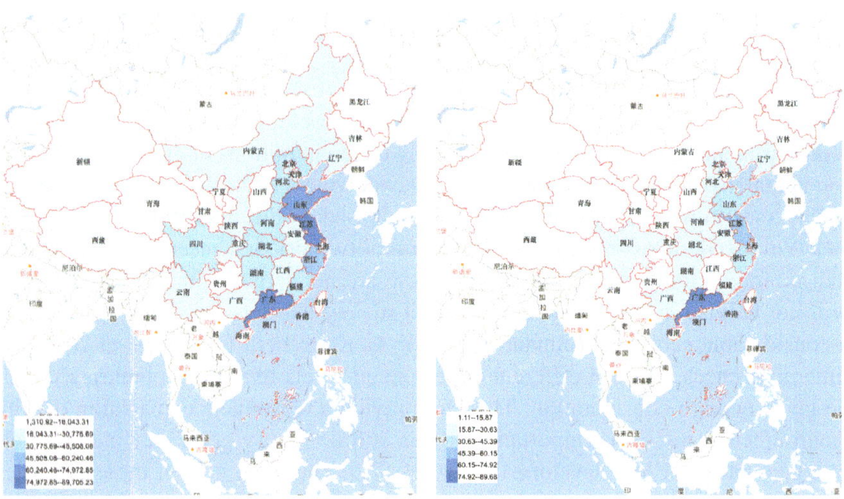

**Fig. 6** Provincial GDP and water supply in 2017. *Source* [75]

### 3.4 Social Dimension

Relying solely on the market or the government control is unable to achieve justice, the role of social capital is important [49]. In another word, shared values, norms and trust formed spontaneously between individuals and among the group [22, 70] is of great importance when realizing justice [49]. Such characteristic determines that the social dimension of environmental justice capital highly depends on the public's consciousness rather than the authority. Consciousness is usually formed in a highly open political system with full recognition of "individual" [54]. However, in China, under the influence of centralized governance, people's "individual" consciousness is awakened with the disappearance of the "working unit" system in recent years, let alone forming shared public norms and trust to solve public affairs and achieve public interest (ibid). Meanwhile, as eastern, urban and economically developed regions in China are often equipped with more transparent and clean governance systems and better education systems, the environmental justice capital of social dimension is with higher density and stock than the western region. This results in that the environmental justice condition in the East is often better than that in the West of China [52]. According to the statistics from National Bureau of Statistics, though the number of CSO has surged in the past two decades, the eastern area still possesses many more CSOs than western region (see Figs. 7 and 8).

As such, it could be observed that the different notions and focus of "justice" have significant impacts on the understanding and application of related environmental policies Additionally, environmental justice capital is not a stable value, which could change with the different socio-economic and social economic levels.

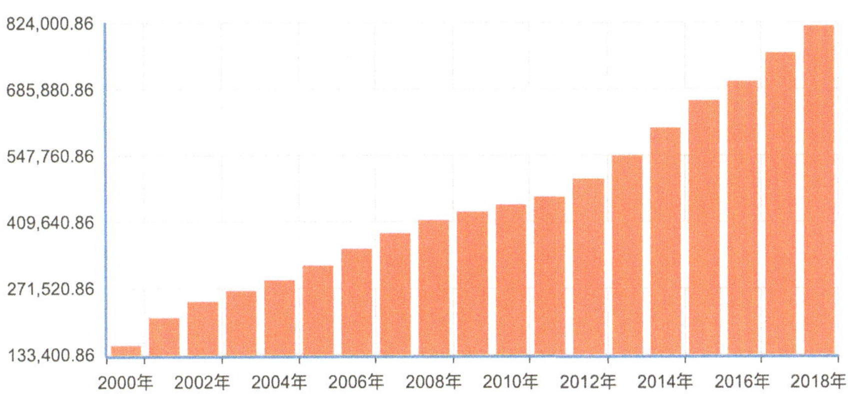


**Fig. 7** Number of CSO from 2000 to 2018. *Source* NBS (2018)

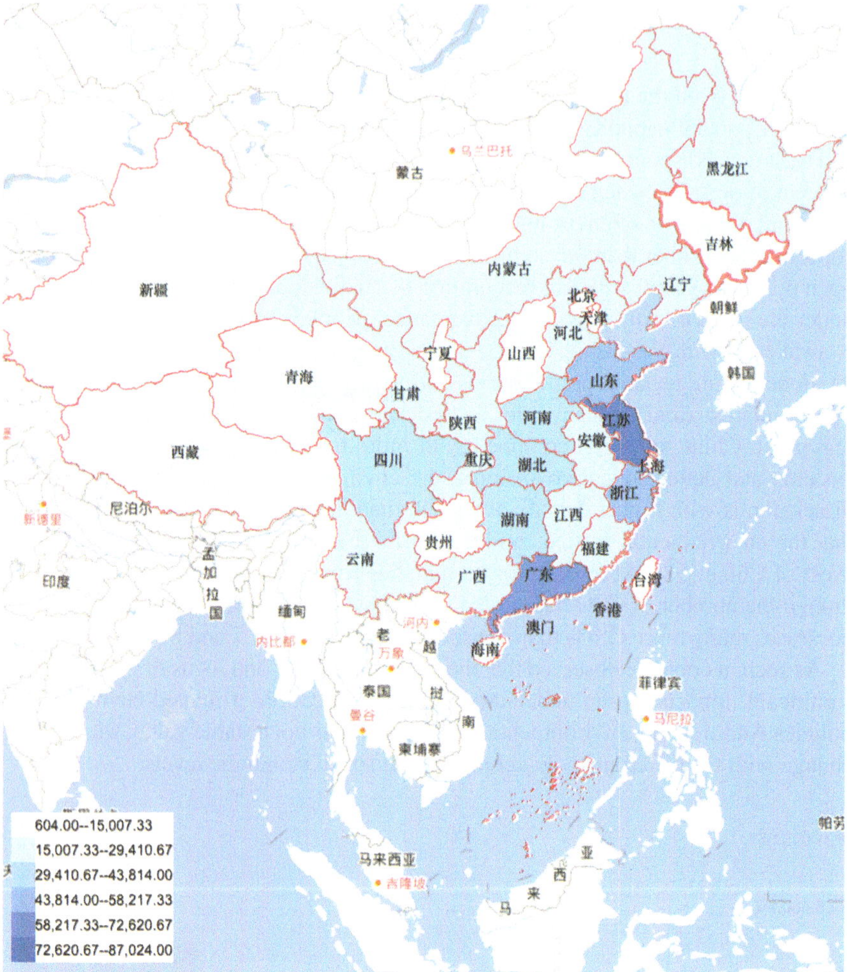


**Fig. 8** Number of CSO by province. *Source* NBS (2017)

## 4 Discussion

Since the Reform and Opening in 1978, China has chosen the path of "taking economic development as the center", which reflects that planned economy is gradually replaced by the market economy. While the development path with the principle of "considering fairness with the priority of efficiency" has prominent limitations that various social problems, such as environmental injustice, began to show up and got worse [34, 98]. Such issues have caused the whole of society's reflections on justice and efficiency [57, 81]. Now, China is becoming a more open and modern society, meanwhile, under the influence of globalization [102], consciousness towards social and environmental justice in China has gradually awakened in recent years [52, 71]. For instance, "lv shui qing shan jiu shi jin shan yin shan" ("lucid waters and lush mountains are invaluable assets") proposed by Chinese President Xi Jinping has already been the new core for development. Moreover, taking the development of EIA as an example, the number of EIA laws and regulations at the central level has surged since ecological civilization has been proposed (see Fig. 9).

In addition to the state's increasing emphasis on environmental protection, the awakening of personal consciousness of "individual" and disenchantment of the western political and economic system may also have a great impact on the future environmental justice in China. According to Xiang's [100] statement of Pacific Paradox, the purpose of today's Chinese students studying abroad is no longer for revitalization of the country, but for self-worth. Moreover, the reason why they choose countries like America for studying abroad is not that they recognize their governance system but they admit the high education level (ibid). As discussed before, an important part of ultrastable system is the ideological identity, while with the "individual" consciousness awakening, what will be the future trend of this social structure and how will the evolving structure affect environmental justice capital remain unknown. One thing we could make sure is that individual rights are important, that is also why China is putting so much effort in poverty alleviation and development of public services.

Last, when we talk about the environmental justice in contexts, besides the time difference, the impacts of other countries should also be took into account. As Premier Li Keqiang has stated that China's accession to the WTO has benefited and meanwhile paid a price at the opening ceremony of the 2013 Summer Davos Forum [20]. Waste-importing and consequent serious environmental pollution might be the

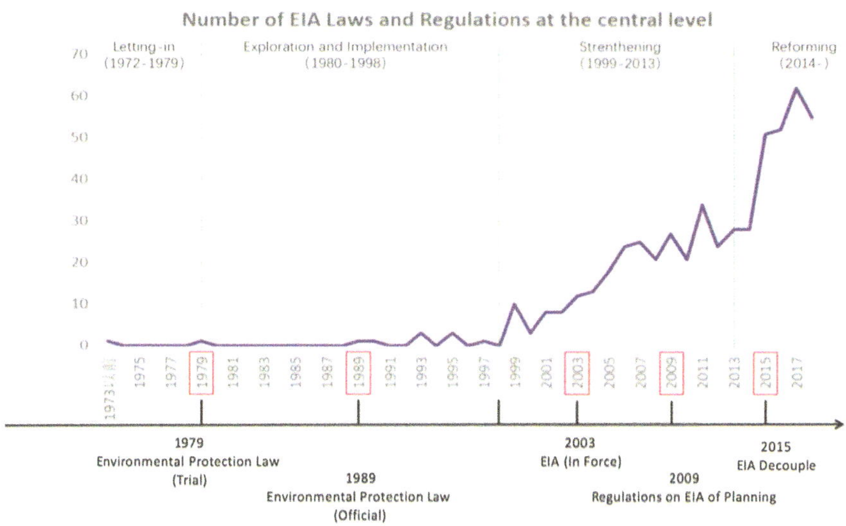


**Fig. 9** Number of EIA laws and regulations on the central level. *Source* [36]

biggest price. For example, 56% of world imports of waste plastics −7.3 Mio. Ton, has been imported to China in 2016 [58]. The waste importing has resulted that the Guiyu County in Guangzhou, one of the biggest informal imported e-waste dismantling bases in the past, has been seriously polluted. That has also brought harm to the health of local people. The physical examination results of the students in one of the villages by the local health center has showed that more than 80% of the primary and secondary school students had respiratory diseases [32]. Until July 18 of 2017, China's Ministry of Environmental Protection submitted a document to the WTO, requesting an urgent adjustment of the list of imported solid waste, which means a closure of importing the foreign waste [13]. China has always blamed for environmental pollution and GHG emission, however, if it is environmental injustice that developed country transfer the harmful waste to less-developed areas and countries so that they could achieve better living environment and environmental justice. The answer is an obvious no, as problems will never disappear because of the shift.

Like the environmental justice research in western countries, getting a more thorough and explicit concept of environmental justice in China requires more empirical and theoretical studies in the future. But again, when we look at the environmental justice in contexts, every dimension of the environmental justice capital and the connections should be understood and interpreted. Only in this way could we see the world with a more equal perspective and we have the opportunity to achieve higher level of justice ladder.

## References

1. Adams, J.S.: Inequity in social exchange. In: Advances in Experimental Social Psychology, pp. 267–299. Elsevier (1965)
2. Adler, P.S., Kwon, S.W.: Social capital: prospects for a new concept. Acad. Manage. Rev. **27**(1), 17–40 (2002)
3. Anheier, H.K., Gerhards, J., Romo, F.P.: Forms of capital and social structure in cultural fields: examining Bourdieu's social topography. Am. J. Sociol. **100**(4), 859–903 (1995)
4. Beck, U., Beck-Gernsheim, E.: Individualization: institutionalized individualism and its social and political consequences. SAGE (2001)
5. Birner, R., Wittmer, H.: Using social capital to create political capital: how do local communities gain political influence? A theoretical approach and empirical evidence from Thailand. In: The Commons in the New Millennium: Challenges and Adaptation, p. 291 (2003)
6. Bourdieu, P.: Economic capital-cultural capital–social capital. the hidden mechanisms of power. Schriften zu Politik and Kultur **1**, 49–78 (1992)
7. Bourdieu, P.: Distinction: a social critique of the judgement of taste. In: Food and Culture, pp. 45–53. Routledge (2012)
8. Bryant, R.L.: Beyond the impasse: the power of political ecology in third world environmental research. Area **29**(1), 5–19 (1997)
9. Bryant, B., Mohai, P.: Race and the Incidence of Environmental Hazards: A Time for Discourse (1992)
10. Bullard, R., Wright, B.H.: Environmentalism and the politics of equity: emergent trends in the black community. Mid-Am. Rev. Sociol. **12**(2), 21–37 (1987)
11. Bullard, R.D.: Dumping in Dixie: Race, Class, and Environmental Quality. Routledge (2018)

12. Bullard, R.: Environmental justice in the 21st century. In: People of color environmental groups. Directory, pp. 1–21 (2000)
13. CCTV: "国正式通知WTO: 不再接收这些'外来垃圾". Accessed 0903 (2017). https://m.news.cctv.com/2017/07/21/ARTIpKDw3jRy2BvSXPoLTlpv170721.shtml
14. Callaghan, E.G., Colton, J.: Building sustainable & resilient communities: a balancing of community capital. Environ. Dev. Sustain. **10**(6), 931–942 (2008)
15. Campbell, K.E., Marsden, P.V., Hurlbert, J.S.: Social resources and socioeconomic status. Soc. Net. **8**(1), 97–117 (1986)
16. Castle, E.N.: Social capital: an interdisciplinary concept. Rural Sociol. **67**(3), 331–349 (2002)
17. Castree, N.: The nature of produced nature: materiality and knowledge construction in Marxism. Antipode **27**(1), 12–48 (1995)
18. Chakraborty, J.: Focus on environmental justice: new directions in international research. Environ. Res. Lett. **12**(3), 030201 (2017)
19. Chan, M.K., So, A.Y.: Crisis and transformation in China's Hong Kong. Routledge (2016)
20. ChinaNews: Li Keqiang: China's accession to the WTO has also paid a price for its gains. Accessed 0903 (2013). https://www.chinanews.com/gn/2013/09-11/5274481.shtml
21. Cohen, P.A.: Discovering history in China: American historical writing on the recent Chinese past. Columbia University Press (2010)
22. Coleman, J.S.: Social capital in the creation of human capital. Am. J. Sociol. **94**, S95–S120 (1988)
23. DfID, UK: Sustainable livelihoods guidance sheets. London: DFID 445 (1999)
24. Di Chiro, G.: Living environmentalisms: coalition politics, social reproduction, and environmental justice. Environ. Polit. **17**(2), 276–298 (2008)
25. Donahue, J.: Water, culture, and power: local struggles in a global context: Island Press (1997)
26. Figueroa, R.M.: Bivalent environmental justice and the culture of poverty. Rutgers JL & Urb. Pol'y **1**, 1 (2004)
27. Follett, M.P.: The new state: group organization. In: The Solution of Popular Government, London (1918)
28. Fombrun, C.J.: Corporate culture, environment, and strategy. Human Res. Manage. **22**(1–2), 139–152 (1983)
29. Fraser, N.: Rethinking recognition. New Left Rev. **3**, 107 (2000)
30. Fraser, N.: Scales of justice: reimagining political space in a globalizing world, vol. 31. Columbia University Press (2009)
31. Green, G.P., Tigges, L.M., Browne, I.: Social resources, job search, and poverty in Atlanta. Res. Commun. Sociol. **5**, 161–182 (1995)
32. Greenpeace: 汕头贵屿电子垃圾拆解业的人类学调查报告 (2003)
33. Grineski, S.E., Collins, T.W., de Lourdes, M., Aguilar, R.: Environmental injustice along the US–Mexico border: residential proximity to industrial parks in Tijuana, Mexico. Environ. Res. Lett. **10**(9), 095012 (2015)
34. Gustafsson, B.A., Shi, L., Sicular, T.: Inequality and public policy in China. Cambridge University Press (2008)
35. Harvey, D.: Social Justice and the City, vol. 1. University of Georgia Press (2010)
36. He J., Bao, C.: Evolution or revolution: where next for SEA in China. In: The 6th China Strategic Environmental Assessment Academic Forum. Haikou, Hainan (2019)
37. Hensengerth, O.: Regionalism, identity, and hydropower dams: the Chinese-built lower sesan 2 dam in Cambodia. J. Current Chin. Affairs **46**(3), 85–118 (2017)
38. Hettne, B.: The new regionalism: a prologue. Globalism and the New Regionalism, vol. 1 (1999)
39. Hicks, A., Misra, J.: Political resources and the growth of welfare in affluent capitalist democracies, 1960–1982. Am. J. Sociol. **99**(3), 668–710 (1993)
40. Huang, G., Hu, X.: Mianzi: Chinese power game. China Renmin University Press (2004)
41. Jackson, S.: Indigenous peoples and water justice in a globalizing world. Oxford Handb. Water Polit. Policy 120–141 (2018)
42. Jacobs, J.: Dark age ahead: vintage Canada (2010)

43. Jin, G., Liu, Q.: The Transformation of Chinese Society (1840–1956): The Fate of Its Ultrastable Structure in Modern Times. China Law Press, Beijing (2011)
44. Kong, F.: Origin of the modern China state. Sanlian (2013)
45. Kroeber, A.R.: China's economy: what everyone needs to know®. Oxford University Press (2020)
46. Kroeber, A.R.: China's Economy: What Everyone Needs to Know®. Oxford University Press (2020)
47. Lake, R.W.: Volunteers, NIMBYs, and environmental justice: dilemmas of democratic practice. Antipode **28**(2), 160–174 (1996)
48. Li, H., Yang, X.: Social capital and social development. Social Sciences Academic Press (2000)
49. Li, Z.: Social capital and the realization of justice. Gansu Theor. Res. **5** (2005)
50. Lin, C., Lu, H.: The country governance in China in the view of social capital. J. China Three Gorges Univ. (Humanities & Social Sciences) **5** (2007)
51. Lin, N.: Inequality in social capital. Contemp. Sociol. **29**(6), 785–795 (2000)
52. Liu, J.: Environmental justice with Chinese characteristics: recent developments in using environmental public interest litigation to strengthen access to environmental justice. Florida A&M University Law Review. Forthcoming, 24–12 (2012)
53. Lyotard, J.-F.: The Postmodern Condition: A Report on Knowledge, vol. 10. U of Minnesota Press (1984)
54. Ma, B.: How is the social justice possible? Jilin University Journal Social Sciences Edition 4 (2006)
55. McCarthy, J.D., Zald., M.N.: Resource mobilization and social movements: a partial theory. Am. J. Sociol. **82**(6), 1212–1241 (1977)
56. McLean, J.: Water injustices and potential remedies in indigenous rural contexts: a water justice analysis. Environmentalist **27**(1), 25–38 (2007)
57. Meng, T.: Chinese people's perception of distributive justice in transitional China: outcome justice and opportunity justice. Chinese Journal of Sociology **32**(6), 108–134 (2012)
58. Miles, T.: China says it won't take any more foreign garbage. Reuters, accessed 0904 (2017). https://www.reuters.com/article/us-china-environment-idUSKBN1A31JI?il=0
59. Mohai, P., Pellow, D., Timmons Roberts, J.: Environmental justice. Annual Rev. Environ. Resour. **34**, 405–430 (2009)
60. Mohai, P., Saha, R.: Racial inequality in the distribution of hazardous waste: a national-level reassessment. Soc. Prob. **54**(3), 343–370 (2007)
61. Mohai, P., Saha, R.: Which came first, people or pollution? A review of theory and evidence from longitudinal environmental justice studies. Environ. Res. Lett. **10**(12), 125011 (2015)
62. Mohai, P., Saha, R.: Which came first, people or pollution? Assessing the disparate siting and post-siting demographic change hypotheses of environmental injustice. Environ. Res. Lett. **10**(11), 115008 (2015)
63. Morello-Frosch, R.A.: Environmental justice and California's riskscape: the distribution of air toxics and associated cancer and non-cancer health risks among diverse communities (1999)
64. NBS (National Bureau of Statistics): Environmental Statistics of 2003. Accessed 0828 (2003). https://www.stats.gov.cn/ztjc/ztsj/hjtjzl/2003/index.html
65. NBS (National Bureau of Statistics): National Statistics. Accessed 0828 (2017). https://data.stats.gov.cn/easyquery.htm?cn=E0103
66. Novotny, P.: Where we live, work, and play: the environmental justice movement and the struggle for a new environmentalism. Greenwood Publishing Group (2000)
67. Pastor, M.: Building social capital to protect natural capital: the quest for environmental justice. In: Natural assets: democratizing environmental ownership, pp. 77–97. Island Press, Washington, DC (2003)
68. Pellow, D.N.: Environmental inequality formation: toward a theory of environmental injustice. Am. Behav. Sci. **43**(4), 581–601 (2000)
69. Pellow, D.N., Brulle, R.J.: Power, justice, and the environment: toward critical environmental justice studies. In: Power, Justice, and the Environment: A Critical Appraisal of the Environmental Justice Movement, pp. 1–19 (2005)

70. Putnam, R.D.: Making democracy work (1993)
71. Quan, R.: Establishing China's environmental justice study models. Geo. Int'l Envtl. L. Rev. **14**, 461 (2001)
72. Rawls, J.: A Theory of Justice (1971)
73. Rescher, N.: Distributive justice (1966)
74. Sandel, M.J.: Liberalism and the limits of justice (1982)
75. Schlosberg, D.: Reconceiving environmental justice: global movements and political theories. Environ. Polit. **13**(3), 517–540 (2004)
76. Schlosberg, D.: Theorising environmental justice: the expanding sphere of a discourse. Environ. Polit. **22**(1), 37–55 (2013)
77. Schlosberg, D., Carruthers, D.: Indigenous struggles, environmental justice, and community capabilities. Global Environ. Polit. **10**(4), 12–35 (2010)
78. Schlosberg, D.: Defining Environmental Justice: Theories, Movements, and Nature. Oxford University Press (2009)
79. Schnaiberg, A., Gould, K.A.: Environment and Society: The Enduring Conflict. Blackburn Press (2000)
80. Schnaiberg, A.: The environment: from surplus to scarcity (1980)
81. Shen, M.: Zhongguo gongminyishi diaocha shuju baogao (Attitudes towards Citizenship in China: Data Report of A National Survey). Beijing: Social Sciences Academic Press (2008)
82. Shrader-Frechette, K.: Environmental Justice: Creating Equality, Reclaiming Democracy. Oxford University Press (2002)
83. Smart, A.: Gifts, bribes, and guanxi: a reconsideration of Bourdieu's social capital. Cult. Anthropol. **8**(3), 388–408 (1993)
84. Stucker, D.: Environmental Injustices, Unsustainable Livelihoods, and Conflict: Natural Capital Inaccessibility and Loss among Rural Households in Tajikistan. In: Environmental Justice and Sustainability in the Former Soviet Union, pp. 215–237 (2009)
85. Sun, T., Zhang, H., Wang, Y., Wang, C.: Approach on treatment after pollution based on environmental kuznets curve of China. Environ. Sci. Manage. **8** (2010)
86. Syme, G.J., Nancarrow, B.E., McCreddin, J.A.: Defining the components of fairness in the allocation of water to environmental and human uses. J. Environ. Manage. **57**(1), 51–70 (1999)
87. Sze, J.: Noxious New York: The Racial Politics of Urban Health and Environmental Justice. MIT press (2006); Sze, J., Jonathan, K.: Environmental justice at the crossroads. Sociol. Compass **2**(4), 1331–1354, London (2008)
88. Sze, J., London, J.K.: Environmental justice at the crossroads. Sociol. Compass **2**(4), 1331–1354 (2008)
89. Tang, B.S., Wong, S.W., Lau, M.H.: Social impact assessment and public participation in China: a case study of land requisition in Guangzhou. Environ. Impact Assess. Rev. **28**(1), 57–72 (2008)
90. Taylor, D.E.: The rise of the environmental justice paradigm: Injustice framing and the social construction of environmental discourses. Am. Behav. Sci. **43**(4), 508–580 (2000)
91. Throsby, D.: Culture, economics and sustainability. J. Cult. Econ. **19**(3), 199–206 (1995)
92. Unesco, Institute of Development Studies, and International Social Science Council.: World social science report 2016: Challenging inequalities: Pathways to a just world. UNESCO Publishing (2016)
93. Viederman, S.: Five capitals and three pillars of sustainability. Newslett. PEGS **4**(1), 5–12 (1994)
94. Walsh, A.: The commodification of the public service of water: a normative perspective. Publ. Reason **3**(2), 90–106 (2011)
95. Walzer, M.: Spheres of justice: a defense of pluralism and equality. Basic books (2008)
96. Wang, F.: A Genealogy of Modern Chinese Thought and Scholarship. Hebei Education Press (2001)
97. Wang, S., Zhou, D.: 生态文明建设与环境法治 (Ecological Civilization Construction and Environmental Law) (2014)

98. Whyte, M.: Myth of the social volcano: Perceptions of inequality and distributive injustice in contemporary China: Stanford University Press (2010)
99. Xiang, B.: Global Body Shopping: An Indian Labor System in the Information Technology Industry. Princeton University Press (2007)
100. Xiang, B.: 把自己作为方法 (Self as Method): 单读 (2020)
101. XinhuaNet.: 91.914 million party members and 46.81 million grassroots party organizations (2020)
102. Yan, Y.X.: The individualization of Chinese society. Shanghai Translation Press (2012)
103. Yin, H.: Reappraisal of cultural change in modern China. Shanghai, China: Sanlian Press (2002)
104. Young, I.M.: Justice and the Politics of Difference (1990); Yu, Y.: Modern Interpretation of Chinese Ideology. Jiangsu People's Publishing House (2003)
105. Yu, Y.: Traditional Chinese Ideology and Its Modern Changes. Guangxi Normal University Press (2004)
106. Yu, Y.: Modern interpretation of Chinese ideology. Jiangsu People's Publishing House (2003)
107. Zhou, X.: The Institutional Logic of Governance in China: An Organizational Approach. SDX-Joint Publishing Company, Beijing, China (2017)
108. Zimmerbauer, K.: From image to identity: building regions by place promotion. Eur. Plann. Stud. **19**(2), 243–260 (2011)
109. Zwarteveen, M.Z., Boelens, R.: Defining, researching and struggling for water justice: some conceptual building blocks for research and action. Water Int. **39**(2), 143–158 (2014)
110. 汪晖: 现代中国思想的兴起, vol. 2: 生活·讀書·新知三联书店 (2004)
111. 邓小平: 邓小平文选, vol. 2: 人民出版社 (1994)

# Spatial-Temporal Characteristics and Driving Factors of Coordination Degree of Ecological Efficiency and Industrial Structure Upgrading in the Yangtze River Economic Belt

**Aoxiang Zhang**

**Abstract** Taking 108 cities in the Yangtze River Economic Belt as research objects, SBM-Undesirable model and coupling coordination degree model were used to calculate the coordination degree between ecological efficiency and industrial structure upgrading in the Yangtze River Economic Belt from 2007 to 2017, and their temporal and spatial evolution was analyzed. Dynamic spatial Dubin model was used to analyze the driving factors of spatial and temporal evolution of coordination degree. The results show that: The coordination degree between ecological efficiency and industrial structure upgrading in the Yangtze River Economic Belt is mainly between moderate imbalance and primary coordination, and the cities with higher coordination level form a "T"-shaped distribution pattern extending from coastal cities in downstream areas to inland areas. The coordination degree has obvious path dependence, and the spatial spillover effect is significant. Increasing the coordination degree of neighboring cities will improve the coordination degree of this city. Increasing population density, optimizing system quality, increasing foreign investment, strengthening environmental regulation and technological progress are all conducive to the increase of coordination degree, while increasing foreign investment and strengthening environmental regulation in neighboring cities will lead to the decrease of coordination degree in this city.

**Keywords** Yangtze river economic belt · Ecological efficiency · Industrial structure upgrading · Coordination degree · DSDM

## 1 Instruction

The Yangtze River, the largest river in China and the third largest river in the world, connecting the eastern coastal and vast inland regions, covering Shanghai, Jiangsu, Zhejiang, Anhui, Jiangxi, Hubei, Hunan, Chongqing, Sichuan, Guizhou and Yunnan,

A. Zhang (✉)
School of Economic Management & Law, University of South China, Hengyang 421001, China
e-mail: 775126196@qq.com

 
F. K. S. Chan et al. (eds.), *Proceedings of the 2020 International Conference on Resource Sustainability: Sustainable Urbanisation in the BRI Era (icRS Urbanisation 2020)*, Environmental Science and Engineering,
https://doi.org/10.1007/978-981-15-9605-6_22

covering an area of about 2.0523 Mio. $km^2$, accounting for 21.4% of the country. In 2018, the total population of the Yangtze River Economic Belt was about 599 Mio. People, accounting for 42.9% of the country, and the regional GDP was about 40.3 trillion yuan, accounts for 44.1% of the country's total. Yangtze River Economic Belt has become the fastest regional growth pole for China's economic development. With the rapid economic development, the ecological environment of the provinces and cities along the Yangtze River is also facing threats. The industries along the Yangtze River are developing rapidly, energy resource constraints are tightening, and environmental problems such as acid rain and smog occur frequently. At present, China's economic development has entered the "new normal", and it has begun to transform from an extensive model that pursues growth speed to a connotative model that pursues structural adjustment and environmental efficiency. In November 2018, the State Council clearly required that ecological priority and green development be the leading, relying on the golden waterways of the Yangtze River, to promote the coordinated development of the upper, middle and lower reaches of the Yangtze River and the high-quality development of the regions along the river. At the same time, the Yangtze River Economic Belt spans the eastern, central and western regions of China. The economic development of each region is different, and the stage of industrial structure upgrading is different. Then, how will the ecological efficiency of each region change in the process of industrial structure adjustment and upgrading? Are ecological efficiency and industrial structure upgrading in a coordinated development? The answers to the above questions have important theoretical and practical significance for promoting the coordinated development of urban economy and environment in the Yangtze River Economic Belt.

## 2 Literature Review

Eco-efficiency was first proposed by Sturm, which refers to the ratio of increased economic value to increased ecological environmental load in a certain period of time [1]. Eco-efficiency comprehensively considers economic growth and ecological environmental protection. The impact of industrial structure on eco-efficiency begins with the study of the relationship between industrial structure and economic growth. Peneder once proposed that the input factors should change from sectors with low productivity or low productivity growth rate. Moving to sectors with high productivity levels or high productivity growth rates can promote the improvement of the overall social productivity level, and the resulting "structural dividend" maintains sustained economic growth [2]. Research by Liu, Qian et al., Han et al. and other scholars have shown that the contribution of industrial structure changes to China's economic growth is very significant, and there is a "structural dividend", but with the improvement of marketization, this "structural dividend" has shown a weakening trend [3–5]. Some scholars noticed the impact of industrial structure adjustment on the ecological environment, introduced the concept of ecological efficiency, and discussed the impact of industrial structure adjustment on the comprehensive benefits of economic growth and environmental improvement. Han et al., Ma et al. established spatial panel data model,studied the driving mechanism and effect of the impact of industrial structure optimization and upgrading on ecological efficiency from the provincial and municipal scales. The research found that industrial rationalization and supererogation have positive local effects and spatial spillover effects between regions or cities, and there is heterogeneity in the strength in different regions [6, 7].

Regarding the reverse effect of ecological efficiency on the upgrading of industrial structure, it is also reflected in both economic growth and ecological environment. The Petty-Clark theorem puts forward the empirical law of industrial structure upgrading. It emphasizes that due to the differences in income elasticity and technological progress in various industries, With the increase in per capita income, national income and labor distribution will gradually transition from the primary industry to the secondary and tertiary industries, reflecting the dynamic transition of industrial upgrading from low to high [8]. The research of Sun et al. and Gao et al. confirmed that consumption demand caused by income increase is an important driving force for the advancement of industrial structure, and it plays a decisive role in the advancement of industrial structure [9, 10]. The effect of environmental pollution on the industrial structure is mainly reflected in the government's environmental supervision policies. The increase in environmental pollution has prompted the government to strengthen environmental regulations, forcing enterprises to relocate or upgrade their industrial structure. Zhong conducted an empirical test using China's inter-provincial panel data, and the results showed that: conducted an empirical test using China's inter-provincial panel data, and the results showed that: environmental regulation, regional industrial transfer, and structural upgrading respectively present a U-shaped change relationship. Only by crossing the threshold of environmental regulation can industrial structure adjustment be promoted [11]. Dou et al. used an intermediary model to empirically analyze the impact of environmental regulations on the transfer of polluting industries in my country, and found that the strengthening of environmental regulations will inhibit the "pollution refuge" effect, and promote the transfer of industries and the upgrading of industrial structure [12]. Qian and Chen studied the dual impact of marine environmental regulations on the upgrading of manufacturing industry structure and the transfer of polluting industries, and found that there is a positive U-shaped relationship between marine environmental regulations and the transfer of polluting industries and the upgrading of industrial structure [13].

In summary, the existing literature has gradually and deeply discussed the relationship between industrial structure upgrading and ecological efficiency. Most of the research perspectives focus on the one-way relationship between the two, based on the coordinated development of the two to study the coordination degree between the two. In addition, most economic geographers believe that cities are the core of coordinated regional development. Provincial-scale research has not met the actual situation in China. Based on this, this article will use the SBM-Undesirable model, coupling coordination degree model measures the coordination degree between ecological efficiency and industrial structure upgrading of 108 cities in the Yangtze River Economic Belt and analyzes its spatial-temporal evolution characteristics, then uses the dynamic spatial Dubin model to analyze the driving factors of the coordination degree, and finally focuses on the Yangtze River economy Propose suggestions for the coordinated development of economy and environment.

# 3 Research Method

## 3.1 Ecological Efficiency

In this paper, SBM-Undesirable model is used to measure the ecological efficiency of cities in the Yangtze River Economic Belt. The model expression is as follows:

$$\rho^* = \min \frac{1 - \frac{1}{M}\sum_{m=1}^{M} S_m^x / x_m^{k'}}{1 + \frac{1}{N+I}\left(\sum_{n=1}^{N} S_n^g / y_n^{k'} + \sum_{i=1}^{I} S_i^b / b_i^{k'}\right)}$$

$$s.t.: \sum_{k=1}^{K} z_k x_m^k + S_m^x = x_m^{k'}, m = 1, \ldots, M; \sum_{k=1}^{K} z_k y_n^k - S_n^x = y_n^{k'}, n = 1, \ldots, N;$$

$$\sum_{k=1}^{K} z_k b_i^k + S_i^b = b_i^{k'}, i = 1, \ldots, I; z_k \geq 0, S_n^g \geq 0, S_i^b \geq 0, S_m^x \geq 0 \tag{1}$$

In Formula (1), $S$ are relaxation variables; $S_m^x$,$S_n^g$ and $S_i^b$ respectively represent input redundancy, expected output deficiency and unexpected output excess. $x_m^{k'}$, $y_n^{k'}$ and $b_i^{k'}$ represent the input vector, the expected output vector and the unexpected output vector of the decision-making unit respectively. $\rho^*$ is the efficiency value, its value range is [0, 1]. When $\rho^* = 1$, the decision-making unit was effective. The selected input-output indicators are shown in Table 1.

## 3.2 Industrial Structure Upgrading

Learn from Linghui [14]: First, take the ratio of the added value of the three industries to GDP as the three coordinates of the spatial coordinate system to form a three-dimensional vector, and calculate the angle between the vector and the industry from low level to high level respectively.

**Table 1** Input-output indicators

| Indicators | Constitute | Representation and unit |
|---|---|---|
| Resource input | Capital input | Investment stock of fixed assets (10,000 yuan) |
| | Labor input | Employees in the unit (10,000) |
| | Water resources consumption | Total annual water supply (10,000 tons) |
| | Energy consumption | Standard coal consumption (10,000 tons) |
| | Consumption of land resources | Urban construction land area (square kilometers) |
| Unexpected output | $SO_2$ emission | Industrial sulfur dioxide emissions (tons) |
| | Smoke emission | Industrial soot emissions (tons) |
| | Wastewater discharge | Discharge of industrial wastewater (10,000 tons) |
| | $CO_2$ emissions | Carbon dioxide emissions (tons) |
| Expected output | GDP | Annual regional real GDP (100 Mio. Yuan) |

$$\theta_j = \arccos\left(\frac{\sum_{i=1}^{3}\left(x_{i,j} \cdot x_{i,0}\right)}{\sqrt{\sum_{i=1}^{3} x_{i,j}^2} \cdot \sqrt{\sum_{i=1}^{3} x_{i,0}^2}}\right), j = 1, 2, 3 \tag{2}$$

After calculating the included angle, calculate the industrial structure upgrading index:

$$\text{Ins} = \sum_{k=1}^{3}\sum_{j=1}^{k}\theta_j \tag{3}$$

The larger the value, the higher the advanced level of industrial structure.The data of each industry proportion comes from China Urban Statistical Yearbook from 2007 to 2017.

## 3.3 Coupling Coordination Degree Model

Coupling coordination degree model is established based on coupling degree model.Coupling comes from engineering physics. Coupling refers to the phenomenon that two or more systems interact with each other, and coupling degree

**Table 2** Classification of coordination degree

| Serial number | Coordination degree | Coordination level | Serial number | Coordination degree | Coordination level |
|---|---|---|---|---|---|
| 1 | 0.01–0.10 | Extreme imbalance | 6 | 0.51–0.60 | Reluctantly coordination |
| 2 | 0.11–0.20 | Severe imbalance | 7 | 0.61–0.70 | Primary coordination |
| 3 | 0.21–0.30 | Moderate imbalance | 8 | 0.71–0.80 | Intermediate coordination |
| 4 | 0.31–0.40 | Mild imbalance | 9 | 0.81–0.90 | Good coordination |
| 5 | 0.41–0.05 | On the verge of imbalance | 10 | 0.91–1.00 | Quality coordination |

indicates the degree of interaction between systems or elements.According to the definition of coupling degree, the coupling degree of ecological efficiency and industrial structure optimization is measured by the following model:

$$C = \left\{ \frac{(X_1 \times X_2)}{(X_1 + X_2)^2} \right\}^{1/2} \quad (4)$$

In Formula (4), it is ecological efficiency.Is the normalized industrial structure upgrading index.Coupling degree explains the degree of interaction and influence between systems, but it can not truly reflect the coordinated development level between system variables. To reflect the coordinated development level of ecological efficiency and industrial structure optimization, the following coupling coordination degree model is established:

$$D = \sqrt{C \times (\alpha X_1 + \beta X_2)} \quad (5)$$

In the Formula (5), D is the coupling coordination degree between ecological efficiency and industrial structure upgrading; $\alpha$ and $\beta$ is the weight of ecological efficiency and industrial structure upgrading index respectively. This chapter holds that ecological efficiency and industrial structure upgrading index are equally important in the system.The value range of coordination degree is [0, 1]. According to the existing research, the coordination degree is divided into 10 grades (as shown in Table 2).

## 3.4 *Spatial Econometric Model*

The spatial weight matrix is incorporated into the econometric model in exploring the driving factors of the spatial and temporal evolution of the coordination degree between ecological efficiency and industrial structure upgrading in the Yangtze River Economic Belt. In identifying the key factors leading to the evolution of coordination degree, the spatial spillover effect of coordination degree and its influencing factors should not be ignored, and the spatial econometric model should be adopted to estimate. Spatial panel measurement models include spatial lag model (SLM), spatial error model (SEM) and spatial Durbin model (SDM). Spatial Dubin model is the general form of SEM and SLM. In order to explore the path dependence of coordination degree, the lag phase of coordination degree is included in the static spatial panel model, and the general form of dynamic spatial panel model is as follows:

$$\begin{aligned} D_{it} &= \alpha W D_t + \varphi D_{it-1} + \beta_0 + \beta_i Z_{it} + \theta W Z_t + \phi_i + \mu_{it} \\ \mu_{it} &= \lambda M \mu_t + \varepsilon_{it} \end{aligned} \tag{6}$$

In Formula (6), $D_{it}$, $D_{it\text{-}1}$ and $WD_{it}$ respectively represent the numerical value of coordination degree of each city in the Yangtze River Economic Belt and its time lag term and space lag term; $Z_{it}$ and $WZ_{it}$ respectively represent explanatory variables and their space lag term; $\alpha$,$\varphi$,$\beta_0$,$\beta_i$ and $\theta$ are coefficient matrices to be estimated;$\phi_i$ and $\mu_{it}$ are individual effect and random disturbance term respectively, and $M$ is the spatial weight matrix of disturbance term and $\lambda$ is the autocorrelation coefficient of random disturbance term. If $\lambda = 0$, the model is a spatial Dubin model (SDM). If $\lambda = 0$ and $\theta = 0$, the model is a spatial lag model (SLM). If $\alpha = \varphi = 0$ and $\theta = 0$, the model is a spatial error model (SEM).In this paper, the spatial panel model will be selected according to LM test and Wald test. If the LM tests of SEM and SLM models are significant, then SDM model is adopted. If LM statistics are not significant, Wald statistics are further used for testing. As for the weight matrix $W$, this chapter chooses the adjacent spatial weight matrix, that is, the adjacent spatial units have significant mutual influence ($W = 1$), while the non-adjacent spatial units basically have no mutual influence ($W = 0$).

Considering the influencing factors of ecological efficiency and industrial structure upgrading index, under the principle of data availability, quantification and comparability, the following variables are selected to analyze the driving factors.The indicators of the variables are selected as shown in Table 3, and the data involved in the column of "Measurement Method" are all from China Urban Statistics Yearbook from 2007 to 2017.

## 4 Empirical Results

### *4.1 Spatial-Temporal Characteristics*

The coordination degree between ecological efficiency and industrial structure upgrading in the Yangtze River Economic Belt was calculated by using the coupling coordination degree model, and its change trend is shown in Fig. 1. From 2007

**Table 3** Qualitative description of each index

| Symbol | Indicators | Measure way | Unit |
|---|---|---|---|
| *PI* | Population density | Urban resident population/Urban area | 10,000 people/km$^2$ |
| *GR* | Expenditure | Fiscal expenditure/GDP | % |
| *FDI* | Dependence on foreign capital | Foreign direct investment/GDP | % |
| *ER* | Environmental regulation | Comprehensive utilization rate of industrial solid waste | % |
| *TEC* | Technical level | Practitioners of scientific research and technology/general practitioners | % |

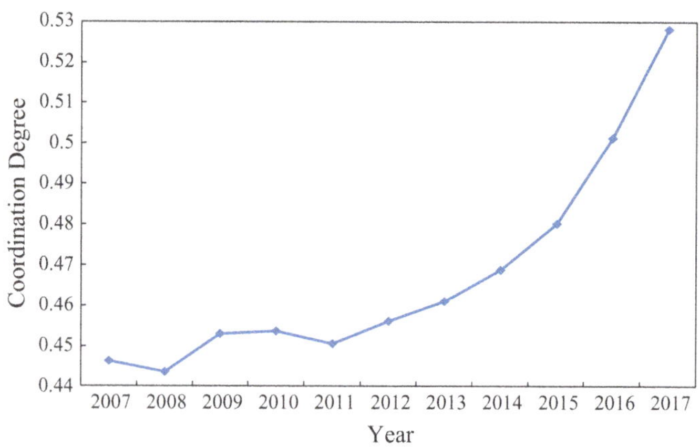


**Fig. 1** Temporal evolution of coordination degree

to 2011, it showed an inverted "N" type with a fluctuating downward trend. The change trend of ecological efficiency and industrial structure index was opposite at this stage, and the rising speed of industrial structure upgrading index was unstable, which led to the fluctuation of coordination degree. After 2012, the coordination degree showed an accelerated upward trend, and increased to 0.528 in 2017. As a whole, it was in a barely coordinated stage, and the coordination degree still needed to be improved. From the spatial distribution of coordination degree types in Fig. 2, in 2007, the coordination degree types were between moderate imbalance and primary coordination. The primary coordination cities were Shanghai, Suzhou, Wuxi, Jinhua, Taizhou, Wenzhou and other cities. The cities with higher coordination degree were mainly concentrated in the coastal areas of the lower reaches of the Yangtze River, while the cities in the middle and upper reaches had lower coordination degree. By 2017, the overall coordination level has been improved. Shanghai, most cities in central and southern Jiangsu, Jiaxing, Hangzhou, Jinhua, Zhoushan, Taizhou and Wenzhou in Zhejiang, Changsha, Zhangjiajie and Jiujiang in the middle reaches of

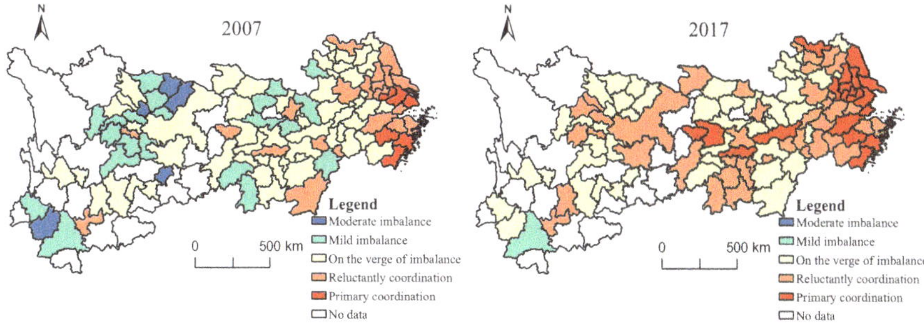


**Fig. 2** Spatial distribution of coordination degree

the Yangtze River have reached primary coordination. Other cities are mainly in the stage of reluctantly coordinating and on the verge of imbalance. Cities with higher coordination degree are distributed in a "T" shape from the eastern coast to the inland. Among them, the coordination degree between Changsha City and Zhangjiajie City in the middle reaches is ahead of the surrounding areas. In 2017, the ecological efficiency of Zhangjiajie City reached 1.28, and the proportion of the tertiary industry reached 68.9%, ranking first and second in the Yangtze River Economic Belt cities, respectively. The rapid development of service industry promoted the coordinated development of industrial structure upgrading and ecological efficiency improvement.

## *4.2 Driving Factors*

Before spatial econometric analysis, it is necessary to discuss whether there is spatial correlation. This chapter uses global Moran's I to detect the spatial correlation of coordination. The results are shown in Table 4. It can be seen that Moran's I index has remained above 0.5 in all years, and passed the test at the significance level of 1%, indicating that there is a significant spatial agglomeration phenomenon in coordination, that is, spatial autocorrelation. Therefore, when analyzing its driving

**Table 4** Global Moran's I value of coordination degree

| Year | Moran's I | Year | Moran's I |
|---|---|---|---|
| 2007 | $0.500^{***}$(9.130) | 2013 | $0.535^{***}$(9.708) |
| 2008 | $0.482^{***}$(8.850) | 2014 | $0583^{***}$(10.580) |
| 2009 | $0.569^{***}$(10.314) | 2015 | $0.618^{***}$(11.155) |
| 2010 | $0.548^{***}$(9.928*) | 2016 | $0.567^{***}$(10.248) |
| 2011 | $0.533^{***}$(9.677) | 2017 | $0.559^{***}$(10.083) |
| 2012 | $0.570^{***}$(9.321) | | |

*Note* Z value is in brackets; $^{***}$ means significant at the level of 1%

factors, it is more reasonable to adopt spatial measurement model. Before parameter estimation, LM test is carried out first, and the results show that both spatial lag LM test and spatial error LM test reject the original hypothesis at 1% significance level, indicating that SEM and SLM models are reasonable, and the general form of both, namely dynamic SDM model, should be adopted. Further Wald test also shows that dynamic SDM model is reasonable (see Table 5). At last, Hausman test was carried out, and the original hypothesis of random effect was rejected at a significant level of 1%, so the dynamic SDM model with fixed effect was selected, and the Return results of the model are shown in Table 5. Next, the Return results are explained.

(1) Time lag term. The coefficient of $\ln D_{t-1}$ is 0.717, which indicates that the coordination degree between urban ecological efficiency and industrial structure upgrading in the Yangtze River Economic Belt has a strong path dependence. The value of the previous period has a significant positive relationship with the current period, and the Matthew effect appears, that is, if the previous period's co-scheduling is higher (lower), the current period's coordination degree will be higher (lower).
(2) Spatial lag term. The coefficient of $W\ln D_t$ is 0.294, which has passed the significance test. According to the first law of geography, the closer the distance between regions, the more similar some characteristic attributes will be. The empirical results of this chapter support this conclusion. There is a significant spatial spillover effect in the coordination degree of cities along the Yangtze River Economic Belt. Specifically, when the coordination degree of neighboring cities increases by 1%, this city increases by 0.294%.
(3) Population density. The coefficient of lnPI is 0.058, which has passed the significance test of 5%, indicating that the population density of cities in the Yangtze River Economic Belt will increase by 1%, and the coordination degree will increase by about 0.06%. On the one hand, cities in the Yangtze River

**Table 5** Return results of Dubin model in dynamic space

| Variable | Coefficient | Spatial lag term | Coefficient |
|---|---|---|---|
| *lnDt-1* | $0.717^{***}$(0.0225) | *WlnD* | $0.294^{***}$(0.0248) |
| *lnPI* | $0.058^{**}$(0.0281) | *WlnPI* | 0.124(0.0931) |
| *lnGR* | $0.017^{*}$(0.0108) | *WlnGR* | −0.006(0.0135) |
| *lnFDI* | $0.004^{**}$(0.0020) | *WlnFDI* | $-0.015^{**}$(0.0038) |
| *lnER* | $0.006^{**}$(0.0029) | *WlnER* | $-0.021^{**}$(0.0106) |
| *lnTEC* | $0.007^{*}$(0.0031) | *WlnTEC* | 0.002(0.0124) |
| $R^2$ | 0.823 | | |
| Obs | 1080 | Log-likelihood | 1682.427 |
| Spatial error LM | $465.50^{***}$ | Spatial lag LM | $70.50^{**}$ |
| Spatial error Wald | $37.41^{***}$ | Spatial lag Wald | $30.51^{**}$ |

*Note* Standard errors are shown in brackets. $^{***}$, $^{**}$ and $^{*}$ are significant at the level of 1%, 5% and 10% respectively

Economic Belt will benefit from the scale effect and agglomeration effect of high urban population density, promote the sharing of urban public infrastructure and services, the overflow of technology and knowledge, and the formation of agglomeration economy such as labor pool, and reduce the energy demand of factor flow, thus reducing resource consumption and pollution emission and improving the ecological efficiency of cities;On the other hand, population agglomeration can promote the upgrading of industrial structure by expanding the customer base of service industry and increasing the accumulation of human capital. The Yangtze River Economic Belt accounts for 21.4% of the whole country, and its population accounts for 42.9% of the whole country, with a high population density. Under the background of the gradual reduction of the "demographic dividend", it ushered in the "second demographic dividend", which promoted the coordinated development of ecological efficiency and industrial structure upgrading through the effects of human capital and economies of scale.

(4) Financial expenditure. The coefficient of lnGR is 0.017, which has passed the significance test of 10%. Fiscal expenditure reflects the government's administrative and regulatory capacity. Government infrastructure expenditure can improve urban infrastructure services and reduce transaction costs and resource consumption. For example, the construction of Zhangjiang High-tech Industrial Park and Suzhou Industrial Park supported by the Yangtze River Delta government has promoted the specialized division of labor and regional integration in the Yangtze River Delta region, produced agglomeration economy, greatly improved the efficiency of resource allocation and utilization, and further promoted the improvement of ecological efficiency. In recent years, the city governments along the Yangtze River Economic Belt have vigorously supported the development of service industry, which has a positive impact on the upgrading of industrial structure. In addition, the government's support for science and education has brought about talent agglomeration and knowledge spillover, which has greatly promoted the development of high-tech industries and service industries, and is conducive to the upgrading of industrial structure.

(5) Dependence on foreign capital. The coefficient of lnFDI is 0.004, which has passed the significance test of 5%. It shows that FDI has produced a "pollution halo" effect in cities in the Yangtze River Economic Belt, which promotes the improvement of economic efficiency, environmental performance and ecological efficiency through technology spillover effect. On the other hand, FDI can not only directly promote the upgrading of industrial structure through the industrial flow of new assets, but also indirectly promote the upgrading of industrial structure through the effects of technology spillover and industrial association. WlnFDI coefficient is negative, and it has passed the significance test of 5%. Under the background of "Chinese decentralization", all local governments are keen to expand the introduction of FDI for the sake of performance evaluation. When neighboring regions attract more FDI inflows by relying on location advantages or policy convenience, the flow or quality of foreign investment in this region will decline, which makes it difficult to play a positive role of FDI in upgrading ecological efficiency and industrial structure.

(6) Environmental regulation. The coefficient of lnER is 0.006, and it has passed the significance test of 5%. It shows that the environmental regulation of cities in the Yangtze River Economic Belt has produced "forced emission reduction effect". Enterprises carry out technological innovation in order to reduce environmental costs. Technological innovation improves factor productivity and reduces pollution emissions per unit output. The increased economic benefits of enterprises make up for the cost of technological research and development, resulting in "innovation compensation effect", thus achieving a win-win situation of economic performance and environmental performance. At the same time, the strengthening of environmental regulations will enhance the public's awareness of environmental protection. Consumers are increasingly inclined to buy environmentally friendly products. Public supervision will play a role in selecting the types of enterprises. Clean technology and environmentally friendly enterprises will eventually gain advantages in the competition, thus promoting the upgrading of the industrial structure in this region. The coefficient of WlnER coefficient is negative, and it has passed the significance test of 5%. At present, the industrial transfer direction of Yangtze River Economic Belt is gradually moving westward, and the strong environmental regulations in the downstream areas increase the cost of some high-pollution and high-energy-consuming industrial enterprises. In order to seek cost savings, the production areas will be moved to the middle and upper reaches with weak environmental regulations. Although it promotes the industrialization process in the upper and middle reaches to a certain extent, it is beneficial to the upgrading of industrial structure, but it has caused damage to the local and ecological environment and adversely affected the ecological efficiency.

(7) Technological progress. The coefficient of lnTec is 0.007, which has passed the significance test at 10% level. On the one hand, technological progress can improve the ecological efficiency by improving the productivity of various factors, reducing the resource consumption and pollution emission per unit GDP; on the other hand, technological progress will have a "return effect" on resource consumption, increasing the consumption of energy and other resources by stimulating economic activities, and offsetting the resources saved by efficiency improvement. From the empirical results, technological progress has a more obvious role in promoting ecological efficiency. At the same time, technological progress is the endogenous direct driving force to promote the adjustment and upgrading of industrial structure. On the supply side, it can promote the improvement of labor and capital productivity, promote the specialized division of production, and improve the quality of labor to provide high-quality labor for emerging industries. On the demand side, technological progress has promoted the emergence of new products, opened up new industrial sectors, stimulated and created new consumer demand, and stimulated the emergence and development of emerging industries, thus promoting the upgrading of industrial structure. Technological progress is an important factor to promote the coordinated development of ecological efficiency and industrial structure upgrading.

## 5 Conclusion

Taking 108 cities in the Yangtze River Economic Belt as research objects, the ecological efficiency from 2007 to 2017 was measured by SBM-Undesirable model, and the coordination degree between ecological efficiency and industrial structure upgrading was measured by coupling coordination degree model, and then the driving factors were analyzed by dynamic spatial Dubin model. The research conclusions are as follows:

(1) In the past 11 year, the ecological efficiency of the Yangtze River Economic Belt has a U-shaped time series evolution trend, and its growth speed has accelerated since 2014. The high-efficiency cities extend inland from coastal cities of Jiangsu, Zhejiang and Shanghai, and are distributed sporadically. The industrial structure upgrading index shows a continuous growth trend, and the industrial upgrading index of the cities in the lower reaches and the provincial capitals in the middle and upper reaches is higher.
(2) The coordination degree between ecological efficiency and industrial structure upgrading in the Yangtze River Economic Belt is on the rise. The type of coordination degree is mainly between moderate imbalance and primary coordination, and the cities with higher coordination level mainly gather in the lower reaches of Jiangsu, Zhejiang, Shanghai and Anhui, forming a "T"-shaped distribution pattern extending from the coastal cities in the lower reaches to the inland.
(3) The coordination degree between ecological efficiency and industrial structure upgrading in Yangtze River Economic Belt has obvious path dependence, and Matthew effect is remarkable. In addition, the spatial spillover effect of coordination degree is obvious, and the increase of coordination degree of neighboring cities will improve the coordination degree of this city. Increasing population density, optimizing system quality, increasing foreign investment, strengthening environmental regulation and technological progress are all conducive to the increase of coordination degree, while increasing foreign investment and strengthening environmental regulation in neighboring cities will lead to the decrease of coordination degree in this city.

## References


1. Schaltegger, S., Sturm, A.: Ökologische Rationalität: Ansatzpunkte zur Ausgestaltung von ökologieorientierten Managementinstrumenten **44**(4), 273–290 (1990)
2. Peneder, M.: Industrial structure and aggregate growth. Struct. Change Econ. Dyn. 14 (2003)
3. Wei, L., Hui, Z.: Changes of industrial structure and technological progress in China's economic growth. Econ. Res. **43**(11), 4–15 (2008)
4. Chunhui, G., Ruogu, Z., Dianfan, Y.: The influence of China's industrial structure change on economic growth and fluctuation. Econ. Res. **46**(05), 4–16+31 (2011)
5. Yonghui, H., Liangxiong, H., Jianhua, Z.: The coming of China's economic structural slowdown era. Stat. Res. **33**(05), 23–33 (2016)

6. Yonghui, H., Liangxiong, H., Xianbin, W.: Has the optimization and upgrading of industrial structure improved the ecological efficiency? Quant. Econ. Tech. Econ. Res. **33**(04), 40–59 (2016)
7. M. Jun, Z. Panchao: The spatial effect of industrial upgrading on improving ecological efficiency of Yangtze river economic belt. J. Nanjing Univ. Technol. (Social Science edition) **19**(02), 73–88+112 (2020)
8. Clark, C.: The conditions of economic progress. London, Macmillan (1940)
9. Jun, S.: Demand factors, technological innovation and industrial structure evolution. Nankai Econ. Res. **5**, 58–71 (2008)
10. Yuandong, G., Weiguo, Z., Qin, Y.: Study on the influencing factors of China's high industrial structure. Econ. Geogr. (6), 96–101 (2015)
11. Zhong, M., Mengjie, L., Weijian, D.: Can environmental regulation force industrial restructuring? An empirical test based on China's provincial panel data. China Popul. Res. Environ. **25**(08), 107–115 (2015)
12. Dou, J., et al.: How does the industry mobility affect pollution industry transfer in China: empirical test on pollution haven hypothesis and porter. J. Clean. Prod. **217**(4), 105–115 (2019)
13. Chen, X., Qian, W.: Effect of marine environmental regulation on the industrial structure adjustment of manufacturing industry: an empirical analysis of China's eleven coastal provinces. Marine Pol. **113**(3), 1–18 (2020)
14. Linghui, F.: An empirical study on the relationship between industrial structure upgrading and economic growth in China. Stat. Res. **27**(08), 79–81 (2010)

# Investigation of the Urban Factors Affecting Microplastic Pollution in Chinese Cities: The Case of Ningbo

**Yuyao Xu, Faith Ka Shun Chan, Matthew Johnson, Thomas Stanton, Jun He, Tian Jia, Jue Wang, Zilin Wang, Yutong Yao, Junting Yang, Yaoyang Xu, Xubiao Yu, and Dong Liu**

**Abstract** Microplastic pollution is an emerging threat to global freshwater ecological security. The emission and discharge of microplastic pollutants is highly associated with human activities and, therefore, cities are particularly at risk of microplastic pollution because they are a concentrated zone of plastic industry and use. Urban rivers may also be significant in transporting microplastic pollution from cities to other areas. Because of rapid urbanization, Chinese coastal cities are potentially at increasing risks of microplastic pollution from freshwater, atmospheric and terrestrial environments. Previous studies discovered that urban factors, including local population density, economic structures, and land-use patterns play decisive roles in microplastic pollution in China's urban catchments. This study builds on past work by analysing the relationship between urban factors and freshwater microplastic pollution along an urban river channel in Ningbo, a megalopolis on the East Coast of China.

Y. Xu · F. K. S. Chan (✉) · T. Jia · J. Wang · Z. Wang · Y. Yao · J. Yang
School of Geographical Sciences, University of Nottingham Ningbo China, Ningbo 315100, China
e-mail: faith.chan@nottingham.edu.cn

M. Johnson (✉) · T. Stanton
School of Geography, University of Nottingham, Nottinghamshire, UK
e-mail: m.johnson@nottingham.ac.uk

J. He (✉)
Department of Environmental Engineering, University of Nottingham Ningbo China, Ningbo 315100, China
e-mail: jun.he@nottingham.edu.cn

F. K. S. Chan
School of Geography, University of Leeds, Leeds LS2 9JT, UK

Water@Leeds Research Institute, University of Leeds, Leeds LS2 9JT, UK

Y. Xu · D. Liu
Institute of Urban Environment, Chinese Academy of Science, Beijing, China

X. Yu
Faculty of Architectural, Civil Engineering and Environment, Ningbo University, Ningbo, China

 
F. K. S. Chan et al. (eds.), *Proceedings of the 2020 International Conference on Resource Sustainability: Sustainable Urbanisation in the BRI Era (icRS Urbanisation 2020)*, Environmental Science and Engineering,
https://doi.org/10.1007/978-981-15-9605-6_23

The microplastic abundance in the Fenghua River, Ningbo, was compared to equivalent measures in other research. This study also considers local urban developments, revealing some of the core factors affecting urban microplastic pollution levels. This manuscript ultimately aims to find countermeasures for controlling China's urban microplastic pollution. These measures will also provide some new perspectives for Chinese cities to deal with the spread and emission of other artificial contaminants in the future, so as to maintain the sustainable development in China, and extensively to other cities in the region.

**Keywords** Microplastic pollution · Chinese urban freshwater environment · River

## 1 Introduction

Urbanization has stimulated the economic development of China; however, this has also brought significant environmental pressures. Plastics materials are integral to numerous industries worldwide. In China, plastic has been widely used in agriculture (as agricultural plastic film) since the early days of the People's Republic of China. Use of plastic products (e.g. plastic bags, tablets, and synthetic fibres) has grown considerably throughout China's modernization and industrial reform [33]. This use has generated similarly considerable quantities of plastic waste, much of which is transported out of urban centres by the rivers that flow through them [3, 10, 19, 30]. As a result, the Chinese government introduced different measures (such as plastic restrictions) to control the use and mismanagement of plastic products in the early 2000s [7]. Plastic pollution is frequently categorised by size (macroplastic > mesoplastic > microplastic), however, microplastic particles (pieces of plastic <5 mm in their largest dimension) are the only category for which particle size is defined, and have been widely discussed in recent years [6, 30, 32].

Plastic debris floating on the surface of ocean was first named 'microplastics' in 2004 [25]. Research in recent years has generally defined microplastics as plastic particles between 1 $\mu$m and 5 mm in diameter and exposure is widespread in ecosystems, including freshwater, atmospheric, soil and marine environments [3, 20, 26, 30]. Microplastics have been documented in diverse environments in China [32]. They are problematic because microplastics can release toxic monomers or additives (such as pigments and flame retardants) used in the production of plastic and can absorb and act as a vector of other pollutants (such as persistent organic pollutants and pathogenic microorganisms) because of their lipophilicity and large surface-to-volume ratio [35]. Given their high mobility, microplastics may therefore amplify the impact of pollutants in different environments.

Research has noted that in the absence of control, microplastic pollution in the environment is likely to grow substantially [27]. Therefore, as more countries (such as the US, Canada and parts of Europe) begin to regulate microplastic pollution, particularly microbeads, China will also enact a new 'plastic limit' in 2021 [16, 18, 30]. This plastic limit is the first national environmental legislation in China that partly covers microplastics (i.e. the plastic microbeads added in personal care products (PCPs) or cosmetics) and highlights the prohibition of the production and sell of plastic bags (<0.025 mm in thickness), polyethylene agricultural films (<0.01 mm in thickness), single-use plastic swabs, and disposable foam plastics [18]. Successful implementation of this legislation, in conjunction with the nationwide promotion of 'garbage classification', will significantly alleviate the plastic pollution in China.

Microplastics in the environment can generally be divided into two categories: primary microplastics produced at the micron or millimetre scale, and secondary microplastics released through the degradation of plastic waste already in the environment [3, 10]. Plastic microbeads that are controlled by legislation in multiple countries make up only a small fraction of all primary microplastics in the environment [16]. Other types of microplastics, such as synthetic fibres and tyre dust, tend to be far more abundant and remain relatively uncontrolled. The lack of suitable management and control strategies of microplastic pollution by the government, is largely due to insufficient understanding of microplastic pollution and, in particular the inadequate consideration of relevant toxicological and pollution mechanisms of microplastics [30]. For this point of view, the investigation of microplastic pollution, which offers important knowledge measured in the field, is necessary right now.

Due to current freshwater management practices in China are insufficient to control the microplastic pollution already in the environment, providing targeted management of microplastic sources is seen as more achievable. Multiple studies have indicated that areas of intensive human activity are the main source of microplastic pollution, particularly urban environments [4, 19, 28]. Rivers also concentrate plastic pollution because they are spatially limited, and tend to accumulate pollutants downstream. For example, Zhao et al. [33] found that the microplastic concentrations in the estuary of the Yangtze River were thousands of times higher than in adjacent seawater, because of dilution effects.

Currently, research of freshwater microplastic pollution in urban watersheds is not extensive [6, 32]. However, investigations in Chinese urban areas (including Wuhan, Guangzhou, Shanghai and Changsha) do suggest that concentrations of microplastics in freshwater are related to population density, economic structure, GDP level and the location of waterbodies spatially in urban environments (i.e. City Centre) [4, 19, 28]. Numerical modelling predictions imply that if China continues its current trajectory without adequate controls, it will be one of the leading emitters of microplastic pollution worldwide [27]. Considering the importance of urban environments (i.e. cities) on microplastic pollution and the volumes of plastic production associated with Chinese economic development, it is essential to better understand how urban factors might affect microplastic pollution.

Therefore, the aim of this study is to explore the role of urban factors in controlling microplastic pollution in urban surface waters under different land-uses, using Ningbo as a case study. We quantify the abundance, size, and morphology of microplastics in an urban freshwater environment, and explore the influence of land-use at each sampling site on microplastic concentrations. We focus predominately on introducing the methodologies used and describing the future direction of research in this area, particularly for controlling microplastic pollution in urban environments.

# 2 Methods

## 2.1 Study Sites

Ningbo is located on the East coast of China and is the most important port city (with in total 8.542 million Long-term population by 2019) in Zhejiang Province Fig. 1. Ningbo-Zhoushan Port's annual container throughput ranked the third (over 275.3 million Standard containers) in the world in 2019 [14]. Rapid urbanization occurred after the "Open door policy" introduced since 1978 in China and continued as Ningbo was promoted as the new economic centre of the south wing of the Yangtze River Delta (YRD) [8]. The land area of Ningbo City reached 9729 km$^2$ in 2016, consisting of 6720 km$^2$ of agriculture (including farmlands, woodlands, gardens, pastureland, agricultural facility lands, rural paths and roads, mainly for primary industry), 1913 km$^2$ of development land (including the land for residential areas, transportation facilities, pubic facilities, special lands and mining lands; mainly for secondary and tertiary industry) and 1095.3 km$^2$ of unused lands (including idle lands, saline-alkali lands, marsh lands, sandy lands bare lands and grasslands not used for husbandry) [23]. Meanwhile, the major economic structure of Ningbo's

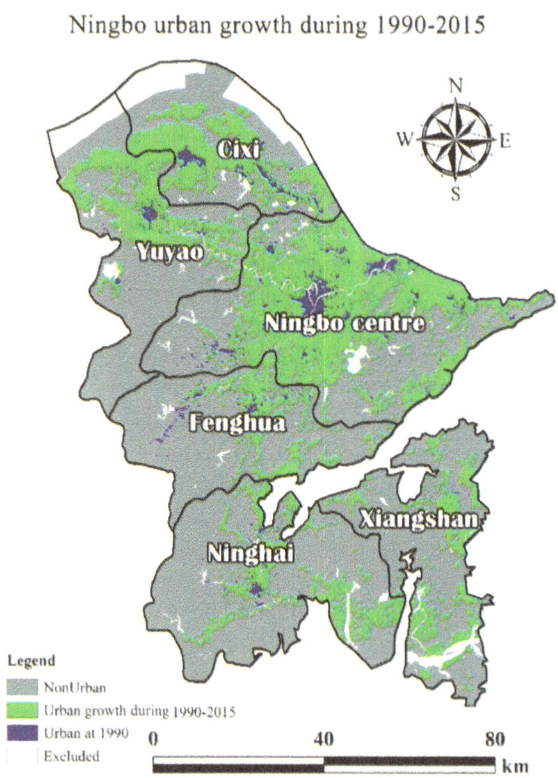


**Fig. 1** Urban development map of Ningbo during 1990–2015 (Purple region is the urban area of Ningbo at 1990 and green area is the urban area at 2015; be edited from source [13])

development land reforms from secondary industry into tertiary industry recently years [23]. In terms of recent economic development, the GDP output of construction land per unit in Ningbo reached 472 million RMB/km$^2$, ranking seventh among the fifteen sub-provincial Chinese cities in 2016 [23]. The development of Ningbo played an important role in the "2010 Regional Planning of the YRD" that also emphasized the construction of development zones along the river (YRD catchment), bay (Hangzhou Bay), and coast (the East China Sea) [17]. The development of Ningbo's central urban area also drives the urbanization development of surrounding satellite areas (including Yuyao and Fenghua) (see Fig. 1. In 2015, the urban areas of the entire Ningbo was classified into six levels according to the size of local population, where the central urban area was classified as a first-level town (>4 Mio. Population size), Yuyao as a second-level town (0.5–1 million Population size) and Fenghua as a third-level town (0.2–0.5 million Population size) [8]. The development of Yuyao and Fenghua is also closely related to Ningbo's major urban rivers.

Three major rivers flow through the main urban area of Ningbo, namely the Yuyao River (YYR), the Fenghua River (FHR), and the Yongjiang River (YJR). YYR starts at Yuyao, flows from the west to east towards the central urban area of Ningbo and experiences a town-rural-urban change of land types along the river (see Fig. 2. FHR flows from south (Fenghua) to north (Ningbo City) with urbanization level increasing along its course (see Figs. 1 and 2. YYR and FHR converge at the Sanjiangkou (SJK) estuary, which is the commercial heart of Ningbo and the source of YJR Fig. 2. YJR flows from west to east into the East China Sea in the Hangzhou Bay area and most of the lands along YJR are highly urbanized (see Fig. 1. Although there are different urbanization patterns along the three rivers, YYR, FHR, and YJR all act as navigation

**Fig. 2** Map of three major rivers of Ningbo urban central area (Google map)

**Table 1** Sampling points along the Fenghua River (FHR) of Ningbo urban central area

| Name of river | Sampling duration | River length | Number of sampling points | Mean distance between each two adjacent sampling sites (m) |
|---|---|---|---|---|
| Fenghua River (FHR) | July, 2019 | 27 km | 43 | 628 |
| | January, 2020 | | 19 | 1588 |

routes, bringing in large amounts of economic income for the development of the riverbank areas, with busy shipping routes.

Four known factors affecting the distribution of microplastics in freshwater environment are summarized in some papers [28], namely, microplastic properties, geographical, hydrological and meteorological conditions of the pollution sites. Considering the complexity of the urban water system, the changeable meteorological conditions, and the variable patterns in urbanization along the rivers in Ningbo, the investigation of microplastic distribution in YYR, FHR and YJR is great significance to understanding urban freshwater microplastic pollution.

Therefore, this study focuses on the concentration of microplastics in the surface water of rivers, considering the influence of urban factors (particularly land-use conditions) through high spatial density sampling along all three major urban rivers of Ningbo City. In this paper, we will report the preliminary findings along the FHR from the Fangqiao estuary (Fenghua) to the SJK estuary (Ningbo City) Table 1.

## 2.2 *Sampling Approaches*

**Fieldwork**. Surface water (0–25 cm depth) was sampled from FHR. A 500 m sampling point interval was initially planned, and conduct the preliminary point selection along one bank of the river (e.g. left bank along FHR) until the end of the river. To ensure the safety and availability of point locations, each point was first investigated to eliminate or replace the non-available sampling sites caused by safety, terrain, land ownership and other issues. The result was a total of 43 sampling sites (628-m mean interval) along FHR (mostly along the left bank) for summer sampling Table 1. To consider the impact of seasonal factors on microplastic pollution, this study selected 19 of those points for a repeated sampling during winter. This sampling strategy has a high spatial resolution relative to other studies, for example, Lin et al. [12] conducted surface water sampling at 14 sites along the Guangzhou of the Pearl River (>50 km). Given the complexity of land-use distribution in Chinese urban areas due to rapid urbanization, high-resolution sampling helps to identify potential sources of microplastics, the influence of adjacent land-use, and current urban factors. In this paper preliminary findings will be presented, consisting of the summer data at three sampling points (A, B and C) and winter data of five sampling points (A, B, C, D and E) Table 1 and Fig. 3. The result of point A, B and C could preliminarily indicate the seasonal factors and the analysis of five winter samples could better explain the influences of different land-use and urbanization conditions.

The surface water was drawn from the bank by a stainless-steel bucket with a rope. 30 L of water was passed through a stainless-steel sieve of 63 μm mesh-size. The residues on the sieve were rinsed into brown glass bottles by deionized water for transport to the laboratory, where they were stored for a short time under low temperature and shaded conditions in the sealed (aluminized paper) containers.

**Lab-work**. Once water samples were brought back to the laboratory, 200 ml of 30% hydrogen peroxide ($H_2O_2$) was added into each sample and heated for five hours in a water bath at 80 °C, digesting bio-organic matter in samples. Each sample was then filtered again through a 63 μm mesh-size stainless-steel sieve and washed into a centrifuge tube by saturated sodium chloride (NaCl) solution (approximately 1.2 g/ml). The centrifuge tube was centrifuged for five minutes at a speed of 4000 rpm to separate the impurities such as sediments and microplastics through the difference in density. The upper liquid in the centrifuge tubes was then filtered through a 0.45-micron pore-size filter membrane with the help of a vacuum pump. Residual materials left on the filter membrane was suspected to be microplastics, pending subsequent identification work.

The filter membrane that completed the filtration was kept in a covered glass dish as a sample for observation. The sample was placed under a 170x maximum magnification microscope (S9D Leica, Germany) equipped with a digital camera (MC170HD Leica, Germany) for manual identification, by using the identification approach from Stanton et al. [20]. At the same time, the size, shape, and number of suspected microplastics observed were recorded. By calculating the amount of microplastics at each point, the concentration of microplastics in the local surface water can be obtained. Combined with the statistics of microplastics' size and shape, local microplastic pollution condition can be assessed.

### 2.3 Land-Use Type

To determine the land-use type of each sampling site, the land-use conditions, within in 1 km of each sampling point, were identified on the map. Considering the complex distribution pattern of land planning around each study site, the land-use condition of the area closest to the sampling point is regarded as the main land-use type of each point, and other land-use situations included within in 1 km of the sampling point are taken as the secondary land-use types of this point Table 2.

## 3 Preliminary Results

From the upstream to the downstream of FHR, the scale of farmlands and villages gradually decreases whilst the residential and commercial areas gradually increase.

**Table 2** Land-use type of five example sampling points along the Fenghua River (FHR)

| Sampling sites | Main land-use type | Secondary land-use types |
|---|---|---|
| Point A | Construction area | Farmland, Industrial area, Village |
| Point B | Industrial area | Farmland, Village, Construction area |
| Point C (watergate) | Farmland | Village, Industrial area |
| Point D (watergate) | Industrial area | Park, Residential area, Farmland |
| Point E | Residential area | Park, Commercial area, Construction area |

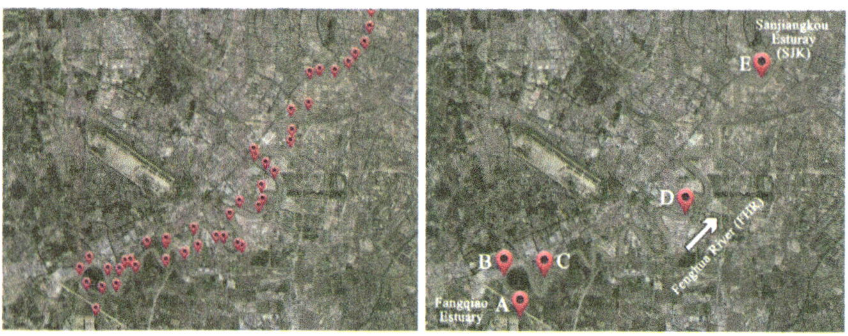


**Fig. 3** 43 sampling points (Left) along Fenghua River (FHR) of this research and the five sampling sites (Right) discussed in following sections

The density of industrial areas first increases and then decreases. The change in pattern in land-use reflects the changing trend of urbanization along FHR, which is low-to-high from upstream to downstream. Three sampling points (Point A, B and C) were located in the upper reaches of FHR, and the land-use types are mainly farmland, villages, low-tech industries (such as small sand and gravel factories) and construction areas, all of which belong to the typical developing suburban economic structure zone. Point D is located at the middle reach of FHR where there is an industrial area (such as Youngor clothing manufacturer). Point E is located in a residential area near the city centre of Ningbo, which is surrounded by business and commercial areas.

Microplastics were detected in the waterbodies at all five sampling sites. During the summer, Point B was found to have the highest concentration of microplastic contamination (9166.76 items/m$^3$), while Point C had the lowest concentration of microplastics (5600 items/m$^3$). Point A and Point B showed no clear difference in surface water microplastic concentration in summer. However, during winter, the

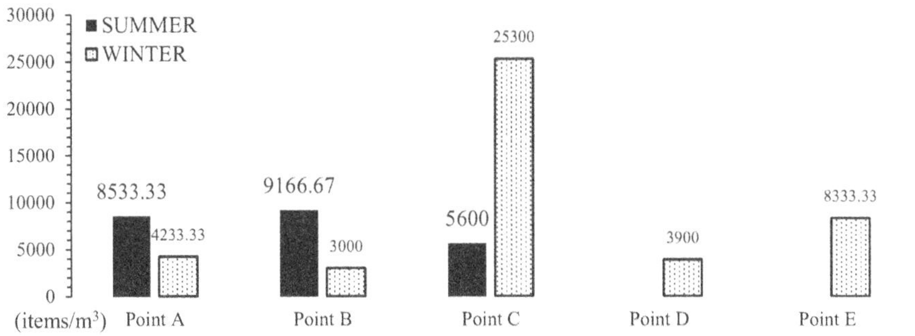


**Fig. 4** Microplastic concentrations in five sampling points (Point A, B, C, D and E) along Fenghua River (FHR) in summer and winter

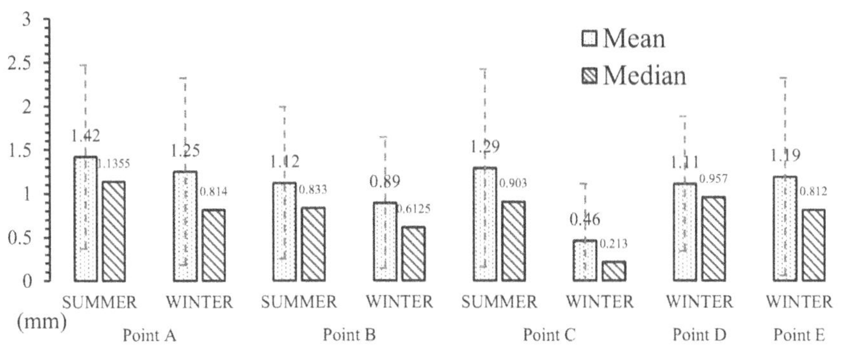


**Fig. 5** Mean and median of the sizes of microplastics detected at five sampling points (Point A, B, C, D and E) along Fenghua River (FHR) in both summer and winter. Error bars represent the standard deviation of the average sizes of microplastics

microplastic concentrations of both points decreased notably, with the microplastic concentration of Point B the lowest recorded Fig. 4. The concentration of Point C rose dramatically in winter and reached 25,300 items/m$^3$, which is over 350% higher than this point in summer Fig. 4. The concentration of microplastics detected at Point E is the second highest among five winter samples Fig. 4.

In terms of microplastic size, the average size of microplastics from Point A (summer and winter), Point B (summer), Point C (summer), Point D (winter) and Point E (winter) are all larger than 1 mm in diameter Fig. 5. The median sizes of Point A (winter), Point B (winter and summer), Point C (summer and winter), Point D (winter) and Point E (winter) were all smaller than 1 mm Fig. 5. Point C in winter has the smallest average size and median size of microplastics Fig. 5. As Fig. 6 shows, compared to the mean size, the median size better reflects the size distribution pattern in microplastic samples, where microplastics smaller than 1 mm account for the majority. According to Fig. 6, all samples are dominated by smaller

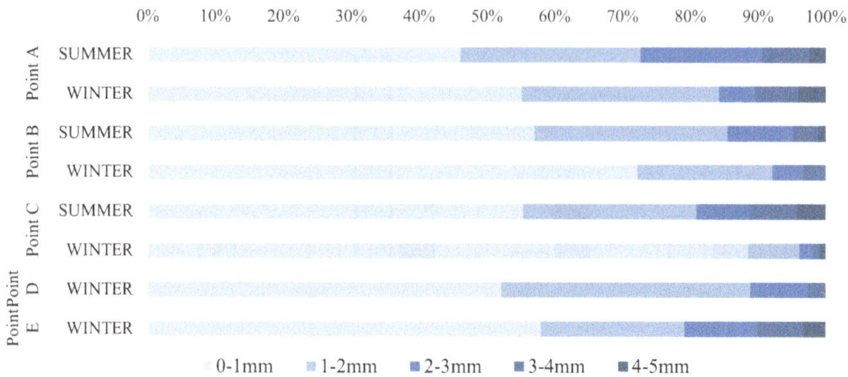


**Fig. 6** The size distribution patterns in the samples from Point A, B, C, D and E along Fenghua River (FHR) in both summer and winter

microplastics (<1mm) and the ratio of smaller microplastics in Point C (winter) is even higher than 85%.

In terms of the shapes of microplastics, this research classified microplastic into four shape classes: globular (spherical plastic particles including microbeads), fibres (wire-like, soft plastic materials), film (soft membranous plastic debris), and fragments (usually irregular hard plastic debris). Fibres dominate all samples except for Point B in winter and Point C in winter Fig. 7. In winter at Point B, fragments dominate but, are not obviously different compared to the number of fibres. The sample at Point C collected in winter is unusual compared to other samples as globular particles dominate the sample but are rare or absent from all other samples. Indeed, in summer at the same site there were no globular microplastics found Fig. 7. Globular

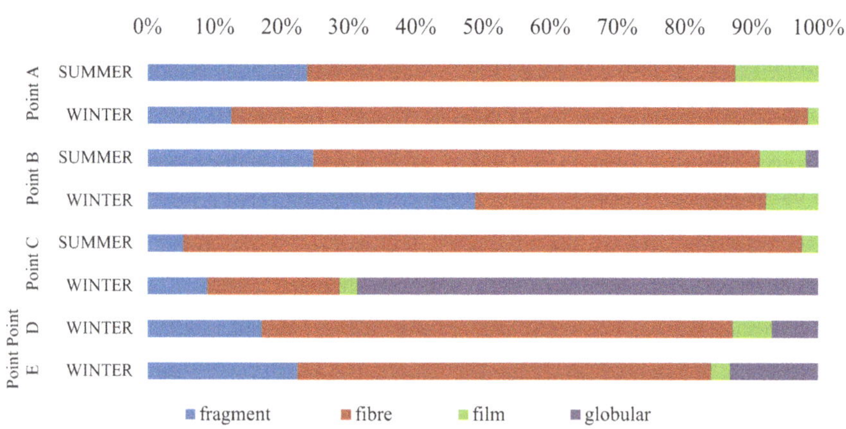


**Fig. 7** The shape distribution pattern in the samples from Point A, B, C, D and E along Fenghua River (FHR) in both summer and winter

microplastics only appear in the samples of Point B (summer), Point C (winter), Point D (winter) and Point E (winter).

# 4 Discussion

## *4.1 Microplastic Concentration*

When the results of the eight samples were compared to previous studies (e.g. [1, 12, 15, 21, 22, 29, 33], it was found that the microplastic concentration of Point C (winter) was extremely high relative to other studies in urban freshwaters Fig. 8. Summer microplastic concentrations at all three points were also above the median of the data from other studies Fig. 8. In contrast, the microplastic concentration of Point A, Point B and Point D in winter were relatively low compared to other data.

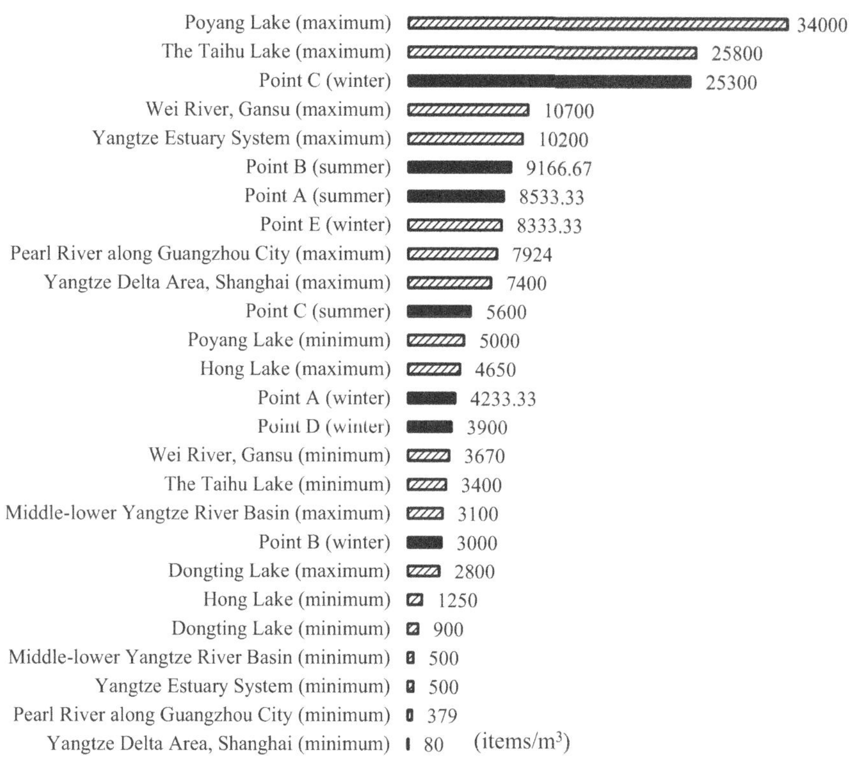


**Fig. 8** Several microplastic concentration data reported in previous publications related to Chinese freshwater microplastic pollution in surface water (*data source* [1, 12, 15, 21, 22, 29, 31, 33], and authors) Black bars are the data from this study

This result indicates that the microplastic pollution in FHR is lower in winter than in summer and, based on the land type and meteorological information, the reason for this may be that rainfall is less frequent in winter than summer, so microplastics are not washed off of adjacent agricultural land in winter. However, this theory cannot explain the peculiarity of Point C in winter.

Point E (winter) has relatively high microplastic concentration (8333.33 items/$m^3$), which is slightly lower than the result of Point A in summer. Point E is located in the central urban area of Ningbo (SJK estuary) and also downstream of FHR and, as such, was expected to have high microplastic concentrations, accumulating materials from upstream and from urban activity. Microplastics usually have higher concentrations in downtown waterbodies due to more frequent human activity [28]. Whilst the comparison in Fig. 8 is useful context, it should be noted that such comparisons between studies is problematic when describing the issue of local microplastic pollution severity. This is because different studies often choose different sampling equipment and analysis approaches, which can result in diverse size ranges or properties of microplastics, which seriously affect the comparability of data between different studies [5]. Therefore, the analysis of microplastic pollution needs to focus on local conditions, as current study approaches are not standardized.

## *4.2 Microplastic Shape*

According to Fig. 7, fibres are the most widespread microplastic shape in most samples. This finding is consistent with the conclusions from many previous studies in the Chinese freshwater environments [30, 32]. Synthetic fibres are one of the most important plastic materials, mainly used in clothing, packaging, and construction making them ubiquitous and hard to mitigate. In Chinese urban areas, the population releases synthetic fibres into domestic wastewater towards wastewater treatment plants (WWTPs) through the washing of clothes. WWTPs can remove synthetic fibres and other pollutants from sewage through filtration, sedimentation, and adsorption [24], but efficiency of removal is dependent on the type of WWTP. In China, removal of microplastics form sewage have been recorded of around 60%, meaning many are likely to be discharged to urban waters, and even the microplastics that are filtered can re-enter the natural environment after landfill [9, 12].

In rural or suburban areas of China, there are still some people washing clothes directly by the river with detergents. Synthetic fibres from clothing and plastic microbeads from detergents therefore go straight into the river. The fibres detected in the samples are likely to have come from a combination of washing clothes and sewage disposal in the urban freshwater environment (i.e. urban river channels and tributaries). Another important source of fibre-plastic is atmospheric deposition and precipitation [20]. Some literature shows that plastic fibres also account for the majority of the atmospheric microplastic pollution in some region of China [2]. Above reasons explain the prevalence of fibres in this study.

Globular microplastics come mainly from microbeads added to PCPs or cosmetics. Plastic microbeads usually enter domestic wastewater after use, going through a similar route as fibres, from WWTPs or other pathways into waterbodies. Fig. 7 shows that the upper reaches of FHR are usually not polluted by globular microplastics, while a small amount of spherical microplastics are usually detected downstream. This result is related to local urbanization degree and population activities. Point C (winter) detected an unusually high concentration of globular microplastics (about 69% in total). Point C is located at the outlet of a watergate. The river network in Ningbo central city is complex. Except for three major urban rivers, all the other areas are covered by a network of urban waterways. These inner urban waterways, which flow through different parts of the urban area, eventually flow into YYR, FHR and YJR at watergates.

The watergates of Ningbo help to adjust the water levels of three major urban rivers for heavy shipping along rivers. There are tens of watergates evenly distributed along both sides of FHR. The water entering the watergates has often accumulated pollution from the urban river network. Here, we hypothesis that the globular microplastics at Point C (winter) were the accumulated emissions from the urban inner river flows including from the living activities of villagers, the emissions of production wastes related to the industrial zone, and the additives in agricultural fertilizers [11]. These were then released when the watergate was operated.

Plastic micro-films had a higher ratio in the samples from upper stream than in downstream Fig. 7. Plastic packaging, bags and agricultural mulch are important sources of thin-film plastics. Through the comparison of upstream and downstream sites, it is suggested that agricultural mulch may be an important source of film microplastics in FHR.

Microplastic fragments usually come from industrial production and mismanaged plastic waste [10]. This explains why Point B, located around the industrial area, has more plastic fragments in both summer and winter than other sampling sites. However, Point D, which is also located around the industrial area and at a watergate, has a lower microplastic concentration and a lower proportion of microplastic fragments Fig. 7. This may be because Point D is in a riverfront park and other studies have found microplastic concentrations that are lower in park zones [4, 19].

### *4.3 Size of Microplastics*

Secondary microplastics are considered to be the major source of current environmental microplastic pollution due to the continuous degradation of plastic wastes in the environment [10]. The distribution of microplastic sizes can therefore reflect the seriousness of local microplastic pollution to some extent [22, 28]. For example, smaller microplastics may indicate that plastic has accumulated for a longer period of time than a site with larger pieces which have not been broken down as much. Considering the fluidity of river water, smaller microplastics may originate from the land areas or atmospheric deposition that were severely polluted by microplastics, upstream input or the macroplastics remain in low flow zone in river. Regardless, microplastic size is ecologically important because smaller microplastics are more likely to be ingested by organisms, potentially affecting ecosystem functioning and food webs.

All eight samples were dominated by smaller microplastics (<1mm), consistent with most studies in China and suggesting that microplastics in FHR have the potential to be ingested by small organisms (e.g. zooplankton). Point C (winter) had the highest proportion of microplastics smaller than 1 mm (88.54%), which was associated with the high abundance of microbeads (usually 0.1–0.5 mm in this study). Winter samples generally have higher ratio of microplastics smaller than 1mm than in summer, which indicates different patterns of microplastic emissions in summer and winter along FHR.

## 4.4 Land-Use and Urban Development

According to the results above, in addition to the seasonal factors, the land-use type as a spatial factor is one of the important factors affecting the local microplastic pollution condition.

First, more microplastics were detected in winter at sampling sites near to the city centre, reflecting the current situation of microplastic pollution in densely populated areas. Second, rural and sub-urban areas have more thin-film microplastics than urban areas, reflecting the role of agricultural mulch in microplastic pollution. Third, more fragments were detected in industrial areas, whereas fewer fragments were emitted in the industrial region with a park. Lastly, plastic microbeads were generally detected near the city centre with large residential areas, which reflects the effect of dense population activities on microbead discharge and emission. But the presence of microbeads in large numbers in rural areas suggests that they may also be emitted from agricultural activities.

Therefore, the changes in land-use type from upstream to downstream along FHR results in a different microplastic pollution pattern along the river, and hints at the underlying pollution mechanisms. In this context, the change of land type associated with future urban development may also change the pattern of local microplastic pollution. From 1991 to 2016, Ningbo's agricultural land (arable land, woodland, garden land, etc.) decreased by 442.6 $km^2$, while the construction land (urban and rural construction land, transportation land, water area and water conservancy facilities and others) increased by 1132.6 $km^2$. At the same time, the major use of construction land is also transferred from the secondary industry to the tertiary industry. If this trend continues, more plastic microbeads and synthetic fibres may be found in FHR in the future. So far, micro-fibres are the most common type of microplastics in Chinese freshwater. Given that China has begun to control microbeads pollution by legislative management, micro-fibres might be the most important part of microplastics water pollution in the future in China.

According to various land types, different methods can be adopted to reduce the emission of certain types of microplastics (e.g. the microbeads and fibres from residential area and micro-fragments from industrial areas). Additionally, the area where the park is located has been found to have lower microplastic pollution in this paper and in previous studies [19]. Green land planning in areas with high pollution levels may therefore be beneficial to reducing microplastic pollution in local waterbodies. In the view of impact of land type on microplastic pollution, future urban planning should consider microplastic pollution and reducing the presence of small microplastics (e.g. <1 mm) in freshwater environments, so as to protect urban water ecology.

## 5 Conclusion

This paper describes the current situations of microplastic pollution in the surface water of local urban catchments by investigating five sampling sites along FHR in Ningbo, China in two seasons (in summer and winter). Smaller microplastics (<1 mm) and synthetic fibres were found to be the most widespread types of microplastics locally.

Seasonal factors and land-use types will influence the microplastic pollution pattern. The pollution of microplastic film and fragments is more serious in suburban areas and rural areas of FHR, while the distribution of microbeads is more widespread in the central urban areas. As population activities and land-use type lead to different types of microplastic emissions, future urban development should take the reduction of microplastic pollution in waterbodies into account.

**Acknowledgement** This study was supported by the National Natural Science Foundation of China (NSFC): International Young Scientist Program (Grant NSFC 41850410497); and the Faculty of Science and Engineering (FoSE) Postgraduate Research Scholarship of University of Nottingham Ningbo China.

## References

1. Ding, L., Mao, R., Guo, X., Yang, X., Zhang, Q., Yang, C.: Microplastics in surface waters and sediments of the Wei River, in the northwest of China. Sci. Total Environ. **667**, 427–434 (2019). https://doi.org/10.1016/j.scitotenv.2019.02.332
2. Dris, R., Gasperi, J., Saad, M., Mirande-Bret, C., Tassin, B.: Synthetic fibres in atmospheric fallout: a source of microplastics in the environment? Marine Pollut. Bull. (2016). https://doi.org/10.1016/j.marpolbul.2016.01.006
3. Eerkes-Medrano, D., Thompson, R.C., Aldridge, D.C.: Microplastics in freshwater systems: a review of the emerging threats, identification of knowledge gaps and prioritisation of research needs. Water Res. **75**, 63–82 (2015). https://doi.org/10.1016/j.watres.2015.02.012

4. Fan, Y., Zheng, K., Zhu, Z., Chen, G., Peng, X.: Distribution, sedimentary record, and persistence of microplastics in the Pearl River catchment, China. Environ. Pollut. **251**, 862–870 (2019). https://doi.org/10.1016/j.envpol.2019.05.056 (2019).
5. Fok, L., Lam, T.W.L., Li, H., Xu, X.: A meta-analysis of methodologies adopted by microplastic studies in China. Sci. Total Environ. **718**, 1–10 (2020). https://doi.org/10.1016/j.scitotenv.2019.135371
6. Fu, Z., Wang, J.: Current practices and future perspectives of microplastics pollution in freshwater ecosystems in China. Sci. Total Environ. **619**, 697–712 (2019). https://doi.org/10.1016/j.scitotenv.2019.07.167
7. GOSC. The Notice of the General Office of State Council on Restricting the Production, Sale and Use of Plastic Shopping Bags. https://www.gov.cn/zwgk/2008-01/08/content_852879.htm. Last accessed 25 Aug 2020
8. GOSC The reply of the State Council on the overall Urban Planning of Ningbo, https://www.gov.cn/gongbao/content/2015/content_2843773.htm. Last accessed 25 Aug 2020
9. He, P., Chen, L., Shao, L., Zhang, H., Lü, F.: Municipal solid waste (MSW) landfill: A source of microplastics? Evidence of microplastics in landfill leachate. Water Res. **159**, 38–45 (2019). https://doi.org/10.1016/j.watres.2019.04.060
10. Horton, A.A., Walton, A., Spurgeon, D.J., Lahive, E., Svendsen, C.: Microplastics in freshwater and terrestrial environments: evaluating the current understanding to identify the knowledge gaps and future research priorities. Sci. Total Environ. **586**, 127–141 (2017). https://doi.org/10.1016/j.scitotenv.2017.01.190
11. Joshi, P.P., van Cleave, A., Held, D.W., Howe, J.A., Auad, M.L.: Preparation of slow release encapsulated insecticide and fertilizer based on superabsorbent polysaccharide microbeads. J. Appl. Polym. Sci. **137**(39) (2020). https://doi.org/10.1002/app.49177
12. Lin, L., Zuo, L., Peng, J., Cai, L., Fok, L., Yan, Y., Li, H., Xu, X.: Occurrence and distribution of microplastics in an urban river: A case study in the Pearl River along Guangzhou City, China. Sci. Total Environ. **644**, 375–381 (2018). https://doi.org/10.1016/j.scitotenv.2018.06.327
13. Liu, Y., Feng, Y.: Simulating the impact of economic and environmental strategies on future urban growth scenarios in Ningbo, China. Sustainability **8**(10), 1045 (2016). https://doi.org/10.3390/su8101045
14. Lloyd's List, One Hundred Ports 2019, https://lloydslist.maritimeintelligence.informa.com/one-hundred-container-ports-2019#comment. Last accessed 25 Aug 2020
15. Luo, W., Su, L., Craig, N.J., Du, F., Wu, C., Shi, H.: Comparison of microplastic pollution in different water bodies from urban creeks to coastal waters. Environ. Pollut. **246**, 174–182 (2019). https://doi.org/10.1016/j.envpol.2018.11.081
16. McDevitt, J.P., Criddle, C.S., Morse, M., Hale, R.C., Bott, C.B., Rochman, C.M.: Addressing the issue of microplastics in the wake of the microbead-free waters act - a new standard can facilitate improved policy. Environ. Sci. Technol. **51**(12), 6611–6617 (2017). https://doi.org/10.1021/acs.est.6b05812
17. NDRC, Circular of the National Development and Reform Commission on the Issuance of Regional Planning for the Yangtze River Delta Region. https://www.gov.cn/zwgk/2010-06/22/content_1633868.htm. Last accessed 25 Aug 2020
18. NDRC: Industrial structure adjustment guidance catalogue of 2019. https://www.gov.cn/xinwen/2019-11/06/5449193/files/26c9d25f713f4ed5b8dc51ae40ef37af.pdf. Last accessed 25 Aug 2020
19. Peng, G., Xu, P., Zhu, B., Bai, M., Li, D.: Microplastics in freshwater river sediments in Shanghai, China: A case study of risk assessment in mega-cities. Environ. Pollut. **234**, 448–456 (2018). https://doi.org/10.1016/j.envpol.2017.11.034
20. Stanton, T., Johnson, M., Nathanail, P., MacNaughtan, W., Gomes, R.L.: Freshwater and airborne textile fibre populations are dominated by "natural", not microplastic, fibres. Sci. Total Environ. **666**, 377–389 (2019). https://doi.org/10.1016/j.scitotenv.2019.02.278
21. Su, L., Cai, H., Kolandhasamy, P., Wu, C., Rochman, C.M., Shi, H.: Using the Asian clam as an indicator of microplastic pollution in freshwater ecosystems. Environ. Pollut. **234**, 347–355 (2018). https://doi.org/10.1016/j.envpol.2017.11.075

22. Su, L., Xue, Y., Li, L., Yang, D., Kolandhasamy, P., Li, D., Shi, H.: Microplastics in Taihu Lake, China. Environ. Pollut. **216**(July), 711–719 (2016). https://doi.org/10.1016/j.envpol.2016.06.036
23. Sun, S., Wang, Y., Zhang, Z., Hong, Q., Yang, Y.: Constructional land use overview of Ningbo City, land and resources information **11**, 26–30 (2018). https://kns.cnki.net/KCMS/detail/detail.aspx?dbcode=CJFQ&dbname=CJFDLAST2019&filename=GTZQ201811005&uid=WEEvREcwSlJHSldRa1FhdXNzY2Z2Vzd3cCtwT3Q0WTZzSjZJbk9rNTRtWT0=$9A4hF_YAuvQ5obgVAqNKPCYcEjKensW4IQMovwHtwkF4VYPoHbKxJw!!&v=MTA2OTk4ZVgxTHV4WVM3RGgxVDNxVHJXTTFGckNVUjdxZllPWnNGeTNuV3I3T0lqblJmN0c0SDluTnJvOUZZWVI=
24. Talvitie, J., Heinonen, M., Pääkkönen, J., Vahtera, E., Mikola, A., Setälä, O., Vahala, R.: Do wastewater treatment plants act as a potential point source of microplastics? Preliminary study in the coastal Gulf of Finland, Baltic Sea. Water Sci. Technol. **72**(9), 1495–1504 (2015). https://doi.org/10.2166/wst.2015.360
25. Thompson, R.C., Olsen, Y., Mitchell, R.P., Davis., Rowland, S.J., John, A.W.G., McGonigle, D., Russell, A.E.: Lost at sea: where is all the plastic? Science **304**(5672), 838 (2004). https://doi.org/10.1126/science.1094559
26. Van Cauwenberghe, L., Devriese, L., Galgani, F., Robbens, J., Janssen, C.R.: Microplastics in sediments: a review of techniques, occurrence and effects. Mar. Environ. Res. **111**, 5–17 (2015). https://doi.org/10.1016/j.marenvres.2015.06.007
27. van Wijnen, J., Ragas, A.M.J., Kroeze, C.: Modelling global river export of microplastics to the marine environment: Sources and future trends. Sci. Total Environ. **673**, 392–401 (2019). https://doi.org/10.1016/j.scitotenv.2019.04.078
28. Wang, W., Ndungu, A.W., Li, Z., Wang, J.: Microplastics pollution in inland freshwaters of China: A case study in urban surface waters of Wuhan, China. Sci. Total Environ. **575**, 1369–1374 (2017). https://doi.org/10.1016/j.scitotenv.2016.09.213
29. Wang, W., Yuan, W., Chen, Y., Wang, J.: Microplastics in surface waters of Dongting Lake and Hong Lake, China. Sci. Total Environ. **633**, 539–545 (2018). https://doi.org/10.1016/j.scitotenv.2018.03.211
30. Xu, Y., Chan, F.K.S., He, J., Johnson, M., Gibbins, C., Kay, P., Stanton, T., Xu. Y., Li, G., Feng, M., Paramor, O., Yu, X., Zhu, Y.: A critical review of microplastic pollution in urban freshwater environments and legislative progress in China: Recommen. Insights (2020). https://doi.org/10.1080/10643389.2020.1801308
31. Yuan, W., Liu, X., Wang, W., Di, M., Wang, J.: Microplastic abundance, distribution and composition in water, sediments, and wild fish from Poyang Lake, China. Ecotoxicol. Environ. Safety. **170**(October 2018), 180–187 (2019). https://doi.org/10.1016/j.ecoenv.2018.11.126
32. Zhang, K., Shi. H., Peng, J., Wang, Y., Xiong, X., Wu, C., Lam, P.K.S.: Microplastic pollution in China's inland water systems: A review of findings, methods, characteristics, effects, and management. Sci. Total Environ. 1641–1653 (2018). https://doi.org/10.1016/j.scitotenv.2018.02.300
33. Zhao, S., Zhu, L., Wang, T., Li, D.: Suspended microplastics in the surface water of the Yangtze Estuary System, China: First observations on occurrence, distribution. Mar. Pollut. Bullet. **86**(1–2), 562–568 (2014). https://doi.org/10.1016/j.marpolbul.2014.06.032
34. Zhou, Q., Zhang, H., Zhou, Y., Li, Y., Xue, Y., Fu, C., Tu, C., Luo, Y.: Separation of microplastics from a coastal soil and their surface microscopic features. Chin. Sci. Bullet. **61**, 1604–1611 (2016). https://doi.org/10.1360/N972015-01098
35. Zou, Y., Xu, Q., Zhang, G., Wang, Y., Liu, C., Zheng, H., Li, M.: Review on the joint toxicity of microplastics and pesticide pollution. Asian J. Ecotoxicol. **12**(4), 25–33 (2017). https://doi.org/10.7524/AJE.1673

# Research on Urbanization Model Based on Residents' Perception

**Lina Zhong, Xiaonan Li, Baolin Deng, Liyu Yang, and Xiangchi Qi**

**Abstract** Driven by many factors, a large number of urbanization models have emerged in China's urbanization development. However, the existing research rarely involves comparing the development of urbanization quality of different models from the perspective of residents' perception. To make up for this research gap, this paper selects 15 non-industrial urbanization rural areas to investigate the development of five types of urbanization: tourism urbanization, real estate urbanization, trade urbanization, science and education urbanization, agricultural urbanization, and transportation urbanization. Through a questionnaire survey, we know the feelings of residents in the process of rural urbanization from three aspects: economic life impact, social life impact, and environmental life impact. By comparing the differences of residents' perceived quality in different urbanization models, the research results show that compared with other urbanization models, traffic urbanization has many shortcomings, and residents are not satisfied with the economy, society, and environment. Residents of agricultural urbanization are satisfied with the local economy, society, and environment. However, in the development of commercial urbanization, environmental protection is neglected. Because tourism has a low peak season, the residents' perception of economic improvement is not obvious. Residents of real estate urbanization are satisfied with the government and income, while residents of science and education urbanization are generally less satisfied with the economy, society, and environment. By summarizing the development characteristics, advantages, and disadvantages of different urbanization models, this paper puts forward some suggestions on how to improve the quality of urbanization development.

**Keywords** Urbanization model · Residents' perception · Variation analysis

---

L. Zhong (✉) · X. Li · B. Deng · L. Yang · X. Qi
Institute for Big Data Research in Tourism, School of Tourism Sciences, Beijing International Studies University, Chaoyang District, Beijing 100020, China
e-mail: zhonglina@bisu.edu.cn

 
F. K. S. Chan et al. (eds.), *Proceedings of the 2020 International Conference on Resource Sustainability: Sustainable Urbanisation in the BRI Era (icRS Urbanisation 2020)*, Environmental Science and Engineering,
https://doi.org/10.1007/978-981-15-9605-6_24

# 1 Introduction

Please note that the first paragraph of a section or subsection is not indented. The first paragraphs that follows a table, figure, equation etc. does not have an indent, either. Under the background of global economic integration, China's social economy has maintained a momentum of rapid development. From 1981 to 2017, the urbanization rate of the country increased from 20 to 57%, with an average annual growth rate of about 1% [1]. This is much higher than the urbanization rate of major developed countries in the world (such as Britain, the United States, and Germany) in the same period [2], making China the country with the greatest contribution to urbanization in the world [3]. Driven by many factors, a large number of urbanization models have emerged in China's urbanization development. Since SPNTU was announced, according to the characteristics and needs of different cities, a large number of driving factors for urbanization have emerged, such as industry, agriculture, commerce, logistics and transportation, tourism and holiday industry, and cultural and creative industries. China's urbanization is in the stage of rapid development. The traditional single industry driving mode can no longer meet the needs of China's urbanization development. China's urbanization is entering a stage driven by many factors. Scholars in the fields of human economic geography, resources and environmental science, urban regional science and so on have made in-depth research on urbanization, but most of the researchers have studied urbanization problems in different types of areas, urbanization models in different areas, and urbanization dynamic mechanism, etc., and few have studied the differences between urbanization quality of different models. Based on this, this paper establishes a comprehensive evaluation index system of urbanization quality from the perspectives of economy, society, and environment, and uses this comprehensive index system to evaluate the urbanization quality of six different models.

# 2 Literature Review

## *2.1 Study on Evaluation of Urbanization Quality*

The evaluation of urbanization quality has attracted more attention from policymakers and scholars [4–11]. The research contents mainly include qualitative evaluation and quantitative evaluation. The qualitative evaluation mainly includes the basic theory [12], urbanization process [13–15], urbanization development [16], and urbanization road [17–21]. For example, Zhang et al. clarified the existing urbanization research and theory and explained their ways to explain the new characteristics of China's urbanization in reform [12]. Feng et al. used fractal theory and urban land use maps to analyze the evolution characteristics of urban form and land use structure in Hangzhou, China [14]. Guana et al. believed that China's traditional land-centered urbanization is a typical "incomplete urbanization" and "low-quality urbanization" [16]. It is increasingly difficult for China's traditional urbanization to continue, and the transformation process should be accelerated as soon as possible. The quantitative evaluation mainly includes a comprehensive evaluation of urbanization quality [13, 22–24], development and application of evaluation index system [25, 26], Dynamic coupling between urbanization and coordinated development of social economy and ecological environment [19, 27–36]. For example, the urban system is a multi-level and complex system, which integrates natural, economic, social, and ecological environments in a specific time and space [37–42]. In the current literature, evaluation indicators can be divided into two categories: (1) single indicators, including infrastructure type [43], environment type [44], land type [19], and space type [42]; (2) Comprehensive indicators, including economy-society-ecology (environment) type [9, 45], population-land-economy-comprehensive type [46], energy, non-agricultural employment, industrial development, and industrial-urban integration [47], economic development-basic public service-ecological environment-urban-rural heterogeneity-population urbanization [48]. For example, Manisha et al. combined statistical analysis and field investigation to clarify and understand the process and characteristics of urbanization in India from infrastructure indicators [43]. Yu et al. pointed out that industrial development, development zone construction, transportation location, and policy measures are the main driving factors to improve the comprehensive urbanization level in Anhui Province [49]. Wu et al. put forward endogenous factors affecting urbanization, such as population growth, industrialization, economic growth, technological progress, material conditions, marketization, and land use policies [50]. Tao et al. evaluated the driving factors of urbanization process with 19 indicators from the perspectives of population urbanization, economic urbanization, social urbanization, and spatial urbanization [51]. Chen et al. classified the latest policies and actions since the release of the New Rural Planning and made a preliminary evaluation of China's new urbanization construction from the aspects of population urbanization, land urbanization, economic urbanization, and integration [46]. Energy, non-agricultural employment, industrial development, and industrial-urban integration are also regarded as important factors of urbanization [46, 47].

Research and analysis methods are more diversified. For example, the establishment of multi-scale Earth observation systems, institutions for collecting cultural, geographical, and economic data, and the creation of the Internet and various databases have greatly promoted the study of urbanization patterns, situations, and development trends [52]. Some studies use satellite images and socio-economic data to assess the degree of land cover change and urban expansion [53] or use remote sensing technology and geographic information system to study and compare geospatial urbanization and landscape change [54]. Bibliometric analysis is also used more and more [55–59]. For example, Zhang et al. conducted literature research on sustainable urbanization based on Science Network and Scopus [58]. Marvuglia et al. used bibliometrics to analyze the English literature on sustainable urbanization on the scientific network to explore the progress and challenges of sustainable urban development [59]. There are also mixed research methods that combine system dynamics with data envelopment analysis to qualitatively and quantitatively evaluate the quality of urbanization [44].

## 2.2 Study on Urbanization Model

In the process of urbanization, according to their economic development foundation, town location factors, town characteristic resources, etc., combined with town positioning and development goals, different towns use different ways and elements to promote urbanization development. For example, with the help of various industries such as industry, agriculture, tourism, real estate, commerce and trade, science and education, transportation hub, etc. Gladstone discovered two different types of tourism urbanization in the United States, one specializing in "sunshine, beach and ocean" tourism, and the other specializing in highly capital-intensive tourist attractions [60]. Liu et al. analyzed the relationship between urbanization and real estate development by combining literature research method [61]. The realization of human resources development of the agricultural transfer population is the inevitable requirement of people-oriented in regional economic development and new urbanization. Wolfgang introduced that accelerating urbanization in Germany is a phenomenon that adapts to industrialization and originates from industrialization [62]. Fu et al. established a comprehensive evaluation index system of regional urbanization from three aspects: economy, society, and transportation [63]. These studies have found many driving factors affecting urbanization. And from the nature of land use, not all urban areas are suitable for a development model. China itself is a big agricultural country, but not all regions focus on developing agriculture and industry. In the initial industrialization, most of the land was occupied as industrial land and construction land, which lost the function of land and destroyed the sustainable long-term development without paying attention to the overall development. The characteristics of urbanization gradually combine the agricultural land and industrial land of the primary and secondary industries with the local tertiary service industry, and promote the development of urbanization with the most suitable leading industry in the local area, and finally achieve the urbanization of people instead of land. From the perspective of population employment, in the process of urbanization, the urban population has gradually increased, and the proportion of the urban population in the total regional population has gradually increased. With the development of industrialization, people poured into cities and joined labor-intensive industries. In the post-industrial era, the tertiary industry plays an increasingly important role in the development of urbanization, with the proportion of employees in the primary industry declining and the proportion of employees in the secondary and tertiary industries increasing.

## 3 Research Method

In this paper, 15 towns around Beijing (capital), Zhuzhou (ordinary city), and Chengdu (provincial capital city), which are representative cities in eastern, central, and western China, are selected as case studies. According to the local town development plan and the specific situation of the high-frequency words "town name, government, urbanization, scenic spot/real estate/industry/agriculture/trade/school/traffic" searched by Baidu, the urbanization ways of each town in the region are determined. 100 questionnaires were distributed in each region, and 1500 questionnaires were distributed in total, and 1459 valid questionnaires were recovered, with an effective rate of 97.27%. The questionnaire uses a Likert five-component scale to measure the impact indicators, and SPSS23.0 software is used to analyze the questionnaire data by one-way ANOVA to compare the residents' perceived quality differences in different urbanization modes. See Table 1 for the specific questionnaire distribution.

## 4 Data Analysis

### *4.1 Reliability and Validity Test of the Questionnaire*

A total of 1459 samples were collected without any missing items. The reliability and validity of the questionnaire were analyzed, and the data were input into SPSS23.0 for testing. The overall Cronbach's coefficient reached 0.867. Generally speaking, an Alpha coefficient of 0.8–0.9 indicates that the reliability of the questionnaire is acceptable. Exploratory factor analysis of the scale shows that the KMO value is 0.914, which is greater than 0.7, and the structural validity is good, which can be used for factor analysis of the questionnaire. To sum up, the established evaluation index system needs no adjustment and can be used as the final evaluation index system.

### *4.2 The Difference in Economic Impact*

It can be seen from Table 2 that in terms of economic impact indicators, there are 8 indicators with significant differences and 2 indicators with slightly significant differences. Among them, the significant difference between "real estate price rise" and "price rise" is the most obvious, indicating that different urbanization models have the highest influence on these two indicators.

Residents of the agricultural urbanization model have the most obvious perception of the indicators of "better development of local property resources or scenery" and "great improvement of your income", while the perception of "per capita arable land has been greatly reduced" is the least obvious. In the process of agricultural

**Table 1** Questionnaire distribution in cities and towns

| Administrative level | Case | Urbanization mode | Town | Sample number |
|---|---|---|---|---|
| Ordinary prefecture-level city | Zhuzhou City (Central) | Tourism urbanization | Xianyu town | 88 |
| | | | Jiubujiang town | 93 |
| | | Real estate urbanization | Mingzhao town | 91 |
| | | Commercial urbanization | Baitutan town | 95 |
| | | | Huangtuling town | 94 |
| Provincial capital city | Chengdu (West) | Tourism urbanization | Jiezi town | 96 |
| | | | Anren town | 97 |
| | | Real estate urbanization | Huayuan town | 85 |
| | | Agricultural urbanization | Wanchun town | 97 |
| | | Science and education urbanization | Liucheng town | 111 |
| Capital city | Beijing (East) | Tourism urbanization | Shidu town | 98 |
| | | Science and education urbanization | Shahe town | 106 |
| | | Agricultural urbanization | Panggezhuang town | 108 |
| | | Real estate urbanization | Dongxiaokou town | 103 |
| | | Traffic urbanization | guan town | 97 |
| Total | | | | 1459 |

urbanization, more local tourism resources and agricultural resources are used for local urbanization to better develop resources, form industrialization, and scale, and better protect resources and environment. In the process of development, other urbanization approaches will greatly change the original features of cities and towns. It can be seen that the development of agricultural urbanization has developed local resources, increased the income of residents, and did not reduce the cultivated land area of farmers in the development of agricultural urbanization.

The residents of commercial urbanization have the highest perception of the indicators, such as "your purchasing power and purchasing desire has increased", "the per capita living area is much larger than before", "you and your friends no longer work in agriculture and choose to work', "the gap between the rich and the poor of residents has widened", and "urbanization has promoted local economic development".

In the process of commercial urbanization, most residents abandon agriculture for business and engage in self-employment. However, due to the different scope of self-employment chosen by different residents, economic income growth is also different. Therefore, commercial urbanization has the strongest perception of the indicator of the "widening gap between the rich and the poor of residents". It can be seen that the development of commercial urbanization mode has improved local purchasing power and income and promoted local economic development.

Tourism urbanization residents are least aware of the indicators of "rising real estate prices", "rising prices", "your purchasing power and purchasing desire have increased" and "urbanization has promoted local economic development". The development of tourism is seasonal, and the low peak season may have a great impact on the local economic situation, so the residents' perception of tourism promoting economic development is low and unstable (standard deviation is 1.31).

Traffic urbanization residents have the highest perception of "rising real estate prices", "rising prices" and "greatly reduced per capita cultivated land area". The perception of "local property resources or scenery have been better developed", "per capita living area is much larger than before" and "your income has been greatly improved" is the least obvious. The common characteristics of commercial towns and transportation towns are superior location conditions, developed commercial logistics, and fast commodity exchange. The residents' perception of price rise and fall is obvious in these areas. In the process of transportation-oriented urbanization, farmers' cultivated land is collected the most, which is not only used to build basic facilities and leading industrial facilities for urbanization development, but also to build major traffic routes. Therefore, the characteristics of traffic urbanization are that the price increases and the cultivated land decrease, but the income and housing are still not optimistic.

Residents of real estate urbanization are least aware of the indicators of "you and your friends no longer work in agriculture, but choose to work" and "the gap between the rich and the poor of residents is widening". The biggest impact of real estate urbanization on residents is the migration of their residence. The original residence and fields are expropriated, and residents are compensated for relocation and moved into new resettlement houses. Therefore, it has little impact on the non-fixed assets of residents, and residents feel relatively weak about the widening gap between the rich and the poor. Compared with other urbanization models, real estate urbanization and agricultural urbanization models have less impact on residents' employment. Real estate urbanization generally occurs in areas with superior location conditions, perfect supporting facilities for surrounding industries, and good ecological environment resources. Many residents choose to work in nearby towns with relatively perfect development, but live in their towns, so their perception of the changes in employment methods brought about by local urbanization is not obvious.

**Table 2** Residents' perception differences of economic impact indicators in different urbanization models

| Sample | Model | | | | | | | | |
|---|---|---|---|---|---|---|---|---|---|
| | Overall mean* | Tourism | Real estate | Commercial | Science and education | Agricultural | Traffic | Overall F | Overall Sig |
| I-HBGI | 3.16 | 3.24 (1.28) | 3.26 (1.20) | 3.20 (1.19) | 3.20 (1.10) | 3.27 (1.24) | 2.78 (1.32*) | 2.600 | 0.024* |
| PP-DTB-H | 3.57 | 3.38 (1.17) | 3.52( 1.13) | 3.76 (1.13) | 3.62 (1.03) | 3.49 (1.21) | 3.65 (1.15) | 3.699 | 0.002** |
| U-HP-LED | 3.75 | 3.55 (1.31*) | 3.81 (1.11) | 3.87 (1.13) | 3.75 (0.99) | 3.7 (1.17) | 3.75 (1.22) | 3.046 | 0.010* |
| REP-HR | 4.05 | 3.74 (1.33) | 3.94 (1.10) | 4.31 (1.11) | 4.09 (1.11) | 3.92 (1.30) | 4.34 (1.21) | 8.608 | 0.000** |
| P-R | 4.06 | 3.77 (1.40) | 3.98 (1.14) | 4.34 (1.10) | 4.06 (1.13) | 3.87 (1.24) | 4.34 (1.24) | 7.873 | 0.000** |
| PCLA-HBL | 3.26 | 3.25 (1.28) | 3.19 (1.30) | 3.56 (1.27) | 3.39 (1.20) | 3.25 (1.37) | 2.91 (1.41*) | 3.991 | 0.001** |
| LPR-S-HBBD | 3.46 | 3.66 (1.21) | 3.42 (1.16) | 3.38 (1.21) | 3.41 (1.00) | 3.67 (1.18) | 3.25 (1.30) | 4.437 | 0.001** |
| RPG-HW | 3.81 | 3.75 (1.19) | 3.62 (1.12) | 3.97 (1.09) | 3.86 (1.05) | 3.70 (1.15) | 3.94 (1.18) | 3.075 | 0.009** |
| PCCLA-HBGR | 3.91 | 3.88 (1.23) | 3.86 (1.12) | 3.96 (1.09) | 3.99 (1.13) | 3.64 (1.29*) | 4.14 (1.20) | 3.115 | 0.008** |
| W-I-F | 3.90 | 3.90 (1.10) | 3.73 (1.04) | 4.05 (0.95) | 3.93 (0.94) | 3.76 (1.16) | 4.03 (1.12) | 3.132 | 0.008** |

*Note* *The significance is below 0.05. Mean measured on a Likert scale of 1–5 where 1 = agree very much and 5 = disagree very much
I-HBGI = "Your income has been greatly improved"
PP-DTB-HI = "Your purchasing power and desire to buy have increased"
U-HP-LED = "Urbanization has promoted local economic development"
REP-HR = "Real estate prices have risen"
P-R = "Price rise"
PCLA-HBL = "The per capita living area has become larger"
LPR-S-HBBD = "The local property resources or scenery have been better developed"
RPG-HW = "The gap between rich and poor has widened"
PCCLA-HBGR = "The per capita cultivated land area has been greatly reduced"
W-I-F = "Work instead of farming"

### 4.3 The Difference in Social Life Influence

As shown in Table 3, 11 indicators are showing significant differences in social life, among which "the government has invested a lot in environmental sanitation, and the sanitary conditions have become better", "the neighborhood relationship is more harmonious than before", "the number of people around you using computers and telephones has increased", "you think the government management department is working well", and "the social security system is better, and you are satisfied with the current social welfare" are all less than 0.001, indicated that these five indicators have been greatly influenced in the process of urbanization. For other indicators of social life, urbanization has less impact.

Agricultural urbanization has the strongest perception of "more people are participating in social security around you". It shows that the government pays more attention to farmers' social security, strengthens the perfection of the social security system, and earnestly safeguards the interests of residents.

The residents of commercial urbanization have the most obvious perception of "more people around you use computers and telephones", "the government has invested a lot in environmental sanitation, and the sanitary conditions have become better", "the neighborhood relationship is more harmonious than before" and "the traffic is more convenient". The development of commercial urbanization has the most obvious effect on stimulating the development of the local regional economy and improving residents' income. Therefore, increasing people's demand for computers and telephones has promoted the improvement of residents' social life quality.

Tourism urbanization residents have the most obvious perception of "the overall quality of residents has improved". The perception of "the noise pollution in life increases and the peace of life is destroyed" is the least obvious. In the process of tourism urbanization, residents have a wide range of social communication, more people and things come into contact with the outside world, and residents' words and deeds represent the image of the town, so residents pay more attention to the improvement of their quality. Although a large number of tourists poured into the town, the residents didn't think it would bring much noise and social security problems. After the tourists left, the town returned to a quiet state. Therefore, it can be concluded that the development of tourism urbanization has improved the quality of residents without increasing the noise of life.

The residents of traffic urbanization have the most obvious perception of "the noise pollution in life increases and the peace of life is destroyed". For "the social security system is better, you are satisfied with the current social welfare", "the government has invested a lot in environmental sanitation, and the sanitary conditions have improved", "you feel that the work of government management departments is in place", "neighborhood relations are more harmonious than before", "the overall quality of residents has improved", "your living environment is safer, and there are fewer crimes", "more people are participating in social security around you", "In terms of compensation for demolition, planning, and construction, etc., the government has listened to the opinions of you and the surrounding residents" and "transportation is more convenient" are the least obvious indicators. In the process of traffic urbanization, although the construction of urban transportation infrastructure has been improved, the local government has invested less in the construction of environmental sanitation. With the improvement of transportation hubs and supporting facilities, noise pollution in life has increased, which has affected residents' lives to a certain extent. Traffic urbanization can promote local economic development, facilitate residents' travel, and improve people's quality of life. But on the whole, the regional economic development around the construction of the transportation hub makes residents leave the original living environment and break up the original neighborhood relations, and the perception of the improvement of the harmony degree of neighborhood relations is not obvious. Therefore, it can be seen that with the development of traffic urbanization, there is much room for improvement in social security, social welfare, health conditions, government management, neighborhood relations, residents' quality, security, and exchange of opinions.

Residents of real estate urbanization have the most obvious perception of "the social security system is better, you are satisfied with the current social welfare", "you think the government management department is working well", "the environment where you live is safer, and there are fewer crimes", which shows that in the process of large-scale development of real estate in real estate urbanization, residents are most satisfied with the evaluation of government management work and can do a good job of communication between the government and residents to a certain extent, and housing problems and social security problems can be properly solved.

The residents of urbanization based on science and education have the most significant perception of "the government has listened to your opinions and the surrounding residents in terms of compensation for demolition, planning, and construction", and the least significant perception of "the number of people around you using computers and telephones has increased". In the aspect of social life impact, the overall satisfaction of urbanized residents based on science and education is low, which is reflected in environmental sanitation, improvement of traffic conditions, noise pollution, neighborhood relations, residents' quality, and social security system. In the process of urbanization based on science and education, for the development and construction of science and technology parks and university campuses, a large amount of land has been expropriated, residents have been placed in new buildings, and the original style of the town has also changed greatly. In the process of building science and technology parks and university campuses, although there is more compensation for residents' demolition, there is a lack of consideration for the sanitary and traffic conditions around the residents' new living environment. When residents move from bungalows to buildings, the original neighborhood relationship is broken up, and the new neighborhood relationship is difficult to establish. With the completion of science and technology parks and university campuses, the number of migrants in cities and towns has increased, which has brought about problems such as noise pollution, great employment competition, and insufficient social security for land-losing farmers. It is worth noting that although it is close to the science and technology park and the university campus, the residents' quality has not improved significantly, which indicates that the improvement of residents' quality has nothing to do with local educational resources, but is more influenced by the improvement of residents' living standards and social communication.

## 4.4 The Differences in Environmental Impacts

As shown in Table 4, in terms of environment, there are 4 indicators with significant differences, among which, "newly built factories and other urbanization facilities have well protected the local environment, and no garbage and wastewater are discharged indiscriminately", "the environment of newly built parks and green spaces is much better than before" and "newly built factories and other urbanization facilities have well protected the local environment" have the most significant differences, which shows that the changes brought by urbanization to the rural environment are compared in different urbanization models.

Residents of agricultural urbanization have the most obvious perception of "many parks and green spaces have been built, and the environment is much better than before". It shows that the development of agricultural urbanization has also protected the local environment.

The residents of commercial urbanization have the most obvious perception of "I hope to continue to build more high-rise buildings here", and the least obvious perception of "the newly-built factories and other urbanization facilities have well protected the local environment, without discharging garbage and wastewater indiscriminately" and "many parks and green spaces have been built, and the environment is much better than before". It shows that in the process of commercial urbanization, more attention is paid to economic development, while the construction of parks and green spaces are neglected, which leads to insufficient environmental protection.

Residents of traffic urbanization have the least obvious perception of "preserving and better protecting and developing our local original scenery in the process of urbanization, and becoming more beautiful".

Residents of real estate urbanization have the most obvious perception of "preserving and better protecting and developing our original local scenery in the process of urbanization, and becoming more beautiful", and "newly built factories and other urbanization facilities have well protected the local environment, and no garbage and wastewater are discharged indiscriminately". In the development process of real estate urbanization, we mainly rely on large-scale real estate projects to build new cities. Most of the newly-built urbanization facilities serve real estate projects, which do less harm to the local environment, and the sewage facilities in the new cities are more perfect. In contrast, the residents of commercial urbanization and traffic urbanization are less satisfied with this index.

Residents of urbanization based on science and education have the least perception of "I hope to continue to build more high-rise buildings here". It shows that the residents of science and education urbanization have lower expectations for the construction of high-rise buildings, and prefer to keep the original urban environment.

**Table 3** Residents' perception differences of social impact indicators in different urbanization models

| Sample | Model | | | | | | | | |
|---|---|---|---|---|---|---|---|---|---|
| | Overall mean* | Tourism | Real estate | Commercial | Science and education | Agricultural | Traffic | Overall F | Overall Sig |
| G-LT-ROC-DPC | 2.78 | 2.68 (1.38) | 2.88 (1.33) | 2.72 (1.24) | 2.99 (1.18) | 2.92 (1.28) | 2.48 (1.23) | 3.553 | 0.003** |
| G-HI-LSC | 3.53 | 3.85 (1.17) | 3.83 (1.02) | 3.87 (1.12) | 3.39 (1.15) | 3.76 (1.16) | 2.48 (1.23) | 6.170 | 0.000** |
| T-MC | 4.00 | 4.04 (1.04) | 4.11 (0.94) | 4.14 (0.94) | 3.97 (1.02) | 4.05 (1.11) | 3.70 (1.45*) | 2.765 | 0.017* |
| YLE-S-FC | 3.34 | 3.49 (1.21) | 3.50 (1.06) | 3.34 (1.16) | 3.38 (1.14) | 3.40 (1.17) | 2.96 (1.29*) | 4.168 | 0.001** |
| NPL-I-LP-D | 3.76 | 3.57 (1.31) | 3.71 (1.11) | 3.73 (1.22) | 3.92 (0.92) | 3.69 (1.30) | 3.96 (1.22) | 3.547 | 0.003** |
| NR-HBMH | 3.48 | 3.64 (1.15) | 3.54 (1.12) | 3.69 (1.05) | 3.38 (1.06) | 3.48 (1.20) | 3.12 (1.29) | 4.893 | 0.000** |
| ROQ-HI | 3.54 | 3.70 (1.14) | 3.63 (1.05) | 3.65 (1.10) | 3.47 (1.11) | 3.59 (1.21) | 3.18 (1.25*) | 4.189 | 0.001** |
| MPAY-UCT | 4.36 | 4.31 (0.92) | 4.18 (1.05) | 4.59 (0.75) | 4.17 (0.97) | 4.33 (0.93) | 4.56 (0.94) | 6.849 | 0.000** |
| GMDW-IP | 2.82 | 2.81 (1.31) | 3.13 (1.28) | 2.64 (1.22) | 2.96 (1.26) | 2.93 (1.33) | 2.44 (1.26) | 6.142 | 0.000** |
| MPAY-PISS | 3.95 | 4.00 (1.07) | 4.09 (0.89) | 3.97 (0.90) | 3.89 (0.99) | 4.12 (0.95) | 3.63 (1.33*) | 4.087 | 0.001** |
| SSS-B-CSW-S | 3.30 | 3.35 (1.21) | 3.61 (1.01) | 3.25 (1.20) | 3.32 (1.15) | 3.47 (1.27) | 2.79 (1.26*) | 7.780 | 0.000** |

*Note* *The significance is below 0.05. Mean measured on a Likert scale of 1–5 where 1 = agree very much and 5 = disagree very much
G-LT-ROC-DPC = "The government listens to residents' opinions on compensation for demolition and planning and construction"
G-HI-LSC = "The government has improved the local sanitary conditions"
T-MC = "Traffic is more convenient"
YLE-S-FC = "Your living environment is safer and there are fewer crimes"
NPL-I-LP-D = "The noise pollution in life increases and the peace of life is destroyed"
NR-HBMH = "Neighborhood relations have become more harmonious"
ROQ-HI = "The overall quality of residents has improved"
MPAY-UCT = "There are more people around you who use computers and telephones"
GMDW-IP = "The work of the government management department is in place"
MPAY-PISS = "more people are participating in social security around you"
SSS-B-CSW-S = "The social security system is better, and you are satisfied with the current social welfare"

**Table 4** Residents' perception differences of environmental impact indicators in different urbanization models

| Sample | Model | | | | | | | | |
|---|---|---|---|---|---|---|---|---|---|
| | Overall mean* | Tourism | Real estate | Commercial | Science and education | Agricultural | Traffic | Overall F | Overall Sig |
| PU-BP-D-OLS-BMB | 3.66 | 3.77 (1.15) | 3.79 (1.03) | 3.56 (1.15) | 3.60 (1.08) | 3.76 (1.12) | 3.50 (1.26) | 2.522 | 0.028* |
| MBF-OUF-HWP-LE-HND-GWI | 3.14 | 3.26 (1.24) | 3.45 (1.14) | 2.80 (1.26) | 3.11 (1.10) | 3.29 (1.24) | 2.95 (1.35) | 8.061 | 0.000** |
| MP-GS-HBB-E-B | 3.56 | 3.60 (1.18) | 3.61 (1.17) | 2.97 (1.36) | 3.60 (1.10) | 3.83 (1.11) | 3.74 (1.24) | 12.295 | 0.000** |
| HTCTB-MHSB | 3.37 | 3.41 (1.35) | 3.52 (1.25) | 3.68 (1.23) | 3.15 (1.20) | 3.30 (1.38) | 3.17 (1.40) | 4.675 | 0.000** |

*Note* *The significance is below 0.05. Mean measured on a Likert scale of 1–5 where 1 = agree very much and 5 = disagree very much
PU-BP-D-OLS-BMB = "The process of urbanization preserves and better protects and develops our original local scenery and becomes more beautiful"
MBF-OUF-HWP-LE-HND-GWI = "Newly built factories and other urbanization facilities have well protected the local environment and have not discharged garbage and wastewater indiscriminately"
MP-GS-HBB-E-B = "Many parks and green spaces have been built, and the environment is much better than before"
HTCTB-MHSB = "I hope to continue to build more high-rise buildings here"

## 5 Research Results

By comparing residents' perceptions of the economy, social life, and environment under different urbanization models, it is found that different urbanization models have different perceptions of different indicators.

The advantage of agricultural urbanization is that it develops resources and increases the income of residents economically, and it does not reduce the cultivated land area of farmers in the development. The government pays more attention to farmers' social security and protects the interests of residents. And protect the environment while developing the economy.

The advantage of commercial urbanization is that it improves local purchasing power and income, promotes local economic development, increases people's demand for computers and telephones, and promotes the improvement of residents' social life quality. The disadvantage is to pursue more economic development while ignoring the construction of parks and green spaces and protecting the environment.

The advantage of tourism-oriented urbanization is that it improves the quality of residents, while it does not increase the noise of life. The disadvantage is that the development of tourism is seasonal, and the low peak season may have a great impact on the local economic situation, so residents' perception of tourism promoting economic development is low and unstable (standard deviation is 1.31).

Under the development background of traffic urbanization, the overall satisfaction of residents is low. Although the construction of urban transportation infrastructure has been improved, the local government has invested less in the construction of environmental sanitation. With the improvement of transportation hubs and supporting facilities, noise pollution in life has increased, which has affected residents' lives to a certain extent. Moreover, in the development process, prices and housing prices rose, and cultivated land decreased, but residents' income did not increase. Housing security, social security, social welfare, sanitary conditions, environmental protection, government management, neighborhood relations, residents' quality, security, and other aspects of traffic urbanization are still not optimistic.

The advantage of real estate urbanization is that residents have the highest satisfaction with the evaluation of government management, and housing problems and social security problems can be properly solved. And the perception of the local gap between the rich and the poor is not obvious, which has less influence on the employment of residents. There is less harm to the environment while developing the economy.

The overall satisfaction of residents of science and education urbanization is low, which is reflected in environmental sanitation, improved traffic conditions, noise pollution, neighborhood relations, residents' quality, and social security system. Although it is close to science and technology parks and university campuses, the quality of residents has not improved significantly, which shows that the improvement of residents' quality has nothing to do with local educational resources, but is more influenced by the improvement of residents' living standards and social communication. Moreover, the residents of science and education urbanization have lower expectations for the construction of high-rise buildings and prefer to keep the original urban environment.

# 6 Suggestion

## 6.1 Agricultural Urbanization

Agricultural urbanization promotes local economic development, protects residents' interests and lives, and protects the environment. Based of these, what agricultural urbanization needs to do is to upgrade the agricultural industry. The industrial upgrading and development of agriculture is the key point to solve many problems in the process of urbanization, such as promoting urban economic development, increasing farmers' income, improving supporting infrastructure, and public service. Therefore, the government should actively promote and comprehensively guide the agricultural industry, promote the transformation and upgrading of the whole agricultural industrial chain, expand the scale of agricultural production, promote agricultural industrialization, continuously broaden the sales channels of agricultural products, improve economic benefits and increase farmers' income.

## 6.2 Tourism Urbanization

Because of relying on single tourism consumption income, the economic income structure of tourism urbanization is weak, and the development of tourism has seasonal characteristics, so the off-season has a great impact on the tourism economy. Therefore, while expanding tourism industry, tourism urbanization must also promote the integration of tourism and other industries, realize the agglomeration of industries, consumption and employment with the tourism industry as the core, promote the overall improvement of urban economy, make urbanization develop in a more benign direction, and better improve the economic income of residents.

## 6.3 Commercial Urbanization

In the development process of commercial urbanization, the income distribution of residents is unequal, and the gap between the rich and the poor is widening. Therefore, the government can influence the distribution of wealth and the industrial distribution of capital and labor using fiscal and industrial policies, and exert the structural adjustment function of credit policies to control the price increase. Increase the input of urban residents' social security work and gradually narrow the gap between urban and rural residents' social welfare security system.

## 6.4 Traffic Urbanization

Residents are less satisfied with the economic changes, social life changes and environmental changes brought about by the development of traffic urbanization. Therefore, it is necessary to plan and build towns from the perspective of "people-oriented", do a good job in land use planning based on the principle of intensive optimization, build a traffic network according to local conditions, scientifically reduce noise pollution, pay attention to ecological environment protection and urban greening construction, and create a green and low-carbon new city environment. In urban management, the government should strengthen regional public security management and strengthen the management of migrants. Focus on improving the educational, cultural, and health conditions in the region, and enhance the quality of life and cultural quality of residents. In terms of social security, it is necessary to implement the integration of urban and rural social security, so that every resident can participate in social security and gradually narrow the social welfare gap with urban residents.

## 6.5 Real Estate Urbanization

Most residents are satisfied with the economic, social, and environmental impacts of real estate urbanization. In addition to paying attention to the current economic development, the development of real estate urbanization should be more forward-looking and make sustainable urban development plans. Make use of the advantages of population agglomeration, actively develop other industrial economies, jointly promote the sustainable growth of the urban economy, and create a "people-oriented" urbanization model.

## 6.6 Science and Education Urbanization

Residents are less satisfied with the social life and environmental impact brought about by the construction of urbanization based on science and education, while they are generally satisfied with the economic impact. Therefore, in line with the principle of "people-oriented", we should speed up the development process of urbanization based on science and education, pay attention to the social life of residents while promoting economic development, actively build a social security system integrating urban and rural areas, and gradually narrow the social welfare gap with urban residents. At the same time, we should combine our advantages to achieve green development and sustainable development. For example, build parks, community centers, and other leisure places.

## 7 Future Research

In this study, 13, 18, and 5 indicators were selected from three aspects: economy, social life, and environment to construct the evaluation index system of urbanization quality. However, this study does not consider the indicators of culture and spirit, and local policies when evaluating the quality of urbanization. In the future, indicators of culture, spirit, and local policies will be added to analyze and compare the urbanization quality differences of different models more comprehensively.

## References

1. National Bureau of Statistics of the People' Republic of China (NBSC). China Statistics Press, Beijing (2018)
2. Yao, S.M., Zhang, P.Y., Yu, C., Li, G.Y., Wang, C.X.: The theory and practice of new urbanization in China. Scientia Geographica Sinica **34**(6), 641–647 (2014)
3. Chen, M., Ye, C., Lu, D., Sui, Y., Guo, S.: Cognition and construction of the theoretical connotations of new urbanization with Chinese characteristics. J. Geog. Sci. **29**(10), 1681–1698 (2019)
4. Bradshaw, Y.W.: Urbanization, personal income, and physical quality of life: The case of Kenya. Stud. Comparative Int. Dev. **23**(4), 15–40 (1988)
5. Liu, H., Weng, Q.: Landscape metrics for analysing urbanization-induced land use and land cover changes. Geocarto Int. **28**(7), 582–593 (2013)
6. Ouyang, T., Zhu, Z., Kuang, Y.: Assessing impact of urbanization on river water quality in the Pearl River Delta Economic Zone. China. Environ. Monit. Assess. **120**(1–3), 313–325 (2006)
7. Pan, J.: From industrial toward ecological in China. Science **336**(6087), 1397 (2012)
8. Qi, Z.F., Ye, X.Y., Zhang, H., Yu, Z.L.: Land fragmentation and variation of ecosystem services in the context of rapid urbanization: The case of Taizhou city, China. Stoch. Env. Res. Risk Assess. **28**(4), 843–855 (2014)
9. Wang, X., Wan, G.: China's urban employment and urbanization rate: A re-estimation. China World Econ. **22**(1), 30–44 (2014)
10. Wei, Y.D., Ye, X.: Urbanization, land use, and sustainable development in China: In: Stochastic Environmental Research and Risk Assessment 2014, pp. 755. Springer, Berlin (2014)
11. Zhang, Z., Su, S., Xiao, R., Jiang, D., Wu, J.: Identifying determinants of urban growth from a multi-scale perspective: A case study of the urban agglomeration around Hangzhou Bay. China. Appl. Geography **45**, 193–202 (2013)
12. Zhang, L.: Conceptualizing China's urbanization under reforms. Habitat Int. **32**(4), 452–470 (2008)
13. Chen, M., Huang, Y., Tang, Z., Lu, D., Liu, H., Ma, L.: The provincial pattern of the relationship between urbanization and economic development in China. J. Geog. Sci. **24**(1), 33–45 (2014)
14. Feng, J., Chen, Y.: Spatiotemporal evolution of urban form and land-use structure in Hangzhou, China: Evidence from fractals. Environ. Plan. B: Plan. Des. **37**(5), 838–856 (2010)

15. Goncalves, R.M., Pacheco, A.d.P., Xavier Tanajura, E.L., da Silva, L.M.: Costal urbanization and shading in Boa Viagem beach, Recife-PE, Brazil. Revista De Geografia Norte Grande (54), 241–55 (2013)
16. Chaolin, G.U., Liya, W.U., Cook, I.: Progress in research on Chinese urbanization. Front. Architect. Res. **1**(2), 101–149 (2012)
17. Bai, X., Shi, P., Liu, Y.: Society: Realizing China's urban dream. Nat. News **509**(7499), 158 (2014)
18. Cao, S., Lv, Y., Zheng, H., Wang, X.: Challenges facing China's unbalanced urbanization strategy. Land Use Pol. **39**, 412–415 (2014)
19. Chen, M., Liu, W., Tao, X.: Evolution and assessment on China's urbanization 1960–2010: Under-urbanization or over-urbanization? Habitat Int. **38**, 25–33 (2013)
20. Hsieh, S.C.: Analyzing urbanization data using rural–urban interaction model and logistic growth model. Comput. Environ. Urban Syst. **45**, 89–100 (2014)
21. Zhao, P.: Too complex to be managed? New trends in peri-urbanisation and its planning in Beijing. Cities **30**, 68–76 (2013)
22. Li, Z.-h., Gao, W., Gao, Z.-q., Shi, R.-h., Liu, C.-s.: A study on assessment of urbanization and ecosystem changes based on MODIS time series in Shanghai municipality from 2000 to 2009: In: Remote Sensing and Modeling of Ecosystems for Sustainability VII 2010, pp. 78090R. International Society for Optics and Photonics (2010)
23. Li, Y., Li, D.: Assessment and forecast of Beijing and Shanghai's urban ecosystem health. Sci. Total Environ. **487**, 154–163 (2014)
24. Yew, C.P.: Pseudo-urbanization? Competitive government behavior and urban sprawl in China. J. Contemp. China **21**(74), 281–298 (2012)
25. Wang, Q., Yuan, X., Zhang, J., Mu, R., Yang, H., Ma, C.: Key evaluation framework for the impacts of urbanization on air environment–A case study. Ecol. Ind. **24**, 266–272 (2013)
26. Ann, T.W., Wu, Y., Zheng, B., Zhang, X., Shen, L.: Identifying risk factors of urban-rural conflict in urbanization: A case of China. Habitat Int. **44**, 177–185 (2014)
27. Al-Mulali, U., Fereidouni, H.G., Lee, J.Y.M., Sab, C.N.B.C.: Exploring the relationship between urbanization, energy consumption, and $CO_2$ emission in MENA countries. Renew. Sustain. Energy Rev. **23**, 107–112 (2013)
28. Gregg, J.W., Jones, C.G., Dawson, T.E.: Urbanization effects on tree growth in the vicinity of New York City. Nature **424**(6945), 183–187 (2003)
29. Liu, Y., Xie, Y.: Asymmetric adjustment of the dynamic relationship between energy intensity and urbanization in China. Energy Econ. **36**, 43–54 (2013)
30. Ponce de Leon Barido, D., Marshall, J.D.: Relationship between urbanization and $CO_2$ emissions depends on income level and policy. Environ. Sci. Technol. **48**(7), 3632–9 (2014)
31. Tu, J.: Spatial variations in the relationships between land use and water quality across an urbanization gradient in the watersheds of northern Georgia, USA. Environ. Manage. **51**(1), 1–17 (2013)
32. Xia, T.-Y., Wang, J.-y., Song, K., Da, L.-j.: Variations in air quality during rapid urbanization in Shanghai, China. Landscape Ecol. Eng. **10**(1), 181–90 (2014)
33. Xiong, Y., Huang, S., Chen, F., Ye, H., Wang, C., Zhu, C.: The impacts of rapid urbanization on the thermal environment: A remote sensing study of Guangzhou, South China. Remote Sens. **4**(7), 2033–2056 (2012)
34. Dongfeng, Y., Chengzhi, Y., Ying, L.: Urbanization and sustainability in China: An analysis based on the urbanization Kuznets-curve. Plan. Theory **12**(4), 391–405 (2013)
35. You, C., Shi, R.: Monitoring urbanization in Shanghai from 1990 to 2005: In: Remote Sensing and Modeling of Ecosystems for Sustainability X 2013, pp. 88690H. International Society for Optics and Photonics (2013)
36. Zhao, J., Chen, S., Jiang, B., Ren, Y., Wang, H., Vause, J., et al.: Temporal trend of green space coverage in China and its relationship with urbanization over the last two decades. Sci. Total Environ. **442**, 455–465 (2013)
37. Benedetti, L., Blumensaat, F., Bönisch, G., Dirckx, G., Jardin, N., Krebs, P., et al.: Systems analysis of urban wastewater systems–two systematic approaches to analyse a complex system. Water Sci. Technol. **52**(12), 171–179 (2005)

38. Kanta, L., Zechman, E.: Complex adaptive systems framework to assess supply-side and demand-side management for urban water resources. J. Water Resources Plan. Manage. **140**(1), 75–85 (2014)
39. Oliver, J.L., Tortosa, L., Vicent, J.F.: A neural network model to develop actions in urban complex systems represented by 2d meshes. Int. J. Comput. Math. **88**(16), 3361–3379 (2011)
40. Zarghami, M., Akbariyeh, S.: System dynamics modeling for complex urban water systems: Application to the city of Tabriz, Iran. . Resources Conserv. Recycling **60**, 99–106 (2012)
41. Zhang, Y., Li, S., Fath, B.D., Yang, Z., Yang, N.: Analysis of an urban energy metabolic system: Comparison of simple and complex model results. Ecol. Model. **223**(1), 14–19 (2011)
42. Zou, Y., Torrens, P.M., Ghanem, R.G., Kevrekidis, I.G.: Accelerating agent-based computation of complex urban systems. Int. J. Geogra. Inf. Sci. **26**(10), 1917–1937 (2012)
43. Jain, M., Korzhenevych, A.: Urbanisation as the rise of census towns in India: An outcome of traditional master planning? Cities **99**, 102627 (2020)
44. Liu, H., Liu, Y., Wang, H., Yang, J., Zhou, X.: Research on the coordinated development of greenization and urbanization based on system dynamics and data envelopment analysis—A case study of Tianjin. J. Clean Prod. **214**, 195–208 (2019)
45. Ghosh, S., Kanjilal, K.: Long-term equilibrium relationship between urbanization, energy consumption and economic activity: Empirical evidence from India. Energy **66**, 324–331 (2014)
46. Chen, M., Liu, W., Lu, D., Chen, H., Ye, C.: Progress of China's new-type urbanization construction since 2014: A preliminary assessment. Cities **78**, 180–193 (2018)
47. Lu, X., Ke, S.: Evaluating the effectiveness of sustainable urban land use in China from the perspective of sustainable urbanization. Habitat Int. **77**, 90–8 (2018).
48. Zhong, L., Li, X., Law, R., Sun, S.: Developing sustainable urbanization index: Case of China. Sustainability **12**(11), 4585 (2020)
49. Hu, Y.: Spatial distribution of comprehensive urbanization level in Anhui Province. Proc. Computer Sci. **131**, 401–408 (2018)
50. Wu, Y., Dong, S., Zhai, J., Huang, D., Huang, Z.: Land management institution as a key confinement of urbanization in Baotou, China application of proposed endogenous urbanization model. Land Use Pol. **57**, 348–355 (2016)
51. Luo, T., Xu, M., Huang, T., Ren, X., Bu, X.: Rethinking the intensified disparity in urbanization trajectory of a Chinese coastal province and its implications. J. Clean Prod. **195**, 1523–1532 (2018)
52. Mingxing, C.: Research progress and scientific issues in the field of urbanization. Geogra. Res. **34**(4), 614–630 (2015)
53. Dewan, A.M., Yamaguchi, Y.: Land use and land cover change in Greater Dhaka, Bangladesh: Using remote sensing to promote sustainable urbanization. Appl. Geogr. **29**(3), 390–401 (2009)
54. Estoque, R.C., Murayama, Y.: Intensity and spatial pattern of urban land changes in the megacities of Southeast Asia. Land Use Pol. **48**, 213–222 (2015)
55. Fu, Y., Zhang, X.: Trajectory of urban sustainability concepts: A 35-year bibliometric analysis. Cities **60**, 113–123 (2017)
56. Li, Y., Commenges, H., Bordignon, F., Bonhomme, C., Deroubaix, J.F.: The Tianjin Eco-City model in the academic literature on urban sustainability. J. Clean Prod. **213**, 59–74 (2019)
57. Cui, X.: How can cities support sustainability: A bibliometric analysis of urban metabolism. Ecol. Ind. **93**, 704–717 (2018)
58. Dan, Z., Jiapeng, X., Yizhu, Z., Jing, W., Siyu, H., Xiao, Z.: Study on sustainable urbanization literature based on Web of Science, scopus, and China national knowledge infrastructure: A scientometric analysis in CiteSpace. J. Clean Prod. 121537 (2020).
59. Marvuglia, A., Havinga, L., Heidrich, O., Fonseca, J., Gaitani, N., Reckien, D.: Advances and challenges in assessing urban sustainability: An advanced bibliometric review. Renew. Sustain. Energy Rev. **124**, 109788 (2020)
60. Gladstone, D.L.: Tourism urbanization in the United States. Urban Affairs Rev. **34**(1), 3–27 (1998)
61. Liu, C.M., Zou, R.C.: Study on the development of Chinese real estate in the context of urbanization: In: Applied Mechanics and Materials 2012, pp. 2284–8. Trans Tech Publications Ltd, (2012)

62. Köllmann, W.: The process of urbanization in Germany at the height of the industrialization period. J. Contemp. History **4**(3), 59–76 (1969)
63. Fu, L.-p., Li, J.-h., He, Y.-j.: The empirical research of urbanization level measure from the spatial economics: On the data of Henan province: In: The 19th International Conference on Industrial Engineering and Engineering Management 2013, pp. 987–97. Springer, Berlin, Heidelberg (2013)

# The Champion of Urban Water Resources Management in the Chinese City—The Case of Ningbo

**Faith Ka Shun Chan, Fangfang Zhu, Lei Li, Miran Lu, Yu-Ting Tang, and James Griffiths**

**Abstract** Ningbo is a coastal megacity located at the East Coast of China and developing rapidly with proactive trading and export economic activities. The city owns a ranked top ten international port and it is the major strategic spot of 21st century maritime Silk Road from the "Belt and Road" policy established for promoting further international trades and developments. In future, populations and economy in Ningbo are expected continuously growing in the next few decades. The demand of quality freshwater resources thus is enormously increasing. Ningbo municipal government has established the "*Five water management*" (五水共治) policy in 2013 that aims to manage (i) sewage discharge; (ii) flooding; (iii) surface water; (iv) water conservation and (v) freshwater supply. Indeed, the municipal government also liaised and initiated the "*Sponge City Program*" after 2015 that Ningbo was selected as one

F. K. S. Chan (✉) · L. Li (✉) · Y.-T. Tang (✉)
School of Geographical Sciences, University of Nottingham Ningbo China, Ningbo 315100, China
e-mail: faith.chan@nottingham.edu.cn

L. Li
e-mail: zy13104@nottingham.edu.cn

Y.-T. Tang
e-mail: yu-ting.tang@nottingham.edu.cn

F. K. S. Chan
Water@Leeds Research Institute, University of Leeds, Leeds LS2 9JT, UK

F. Zhu (✉)
Department of Civil Engineering, University of Nottingham Ningbo China, Ningbo 315100, China
e-mail: Fangfang.ZHU@nottingham.edu.cn

M. Lu (✉)
Department of Mathematical Sciences, University of Nottingham Ningbo China, Ningbo 315100, China
e-mail: smyml1@nottingham.edu.cn

J. Griffiths (✉)
National Institute of Water & Atmospheric Research Ltd (NIWA), 10 Kyle Street, Riccarton, Christchurch, New Zealand
e-mail: james.griffiths@niwa.co.nz

 
F. K. S. Chan et al. (eds.), *Proceedings of the 2020 International Conference on Resource Sustainability: Sustainable Urbanisation in the BRI Era (icRS Urbanisation 2020)*, Environmental Science and Engineering,
https://doi.org/10.1007/978-981-15-9605-6_25

of the pilot city; these policies and practices are successful up to now. This article adopts the case study of Ningbo to investigate the reasons of municipal government to promote the policy, to understand the public perception of this water management policy in Ningbo through conducted semi-structured interviews. During the 2017 and 2019, we conducted a questionnaire (N = 110) and interviews (N = 10) that follow up for justification of the public perception with the local communities. Our findings indicated that the communities had not been engaged closely with these practices, but generally supporting these two urban water management practices; and agreed that the urban water conditions (urban floods and pollution) had been improved. Also, the article discusses whether these (5 Water and SCP) practices can be extensively applied in other Chinese cities. We will provide recommendations at the end of the article.

**Keywords** Five water management · Sewage · Water supply · Flooding · Surface water and water conservation

## 1 Introduction

In recent years, China has undergone rapid economic and social development, and one of the negative consequences is the severe environmental pollution [1]. This is especially the case in Ningbo, which is one of the most developed areas and rich in water resources. It has been reported that the water qualities in many rivers in Ningbo are Grade 5, and water scarcity problems also occur because of water quality and fast development [2]; these lead to deterioration of ecological system. Furthermore, as a coastal city, Ningbo suffers from typhoons which can cause severe flooding. The problems of water scarcity, water pollution, and deterioration of ecological system have become an increasingly prominent concern of Ningbo residents, also the municipal government at Ningbo and extensively to the provincial level for Zhejiang province. For instance, there was a TV program called "*Find a river where you can swim in*", which caused a huge impact, and also gained the attention of the government [3].

Considerable effort has been made for better water management in Zhejiang province. For example, water supply project for tens of millions of farmers, water resources protection project, and seawall project. These have launched a solid foundation for the proposal and deployment of the '*Five Water Management*'. The government has fully understood the weakness of existing water management. Water problems are complex and not independent. The main problem in the water management is the multi-sector involvement, which may cause contradictions among institutions and stakeholders [4]. The functions of rivers and hydraulic structures are regulated by different departments for various purposes. For example, Water Conservancy Department is concerned with the flood control; however, environmental protection authorities are concerned with water qualities. As the most economically developed area in China, Zhejiang province has received high expectations from the central government. The Zhejiang provincial Party committee put forward the strategic deployment of '*Five Water Management Plan*', which includes (i) sewage discharge treatment; (ii) flood control; (iii) waterlogging discharge; (iv) freshwater supply and (v) water conservation at the fourth plenary session of the thirteenth central committee. It should be noted that the 'Five Water Management Plan' is not a single administrative activity, but a public policy with comprehensive management tasks including a series of economic development, environmental protection, and city transformation that aligns with the Sponge City Program (SCP) in Ningbo which has been selected as one of the pilot cities for adapting the sustainable urban stormwater management practice by Sponge infrastructure and achieve multiple benefits [5], which were supported by relevant legislation practices to deliver better urban water management strategies.

The aim of '*Five Water Management*' is to reach the harmonious state of human and water for sustainable development. This will further force transformation and upgrade of industry. The implementation schedule of the '*Five Water Management*' plan in different time scales is: solving urgent and prominent problems in 2014–2016, and achieving some improvement; solving the fundamental problems in 2014–2018, and achieving significant improvement; solving all the problems and achieving a qualitative change in 2014–2020. The '*Five Water Management*' shows the determination of Zhejiang government to solve these water environment problems. Zhejiang province serves as a pilot for other provinces in China.

Therefore, this paper aims to investigate the current five water management plan and SCP at the case of Ningbo; this aim will be followed by some specific objectives. First, we are investigating the detailed background of Fiver water management and reviewing the current progress of the '*Five Water Management*' and SCP at Ningbo. Second, we would like to review whether these policy plans are successful and whether the other cities can learn from this plan. Third, we would like to provide justification from our interviews and questionnaire to triangulate and validate current practices and offer recommendations for integrated water management in China.

## 2 Literature Review and Background

### *2.1 Fiver Water Management*

Ningbo Municipal Government has thoroughly implemented the provincial and municipal 'Five Water Management', encouraging all counter parts and residents to engage, boosting innovation and establishing a more comprehensive urban water supply and wastewater treatment system. The final goal is to force economic and social transformation and development by water management, and remarkable results have been achieved [6].

### 2.1.1 Infrastructure Reinforcement (治水强基 in Chinese)

After Typhoon Fitow, Ningbo government has determined to reinforce the infrastructure to protect the city from flooding. Ningbo has invested a large amount of money for engineering measures such as the construction of riverbanks and reinforcement of hydraulic structures. For the flooding problem, the main approaches include construction and reinforcement of river banks, drainage channels, and also hydraulic structures such as reservoirs, gates and pumps. Hundreds of kilometres of river banks of Yong Jiang and other rivers that go to sea directly have been constructed. Besides, flooding discharge channel networks ranging 28 km were also constructed. The obstacles (e.g. garbage, pollutants from domestic and industrial sources) in the river, which affect the flood discharge, were removed [7, 8].

### 2.1.2 Zero Sewage Discharge Into Rivers

For the environmental pollution, Ningbo government has tried to reduce the pollution from the source. Factories that discharge exceeded polluted water or waste has been required to upgrade and introduce new technology to reduce pollution. Major polluting industries were commanded to close and move. Pilots of zero sewage discharge areas have been established, and zero sewage discharge strategy has been promoted in Ningbo. The Municipal Ningbo Government worked to identify the characteristics of rivers including the back utilization, bridge information, sewer discharge arrangement, fishery, and rubbish disposal along the river, the green plant covering, livestock, and reclamation depot. Comprehensive measures were taken to control rural pollution from non-point sources and pollution from livestock and poultry farming. Furthermore, Ningbo has set pipelines to collect the foul water especially in rural areas [7].

### 2.1.3 Existing Pollution Control

For seriously polluted rivers, dredging, water transfer and ecological remediation and ecological restoration has been used. Ecological revetment has been utilized in rivers in rural areas [9, 10].

### 2.1.4 Policy Innovation

In July, 2018, China's Ministry of Water Resources announced a new measure-River chief mechanism, aiming to improve China's water governance by overcoming the drawbacks of multi-sector management. It had been successfully implemented across most provinces in China. The "*River Chief*" mechanism places responsibility for protecting bodies of water squarely on the shoulders of government officials. According to China Daily [11], over 300,000 officials have been appointed as river chiefs around the country, operating on the provincial, city, county, and township levels, and another 760,000 have been designated at the village level.

River chief policy is well implemented, and the performance of the river chief is an important indicator for their career promotion is their job [11]. River Chiefs are well-placed to coordinate various governmental departments to improve the efficiency and efficacy of a multitude of water-resource management efforts, and now to face clear consequences if they fail to maintain or improve water quality in their assigned areas.

This policy does address part of the current problem with water governance in China, which is the question of inter-jurisdictional cooperation. River Chief policy is an "ingenious" method of using Party hierarchy as a lever to improve coordination between various government organs [12]. The system is a massive undertaking, with the potential to dramatically improve the quality and integrity of China's water resources, and provides effective water management tackling collaborative issues in the Chinese context, but its long-term effects and sustainability remain to be determined [12, 13].

## *2.2 Sponge City Program*

The Sponge City Program (SCP) was initiated in 2013 and the function is parallel to some global best practices, such as the "*Low Impact Developments*" (LID) in the USA [14]; Sustainable Urban Drainage Systems (SuDs) [3, 15] and the Blue-Green Cities (BGCs) [16] in the UK; and Water Sensitive Urban Design (WSUD) in Australia [17] or Low Impact Developments Urban Design (LIDUD) in New Zealand. The SCP is required to (i) adopt BGI and SuDs such concepts which improve the effective control of urban peak runoff, and storing and filtering stormwater; (ii) upgrade traditional engineered drainage systems (e.g. construction of tanks and tunnels); and improve current land-drainage protection standards to balance peak discharges and to alleviate stormwater (to achieve 1-in-30 year return period protection standard); and (iii) reassure multi-functional objectives in drainage design and in achieving good ecosystem services [18, 19]. The SCP infrastructures in Chinese pilot cities (selected pilot 30 cities) that adopted, for example, artificial wetlands, ponds, green-roofs, bio-swales, rain gardens, pervious pavements, these infrastructures were normally often designed to enhance natural hydrological response and improve natural soil infiltration, stormwater retention, storage, purification, recharge groundwater and improve water quality of the runoff [20, 21]. Certainly, Xia, Zhang [22] indicated that the SCP development is able to deliver better urban water management and achieve sustainable development goals by encouraging ecological, social and economic perspectives to integrate with the urban drainage and master plans in Chinese cities. Ningbo is selected as one of the pilot city in the SCP in 2015 and constructed several SCP infrastructures sites (including Eco-corridor in New East Town (NET), and Cicheng Park, etc.) (see Fig. 1).

**Fig. 1** SCP site locations in Ningbo as interviews and questionnaire sites for this study (Locations A: New East Town; B: Ci Cheng Park; C: Sanjiangkou Park; D: Yinzhou Park; E: Nan Tang Old Street) (*Source* authors)

# 3 Methods

## *3.1 Questionnaire*

Questionnaire is an effective and popular method to study perceptions, attitudes, viewpoints and foresights [6]. In this study, a questionnaire was conducted to gain an initial understanding on public perspectives on the 'Five Water Management' and SCP in Ningbo. The questionnaire was developed according to the research questions and targeted understanding about the opinions, perceptions and viewpoints of the development of both urban water management strategies (Five Water Management and SCP). Since both water management practices are related to addressing urban water issues. The questionnaire consisted of the following parts:

a. General questions (e.g. the information from the communities: age, education background, residential location, gender);
b. Public perceptions on SCP
c. Public perceptions on 'Five Water Management' (e.g. urban river restoration, SCP and Five Water Management)
d. Sustainability issues (e.g. social-economic, ecology and wellbeing) in SCP
e. Future foresights on climate change and public participation

Key Question 1: Interviewees' perception on connections between 'Five Water Management Plan' with SCP projects?

Key Question 2: Interviewees' viewpoints on future SCP should be continued?

The survey sites for questionnaires were selected from the SCP projects location in Ningbo (see Fig. 1) as for investigating the communities who are living in that area or commonly using these SCP facilities (e.g. Sponge urban parks, Sponge healthy running tracks, Sponge artificial ponds and wetlands.). The survey was conducted by paper-style format, the data collection and sampling period was from June to August 2017. Due to the limited labour resources and weather conditions, this study managed to conduct 110 questionnaires. All survey data were statistically analysed using Excel and R software.

## 3.2 *Interview*

Semi-structured Interviews in this study were conducted by face-to-face in the public space (inside the SCP parks, coffee shops, café, etc.) [23]. The interviews were conducted to help understand more in-depth perspectives from the communities to Five Water Management and SCP, because the interviewees were picked randomly and each interview had gone through rather a short interview with about 10 min each.

Total 10 interviewees were invited for interviews who all are living around the SCP sites (within 3 km radius) in Ningbo during November 2019 (see Table 1) and aiming to articulate the perception issues specifically from the 2017 questionnaire data set. All interviewees have been informed and agreed the consent of ethical

**Table 1** Interviewees' information

| Participant ID (Interviewees) | Age, gender and occupation | Interviewees' home location (within 3 km radius at the relevant SCP sites) |
|---|---|---|
| A | Elderly (F), 60, retired | NET Eco-corridor |
| B | Flower sales-person in South Business District, 35 | Yinzhou Park |
| C | Elderly (F), 55, retired | Sanjiangkou Park |
| D | Housewife (F), 40 | Cicheng Park |
| E | Primary school teacher (M), 30 | Cicheng Park |
| F | Student (F), 20 in South Business District | Yinzhou Park |
| G | Waitress (F), 35 in Nantang Old Street | Nantang Old Street, Sanjiangkou Park |
| H | Elderly (M), 60, retired | Nantang Old Street, Sanjiangkou Park |
| I | Businessman (M), 55, at reside around the Tianyi Square | Nantang Old Street, Sanjiangkou Park |
| J | Student (M) 20, reside around the Tianyi Square | Nantang Old Street, Sanjiangkou Park |

practice (approved by the ethics application at the affiliated institution, University of Nottingham). The data was transcribed and translated. The analyses based on the ground theory, and the researchers undertook manual coding that match with the key research themes in this study.

# 4 Results and Discussion

## *4.1 Distribution of the Respondents*

According to the general information, the respondents were evenly distributed in gender with Male (47.3%) and Female (51.8%) and 1 participant did not disclose on the gender. Most participants are at the age group of 20–29 (with 56.4%), and majority of participants in this questionnaire set were living in Yinzhou district (with 73.6%), and dominated with the education level at the Tertiary/higher education/university level (with 83.6%) (Table 2).

**Table 2** Demographic information of participants in this questionnaire survey

| Independent variables | Information | Number | Percentage (%) |
|---|---|---|---|
| Total | | 110 | |
| Age | Below 20 | 20 | 18.18 |
| | 20–29 | 62 | 56.36 |
| | 30–39 | 15 | 13.64 |
| | 40–49 | 8 | 7.27 |
| | 50 above | 5 | 4.55 |
| Education | Primary | 1 | 0.91 |
| | Secondary | 7 | 6.36 |
| | Tertiary/higher education/university | 92 | 83.64 |
| | Postgraduate or above | 10 | 9.09 |
| Districts | Yinzhou | 81 | 73.64 |
| | Haishu | 11 | 10.00 |
| | Jiangbei | 9 | 8.18 |
| | Others | 9 | 8.18 |
| Sex | Male | 52 | 47.27 |
| | Female | 57 | 51.82 |
| | Not disclose | 1 | 0.91 |

## 4.2 *Perceptions on Five Water Management Plan and SCP Projects*

In this questionnaire, we have conducted some other issues such as general information, perception on Fiver Water Management Plan, SCP, sustainability issues and future foresights about these programs. In this section, we particularly focused on illustrating the results that responding our major research questions. In this section, we particularly demonstrated that conducted five specific questions in the questionnaire set on the perception issue (see Fig. 2) and listed here:

> *"11. Ningbo has done a lot of works on urban river restoration under the 5 water management plan, are you familiar of this policy/what are the 5 water management about?*
>
> *a. Not at all, b. little, c. modest, d. fully understood*
>
> *12. Do you think it is useful to promote urban river restoration in Ningbo?*
>
> *a. Not at all, b. a little, c. modest, d. very useful*
>
> *13. Do you think there are any connections between Five Water Management Plan with sponge city projects?*
>
> *a. Not at all, b. a little, c. modest, d. a lot of connections*
>
> *14. Do you think more areas in Ningbo other than East town (Jiangdong) and North district (Jiangbei) need to be undertaken sponge city and urban river restoration projects?*
>
> *a. No, b. Yes*
>
> *15. Do you think current awareness and participation are enough to public for these projects?*
>
> *a. Not at all, b. a little, c. modest, d. enough"*

We have found the results (see Fig. 2) demonstrated the communities were not too familiar about the urban river restoration under Five Water Management Plan and urban river restoration (Q11) as 33.64% participants responded unfamiliar and another 44.55% respondent expressed with little understanding on the Five Water Management Plan.

Also, we found that the communities also have rather conservative viewpoints on the usefulness of promoting urban river restoration in Ningbo (Q12) as only at about 9% participants expressed that promoting urban river restoration in Ningbo was very useful, but majority were expressed not useful or only little useful. Interestingly, we also found that majority of participants found no connection (3.64%) or only with little connection (about 27%) between Five Water Management Plan with SCP in Ningbo (Q13).

Participants believed it was not too useful to promote the urban river restoration and expressed only a little connection between Five Water Management Plan, urban river restoration and SCP as according to the analysis of the connection among these practices (Q12 and Q13) (see Fig. 3), which reflected the public perception to the functions and benefits in urban ecology, water management and urban flood risk management and climate change adaption with these three main strategies from the data set in 2017 were yet understood the Science and technical aspects well. That findings were implied by Q14 and Q15, which majority of community residents expressed that the SCP and urban river restoration should still be promoted and continuously expanded Five Water Management should still be undertaken (over 65%), but majority of participants admitted that their own awareness and participation were not enough (with about total 58% expressed not at all or a little awareness and participation) for these practices. That reflected the awareness and participation issues might influence the perceptions on the urban water management issues in Ningbo (include river restoration, Five Water Management and SCP) from the 2017 questionnaire data set.

For such findings on public perceptions, we justified with the 2019 interview data with the local community and found that there were some evident afterwards. For some reasons, such as lacking of education and promotion, especially to elderlies, an elderly interviewee living around the Yinzhou Park responded (interviewee A): "…*I have not heard too much on these programs, because I am getting old at this age, do not pay attention*…" Another elderly interviewee (interviewee H) also responded: "…*read and understood it (these programs/strategies) via the newspaper but can't remember it clearly, found these are helpful policies*".

Furthermore, we interviewed some interviewees who understood these schemes via social media channels, such as via the mobile apps that the elderlies might found their difficulties to adapt the implications from technology. Two students expressed (interviewee F, resided at South Business district close to Yinzhou Park) "*Heard it in*

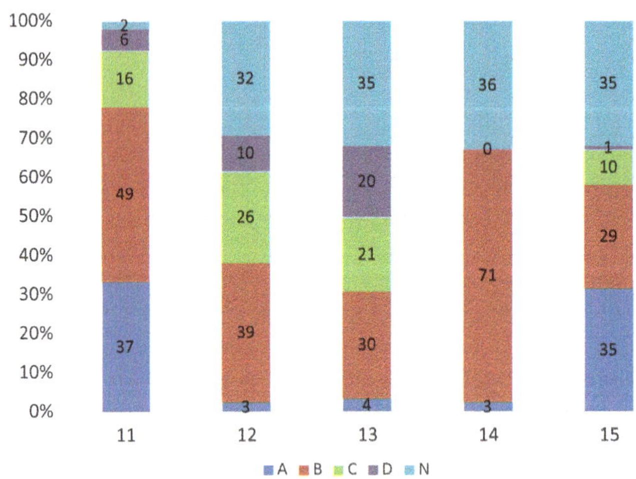


**Fig. 2.** Public perception on urban river restoration, Sponge City Program and Five Water Management (note perception on these questions refer as above: Blue: Not at all (for Q11–13, 15)/Yes (Q14), Orange: a little (for Q11–13, 15)/No (Q14), Grey: Modest (Q11–13, 15), Yellow: fully understood (Q11)/very useful (Q12)/lots of connection (Q13)/enough (Q15), Green: Answer not valid/incomplete/not answered

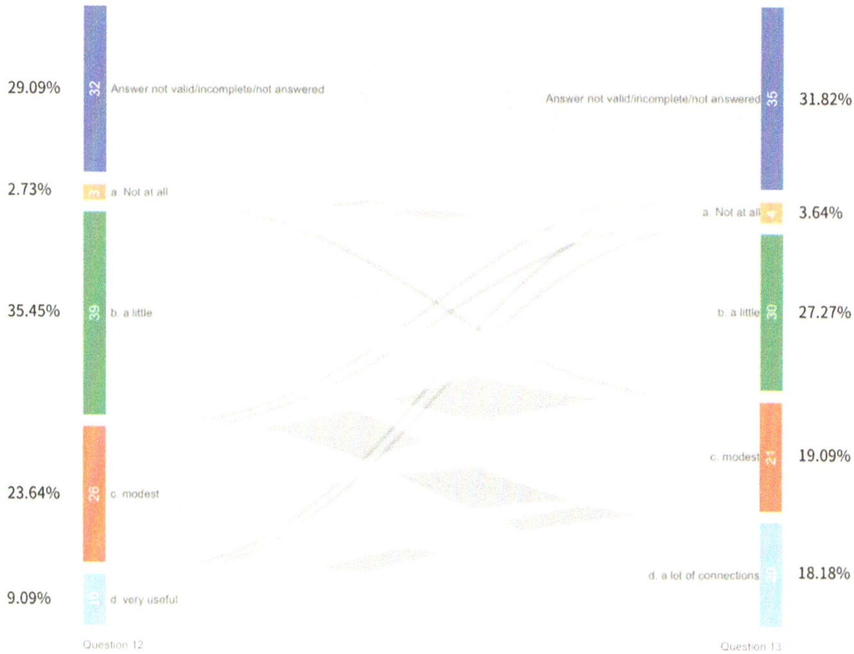


**Fig. 3** Sankey diagram with Q12 and Q13 on the relationship about the urban river restoration (Q12) and Five Water Management and SCP (Q13)

*the car radio and received relevant publicity via SMS.*"; another student (interviewee J, living around the Tianyi Square close to Sanjiangkou Park) expressed that "…*Saw (these programs) on app news*…" Also, for the work professional age that a flower-seller operated the shops at a shopping mall around the Yinzhou Park (interviewee B) expressed that: "…*saw related information on mobile phone text messages*…"; another interviewee working as a primary school teacher (interviewee E, Ci Cheng Park) expressed he understood about these urban water management strategies via television channels "…*understanding (these programs) from TV news, three out of five water management are remembered: sewage, drainage water and flood.*"

From these justifications in the 2019 interviews, we can understand that the Municipal Government has actually used and adopted multiple channels on these urban water management strategies by TV, newspaper and lately via mobile apps, internet sources for enhancing promotion and improvement on public awareness and participation.

However, the government may take more attention for the elderly group as they normally have difficulties to access media source and similar finding on the climate risk information from typhoon in Tai O town, Hong Kong that the elderly people normally difficult to access technology [24], thus the perception of these programs are understandably lower than the other age groups. We recommend the government may engage with the communities in several age groups for gathering them for some

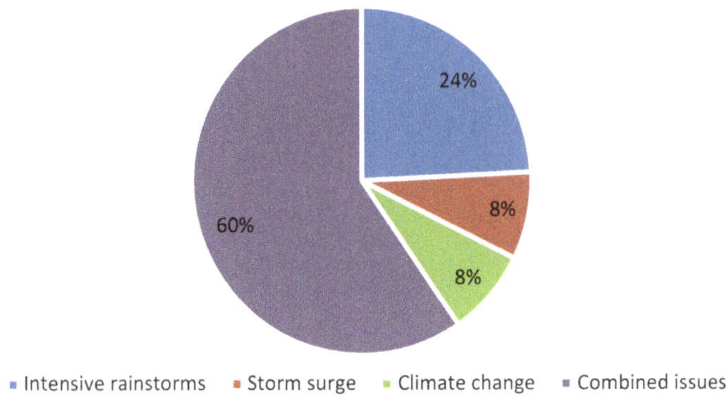


**Fig. 4** The percentages of participants understanding on the causes of urban floods in Ningbo

local education activities or community engagement activities to enhance further engagement.

On the other hand, we also found that the public understand the cause and consequences of urban floods quite well as that reflected from Q8 here:

*"8. Why urban flooding is occurring in Ningbo?*

*a. Intensive rainstorms, b. storm surge, c. climate change, d. combined issues"*

Over half of respondents (60%) understood that the urban flood issues in Ningbo are the combined issues that connected with the climate change, storm surges from typhoons and cyclonic effects (as Ningbo frequently have typhoons from West Pacific during August to October) and intensive rainstorms that reflected the public have good scientific understanding about the meteorological effects on flooding (see Fig. 4).

## *4.3 Future Foresight*

**Sustainability issues on social, economic, ecology and wellbeing**

*16. Do you think the sponge city and urban river restoration projects are able to improve the living conditions/quality?*

*a. Not at all, b. a little, c. modest, d. very helpful*

*17. Do you think the sponge city and urban river restoration projects are able to enhance greening to urban spaces and improve biodiversity?*

*a. Not at all, b. a little, c. modest, d. very helpful*

*18. Do you think there are any economic benefits from these projects, for example the reflections from property prices?*

*a. Not at all, b. a little, c. modest, d. very relevant*

*19. Do you think you feel happier and living with wellbeing with urban green and sponge infrastructures (e.g. swale, rain garden, green roofs, etc.) and restored river?*

*a. Not at all, b. a little, c. modest, d. very happy*

*20. Will you be satisfied if the municipal bureaus continue to implement the sponge city and river restoration practices alongside with current urban planning, water resources and flood management plans in Ningbo?*

*a. Not at all, b. a little, c. modest, d. very satisfied*

In the section of 'Sustainability issues on social, economic, ecology and well-being' (Q16 to Q20) (see Fig. 5), it can be found that majority participants perceived 'a little' as the effect level of SCP and river restoration projects, while 'not at all' and 'very helpful/happy' were considered by the least participants. Over 60% participants believed such projects have 'a little' and 'modest' values in terms of the social benefits (Q16. 70% (77 people) improvement of the living quality; Q19. 62.7% (69 people) feeling happier and living with wellbeing), ecology benefits (Q17. 64.5% (71 people) enhancing of greening to urban spaces and improving of biodiversity), and economic benefits (Q18. 70.9% (78 people) increasing property prices). Regarding the satisfaction on the municipal bureaus continuing to implement the sponge city and river restoration practices alongside with current urban planning, water resources and flood management plans in Ningbo, 13.6% participants (15 people) felt unsatisfied on it compared to fewer percentages of other four questions, which might be a reminder for relevant municipal bureaus when designing and implementing such projects to consider public perceptions and opinions. That is vitally important for the current and future urban planning and related to the local communities whether they really engage with the SCP and Five Water Management Plan and relevant programs in Ningbo. It reflected not only their perception and understanding of the functionality of these infrastructures and practices whether are beneficial to their daily life and well-being, but also the level of understanding to these water programs, which might help the Municipal Government to promote the programs further extensively to wider communities and districts in the city. Their early involvement (awareness and perception) will lead to stronger public engagement and participation, deliver of successful outcomes, and help stakeholders to make decisions that based on the needs from communities.

Specifically for SCP, we also found that despite the public not being very aware and engaged with these urban water management practices from Sect. 4.2, interestingly the communities that live nearby the SCP sites still support that future SPC projects should be continuously carrying out in Ningbo and other Chinese cities at Q10, see here:

*"10. Do You Think More Future Sponge City Projects Should Be Continue in Ningbo and Chinese Cities?"*

We found that there were over 95% respondents expressed the full support that the SCP should be carried on in Ningbo and Chinese cities and interestingly as we noted that majority of respondents awareness and perception remained at the inadequate level, but witnessed that the public and communities were actually understanding quite well about the cause of floods and situation as discussed in Sect. 4.2.

**Future foresights, climate change and public participation/engagement**

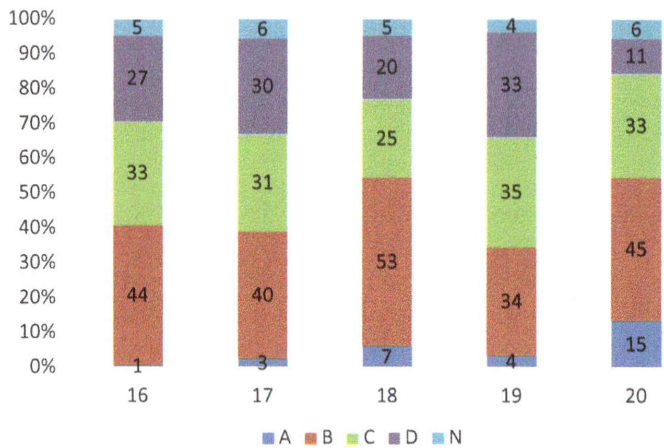


**Fig. 5.** Sustainability issues on social, economic, ecology and wellbeing. The number labelled in the histogram is the participant number of each question option (the total number is 110)

*21. Will you support the municipal government continue to finance the sponge city and relevant projects including urban river restoration projects?*

*a. Not at all, b. a little, c. support, d. very supportive*

*22. Will you continue supporting 5 water management plan in Ningbo?*

*a. Not at all, b. a little, c. support, d. very supportive*

*23. Do you think joint venture or Public Private Partnership (PPP) will be a better choice/more economic sustainable to finance these projects?*

*a. Not at all, b. a little, c. useful, d. very useful*

*24. Do you think Sponge city project can address the climate change?*

*a. Not at all, b. a little, c. useful, d. very useful*

*25. Do you think public participation is enough on sponge city and 5 water management plan projects at the moment?*

*a. Not at all, b. a little, c. modest, d. enough*

In this section (see Fig. 6), 85% participants (94 people) supported the Five Water Management Plan (with 53% was very supportive) (Q22), which was similar to the support level of SCP with over 86% (95 people) (Q10). The continued financial support by the municipal government on SCP and relevant projects such as river restoration was supported by 85% participants (93 people) (Q21). Specifically, PPP for such projects was considered by 76% participants (40% useful and 25% very useful) as an economic sustainable mechanisms (Q23).

The lack of public participation in SCP and Five Water Management projects was expressed by 29.1% participants (32 people) with ‘not at all’ and 39.1% (43 people) ‘a little’ (Q25). Regarding the function of SCP on the climate change (Q24), over half participants perceived a weak link between them that 10% participants (11 people) have offered the answer of ‘not at all’ and 40.9% (45 people) of ‘a little’.

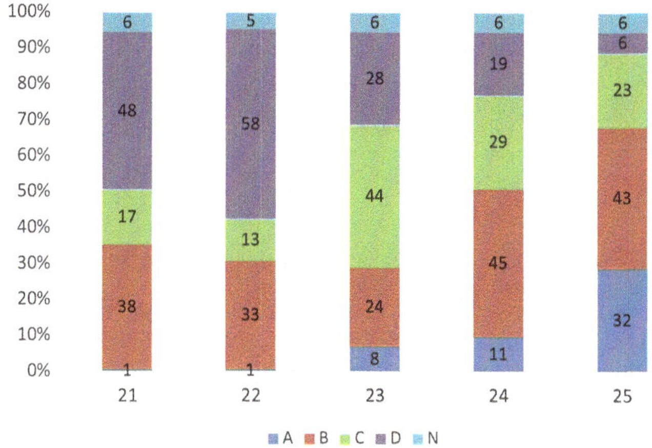


**Fig. 6.** Future foresights, climate change and public participation/engagement. The number labelled in the histogram is the participant number of each question option (the total number is 110)

## 5 Conclusion

Ningbo Municipal Government has been witnessed and evidently spent tremendous efforts on improving urban water issues. The government has been addressing these issues by implementing SCP, Five Water Management and undertaking the large scale urban river restoration schemes during the last few years, despite the time needed to see the performance and effects based on the regular monitoring and review. The positive effects at this early stage is reflected in the support of these schemes in 2017 from the communities and public. These showed the current the limitation of reflecting the latest situation and phenomenon of these schemes as the data has been finalised after only 3 year of implementation.

Public perception can be improved by using several channels and adopt "*fit-for-purpose*" approach such as looking after different age groups' understanding and learning of these strategies (i.e. elderly and other age groups). These lessons will be vitally important for other cities in Zhejiang Province to improve the urban water management strategies holistically and extensively providing good lessons for other Chinese cities or even Asian cities in the region that faced potential water risks from climate change and rapid urbanisation factors.

**Acknowledgement** This work was supported by the Research grant on the Strategy and Practice of the Development of Sponge City in Ningbo New Eastern Town funded by the Ningbo Housing and Urban-Rural Development Bureau. We would like to send our gratitude for the support from the National Natural Science Foundation of China (NSFC) (Grant numbers: 41850410497); the National Key R&D Program of China (Grant Number: 2019YFC1510400); and the Faculty of Science and Engineering (FoSE) Postgraduate Research Scholarship of University of Nottingham Ningbo China. Lastly,we also appreciated the research assistants Shuyang Xu, Liuge Yu, Wanying

Gu, Ming Yuan, Rui Jiang, Yiming Wang for their support in 2017 summer data collection and GEOG 3042 module students in 2019 Winter data collection.

## References


1. Fu, C., et al.: Freshwater fish biodiversity in the Yangtze River basin of China: Patterns, threats and conservation. Biodivers. Conserv. **12**(8), 1649–1685 (2003)
2. Tang, Y.T., Chan, F.K.S., Griffiths, J.: City profile: Ningbo. Cities **42**, 97–108 (2015)
3. Mitchell, G.: Mapping hazard from urban non-point pollution: A screening model to support sustainable urban drainage planning. J. Environ. Manage. **74**(1), 1–9 (2005)
4. Tseng, C.P., Penning-Rowsell, E.C.: Micro-political and related barriers to stakeholder engagement in flood risk management. Geograph. J. **178**(3), 253–269 (2012)
5. Tang, Y.-T., et al.: Aligning ancient and modern approaches to sustainable urban water management in China: Ningbo as a "Blue-Green City" in the "Sponge City" campaign. J. Flood Risk Manage. **11**(4), e12451 (2018)
6. Wang, Y., Sun, M., Song, B.: Public perceptions of and willingness to pay for sponge city initiatives in China. Resour. Conserv. Recycl. **122**, 11–20 (2017)
7. Ningbo Daily, Regulations of Ningbo Municipality on the Administration of River Channels. https://gtog.ningbo.gov.cn/art/2019/9/14/art_389_1021227.html. Last accessed 26 July 2020
8. Yu, L.: Study of 'Five Water Co-governance' from the Perspective of Multiple Governance: a Case Study of Yuhang District of Hangzhou. Northwest University (2018)
9. Ningbo Daily, Ningbo Water Control Pilot Trial Results and by 2020 Plans to Arrange 3064 Projects. https://zt.cnnb.com.cn/system/2019/11/28/030105097.shtml. Last accessed 29 July 2020
10. China Ningbo Network, Membrane Technology is used to Treat Domestic Sewage and How to Use the Reclaimed Water after the Tender. https://zt.cnnb.com.cn/system/2019/10/21/030094662.shtml. Last accessed 30 July 2020
11. China Ningbo Network, Control of Small Urban Lakes and Yuehu Lake Has Words. https://zt.cnnb.com.cn/system/2019/10/24/030095664.shtml. Last accessed 30 July 2020
12. China's River Chiefs—An "Ingenious" Approach to Water Governance in China? International Rivers Organization. Available at: https://www.internationalrivers.org/blogs/1126/china%E2%80%99s-river-chiefs-an-%E2%80%9Cingenious%E2%80%9D-approach-to-water-governance-in-china. Last accessed 14 Aug 2020
13. Wang, Y., Chen, X.: River chief system as a collaborative water governance approach in China. Int. J. Water Resour. Dev. **36**(4), 610–630 (2020)
14. Pyke, C., et al.: Assessment of low impact development for managing stormwater with changing precipitation due to climate change. Landscape Urban Plan. **103**(2), 166–173 (2011)
15. Griffiths, J.A.: Sustainable Urban Drainage, in Reference Module in Earth Systems and Environmental Sciences. Elsevier (2016)
16. Thorne, C.R., et al.: Overcoming uncertainty and barriers to adoption of Blue-Green Infrastructure for urban flood risk management. J. Flood Risk Manage. n/a-n/a (2015)
17. Morison, P.J., Brown, R.R.: Understanding the nature of publics and local policy commitment to Water Sensitive Urban Design. Landscape Urban Plan. **99**(2), 83–92 (2011)
18. Voyde, E., Fassman, E., Simcock, R.: Hydrology of an extensive living roof under sub-tropical climate conditions in Auckland, New Zealand. J. Hydrol. **394**(3–4), 384–395 (2010)
19. van Roon, M.: Water localisation and reclamation: Steps towards low impact urban design and development. J. Environ. Manage. **83**(4), 437–447 (2007)
20. Everett, G., et al.: Sustainable drainage systems: helping people live with water. In: Proceedings of the ICE-Water Management (2015)

21. Qin, H.P., Li, Z.X., Fu, G.: The effects of low impact development on urban flooding under different rainfall characteristics. J. Environ. Manage. **129**, 577–585 (2013)
22. Xia, J., et al.: Opportunities and challenges of the Sponge City construction related to urban water issues in China. Science China Earth Sciences p. 1–7 (2017)
23. King, N., Horrocks, C.: Interviews in qualitative research (2010)
24. Lo, A.Y., et al.: Contested transformations: sustainable economic development and capacity for adapting to climate change. Ann. Am. Assoc. Geographers **110**(1), 223–241 (2020)

# Online Social Media—A Vehicle for City Branding in China: The Case of Sponge City Program (SCP)

**Dimple Thadani, Lei Li, and Faith Ka Shun Chan**

**Abstract** The shift of academic attention from city marketing to city branding heralds a new era of representation and signification of cities as brands. Conscious practices of nationwide program implementation could be used not only for improving urban resilience but for city brands promotion. In China, to cope with the frequent occurrence of urban water issues, a nationwide initiative called Sponge City Program (SCP) has been launched since 2013. Given the significant impact of online social media on brand image, various levels of the Chinese Governments have begun to use online social media for city program promotion. In particular, brand community formed by stakeholders on online social media can significantly influence the brand image by co-creating contents. In this paper, we first analyze the current application of online social media for city marketing in China based on the SCP implementation. Next, we propose a framework/typology to inform how Chinese' social media could be used for effective city branding. Finally, we conclude by discussing the implications of online social media use in China.

**Keywords** Sponge city program · City branding · Online social media · Urban water management

D. Thadani (✉)
Nottingham University Business School, University of Nottingham Ningbo China, Ningbo 315100, China
e-mail: Dimple.thadani@nottingham.edu.cn

L. Li · F. K. S. Chan (✉)
School of Geographical Sciences, Faculty of Science and Engineering, University of Nottingham Ningbo China, Ningbo 315100, China
e-mail: Faith.Chan@nottingham.edu.cn

F. K. S. Chan
School of Geography, University of Leeds, Leeds LS2 9JT, UK

Water@Leeds Research Institute, University of Leeds, Leeds LS2 9JT, UK



F. K. S. Chan et al. (eds.), *Proceedings of the 2020 International Conference on Resource Sustainability: Sustainable Urbanisation in the BRI Era (icRS Urbanisation 2020)*, Environmental Science and Engineering,
https://doi.org/10.1007/978-981-15-9605-6_26

# 1 Introduction

The rise of globalization and its consequent urban planning restructuration have been accompanied by cutthroat intercity competition [11, 12, 27]. The accent of financialization—capturing investments and capitals–has made city marketing and city branding integral strategies for competitiveness development and sustainable urban growth [8]. The shift of attention from city marketing to city branding heralds a new era of representation of cities as brands. In this perspective, "*city, as the object of branding, becomes of locus of value, embedded in social practices and situated in a spatial location*" [3] (p. 148). Positive city branding is regarded as an economic and social instrument to enhance city's welfare [13]. Although effective city branding is a prolonged management process, conscious practices of nationwide urban program implementation (e.g. utilizing appropriate communication channels) facilitates city brands promotion.

In China, to cope with the frequent occurrence of urban water issues (i.e. floods and drought), a nationwide initiative called Sponge City Program (SCP) has been launched since 2013. Community perception, acceptance and financial challenges are considered to be the strongest barriers against the adoption and implementation of SCP [21]. Public-Private Partnership (PPP) is one of the best ways to make the program viable [14, 37], which has greater opportunities to be achieved via city marketing and branding. Given the tremendous impact of online social media on brand image, various levels of the Chinese Governments have begun to use online social media for city program promotion. In particular, brand community formed by stakeholders on online social media could significantly influence the brand image by co-creating contents.

The goal of this article is twofold. First, we analyze the current application of online social media for city marketing in China based on the case of SCP implementation. Then, we propose a framework/typology to inform how Chinese social media could be used for effective city branding. Implications on online social media use in China are discussed.

# 2 Current Application of Online Social Media: The Case of SCP Implementation

## *2.1 City Marketing and City Branding*

Cutthroat competition among cities motivates city government to utilize marketing and branding strategies for city promotion. City marketing is a process of planning and promoting cities to satisfy the needs of city target markets [9, 17–19]. The main entities in city marketing implementation are local government, organizations and residents [4, 9, 25]. City brand is considered to be one of the essential city products [9, 17, 25, 38] that provides a conscious and reliable commitment to stakeholders (i.e. local government, residents and organizations) in promoting city images [31, 36, 38]. City branding can thus be considered as one of the city marketing strategies to strengthen long term relationships between a city and its stakeholders so as to establish good city image [17]. Specifically, city branding is process of identity building in which geographical, social, and cultural attributes are integrated [31, 36, 38], and thrives as a means for promoting competitive advantages [2]. Utilizing online social media for city marketing and city branding have been applied in various context [e.g. 23, 32, 34, 35] including nationwide program implementation.

## *2.2 Online Social Media*

With rapid propagation of online social media, webs/apps users including the government, organizations and city residents can easily create and disseminate media content. Hence, online social media plays an important role in shaping public perceptions and attitude [6], which impacts community acceptance and financial resources for nationwide program implementation. In particular, online social media is broadly defined as online technologies and platforms that allow users to connect, communicate, and interact with each other, often by posting, sharing, or co-producing information online [7]. Different typologies of online social media depending on their type and characteristics have been proposed by prior researchers [e.g. 20, 29]. The most widely used online social media typology was proposed by Kaplan and Haenlein [16] based on two dimensions: the degree of social presence/ media richness, and self-presentation/ self-disclosure. Table 1 summarizes six types of online social media with global examples.

Stockmann and Luo [30] classified Chinese cyberspace according to two dimensions—social interaction and information sharing, and suggested that patterns of public opinion are different for different types of online social media.

For instance, the *twitter-like Weibo* offers users the power to become the source of information, including individual, organizations and the government. *Baidu Tieba* is a BBS-like chat forum with an average of 50 million. New post per day. Indeed, the emergence of online media has provided the government with more opportunities to better communicate with the public. Previous studies have addressed the role of online media in terms of government transparency and interactivity [10, 26, 28]. It is believed social media could be an effective vehicle for city marketing and city branding [39].

**Table 1** Six types of online social media [16]

| Type of online social media | Descriptions | Examples |
|---|---|---|
| 1. Collaborative projects | Platforms that allow users to add, remove, and change text-based content | *Wikipedia* |
| 2. Blogs and micro-blogs | Platforms that come in a multitude of different variations, from personal diaries describing the author's life to summaries of all relevant information in one specific content area | *Twitter; Weibo* |
| 3. Content communities | Platforms that aim at sharing content among users including news site | *Youtube; Youku* |
| 4. Social networking services | Platforms that aim at connecting users by creating personal profiles | *Facebook; WeChat* |
| 5. Virtual game worlds | Platforms that replicates three dimensional environment in which users can appear in the form of avatars or virtual characters with game rules restricting the range of interactions | *World of Warcraft* |
| 6. Virtual social worlds | Platforms that replicates three dimensional environment in which users can appear in the form of avatars or virtual characters, but no rules restricting the range of possible interactions | *Second Life* |

## *2.3 Online Social Media for City Marketing in the Current SCP Implementation*

In the past, the only active entity in city marketing and branding is the city government [25]. City program promotion was often done via traditional media such as TV advertisement and magazines. These traditional promotion approaches are high-cost in operation and lack feedback from the audience. Effective city marketing implementation needs multi-directional interactions between the city government, organizations and city residents.

The contemporary online social media channel provides a bilateral beneficial vehicle for mutual communication between city's stakeholders because of its participative, interactive and ubiquitous nature [23, 32, 34, 35]. Given the benefits of online social media, various levels of the Chinese Governments have begun to utilize online social media for city program promotion. In this article, we are focusing on the use of social media for one of the nationally significant program of urban water management– The *Sponge City Program* (SCP).

In China, a nationwide initiative namely, *Sponge City Program* (SCP), intended to improve urban water management [5]. Rapid urbanization, socio-economic development and population growth coupled with global climate change, have recently caused severe urban floods and other water issues in Chinese cities [15]. In response, the SCP strategy was put forward by the Chinese National Government in 2013, intended to increase urban resilience in adapting to environmental changes and dealing with disasters caused by water. SCP integrates a new urban planning and development mode [5, 22], which envisions Chinese cities cope with sustainable urban water systems that operates like a sponge to absorb, store, purify and release urban stormwater spontaneously [24].

## 2.4 The Current Online Social Media Use for SCP Implementation in China

To understand the current application of online social media for nationwide program promotion in China, we have done a systematic search by identifying SCP related articles published in online social media between 2013 and 2019. We followed the guidelines provided by Webster and Watson [33] to identify relevant articles. Given the variety of terminology describing SCP, we based the articles search on a range of Chinese and English keywords including "*Sponge City*", "*Sponge City Program*", "*SCP*", "*海绵城市*" using *Baidu* and *Google* search engines. In addition, the top 5 most popularly international and Chinese online social media were manually searched.

The initial searches returned 1651 articles in total. Following the guidelines of the systematic review methodology, we applied inclusion and exclusion criteria to the initial set of articles. The inclusion criteria were (1) SCP was the main focus of investigation and comments; (2) the article was written in Chinese or English; (3) article published between 2013 and 2019. The exclusion criteria were applied to (1) articles in which SCP was referenced merely as the main topic; (2) duplicate or linguistically unusable articles; (3) photos or videos without SCP related hashtags. 890 articles were excluded from analysis because they failed to meet the criteria stipulated in the current research design. At the end of the review and screening process, 418 articles published between 2013 and 2019 were identified.

**Preliminary Findings**

418 SCP articles published on online social media between 2013 and 2019 were identified at the end of the review and screening process. Table 2 provides an overview of the number of SCP articles by year and online social media outlet categorized using Kaplan and Haenlein [16] typology.

Referring to Table 2, SCP articles first appeared in news website (i.e., *WeChat News*) in 2014, and majority of these articles (90%) were found to publish in news websites. Surprisingly, among the 407 published articles in the content communities, only 63 of them specified *City* information; and only 27 of these articles were involved in the interactivity activities (e.g. forwarded/liked/commented). SCP articles first published on video/audio community in 2015, and has been steadily increasing ever since.

**Table 2** Online social media outlets for SCP articles

| Online social media outlets | Total | 2019 | 2018 | 2017 | 2016 | 2015 | 2014 | 2013 |
|---|---|---|---|---|---|---|---|---|
| Content communities | | | | | | | | |
| – News website (e.g. *Sina News; Business insiders; Wechat News*) | 377 | 68 | 65 | 90 | 99 | 54 | 1 | 0 |
| – Dynamic video/audio community (e.g. *Tiktok, Youku, Voice tube 影音號*) | 30 | 13 | 10 | 6 | 1 | 0 | 0 | 0 |
| Blogs and micro-blogs (e.g. *Weibo*) | 9 | 3 | 4 | 2 | 0 | 0 | 0 | 0 |
| Social networking services (e.g. *WeChat*) | 2 | 2 | 0 | 0 | 0 | 0 | 0 | 0 |
| Total | 418 | 86 | 79 | 98 | 100 | 54 | 1 | 0 |

## 3 Framework of Chinese Social Media for City Branding

Although it is believed that online social media could be an effective vehicle for city marketing and city branding [39], our preliminary findings suggest that the current SCP implementation have not fully utilized the interactive and participative nature of online social media. Majority of SCP articles were published in static news websites with emphasizes on technical content which made them difficult to be commented. Among all these news websites, 95% of interactive actions such as forwarding, commenting and liking were done on Chinese SCP articles.

More importantly, there is a lack of two-way communications between the city government, organizations and city resident, which are considered to be essential components of city branding. Recognizing that interactivity, source of information, and media richness are important dimensions that could facilitate city branding, we propose the following framework by integrating typologies suggested by Kaplan and Haenlein [16] and Stockmann and Luo [30].

The first dimension relates to the degree to which individual users/organization/government users disseminate information on the social media platform. The second dimension relates to the degree to which the functions of the platform support two-way interactions between users. The third dimensions relates to the degree of media richness provided by the platform. The top 5 most popular Chinese social media platforms in 2019 [1] are mapped onto a three-dimensional construct of online social media as shown in Fig. 1.

Although city government has been using online media that enhances the openness and transparency of city marketing and branding, it remains a government-lead marketing approach that lacks interaction and participation from other stakeholders.

In addition, from our preliminary results, we found that Chinese SCP articles published on Chinese online social media received far more attention from the audiences than those published in non-Chinese social media. Thus, we believe that we should put more focus on utilizing these Chinese oriented social media platforms. Fig 2. Suggests a roadmap for future online social media use for SCP implementation.

To strengthen long term relationships between a city and its stakeholders, it is suggested that platforms with high media richness and high social interaction

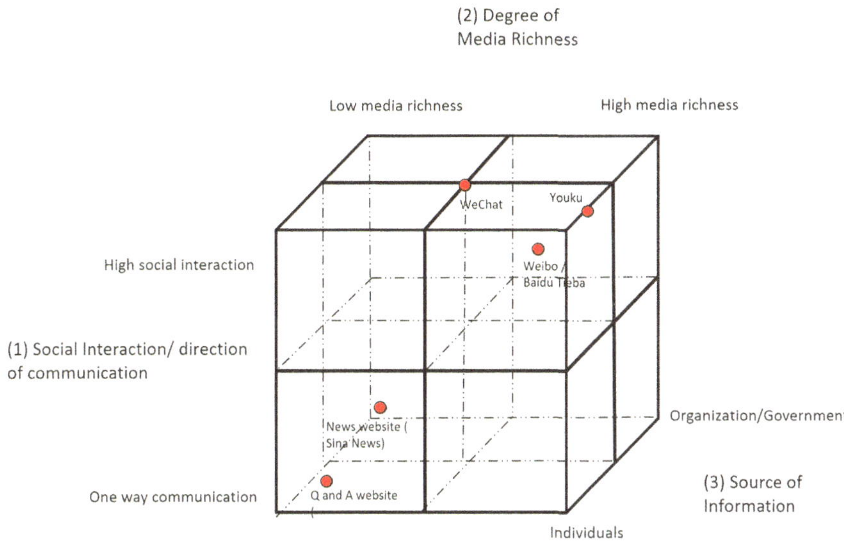


**Fig. 1** The three-dimensional mapping of various Chinese online social media

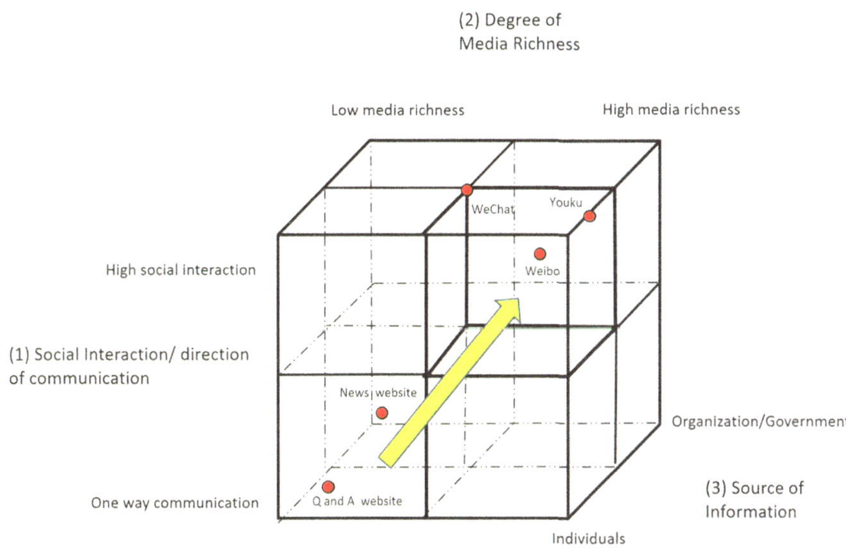


**Fig. 2** A roadmap for future online social media use for SCP implementation

such as *Youku, Weibo and WeChat* groups should be utilized. The participative and interactive, nature of these platforms serve as effective channel for city marketing, particularly for implementing a city branding strategy to build city uniqueness and commitment among its audience.

## 4 Conclusion

This article explores the current application of social media use in SCP implementation, and presents an integrated framework of online social media for city branding in China. Compared with traditional media, online social media provides a more participative, and interactive dialogue platform for city stakeholders to voice out their ideas.

Despite the fact that microblog and video platforms are very widespread social media categories in china, they have not been fully utilized for city marketing and branding in the case of SCP implementation, implying that government-lead city promotion is still the dominant marketing approach. Official government website and authoritative news website are fulfilling the purpose of information dissemination, the content is too technical for city stakeholders to engage.

Future studies should further explore the integration framework of both traditional and online social media in the aspects of city marketing and branding. It is also necessary to conduct a more robust empirical analysis on the mapping of the characteristics content to types of online social media platforms in China.

**Acknowledgement** The research presented in this paper was funded by Institute of Asia Pacific Studies (IAPS) research grant 2019/20; and the Faculty of Science and Engineering Postgraduate research grant 2018/21– University of Nottingham Ningbo China. We would like to send our gratitude for the support from the National Natural Science Foundation of China (NSFC) (Grant numbers: 41850410497); the National Key R&D Program of China (Grant Number: 2019YFC1510400).

## References

1. 10 Most Popular Social Media Sites in China (2019 Updated). https://topdigital.agency/10-most-popular-social-media-sites-in-china-2019-updated/ (2019). Accessed 9 May 2020
2. Ashworth, G., Voogd, H.: Selling the City: Marketing Approaches in Public Sector Urban Planning. Belhaven Press, London, UK (1990)
3. Bonakdar, A., Audirac, I.: City branding and the link to urban planning: theories, practices, and challenges. J. Plan. Literature **35**(2), 2020 (2020)
4. Cao, B.L.: Social media: definition, history of development, features and future—the ambiguous cognition of social media. J. Hunan Radio Television Univ. **47**(3), 47 (2011)
5. Chan, F., et al.: "Sponge City" in China—a breakthrough of planning and flood risk management in the urban context. Land Use Policy **76**(76), 772–228 (2018)
6. Cheung, M.K., Thadani, D.T.: The impact of electronic word-of-mouth communication: a literature analysis and integrative model. Decis. Support Syst. **54**(1), 461–470 (2012)
7. Correa, T., Hinsley, A.W., de Zúñiga, H.G.: Who interacts on the Web? The intersection of users' personality and social media use. Comput. Hum. Behav. **26**(2), 247–253 (2010)
8. Friedmann, J.: The world city hypothesis. Development Change **17**(1), 69–83 (1986)
9. Guo, G.Q., Liu, Y.P.: The new development of the research on city marketing. Contemp. Econ. Manage. **28**(2), 5–12 (2006)
10. Harris, J., McKenzie, K., Rentfro, R.: Building trust or mistrust: the search for performance reporting from a citizen's perspective. J. Govern. Finan. Manage. **58**(4), 40–46 (2009)
11. Harvey, D.: From Managerialism to entrepreneurialism: the transformation in urban governance in late capitalism. Human Geography **71**(1), 3–17 (1989)

12. Harvey, D.: A Brief History of Neoliberalism. Oxford University Press, Oxford, UK (2007)
13. Hernandez, J., Lopez, M.C.: Is there a role for informal settlements in branding cities? J. Place Manage. Development **4**(1) (2011)
14. Hodge, G.A., Greve, C.: Public–private partnerships: an international performance review. Public Administration Rev. **67**(3), 545–558 (2007)
15. Jiang, Y.: China's water scarcity. . J. Environ. Manage. **90**(11), 3185–3196 (2009)
16. Kaplan, A.M., Haenlein, M.: Users of the world, unite! The challenges and opportunities of social media. Bus. Horiz. **53**(1), 59–68 (2009)
17. Kavaratzis, M.: From city marketing to city branding: towards a theoretical framework for developing city brands. Place Branding **1**(1), 58–73 (2004)
18. Kavaratzis, M., Ashworth, G.J.: City branding: An effective assertion of identity or a transitory marketing trick? Tijdschrift voor Economische en Sociale Geografie **96**(5), 506–514 (2005)
19. Kotler, P., Asplund, C., Rein, I., Heider, D.: Marketing places Europe: attracting investments, industries, residents and visitors European cities, communities, regions and nations Pearson Education, London (1999)
20. Krishnamurthy, S., Dou, W.: Advertising with user-generated content: a framework and research agenda. J. Interactive Advertising **8**(2), 1–7 (2008)
21. Liang, X.: Integrated economic and financial analysis of China's Sponge City Program for water-resilient urban development. Sustainability **10**(3) (2018)
22. Liu, H., Jia, Y., Niu, C.: Sponge city" concept helps solve China's urban water problems. Environ. Earth Sci. **76**(14) (2017)
23. Mangold, W.G., Faulds, D.J.: Social media: The new hybrid element of the promotion mix. Bus. Horiz. **52**(1), 357–365 (2009)
24. MHURD: Sponge city construction in Xiamen has been showing results. https://www.mohurd.gov.cn/dfxx/201608/t20160816_228553.html (2016). Accessed 5 Sept 2020
25. Qiu, Q.: Role analysis in city marketing. https://www.paper.edu.cn (2005). Accessed 9 Aug 2020
26. Sadeghi, L. (ed.) Web 2.0. CRC Press, Boca Raton, FL (2012)
27. Sassen, S.: Global Networks Linked Cities. . Routledge, New York (2016)
28. Searson, E.M., Johnson, M.A.: Transparency laws and interactive public relations: an analysis of Latin American government Web sites. Public Relations Rev. **36**(2), 120–126 (2010)
29. Shao, G.: Understanding the appeal of user-generated media: a uses and gratification perspective. Internet Res. (19), 1 (2009)
30. Stockmann, D., Luo, T.: Which social media facilitate online public opinion in China? Probl. Post-Communism **64**(3–4), 189–202 (2017)
31. Trueman, M.M., Klemm, M., Giroud, A., Lindley, T.: Bradford in the premier league? A multidisciplinary approach to branding and re-positioning a city. In: Working Paper 01/04. Bradford University School of Management, Bradford, UK, (2001)
32. Wang, L.: Marketing in social media era. Management & Technology of SME 8, 202–203 (2012)
33. Webster, J., Watson, R.T.: Analyzing the past to prepare for the future: writing a literature review. MIS Q. **26**(2), 13–23 (2002)
34. Wigmo, J., Wikström, E.: Social media marketing: what role can social media play as a marketing tool? Linnaus University (2010)
35. Yu, X.: Analysis of brand communication strategies in social media era. The Press **4**, 122–133 (2011)
36. Zhang, L., Zhao, S.X.B.: City branding and the Olympic effect: A case study of Beijing. Cities **26**(5), 245–254 (2009)
37. Zhang, S., Zevenbergen, C., Rabé, P., Jiang, Y.: The influences of sponge city on property values in Wuhan, China. Water **10**(6) (2018)
38. Zhao, Q.S.: The interactive mechanism between city brand and city customers. Mod. Econ. Res. **9**, 36–39 (2010)
39. Zhou, L., Wang, T.: Social media: A new vehicle for city marketing in China. Cities **37**(C), 27–32 (2014)

# The Development and Emerging Trend of Closed-Loop Supply Chain Research in China: Visual Analysis of CSSCI Documents of CNKI

Chunyan Zhu and Xiaochen Wang

**Abstract** In recent years, under the dual pressure of ecological environment and sustainable development, the reform of closed-loop supply chain has been paid more and more attention. The closed-loop supply chain provides guarantee for the development of circular economy and the enhancement of ecological environment sustainability. The research on the closed-loop supply chain has become a hot issue widely concerned by domestic scholars. In order to clearly understand the research status and hot spot distribution of closed-loop supply chain, the SSCI literature in CNKI database was taken as the research object, and the descriptive statistical analysis and cluster analysis were carried out by CiteSpace. This paper summarizes the evolution process, research context and other information of closed-loop supply chain research, and uses the visualization map to mine the latest research trends and trends of closed-loop supply chain, trying to provide a reference for the construction of macro knowledge map in the field of closed-loop supply chain, and provide a basic understanding for domestic and foreign scholars on the research of closed-loop supply chain.

**Keywords** Closed loop supply chain · Sustainable development · Visual analysis

## 1 Introduction

The supply chain refers to a functional network chain structure around the core enterprise, starting from each required accessory parts, and then making intermediate products and final products by the manufacturer, and finally sending the products to the final consumers through the whole sales network. Due to multiple pressures, such as continuous deterioration of the environment, shortage of resources and restrictions of laws and regulations, the term "green" was put forward in the "guiding opinions of the general office of the State Council on actively promoting the innovation and application of supply chain" issued in 2017, which reminds the development of

C. Zhu · X. Wang (✉)
School of Management, University of Science and Technology, Xi'an 710054, China
e-mail: wxccc0918@163.com



F. K. S. Chan et al. (eds.), *Proceedings of the 2020 International Conference on Resource Sustainability: Sustainable Urbanisation in the BRI Era (icRS Urbanisation 2020)*, Environmental Science and Engineering,
https://doi.org/10.1007/978-981-15-9605-6_27

supply chain needs reform. On the basis of the traditional supply chain, a series of operation links such as detection recovery, feedback reprocessing and redistributing are added to form a nearly closed ring structure, which can close the flow of materials, realize product recovery and effective management of product life cycle, reduce production waste and reduce the adverse impact of supply chain activities on the environment-CLSC.

In 1999, Qunyan [1] mentioned green supply chain for the first time in this paper, and thought that ecological environment management should be paid attention to; in 2001, Hesselbach, a foreign scholar, first proposed the concept of closed-loop supply chain in the field of mechanical and electrical equipment logistics. Although the research on closed-loop supply chain started late, domestic scholars all think that ecological problems, pollution reduction and resource optimization should be considered in the process of logistics development. In order to accurately grasp the research and development status of closed-loop supply chain and the emerging trend in the future, this paper takes the literature in CNKI database as the research basis, uses CiteSpace 5.5.r2 to sort and analyze the literature of closed-loop supply chain, and understands the development process and the most active research direction of this field from multiple perspectives through the visual knowledge map, which is the research direction of this field Scholars find new directions and topics for reference.

## 2 Research Design

Based on CiteSpace, the research of closed-loop supply chain is mainly divided into three parts: literature search, data entry and visual analysis, which are finally represented in the form of atlas and text. The first is to determine the research content, and determine the research topic for specific problems; the second is to determine the data, determine the data source according to the research topic, and store the processed data in the database; the third is to draw the visualization map, and make the map from the "table" and "inside" levels respectively, so as to deepen the understanding of the researchers on the visualization map; the fourth is the conclusion, and analyze the results of the visualization map The conclusion can be drawn and the feasible suggestions can be put forward.

### *2.1 Research Method*

Bibliometrics is a kind of interdisciplinary science, which integrates mathematics, statistics and Philology, and is a comprehensive method for quantitative analysis of knowledge carrier. Another popular method in academic circles is visualization chart analysis, which mainly reflects the importance of all kinds of information in the research through the size of nodes in the graph, and reflects the relationship between nodes by using text clustering, so as to obtain valuable information by studying core journals, so as to grasp the research frontier and direction. In CiteSpace, descriptive statistical analysis and cluster analysis are usually used to study the development status and future trend of a certain field. Based on this, this paper uses CiteSpace software system to conduct bibliometric analysis in the field of closed-loop supply chain, aiming to show the status quo, hot spots and research trends of the research through data analysis and information mining.

### *2.2 Data Sources*

The data of this paper comes from CNKI full-text database, which was collected on February 20, 2020. In CNKI database, theme retrieval is used to give consideration to both precision and recall. The retrieval conditions are "theme: closed-loop supply chain" or "theme: green logistics". The literature type is limited to academic journals, and the source category is limited to CSSCI. The retrieval year span is 2000–2020 (data in 2020 is selected to February 19). A total of 713 journal literatures are retrieved. The reasons for choosing the retrieval scheme are: first, the number of Chinese journals in CNKI journal database is more comprehensive; two, the literature quality of CSSCI source journals is higher, which can better reflect the research status of domestic closed-loop supply chain.

## 3 Visual Analysis

In this paper, CiteSpace software visual analysis of closed-loop supply chain, explore the development process of closed-loop supply chain and future research hotspot, according to which the introduction of "Outside" and "Inside" to distinguish the two levels of this study. "Outside" mainly includes the analysis of the volume, key words, author and organization of the document; "Inside" mainly includes cluster analysis and trend analysis.

### *3.1 "Outside" of Visual Analysis of Closed-Loop Supply Chain*

**Energy analysis of closed-loop supply chain**. In order to understand the quantity and quality of literature in the field of closed-loop supply chain, this paper makes a statistical analysis of CSSCI publications and total publications of secondary technical journals in CNKI database. It is found that the total number of literatures in the related journals of closed-loop supply chain research has reached 5113 since 2000, and the overall publications have remained at a high level. After 2006, the number of publications has been greatly increased since 2007 at the same time, the popularization of closed-loop supply chain and green logistics concept has been greatly improved. The distribution of the number of sample documents in 2000–2020 is shown in Fig. 1. Through the analysis, the volume of papers issued by all journals is more consistent with the power function model, which can be expressed as follows:

$$y = \alpha x^{\beta} \tag{1}$$

In Formula (1), y is the cumulative number of documents; X is the publication year of documents; α and β are parameters.

During the research period, it can be seen that the total amount of documents issued approximately presents a power function growth (R2 = 0.56 reflects the fitting degree between the estimated value of the trend line and the corresponding actual data), which indicates that the research of closed-loop supply chain is still in the growth stage, and may appear a sharp increase trend with the continuous innovation of the research. It can be seen from the broken line chart that since 2000, the number of documents issued has grown in three stages, with 2004 and 2007 as the boundary: 2000–2004 is the initial stage of closed-loop supply chain research, and the number of documents issued is in a steady growth stage; 2004–2007 is the rapid growth stage of closed-loop supply chain research, with an average annual growth rate of 72.8%; 2007–2019 is the stable development stage of closed-loop supply chain research, reaching All 357. Although the number of CSSCI journals is not much, it is steadily increasing, with 59 papers per year in the past six years.

**Hot spot analysis**. Key words are words or professional terms extracted from the literature to directly reflect the theme of the literature and describe the key points of the article. They are the author's extraction and generalization of the information

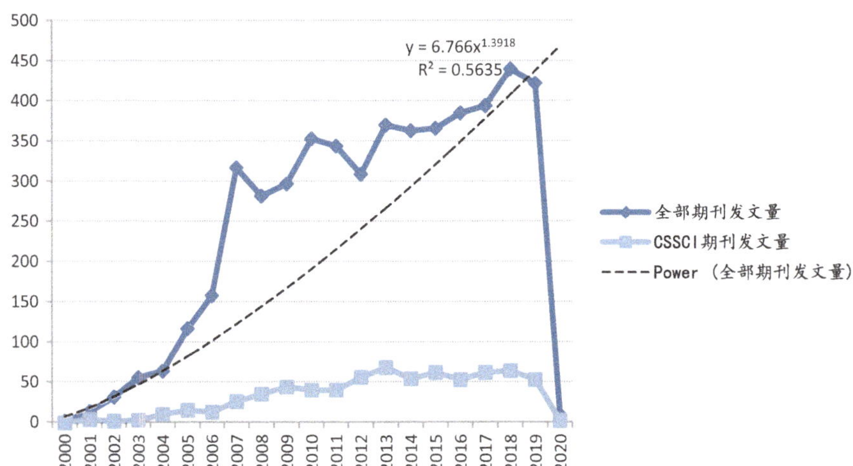


**Fig. 1** Distribution of research literature on closed-loop supply chain in 2000–2020

**Table 1** Top 10 high frequency keywords in sample literature in 2000–2020

| Order | Count | Centrality | Year | Keywords |
|---|---|---|---|---|
| 1 | 441 | 0.39 | 2000 | Closed loop supply chain |
| 2 | 102 | 0.34 | 2004 | Green Logistics |
| 3 | 100 | 0.40 | 2008 | Remanufacturing |
| 4 | 56 | 0.17 | 2006 | coordinate |
| 5 | 47 | 0.05 | 2000 | Consumer preferences |
| 6 | 39 | 0.19 | 2008 | Game theory |
| 7 | 34 | 0.07 | 2010 | Stackelberg game |
| 8 | 24 | 0.05 | 2011 | Closed loop supply chain network |
| 9 | 22 | 0.12 | 2005 | Reverse logistics |
| 10 | 22 | 0.05 | 2004 | sustainable development |

content of the article and reflect the theme and central idea of the article. The research on key words is helpful to infer the hot issues in this field. Therefore, this paper lists and sorts the top 10 keywords with the highest frequency in the sample literature, and understands and analyzes the centrality of these words, as shown in Table 1.

Combined with the frequency of keywords, the research focus of closed-loop supply chain mainly focuses on "closed-loop supply chain", "green logistics", "remanufacturing" and "coordination", and forms a central co-occurrence relationship structure with the analysis of high-frequency keywords as the core. Centrality is an important indicator to detect the influence of nodes in the network, and the nodes with high centrality are often the key hubs connecting various research fields. There are six nodes with centrality greater than 0.1, which indicates that the key words of "closed-loop supply chain", "green logistics", "remanufacturing", "coordination", "game theory" and "reverse logistics" play a key role and influence in this field, leading the research direction of closed-loop supply chain.

**Cooperation analysis of closed-loop supply chain**. The author is the main body of the article creation. By analyzing the author's network structure in the visualization chart, we can reveal the cooperation between the core authors in this field. As shown in Fig. 2, the node size represents the frequency of their occurrence, and the connection between nodes indicates the co-occurrence strength. In terms of frequency, the top four authors are Da Qingli, Li Bangyi, Nie Jiajia and Li Xinran, followed by SUN Hao, Gong Yande, Xie Jiaping and Liang Ling. From the perspective of cooperative network, the network structure of closed-loop supply chain research shows the characteristics of small concentration and large dispersion. It can be seen that in the process of research in this field, the author spontaneously formed his own core research team, such as the team led by Da Qingli, SUN Hao, Huang Zuqing, Wang Wenbin, etc., and made achievements in the competition, remanufacturing, recovery rate and fair preference of closed-loop supply chain The gongyande team, led by Li Bangyi, has its own views on the supply chain fund subsidy, government subsidy and recycling channels. Meanwhile, gongyande serves as a cooperation bridge between

**Fig. 2** Cooperation network of authors of sample documents in 2000–2020

the two teams to form a new author cooperation network. However, at present, although each team has made some achievements in the specific fields of closed-loop supply chain, due to the great differences in research directions between each team, the overall strength of the connection is weak, and the research atmosphere of integration and cooperation between scholars is not strong enough.

By using CiteSpace 5.5.r2 software and taking the issuing organization of sample literature as the node type, the cooperative network analysis (as shown in Fig. 3) is carried out to understand the research status and actual contribution of the issuing organization to the field of closed-loop logistics. As can be seen from Fig. 3, the institutions with a large number of papers are Southeast University, Southwest Jiaotong University, Nanjing University of Aeronautics and Astronautics, Dalian University of technology, etc., which form the main institutions of closed-loop supply chain research in China. Among all the publishing institutions, China University of mining and technology, Dalian University of technology and other universities led by Southeast University have a close communication and cooperation relationship, forming a cooperation network. At the same time, from the cluster diagram, we can find that most of the cooperative institutions are located in the same region, such as Huazhong

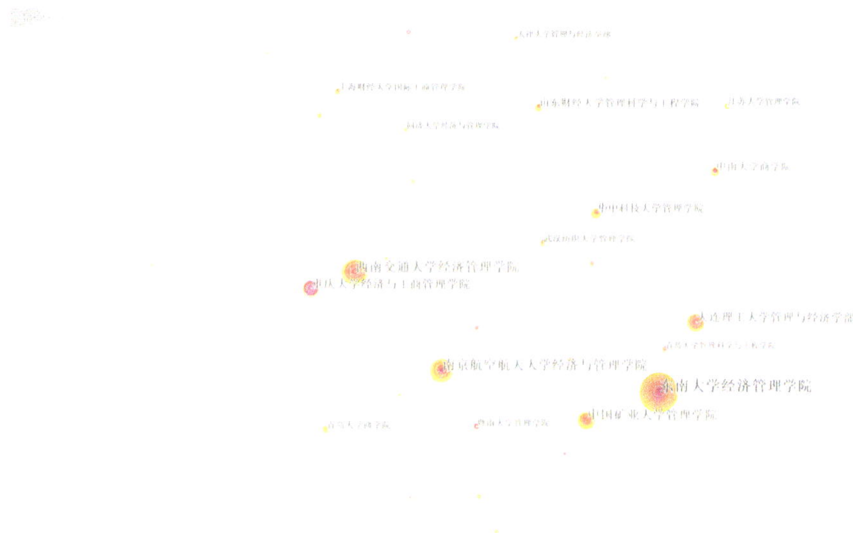


**Fig. 3** Network of authors' cooperative organizations of sample documents in 2000–2020

University of science and technology and Wuhan Textile University, Southwest Jiaotong University and Chongqing University, and there are few cross province cooperative research institutions. It can also be seen from Fig. 3 that among all the document issuing institutions, the major universities dominate the document issuing, which shows that universities play an important role in the research of closed-loop supply chain with their excellent research capabilities. When analyzing the key issuing organizations, we found that most of them are coastal areas along the river, which is consistent with the rapid development of domestic logistics. Coastal areas along the river have convenient transportation, suitable environment and dense population distribution, which are the best areas for the development of logistics supply chain, occupying a unique geographical position.

## 3.2 *"Inside" of Visual Analysis of Closed-Loop Supply Chain*

**Cluster analysis of closed-loop supply chain**. Key words are the core of the paper. There must be some relationship among the key words given in a paper, which can be expressed by the frequency of co-occurrence. By counting the frequency of the two subject words in a group of documents in the same document, a coinage network composed of these word pairs can be formed. As shown in Fig. 4, it is a keyword co-occurrence network generated by CiteSpace. The clustering view focuses on the structural features and key nodes between clusters. First of all, we can see from the parameters in the upper left corner that the number of nodes in the network is 133, the number of sides is 375, and the network density is 0.0427. The number of nodes

**Fig. 4** Clustering network of closed-loop supply chain research literature in 2000–2020

is the number of keywords in the graph, and the number of sides is the number of connections between keywords. As long as the keywords appear in the same literature, a line will appear between two nodes. The node size represents the frequency of keywords. The larger the node, the more indirectly reflects the research front content. Cluster analysis uses statistics, taking the frequency of coinage as the research object, and simplifies the network relationship of coinage into the relationship between several groups, which can directly represent the clustering process. As shown in Fig. 4, there are seven categories of clustering network, which represent the seven main directions of closed-loop supply chain research in China: #0 "coordination", #1 "green logistics", #2 "moral hazard", #3 "Stackelberg game", #4 "closed-loop supply chain network", #5 "remanufacturing" and #7 "supply chain stability". The clustering view gathers keywords and emergent words with the same or similar meanings into one category, It reveals the internal relationship between the papers. It can be seen from Fig. 4 that the research focus of domestic scholars on closed-loop supply chain is mainly to keep the overall coordination and stability of closed-loop supply chain while taking into account green logistics, so as to realize the green and healthy development of closed-loop supply chain.

Based on the above clustering keywords and related literature analysis, the following research directions can be obtained:

*Combined with green logistics to talk about closed-loop supply chain.* From Fig. 4, we can see that the words "green", "coordination", "recovery rate" occupy a very important research position. Today, with the economic development booming, the state attaches more and more importance to sustainable development. The government proposes the development plan of green logistics system: dominated by the government's legal system and administrative means, implemented by relevant economic means, extended producer responsibility The system is auxiliary. Through environmental legislation, pollution charge system, permit system and the establishment of green logistics standards to intervene in the external diseconomy of logistics activities [2]; through the construction of a perfect logistics information system to improve the utilization and economy of logistics resources, and comprehensively improve the level of enterprise information management [3]; promote the healthy development of green logistics by formulating green subsidy policies, tax support policies and other behaviors to encourage and guide the main body of logistics [4]; at the same time, vigorously develop the third-party logistics, further optimize the overall structure and industrial upgrading, which will help the enterprise to obtain greater profits and contribute to the sustainability of the ecological environment. To sum up, we should not only stabilize the development of closed-loop supply chain, but also focus on green sustainability to form a green and healthy closed-loop supply chain circle.

*Combined with the game theory model to discuss the closed-loop supply chain.* The incentive structure of the same research problem is formulated, and the method to study its interaction is called game theory. This is a kind of mathematical theory and method to study the phenomenon of struggle or competition. Game theory is mainly used to predict the behavior of the same thing and optimize its behavior strategy. It is often used to select the best in economics. For example [5], etc. proposed to consider the coordination mechanism of green supply chain under the condition of government incentive by building a game model. The research shows that the higher the green degree of closed-loop supply chain, the higher the government subsidy, and the higher the profit of each organization [6]. Established a game model to determine how manufacturers, retailers and third-party recycling models should conduct pricing and channel selection [7]. Used Stackelberg game model to select financing mode of closed-loop supply chain under the condition of capital constraints of recyclers. The research shows that under certain conditions, internal financing can achieve win-win between manufacturers and recyclers [8]. Studied the decision-making of transnational green supply chain by building a game model, and found that both production cost subsidy and green input cost subsidy can increase the economic and environmental benefits of the supply chain. Using game theory model, we can better use formula and data to analyze the strategic competition problems in the research of closed-loop supply chain, so as to get different solutions in different situations.

*Combined with information sharing to talk about closed-loop supply chain.* The real-time sharing of information is essential for the supply chain. Effective information sharing is conducive to the organic connection between supply and demand of enterprises, improve the response capacity of the supply chain, and at the same time, it is conducive to the accurate management of the supply chain, reduce logistics costs, improve resource utilization, and achieve reliable delivery of the supply chain. Therefore, there are [9] and other studies on the impact of information sharing on differential pricing under government regulation. The research shows that improving the level of government regulation in information sharing is conducive to improving the production process and recovery level of manufacturers [10] and other studies consider the impact of whether the retailer carries out predictive information sharing on the profits of each member of the closed-loop supply chain under the condition of government reward and punishment mechanism. The research shows that when the manufacturer's recovery efficiency is high, voluntary sharing of demand information can get the balance of the whole closed-loop supply chain [11]. Constructed a closed-loop supply chain information service system based on the Internet of things technology in order to realize resource conservation and promote the application of information in the closed-loop supply chain, exemplifying the importance of demand information sharing in the closed-loop supply chain.

To sum up, through cluster analysis and combined with specific literature, the advantages of closed-loop supply chain in different perspectives and the huge advantages it brings provide continuous momentum for the coordinated development of China's supply chain economy.

**Trend analysis of closed-loop supply chain**. CiteSpace can provide the rendering mode of timezone view, real-time area view. Time zone diagram focuses on the time dimension to represent the view of knowledge evolution, and clearly shows the update and interaction of literature. The amount of literature in a certain time zone reflects the development level of the field. The more literature, the more prosperous it is. On the contrary, the field is in a low period. As shown in Fig. 5, in the early stage of closed-loop supply chain research, the theme of closed-loop supply chain and consumer preferences occupied the main part of the research, and in the past 20 year, it has continuously influenced the trend of subsequent literature, almost covering all literature directions in this field; with the passage of time, sustainable development and green logistics were placed in the economic background at that time, which was generally recognized by researchers With the development of economy for many years, the coordinated development of closed-loop supply chain, multi interest game and risk control have become more and more important, followed by differential pricing strategy and government reward and punishment mechanism; there are

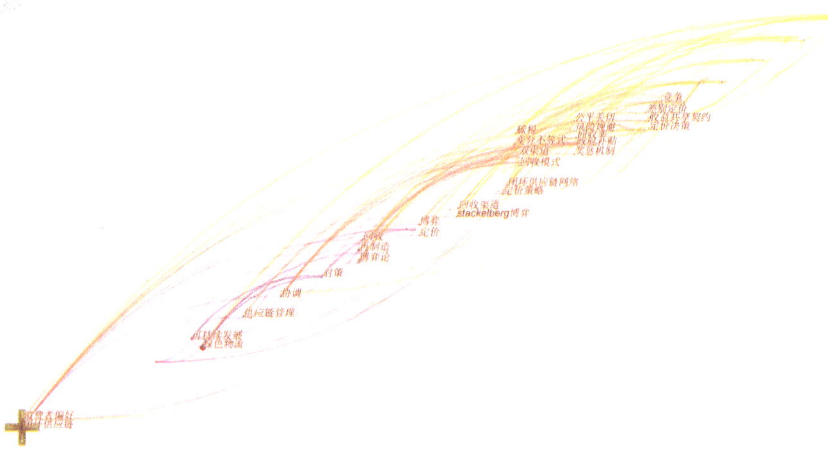


Fig. 5 Development trend of closed-loop supply chain research in 2000–2020

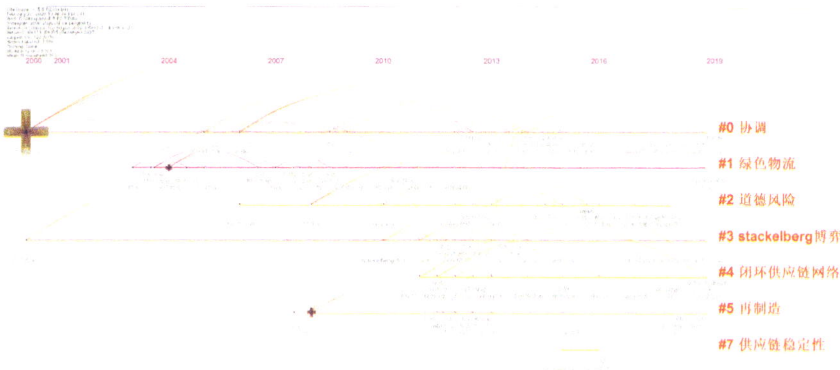


**Fig. 6** Timeline development of closed-loop supply chain research in 2000–2020

many connections between recycling and remanufacturing nodes and dual channel and government subsidy nodes, which shows the research in these two periods The hot spot has strong inheritance; in line with the development trend of the closed-loop supply chain, more and more literatures discuss the combination of income, risk, cooperation and competition, aiming to determine an overall balance point that conforms to the interests of multiple parties and achieves the balance of income of all parties in the closed-loop supply chain.

CiteSpace also provides the rendering mode of Timeline view, namely Timeline view. As shown in Fig. 6, the module value of clustering $q = 0.5101$ (generally, $Q > 0.3$ means significant clustering structure), and the average contour value of clustering silhouette $S = 0.766$ (it is generally believed that $s > 0.5$ clustering is reasonable and $s > 0.7$ clustering is convincing). We can do the following analysis: from 2000, coordination and Stackelberg game clustering began to appear, that is, the first reference of this clustering respectively; in 2004, green logistics clustering appeared and began to increase, and this focus entered the eyes of the writers, and in the process of the development of green logistics clustering, it is constantly cited Lead the development direction of other clusters, and connect with multiple clusters; around 2008, the topic of remanufacturing began to increase dramatically, and it was related to coordination, green logistics, Stackelberg game clustering, covering the research of multiple clusters, especially after 2018, the topic of remanufacturing became more and more intensive, which is likely to become the main focus of future literature development; in 2018, moral hazard clustering started The research focus in 2019 is mainly focused on Coordination clustering and closed-loop supply chain network clustering, which are also based on the previous results, and also have deep relationship with other clusters.

## 4 Conclusion

Environmental protection and green sustainability play an important role in all levels of production and transportation. While reducing the harm to the environment in the process of logistics operation, making full use of resources is the general trend of logistics development. The purpose of closed-loop supply chain management is to realize the harmonious unity of "economy and environment" and the maximization of benefits. This concept is conducive to the sustainable development of enterprises and the whole society. Improving the closed-loop supply chain system will bring great benefits to the ecological environment and economic development. The purpose of this study is to determine the evolution process and emerging development trend of closed-loop supply chain through the descriptive statistical analysis and cluster analysis in CiteSpace, so as to get the following conclusions.

The knowledge map represented by coordinated development, green logistics, remanufacturing and supply chain stability is forming, and the basic thought system of closed-loop supply chain research is gradually being constructed. Since the emergence of closed-loop supply chain in the field of vision of researchers in 2000, it has reached its first peak in 2007. Since then, it has maintained a high trend in the volume of publications, but the quality of research has not been significantly improved. The annual volume of publications on CSSCI is only 45–65, which is related to the lack of cooperation and communication between domestic scholars and research institutions and the lack of in-depth research.

At present, the focus of domestic researchers on closed-loop supply chain research is mainly to use game theory to analyze the benefits of each member of the supply chain, without giving specific operation steps. The research has limitations and lacks in depth and breadth. The specific links of closed-loop supply chain research need to be optimized and improved, which can be further explored in combination with the connotation and significance of closed-loop supply chain Dig.

In the future, the research trend of closed-loop supply chain is mainly reflected in market pricing and benefit distribution of supply chain members. The stable development of the closed-loop supply chain network is inseparable from the interest balance among the members and the overall sustainability of the supply chain. How to maintain the coordinated development of the whole closed-loop supply chain while maximizing its interests is the focus of future research.

The shortcomings of this paper and the problems that can be further explored are as follows: the range of keywords selected in cluster analysis is not wide enough, the retrieval of relevant literature is relatively limited, and future research will strengthen the exploration of relevant topics; because the data source is mainly Chinese literature, which can not involve foreign journals, it is unable to make a co citation analysis, and it is unable to give a representative point of view in this part. The future research will further involve the domestic and foreign literature, bring more information to the researchers and workers of closed-loop supply chain, and then effectively promote the development of closed-loop supply chain research.

## References


1. Qunyan, Z., Dequn, Z.: Analysis on environmental management behavior of modern enterprises. Shanghai Environ. Sci. **11**, 515–516 (1999)
2. Gang, L.: Analysis on the restrictive factors and realization path of low carbon logistics development in China. Econ. Res. Guide **08**, 45–46 (2015)
3. Shenghua, N.: On the development strategy of green logistics. Industrial Sci. Tribune **8**(08), 61–62 (2009)
4. Guochuan, Y.: Restrictive factors and Countermeasures in the development of green logistics in China. J. Bus. Econ. **02**, 18–23 (2010)
5. Xinlin, W., Shengqiang, H., Xiaobin, L. Game model and contract coordination of green supply chain considering government incentive policy. Industrial Eng. **22**(06), 17–26 (2019)
6. Yande, G., Yuwei, J.: Research on pricing and channel selection of hybrid recycling mode in closed loop supply chain. Soft Sci. **32**(05), 127–131+144 (2018)
7. Tianhui, Y., Chunyi, L., Bingbing, C.: Financing mode selection strategy of closed-loop supply chain with manufacturer led and recycler capital constraint. Chin. J. Manage. **17**(01), 139–147 (2020)
8. Mingwu, L., Mengyue, Q., Xiang, C.: Research on decision making of transnational green supply chain from the perspective of tariff and subsidy. Soft Sci. 33(09), 21–26 (2019)
9. Fang, L., Xin, M., Jia, H.: Influence of asymmetric information on differential pricing of closed-loop supply chain under government regulation. Chin. J. Manage. Sci. **27**(07), 116–126 (2019)
10. Pan, Z.: Research on demand forecasting information sharing in closed-loop supply chain with government rewards and punishments. Chin. J. Manage. Sci. **27**(02), 107–118 (2019)
11. Yang, G., Jian, Li.: Research on information service system of remanufacturing closed loop supply chain based on Internet of things. Sci. Technol. Progress Policy **31**(03), 19–25 (2014)

# Research on Carbon Emission Characteristics and Influencing Factors of Logistics Industry: Based on EKC Model and LMDI Decomposition Method

**Chunyan Zhu and Xu Guo**

**Abstract** Based on the energy consumption data of logistics industry in Shaanxi Province from 2009 to 2018, the carbon emission characteristics and influencing factors of logistics industry were studied. Firstly, the carbon emission coefficient method is used to measure the total carbon emissions of the logistics industry in Shaanxi Province, and it is found that the carbon emissions of the logistics industry have been fluctuating up and down in the past ten years. Then, the EKC model and LMDI factor decomposition method are used to analyze the carbon emission characteristics and influencing factors of logistics industry in Shaanxi Province, the results show that the relationship between the carbon emissions of the logistics industry and economic development is a general inverted U-shaped curve. Among the driving factors of carbon emissions of logistics industry in Shaanxi Province, the economic development level has the strongest pulling effect, followed by energy structure and demographic factors; among the restraining factors, energy efficiency has the strongest inhibitory effect, followed by industrial structure. Shaanxi Province needs to continuously improve its energy consumption structure, promote clean energy and low-carbon technologies, realize energy conservation and emission reduction, and promote the green development of logistics industry.

**Keywords** Logistics industry · Carbon emission · EKC model · Influencing factors · LMDI decomposition method

## 1 Introduction

In recent years, the implementation of strategies such as the great western development strategy and the Belt and Road Initiative has brought favorable conditions for the economic development of Shaanxi Province. However, the ecological environment is still a constraint and a short board for the development of Shaanxi Province.

C. Zhu · X. Guo (✉)
School of Management, Xi'an University of Science and Technology, Xi'an 710054, China
e-mail: guxee@qq.com

 
F. K. S. Chan et al. (eds.), *Proceedings of the 2020 International Conference on Resource Sustainability: Sustainable Urbanisation in the BRI Era (icRS Urbanisation 2020)*, Environmental Science and Engineering,
https://doi.org/10.1007/978-981-15-9605-6_28

According to the 2019 China urban air quality rankings, Weinan, Xianyang and Xi'an are among the top 10 cities with poor air quality. Shaanxi Province is still facing serious ecological and environmental problems. In January 2018, the 13th Five-Year Plan of Shaanxi Province proposed that by 2020, Shaanxi Province will achieve a target of reducing carbon emissions per unit output value by 18% and energy consumption below 139 million tons of standard coal. The logistics industry is very important to the development of the national economy, and its energy consumption is gradually increasing with the development of the economy. In 2018, the logistics industry in Shaanxi Province consumed 10.96 million tons of standard coal, accounting for 8.5% of the total energy consumption. Compared with other industries, the logistics industry is currently in the economic development mode of excessive dependence on resources, with great pressure on carbon emission reduction. Therefore, studying the carbon emission characteristics and influencing factors of the logistics industry in Shaanxi Province is of great significance for energy conservation and emission reduction and the green development of the logistics industry in Shaanxi Province.

The rest of the paper is divided into five sections. Section 2 reviews the related research of scholars. Section 3 describes the models and data sources. Section 4 conducts the empirical analysis and discusses the analysis results. Finally, Sect. 5 summarizes the main points of the conclusion.

## 2 Literature Review

Carbon emission is a hot issue that many scholars have paid attention to in recent years. Scholars at home and abroad have done extensive research on regional and industrial carbon emissions, mainly focusing on the following three aspects.

In terms of carbon emission measurement, there are mainly life cycle method, carbon emission coefficient method and input-output method, etc. Sim and Sim [1] used the life cycle method to measure the carbon emissions of the Korean automotive industry and evaluated the environmental impact of vehicle logistics. Liu and Li [2] estimated the carbon emissions of the logistics industry in 30 provinces in China from 2007 to 2013 according to the carbon emission coefficient method provided by the IPCC. Wang and Liu [3] respectively calculated carbon emissions of China's logistics industry by using carbon emission coefficient method and input-output method. The study found that the carbon emission coefficient method was low and could not reflect the real carbon emission level of China's logistics industry.

In terms of carbon emission characteristics, many scholars use the EKC model to study the relationship between environmental pollution and economic growth. However, there is no final conclusion on the form of EKC in China's academic circles. Due to the different index selection and calculation methods, the shape of the EKC curve is also different [4]. Zhao et al. [5] established a cubic EKC model to regress energy consumption and per capita GDP, and found that there was an n-shaped curve relationship between energy consumption, energy intensity and per capita GDP. Luo and Liu [6] believe that the demand structure will affect economic growth and thus the environment. Through EKC model, the relationship between demand structure and environmental quality of 31 provinces and cities was analyzed, and it was found that there was a U-shaped curve relationship between the two. Shahbaz et al. [7] studied the relationship between globalization and carbon emissions in high, middle and low-income countries by using EKC model. The results show that 18% of the high-income and middle-income countries conform to the inverted U-shaped curve, 8% of the countries show a U-shaped relationship, while other countries have no such curve relationship.

In terms of influencing factors of carbon emissions, scholars use LMDI factor decomposition method to study this issue based on different decomposition methods. Sun et al. [8] divided the industry into three types of industries: high, medium, and low pressure, and used the LMDI factor decomposition method to decompose the factors that affect the energy carbon footprint of industrial enterprises, which were specifically decomposed into energy structure, energy intensity, industrial structure Europe, economic development and population pressure of forest and grassland. Achour and Belloumi [9] used the LMDI method to decompose the driving factors of energy consumption in Tunisia's transportation industry, decomposing energy consumption into energy intensity, transportation intensity, transportation structure, population scale and economic output effect. Liu [10] used the LMDI factor decomposition method to analyze the driving factors of carbon emissions of logistics industry in Shaanxi Province, and decomposed them into energy structure, energy intensity, energy efficiency and logistics industry output value.

To sum up, in the existing literature research, most scholars use more concentrated research methods and more comprehensive research. However, there are few studies on carbon emissions in the logistics industry in Shaanxi Province. Although some scholars have studied the carbon emissions and influencing factors of logistics industry, there is no relevant research on the relationship between economic development of logistics industry and carbon emission in Shaanxi Province. Therefore, this paper takes the logistics industry in Shaanxi Province as the research object, measures its carbon emissions, analyzes the relationship between environmental quality and economic development of logistics industry by EKC model, uses LMDI factor decomposition model to study the influencing factors of carbon emission of logistics industry, in order to obtain some valuable research conclusions.

# 3 Model Construction and Data Sources

## *3.1 Carbon Emission Measurement Method*

According to the literature review, the current academia mainly adopts the carbon emission coefficient method for carbon emission measurement. Therefore, this article refers to the calculation method in Cao [11] to estimate the carbon emission of the logistics industry in Shaanxi Province. The specific calculation method is as follows:

$$C_t = \sum_i C_{it} = \sum_i E_{it} \times V_i \times F_i \tag{1}$$

In Formula (1), $C_t$ is the total carbon emission of logistics industry in Shaanxi Province in the $t$ year, $C_{it}$ is the carbon emission of $i$ energy consumed by the logistics industry in Shaanxi Province in the $t$ year, $E_{it}$ is the amount of $i$ energy consumed by the logistics industry in Shaanxi Province in the $t$ year, $V_i$ is the conversion standard coal coefficient of $i$ energy, $F_i$ is the carbon emission coefficient of $i$ energy, and the latter two are fixed values.

## 3.2 EKC Model

This paper chooses the EKC model to study the carbon emission characteristics of the logistics industry in Shaanxi Province. The EKC model is a commonly used model to study the relationship between environmental quality and economic growth, which generally presents an inverted U-shaped curve. With the use of more and more scholars, the research results are diversified. Many scholars have discovered that different parameter values to be estimated will present different curve shapes. In addition to the inverted U-shaped curve relationship, there will be U-shaped, N-shaped, and inverted N-type and linear relationship. This paper selects a more comprehensive and flexible cubic regression curve model to study the carbon emission characteristics of the logistics industry in Shaanxi Province, the formula is as follows:

$$Y_t = \alpha + \beta_1 X_t + \beta_2 X_t^2 + \beta_3 X_t^3 + \gamma \tag{2}$$

In Formula (2), $Y_t$ is the index value reflecting the environmental quality in the $t$ year, $X_t$ is the index value reflecting the economic growth in the $t$ year, $\alpha$ is the fixed effect which does not change with the period, $\beta_1$, $\beta_2$ and $\beta_3$ are the parameters to be estimated, and $\gamma$ is the random error term.

## 3.3 Kaya Identity and LMDI Decomposition Method

Kaya Identities believe that carbon emissions are related to factors such as energy, economy, and population. It is a common method for studying the factors affecting carbon emissions. The Kaya identity is deformed, the LMDI decomposition method is used to analyze the influencing factors of carbon emissions, and the following decomposition model is established [12]:

$$C_t = \sum_i C_{it} = \sum_i \frac{C_{it}}{E_{it}} \times \frac{E_{it}}{E_t} \times \frac{E_t}{L_t} \times \frac{L_t}{G_t} \times \frac{G_t}{P_t} \times P_t \tag{3}$$

In Formula (3), $E_t$ is the total energy consumption of Shaanxi Province in the $t$ year; $L_t$ is the GDP value of logistics industry in the $t$ year of Shaanxi Province; $G_t$ is the total GDP of Shaanxi Province in the $t$ year; $P_t$ is the total population of Shaanxi Province in the $t$ year.

$$C_t = \sum_i C_{it} = \sum_i F_i \times S_{it} \times K_t \times R_t \times Q_t \times P_t \tag{4}$$

In Formula (4), $F_i = C_{it}/E_{it}$ stands for carbon emission intensity and refers to carbon emission coefficient; $S_{it} = E_{it}/E_t$ represents energy structure, refers to the ratio of $i$ energy consumption to total consumption of logistics industry in Shaanxi Province in the $t$ year; $K_t = E_t/L_t$ represents energy efficiency, refers to the amount of energy consumed per unit of GDP of logistics industry in Shaanxi Province in the $t$ year; $R_t = L_t/G_t$ represents industrial structure, which refers to the ratio of logistics industry GDP to GDP of Shaanxi Province; $Q_t = G_t/P_t$ represents the level of economic development and refers to the per capita GDP of Shaanxi Province.

Using LMDI decomposition method, the carbon emission change of logistics industry in Shaanxi Province from the base period to the $t$ year is decomposed as follows:

$$\Delta C = C^t - C^0 = \Delta C_F + \Delta C_S + \Delta C_K + \Delta C_R + \Delta C_Q + \Delta C_P \tag{5}$$

In Formula (5), $C_t$ and $C_0$ respectively represent the carbon emissions from the base period to the $t$ year of the logistics industry in Shaanxi Province, and their difference $\Delta C$ represents the carbon emission change from the base period to the $t$ year of the logistics industry in Shaanxi Province, and $\Delta C_j$ ($j = S, K, R, Q, P$) represents the contribution value of each influencing factor. Since the carbon emission coefficient is fixed, $\Delta C_f = 0$.

$$\Delta C_j = \frac{C^t - C^0}{\ln C^t - \ln C^0} \times \ln \frac{j^t}{j^0} \quad (j = S, K, R, Q, P) \tag{6}$$

## 3.4 Data Sources

The logistics industry is an emerging compound industry, and its business scope covers many industries. Since the National Bureau of Standards has not defined the logistics industry separately, and considering that the transportation, warehousing and postal industry have a great contribution to the development of the logistics industry, the relevant data of the logistics industry in Shaanxi Province used in this paper are all from the transportation, warehousing and postal industry of Shaanxi Province [13].

The energy consumption and GDP of logistics industry used in this paper come from the energy consumption and GDP added value of transportation, storage and postal industry in the 2010–2019 "Shaanxi Statistical Yearbook"; freight turnover, Shaanxi GDP and total population of Shaanxi Province are from the 2010–2019 "Shaanxi Statistical Yearbook"; the conversion coefficients and carbon emission coefficients of various energy sources are from the "China Energy Statistical Yearbook".

# 4 Empirical Analysis

## *4.1 Carbon Emission Measurement Results*

The main energy balance sheet in the "Shaanxi Statistical Yearbook" shows that the final energy consumption of the logistics industry is mainly divided into the following seven types of energy. Table 1 lists the seven types of energy conversion factor and carbon emission factor. The unit of conversion coefficient of natural gas into standard coal is kg standard coal/$m^3$, and the unit of conversion coefficient of electric power is kg standard coal/kWh.

According to Formula (1) and Table 1, the total carbon emission of logistics industry in Shaanxi Province can be calculated. The results are shown in Table 2. The unit of carbon emission is 10,000 tons, accounting for %, and TCE represents the total carbon emission, CE represents the carbon emission, and P represents the proportion.

**Table 1** Standard coal conversion factor and carbon emission factor of seven energy sources

| | Raw coal | Natural gas | Electric power | Gasoline | Kerosene | Diesel oil | Fuel oil |
|---|---|---|---|---|---|---|---|
| Conversion coefficient of standard coal (kg standard coal/kg) | 0.7143 | 1.2143 | 0.1229 | 1.4714 | 1.4714 | 1.4571 | 1.4286 |
| Carbon emission factor (T carbon/t standard coal) | 0.7559 | 0.4483 | 2.2132 | 0.5538 | 0.5714 | 0.5921 | 0.6185 |

**Table 2** Total carbon emission and carbon emission proportion of seven energy sources

| Years | TCE | Raw coal | | Natural gas | | Electric power | | Gasoline | | Kerosene | | Diesel oil | | Fuel oil | |
|---|---|---|---|---|---|---|---|---|---|---|---|---|---|---|---|
| | | CE | P | CE | P | CE | P | CE | P | CE | P | CE | P | CE | P |
| 2009 | 533.04 | 20.52 | 3.85 | 19.33 | 3.63 | 100.37 | 18.83 | 107.56 | 20.18 | 5.31 | 1.00 | 279.96 | 52.52 | 0.00 | 0.00 |
| 2010 | 580.64 | 28.08 | 4.84 | 23.68 | 4.08 | 105.05 | 18.09 | 111.23 | 19.16 | 5.47 | 0.94 | 307.14 | 52.90 | 0.00 | 0.00 |
| 2011 | 631.14 | 30.18 | 4.78 | 25.04 | 3.97 | 118.24 | 18.73 | 120.07 | 19.02 | 5.97 | 0.95 | 331.64 | 52.55 | 0.00 | 0.00 |
| 2012 | 654.68 | 31.91 | 4.87 | 26.08 | 3.98 | 129.15 | 19.73 | 121.75 | 18.60 | 6.39 | 0.98 | 339.41 | 51.84 | 0.00 | 0.00 |
| 2013 | 679.31 | 33.25 | 4.89 | 27.87 | 4.10 | 134.56 | 19.81 | 127.49 | 18.77 | 6.56 | 0.97 | 349.59 | 51.46 | 0.00 | 0.00 |
| 2014 | 568.32 | 15.43 | 2.72 | 14.21 | 2.50 | 144.54 | 25.43 | 74.85 | 13.17 | 27.94 | 4.92 | 282.43 | 49.70 | 8.96 | 1.57 |
| 2015 | 559.89 | 14.71 | 2.63 | 17.75 | 3.17 | 145.17 | 25.93 | 81.60 | 14.57 | 28.56 | 5.10 | 254.63 | 45.48 | 17.48 | 3.12 |
| 2016 | 524.83 | 12.13 | 2.31 | 15.19 | 2.89 | 165.05 | 31.45 | 78.18 | 14.90 | 25.06 | 4.77 | 224.99 | 42.87 | 4.24 | 0.81 |
| 2017 | 537.50 | 10.14 | 1.89 | 15.57 | 2.90 | 177.13 | 32.95 | 92.04 | 17.12 | 41.85 | 7.79 | 198.73 | 36.97 | 2.03 | 0.38 |
| 2018 | 594.44 | 6.25 | 1.05 | 18.56 | 3.12 | 196.69 | 33.09 | 96.50 | 16.23 | 65.02 | 10.94 | 210.25 | 35.37 | 1.18 | 0.20 |

It can be seen from Table 2 that the carbon emission of logistics industry in Shaanxi Province shows a fluctuating trend from 2009 to 2018. In phases, from 2009 to 2013, the total carbon emissions increased continuously, reaching a maximum of 6.79 million tons in 2013; starting to decline from 2014 to 2016, reaching a minimum of 5.25 million tons in 2016; from 2017 to 2018, it rose again, reaching 5.94 million tons in 2018, but it is still less than the 2013 maximum. Analysis of the reasons shows that before 2013, the logistics industry in Shaanxi Province mainly relied on energy sources with higher carbon content such as gasoline and diesel. Since 2014, the increase in electricity and kerosene consumption and the use of fuel oil have made the energy consumption structure more reasonable. Therefore, the total carbon emissions in 2014 decreased by more than 1 million tons compared with 2013.

Among the seven energy sources, diesel has the largest carbon emissions, accounting for more than 35%; followed by electricity and gasoline, with fuel oil accounting for the smallest proportion, reaching only 3.12%. From the perspective of the trend, the carbon emissions of raw coal and diesel showed a downward trend. The carbon emissions of diesel has dropped from 52.52% to 35.37% in ten years; the carbon emissions of electricity and kerosene have shown an overall upward trend; the carbon emissions of electric power increased from 18.83% to 33.09%; the carbon emissions of fuel oil increased first and then decreased; the carbon emissions of gasoline and natural gas were generally in line with the trend of total carbon emissions.

## *4.2 Carbon Emission Characteristics Analysis*

In order to eliminate the unreasonable factors, this paper uses the carbon emissions per unit freight traffic turnover of the logistics industry as the index to reflect the environmental quality, and uses the GDP per unit freight traffic turnover of the logistics industry as the indicator to reflect the economic development of the logistics industry. Carbon emissions per unit freight traffic turnover is equal to the ratio of total carbon emissions to freight traffic turnover; GDP per unit freight traffic turnover is equal to the ratio of logistics GDP to freight traffic turnover [14]. This paper uses SPSS to regress the carbon emissions per unit freight traffic turnover and GDP per unit freight traffic turnover of logistics industry in Shaanxi Province with linear regression, quadratic curve and cubic curve models respectively. The results are shown in Table 3.

From the model regression results, the *p*-values of the linear, quadratic and cubic models are all equal to 0.001 and less than 0.05. Therefore, all three models have passed the test and are significant. The larger the model $R^2$, the better the fitting result, so the best fitting model is selected by comparing the $R^2$ of the three models. Because the $R^2$ of the quadratic curve model is the largest, which is 0.854, and the fitting effect is good, it can be judged that the quadratic curve model can better describe the relationship between the economic development of the logistics industry and environmental quality in Shaanxi Province. According to the parameter evaluation

**Table 3** Model regression results and parameter evaluation

| Equation | Model summary | | | | | Parameter evaluation | | | |
|---|---|---|---|---|---|---|---|---|---|
| | R squared | F | df1 | df2 | Significance | Constant | b1 | b2 | b3 |
| Linear | 0.744 | 23.306 | 1 | 8 | 0.001 | 0.512 | −1.606 | | |
| Quadratic curve | 0.854 | 20.526 | 2 | 7 | 0.001 | −1.596 | 19.491 | −52.407 | |
| Cubic curve | 0.851 | 20.022 | 2 | 7 | 0.001 | −0.890 | 8.940 | 0.000 | −86.460 |

of the quadratic curve model in Table 3, the following regression equation can be obtained:

$$Y = -1.596 + 19.491X - 52.407X^2$$

Using SPSS to obtain the fitting graph of the quadratic curve model, it can be seen from the regression equation and fitting graph that there is an inverted U-shaped relationship between the carbon emission per unit freight traffic turnover of Shaanxi's logistics industry and the GDP per unit freight traffic turnover. By deriving the regression equation once and making its derivative equal to 0, the vertex coordinates of the quadratic curve can be obtained as (0.186, 0.216), that is, the GDP of unit freight traffic turnover is RMB 0.186/ton-kilometer, which is similar to the GDP value of logistics industry in Shaanxi Province in 2010. For the years where the GDP per unit freight traffic turnover is less than RMB 0.186/ton-kilometer, that is, in 2009 and 2013, they were RMB 0.180/ton-kilometer and RMB 0.175/ton-kilometer respectively, the carbon emissions of logistics industry in Shaanxi Province will continue to increase with the development of logistics industry; From a point of view, the GDP per unit freight traffic turnover in the remaining years is greater than RMB 0.186/ton-kilometer, and the carbon emissions of the logistics industry in Shaanxi Province will continue to decrease with the development of the logistics industry, which means that the carbon emissions per unit output value of the logistics industry in Shaanxi Province is on a downward trend. The economic growth of the logistics industry is less and less dependent on energy consumption, and energy efficiency is gradually increasing (Fig. 1).

## *4.3 Analysis of Influencing Factors of Carbon Emission*

Using the LMDI decomposition formula, using 2009 as the base period to analyze the factors affecting the carbon emissions of the logistics industry in Shaanxi Province from 2010 to 2018, the results are shown in Table 4. The unit is 10,000 tons.

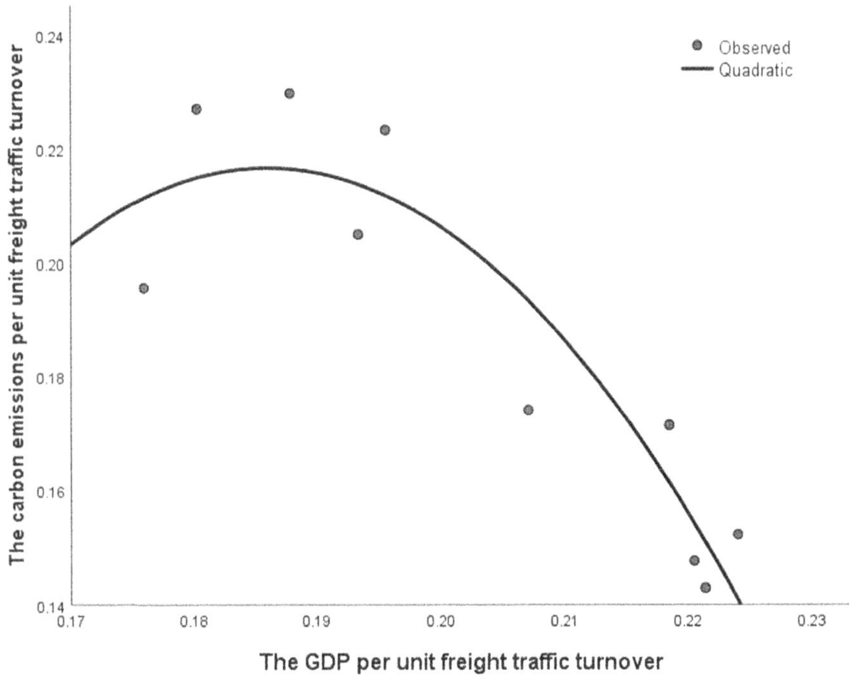


**Fig. 1** Fitting diagram of quadratic curve model

**Table 4.** Carbon emission decomposition results of logistics industry in Shaanxi Province

| | $\Delta C$ | $\Delta C_S$ | $\Delta C_K$ | $\Delta C_R$ | $\Delta C_Q$ | $\Delta C_P$ |
|---|---|---|---|---|---|---|
| 2010 | 47.581 | −2.504 | −13.618 | −55.566 | 118.091 | 1.193 |
| 2011 | 98.077 | 0.713 | −57.382 | −92.705 | 244.989 | 2.487 |
| 2012 | 121.589 | 5.958 | −107.670 | −114.136 | 333.371 | 4.112 |
| 2013 | 146.210 | 6.184 | −81.412 | −191.485 | 407.026 | 5.957 |
| 2014 | 34.646 | 25.717 | −252.241 | −164.789 | 410.757 | 6.918 |
| 2015 | 26.784 | 26.041 | −292.338 | −146.912 | 421.594 | 9.566 |
| 2016 | −8.159 | 50.506 | −383.465 | −138.830 | 442.679 | 11.994 |
| 2017 | 4.379 | 55.585 | −416.100 | −162.786 | 503.814 | 15.032 |
| 2018 | 59.649 | 58.301 | −410.975 | −194.339 | 579.749 | 19.751 |

On the whole, the growth rate of carbon emission of logistics industry in Shaanxi Province showed a trend of first increasing, then decreasing and then increasing. The growth rate of carbon emissions was relatively rapid from 2010 to 2013. In the following four years, the growth rate of carbon emission slowed down and became negative in 2016. Compared with 2009, the carbon emission in 2016 decreased by 81,590 tons; after that, the growth rate of carbon emissions accelerated again from

2017 to 2018. Among the five influencing factors, energy structure, economic development level and population factors play a driving role in the growth of carbon emission of logistics industry in Shaanxi Province, while energy efficiency and industrial structure play a restraining role in the growth of carbon emission of logistics industry in Shaanxi Province. The following five factors are analyzed respectively.

1. Energy structure factor has a positive effect on the carbon emissions of the logistics industry. From 2010 to 2018, the contribution value of energy structure factors gradually increased, reaching 583,100 tons in 2018, indicating that the energy consumption structure of the logistics industry in Shaanxi Province is unreasonable. Specifically, the transportation industry accounts for a large proportion in the logistics industry, and the consumption of gasoline and diesel is high, which is also the main reason for the unreasonable energy consumption structure of the logistics industry.
2. The energy efficiency factor has a restraining effect on the growth of carbon emissions of logistics industry. From 2010 to 2018, the contribution value of energy efficiency factor showed a downward trend, from −136,180 tons to −4,109,750 tons, which was the strongest restrictive factor. Energy efficiency has a good inhibition effect, which shows that the carbon emissions per unit output value of logistics industry in Shaanxi Province has been reduced, and the energy use efficiency has gradually increased.
3. The factor of industrial structure play an inhibitory role in the growth of carbon emissions of logistics industry in Shaanxi Province. Similar to the energy efficiency factor, the contribution value of industrial structure factor also showed a downward trend, from the numerical point of view, its inhibitory effect was not as strong as that of energy efficiency factor, but it also exceeded the pulling effect of energy structure and population factors. The factor of industrial structure decreases year by year and is negative, which shows that the logistics industry in Shaanxi Province is well developed and the industrial structure is more reasonable.
4. The level of economic development plays an important role in promoting carbon emissions of logistics industry in Shaanxi Province. In 2010, the contribution value of economic development level factor reached 1,180,910 tons, and has been growing steadily since then, reaching 5,797,490 tons in 2018. The development of logistics industry will inevitably lead to the increase of carbon emissions of logistics industry, especially in recent years, with the development of e-commerce and the gradual increase of commodity trade, the logistics industry has been rapid development.
5. Population factor also plays a driving role in carbon emission of logistics industry in Shaanxi Province. The contribution of population factor to the growth of carbon emissions of logistics industry is relatively small, but it has been on an upward trend year by year. With the rapid growth of population, the demand for logistics is also increasing, resulting in more logistics activities, leading to the growth of carbon emissions in the logistics industry.

## 5 Conclusion and Suggestion

Based on the energy consumption data of logistics industry in Shaanxi Province from 2009 to 2018, this paper uses the carbon emission coefficient method, the EKC model and the LDMI factor decomposition method to study the carbon emissions, carbon emission characteristics and influencing factors of logistics industry in Shaanxi Province. The main conclusions are as follows:

(1) From 2009 to 2018, the total carbon emissions of logistics industry in Shaanxi Province showed a fluctuating trend. Among the seven energy sources, diesel has the largest carbon emissions, followed by electricity and gasoline. The carbon emissions of raw coal and diesel showed a downward trend as a whole, while electricity and kerosene showed an overall upward trend. (2) There is an inverted U-shaped relationship between the carbon emissions of logistics industry and economic growth in Shaanxi Province. For the years when the GDP of unit freight traffic turnover is greater than RMB 0.186/ton-kilometer, the carbon emissions of logistics industry will continue to decrease with the development of the logistics industry. (3) The economic development level factor has the strongest pulling effect on carbon emission of logistics industry in Shaanxi Province, followed by energy structure and population factor; energy efficiency factor has the strongest inhibitory effect, followed by industrial structure factor.

To sum up, the energy efficiency of the logistics industry in Shaanxi Province has been continuously improved, the industrial structure is relatively reasonable, and the logistics industry is developing healthily, but there are still some problems such as rising carbon emissions and irrational energy structure. Therefore, Shaanxi Province needs to constantly improve the energy consumption structure of the logistics industry, promote the use of clean energy, reduce the use of high-carbon diesel and gasoline, improve energy utilization efficiency, and reduce the dependence of economic development of logistics industry on energy consumption. At the same time, it is necessary to integrate logistics resources, promote low-carbon technology, improve the operation efficiency and service level of the logistics industry, gradually reduce the carbon emission of the logistics industry, and promote the green development of the logistics industry [15].

## References

1. Sim, J., Sim, J.: Air emission and environmental impact assessment of Korean automotive logistics. J. Clean. Prod. **159**, 130–140 (2017)
2. Liu, Y., Li, L.: Research on decoupling and its influencing factors of the carbon emission for logistics industry in China. Environ. Sci. Technol. **41**(05), 177–181 (2018)
3. Wang, L.P., Liu, M.H.: Carbon emission measurement for China's logistics industry and its influence factors based on input-output method. Res. Sci. **40**(1), 195–206 (2018)
4. Chang, J.: Judgment and countermeasures on the environmental Kuznets curve of China's foreign trade. J. Commercial Econ. **27**, 99–101 (2015)

5. Zhao, A.W., He, Y., Wang, S.Y., Li, D.: EKC Test for China's energy consumption and the impact factors analysis. J. Syst. Manag. **23**(03), 416–422 (2014)
6. Luo, N.S., Liu, T.: Empirical study on influence of demand structure on the quality of environment. China Soft Sci. **12**, 38–47 (2014)
7. Shahbaz, M., Mahalik, M., Shahzad, S.: Testing the globalization-driven carbon emissions hypothesis: international evidence. Int. Econ. **158**, 25–38 (2019)
8. Sun, L.W., Wang, D.F., Du, J., Han, Y.: A study of ecological pressure intensity of carbon footprint's factor decomposition of chinese industry based on LMDI. Ecol. Econ. **35**(01), 13–18 (2019)
9. Achour, H., Belloumi, M.: Decomposing the influencing factors of energy consumption in Tunisian transportation sector using the LMDI method. Transp. Policy **52**, 64–71 (2016)
10. Liu, Y.: The research on decomposition and decoupling effect of driving factors of carbon emission from logistics industry under open economy: taking Shaanxi Province as an example. Ecol. Econ. **34**(11), 84–89 (2018)
11. Cao, H.G., Tong, X., Chen, K., Li, G.: An empirical study of regional disparity of China's carbon emission. J. Industr. Technol. Econ. **11**, 84–96 (2015)
12. Wang, X., Wan, M.: Research on carbon emission scale and influencing factors of logistics industry in anhui province: based on the comparative perspective of four provinces and cities in the Yangtze River Delta. J. Anhui Univ. (Philos. Soc. Sci. Ed.) **41**(03), 148–156 (2017)
13. Wang, L.P.: An empirical study on the relationship between carbon emission and energy consumption and economic growth in the logistics industry—a case study of Henan Province. Chin. J. Syst. Sci. **25**(02), 112–116 (2017)
14. Zhou, Y., Guo, L.J., Tang, E.B.: An EKC-based study on carbon emission characteristics of the provincial logistics industry: evidence from four provinces in China. J. Beijing Jiaotong Univ. (Soc. Sci. Ed.) **17**(01), 85–93 (2018)
15. Liang, W., Fang, S.H.: Research on the dynamic relationship between logistics industry growth, urbanization and carbon emissions. Jianghan Acad. **38**(04), 73–81 (2019)

# Towards the Low Emission Roadmap of China's Commercial Building: Mitigation Retrospection and Peaking Simulation

**Minda Ma and Weiguang Cai**

**Abstract** This study is the first to assess the historical carbon mitigation and simulate the energy and emission peaks of China's commercial building sector using a dynamic emission scenario. It shows that the emission mitigation of the commercial building sector during 2000–2016 is 1221.50 (±486.89) million tons of carbon dioxide ($MtCO_2$), and the scenario simulation demonstrates that the commercial building sector will achieve its carbon emission peak in year 2039 (±8) with a peak value of 1154.88 (±191.05) $MtCO_2$. The sensitivity analysis reveals that the impacts of emission factor and GDP per capita are the most significant for the uncertainty of emission peaks. A strict energy demand benchmark of the commercial building sector suggests a control at 465.99 million tons of standard coal equivalent (Mtce), and its peaking year is estimated for 2035, which is 13 years ahead of the business-as-usual scenario, with energy savings of 112.90 Mtce. For the earliest peaking time, if the commercial building sector aims to achieve its emission peak before 2030, the emission peak should be controlled at 960 $MtCO_2$. Overall, this paper can assist the government in more accurate and feasible building emission mitigation strategies. Moreover, the results provide a more powerful decision-making reference in issuing targeted and feasible strategies for future commercial building emission mitigation.

**Keywords** Commercial building · Emission mitigation · Energy and emission peaks · Decomposition analysis · Dynamic scenario analysis

M. Ma (✉)
Department of Earth System Science, Tsinghua University, Beijing 100084, PR China
e-mail: maminda@tsinghua.edu.cn

W. Cai
Building Energy Big Data Group, International Research Center for Sustainable Built Environment, Chongqing University, Chongqing 400045, PR China



F. K. S. Chan et al. (eds.), *Proceedings of the 2020 International Conference on Resource Sustainability: Sustainable Urbanisation in the BRI Era (icRS Urbanisation 2020)*, Environmental Science and Engineering,
https://doi.org/10.1007/978-981-15-9605-6_29

## Nonmenclature

### Abbreviations

| | |
|---|---|
| BAU | Business as usual |
| LMDI | Log-Mean Divisia Index |
| Mtce | Million tons of standard coal equivalent |
| $MtCO_2$ | Million tons of carbon dioxide |

### Symbols

| | |
|---|---|
| $C$ | Carbon emissions released by commercial buildings |
| $CM$ | Carbon mitigation of commercial buildings |
| $E$ | Energy demand of commercial buildings |
| $e$ | Energy intensity of commercial buildings |
| $F$ | Commercial building stocks |
| $G$ | Gross Domestic Product (GDP) |
| $g$ | GDP per capita |
| $G_s$ | Output value of the service industry |
| $I$ | Industrial efficiency of the service industry |
| $K$ | Energy-related carbon intensity of commercial buildings (i.e., the emission factor) |
| $P$ | Population size |
| $S$ | Industrial structure |

### Greek Letters

| | |
|---|---|
| $\omega$ | Random value |
| $\sigma$ | Standard deviation |

## 1 Introduction

25th Conference of the Parties announced that the global building sector and construction industry account for nearly 40% of the total carbon emissions and 36% of the end-use energy [1]. Growing population size as well as rapid development of service industry in emerging economies and developing countries lead to an energy demand in buildings that could increase by 60% in 2070 [2]. Besides, building stock is expected to double by 2050, contributing to the continuous increase of global construction emissions [3]. As the largest developing economy, China is facing a fast-growing energy demand pressured by the service industry, which requires massive fossil energy use and further increases the emissions of carbon dioxide ($CO_2$) [4]. Carbon emissions driven by the commercial buildings have increased significantly at a 6.83% yearly rate over the past decade, and the total emissions in 2017 were 767.86 million tons of CO2 ($MtCO_2$) [5].

On the other hand, the building sector provides the largest cost-effective emission mitigation potential [6], with possible net cost savings and economic gains via existing technologies and policies [7]. Lawrence Berkeley National Laboratory [8] developed a bottom-up scenario analysis via long-range energy alternatives planning model to illustrate the prospective emission roadmaps of China's building sector for 2050. The most positive emission mitigation scenario showed that the commercial building would achieve the goal in 2040, with a total primary energy consumption of 600 million tons of standard coal equivalent (Mtce). However, the above studies conducted the emission peak analyses through static scenario models, and the impact of uncertainty on the emission peaks was rarely considered. Therefore, the above peak schemes may discourage the government to accurately and feasibly conduct the emission mitigation goals. Here, three issues should be addressed for the Chinese commercial building sector as follows:

- How is the carbon mitigation potential of commercial buildings over the last decades?
- What are the prospective building emission pathways considering the uncertainty impacts?
- How to reinvent the future commercial building and achieve its emission peak goals the earliest?

To address these issues, this study is, to the best of our knowledge, the first to propose a historical carbon mitigation assessment, develop a dynamic emission scenario analysis, and identify low carbon pathways of the commercial building sector in China over the period of 2000–2070. Specifically, this work establishes the emission model by Kaya identity, assesses the carbon mitigation during 2000–2016, and analyzes the prospective energy and emission scenarios up to 2070 via Monte Carlo simulation. Thereafter, the strict schemes of the energy demand of the commercial building sector are proposed by applying the uncertainty measured by the scenario analysis. Furthermore, the emission benchmarks for the future commercial building sector were discussed to achieve its earliest emission target. The flowchart of this study is expressed in Fig. 1.

The most significant contribution of this study is the simulation of energy and emission peaks of the commercial building sector considering the impact of uncertainty. Currently, there are few studies on such topic. To the best of our knowledge, the existing studies only performed the emission peak analysis using a static scenario model, and the uncertainty of the emission peaks caused by the model parameter changes was not considered. Therefore, the proposed peak scheme may

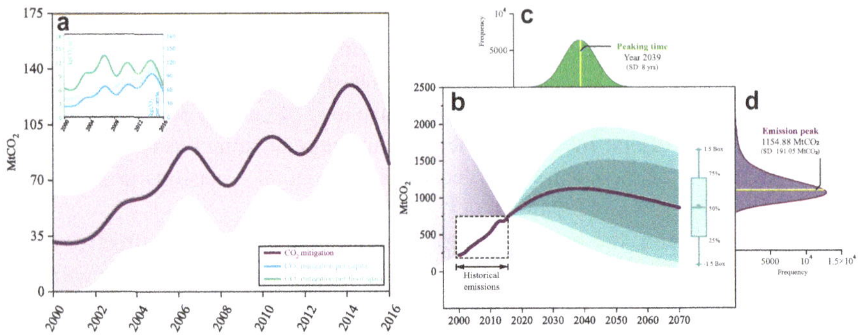


**Fig. 1** Flowchart of this study: **a** historical carbon emission mitigation assessment since 2000, **b** dynamic scenario simulations of the prospective emission peak up to 2070, and **c–d** distributions of the emission peak and its peaking time of the commercial building sector

not be adequate to accurately and feasibly implement emission mitigation goals, which will be further explained in the literature review.

In this paper, Sect. 2 focuses on the materials and methods, including the emission model, historical carbon mitigation assessment, static scenario design, and dynamic scenario simulation. In addition, dataset is also introduced. Section 3 shows the historical carbon mitigation trend, and illustrates the prospective carbon emission pathways in the commercial building sector. Section 4 summarizes the core findings and proposes upcoming studies.

# 2 Materials and Methods

## *2.1 Emission Mitigation Model*

To assess the historical carbon mitigation of commercial buildings and establish different carbon emission scenarios of the future commercial building sector, this study considered a series of impact factors to set the building emission model via the Kaya identity, as expressed and refined in Eqs. 1 and 2, respectively.

$$C = P \cdot \frac{G}{P} \cdot \frac{G_s}{G} \cdot \frac{F}{G_s} \cdot \frac{E}{F} \cdot K \tag{1}$$

$$C = P \cdot g \cdot S \cdot I \cdot e \cdot K \tag{2}$$

Due to the space limitation, all the parameters in Eqs. 1 and 2 are explained in the Nomenclature. After establishing the building emission model, the Log-Mean Divisia Index (LMDI) decomposition was applied to analyze the contribution of each factor to the carbon emissions of commercial buildings. Furthermore, the carbon mitigation of the commercial building sector was estimated based on the decomposition result. LMDI is a typical approach to estimate the contribution of each factor to the carbon emissions [9]. Combined with the Kaya identity, the change of carbon emissions of commercial buildings in Period $\Delta T$ ($\Delta C|_{0\to T}$) is decomposed as:

$$\Delta C|_{0\to T} = C|_T - C|_0 = \left(\Delta C_P + \Delta C_g + \Delta C_S + \Delta C_I + \Delta C_e + \Delta C_K\right)|_{0\to T} \tag{3}$$

where

$$\Delta C_K = \frac{C|_T - C|_0}{\ln C|_T - \ln C|_0} \Delta \ln\left(\frac{K|_T}{K|_0}\right) \tag{4}$$

Based on Eq. 3, the carbon mitigation of the commercial building sector ($CM$) is expressed as:

$$CM|_{0\to T} = -\sum \Delta C_x|_{0\to T} \tag{5}$$

where

$$\Delta C_x|_{0\to T} \in \left(\Delta C_P, \Delta C_g, \Delta C_S, \Delta C_I, \Delta C_e, \Delta C_K\right), \Delta C_x|_{0\to T} < 0. \tag{6}$$

## 2.2 *Emission Scenario Analysis*

Scenario analysis is a typical approach to illustrate the prospective roadmap of carbon emissions, and it has been widely applied for different emission sectors. The key to design different emission scenarios is to obtain a set of parameters that affect the carbon emissions (i.e., impact factors). The final carbon emissions should produce a significant feedback based on changes of key parameters. Moreover, the essence of scenario analysis is not to reliably simulate the prospective emissions following the hypothetical socioeconomic and technological levels, but to reveal how will the emission develop upon different strategies and then identify the emission mitigation pathway for the future.

In this section, the static emission scenario at the business-as-usual (BAU) level was developed. For that, the BAU level of emission trend and socioeconomic progress was considered over the period of 2000–2070. The BAU scenario is the most likely to occur, and its emission trend and socioeconomic progress follow the existing background. Furthermore, the potential change degrees of all parameters (see Eq. 2) in this scenario are at the BAU level, which is the benchmark for designing the other emission scenarios.

After setting the static emission scenarios for the commercial building sector over the period of 2000–2070, the dynamic BAU scenario was modeled via Monte Carlo simulation. This model covered the uncertainties caused by parameter changes in the emission model. The simulation illustrates the possible ranges of prospective energy demand and carbon emissions with different probabilities. Monte Carlo simulation is a useful tool to project risk analysis, and it has been recommended to model prospective carbon emission roadmaps considering various uncertainties [10]. This study presented the dynamic scenario simulation in three steps. First, taking Eq. 2 as the emission model, random values ($\omega$) following different distributions were set to express the potential range of factors change affecting the carbon emissions in the commercial building sector. Then, the static BAU scenario changed into the dynamic BAU scenario based on the uncertainties (see Eq. 7). Second, various times of simulations were deployed to model the dynamic BAU scenario by randomly sampling the parameters following the probability distributions defined in Step 1. A total number of 100,000 Monte Carlo simulations were applied to ensure the simulation obtained reliable results. Finally, the simulation results (i.e., frequency distributions) of the potential carbon emission peak and respective peaking time were fitted via the suitable distributions. Then, the emission peak and peaking time were identified.

$$C^{Dynamic} = C^{Static}\Delta\left(1 + \omega\Delta\frac{T - 2016}{2070 - 2016}\right), \quad \omega \sim N(0, \sigma) \tag{7}$$

Using Monte Carlo simulation, this study drew the possible ranges of carbon emission peak and peaking time of the future commercial building sector at different probabilities. Therefore, the results can support the government to conduct emission mitigation strategies more accurately and feasibly for the commercial building sector. Moreover, a more powerful decision-making reference is provided for targeted and feasible strategies to mitigate building emissions in the future.

## 2.3 Dataset

This study collected historical data (2000–2016) of energy, emissions, stocks, and economy of the Chinese commercial building sector from https://www.researchgate.net/project/China-Building-Energy-and-Emission-Database-CBEED. Historical data on population size were accessed from https://population.un.org/wpp/. For the data in the period of 2016–2070, this study referred to predictions issued by the UN World Population Prospects and CBEED.

# 3 Results and Discussion

## 3.1 Historical Carbon Mitigation of Commercial Buildings

Figure 2 shows the carbon emission mitigation of the commercial building sector over the period of 2000–2016, which was assessed via Eqs. 5 and 6. Regarding the uncertainty of the emission mitigation result, the error bars at different levels should be considered. For the total $CO_2$ mitigation, $CO_2$ mitigation per capita, and per floor space (i.e., $CO_2$ mitigation intensity), the uncertainty values are: ±30.43 $MtCO_2$ per year, ±21.80 $kgCO_2$ per capita, and ±2.52 kgce $m^{-2}$, respectively.

In general, the total carbon mitigation of the commercial building sector in 2000–2016 is 1221.50 (±486.89) $MtCO_2$ (see Fig. 2a). Specifically, the total emission mitigation values during the following three periods are: 211.13 (±152.15, 2000–2005), 391.40 (±152.15, 2006–2010), and 618.97 (±182.58, 2011–2016) $MtCO_2$, respectively. Furthermore, the intensity of $CO_2$ mitigation during the above three periods is: 7.69 (±2.52), 11.18 (±2.52), and 10.41 (±2.52) kgce $m^{-2}$ yr, respectively (see Fig. 2b). As the historical emission mitigation trends in Fig. 2a, b follow the W-shaped curves, the mitigation trend is not monotonic. After fitting the mitigation results, it can be observed that the continuous increase of carbon mitigation at three different scales is significant. Let the total emission mitigation be an example, the accumulative mitigation value in 2000–2016 increased consistently, at a 74.01 $MtCO_2$ yearly level.

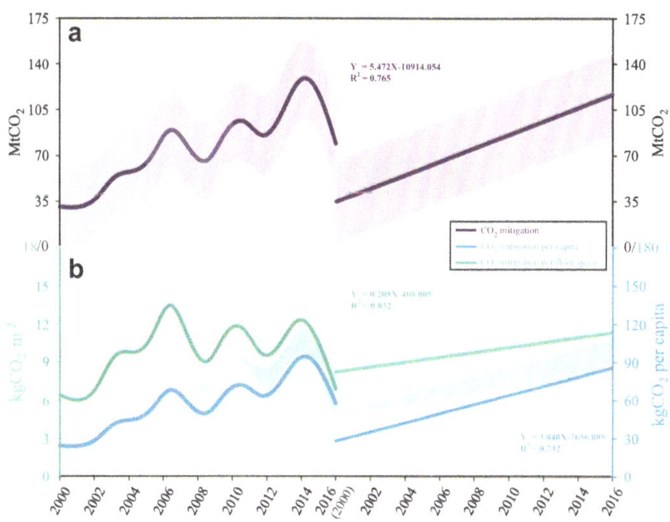


**Fig. 2** Historical carbon emission mitigation of commercial buildings at the scales of **a** total $CO_2$ mitigation and **b** $CO_2$ mitigation per capita and per floor space

## 3.2 *Emission Paths of the Future Commercial Buildings*

Figure 3a illustrates the static scenario at the BAU level of the carbon emissions in the commercial building sector and the potential ranges of the dynamic scenario simulations with different probabilities. Following different levels of emission trend and socioeconomic progress, the static and dynamic emission scenarios show the pathways of carbon emission development up to 2070. It was observed that the emission trend of the commercial building sector in the static scenario would follow the inversed U-shaped curve. Specifically, commercial buildings would achieve their emission peak in 2038 (emission peak: 1128.95 $MtCO_2$).

Figure 3b, c briefly shows the results of the BAU scenario analysis of emissions in the commercial building sector based on 100,000 Monte Carlo simulation runs. The commercial building sector would achieve the emission peak in 2039 in all probability. Regarding the uncertainty of the simulation result, one standard deviation value of 8 years for the peaking time should be considered. Here, the interval of peaking year was expressed as 2039 (±8). As for the peak value of carbon emissions of the commercial building sector, it would be 1154.88 (±191.05) $MtCO_2$. It was observed that the results of static scenario analysis (see Fig. 3a) were basically consistent with the ones of dynamic scenario simulation (see Fig. 3b, c).

Regarding the peak and peaking year of the energy demand of the commercial building sector, Fig. 4a illustrates that commercial buildings would achieve the peak in year 2048 (±13) with the value of 578.89 (±112.90) Mtce. To analyze the difference between energy and emission peaks, the distributions of emission peak with its peaking year were illustrated in Fig. 4b, it reflects that the peaking

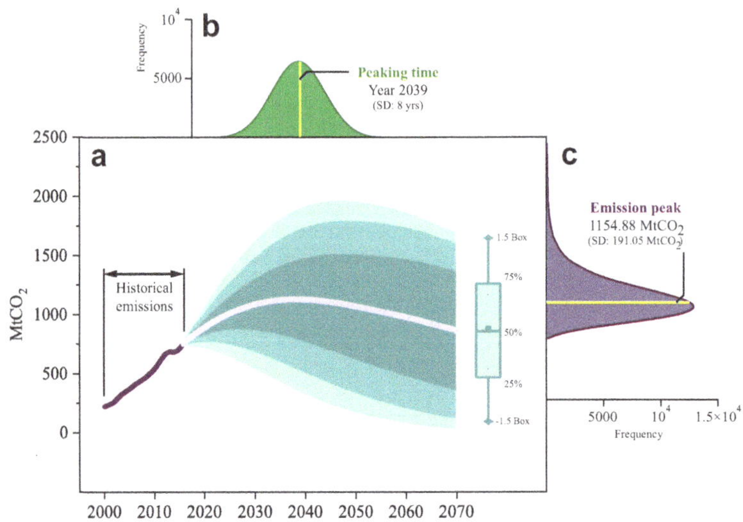


**Fig. 3** Dataset of **a** static scenarios of carbon emissions in the commercial building sector and **b–c** potential ranges of dynamic scenario simulations with different probabilities in 2000–2070

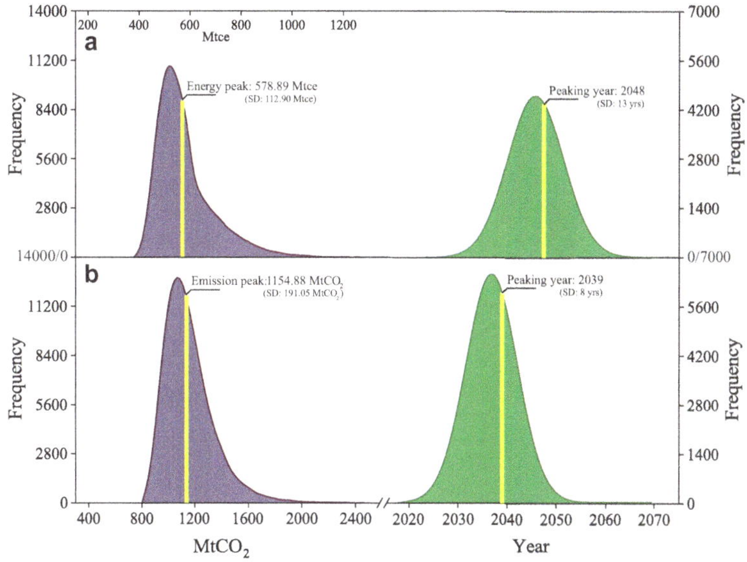


**Fig. 4** Distributions of **a** energy peak with its peaking year and **b** emission peak with its peaking year of the commercial building sector

year of carbon emissions usually appears earlier than the peaking year of energy demand in the commercial building sector. Furthermore, by applying the uncertainty measured by the scenario analysis, to achieve the energy peak goal the earliest, a strict energy demand benchmark of the commercial building sector suggests a control at 465.99 Mtce, and its peaking year is estimated for 2035, which is 13 years ahead of the BAU scenario, with energy savings of 112.90 Mtce. For the earliest peaking time of emissions, if the commercial building sector aims to achieve its emission peak in 2030 (one negative standard deviation value scenario compared to the BAU scenario), the emission peak should be controlled at 960 $MtCO_2$ approximately.

## *3.3 Sensitivity Analysis for the Emission Uncertainty*

The scenario analyses shown in Figs. 3 and 4 indicate the existence of the uncertainty of carbon emission peak and its peaking time based on the Monte Carlo simulation. Therefore, a sensitivity analysis is proposed for the uncertainty of the peak and peaking time of carbon emissions from the commercial building sector, as shown in Fig. 5. It can be observed that the impact of emission factor of commercial buildings on the uncertainty of the emission peak was the most significant, and its contribution reached 31.7%. GDP per capita and the industrial structure also positively contributed to the uncertainty of the emission peak, and their contributions reached 31.1 and 22.6%, respectively. The above factors contributed to approximately 85.4% (i.e.,

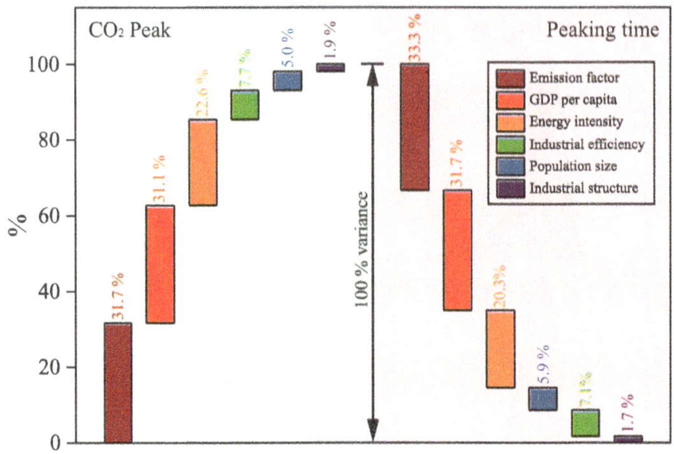


**Fig. 5** Sensitivity analysis in the commercial building sector for the uncertainty of **a** carbon emission peak and **b** the peaking time

±176.55 $MtCO_2$) of the future carbon emission peak of the commercial building sector.

Regarding the uncertainty of peaking time, the sensitivity analysis result was similar to the one shown in the emission peak. The top three factors affecting the uncertainty of peaking time in decreasing order are as follows: the emission factor (33.3%), GDP per capita (31.7%), and the industrial structure (20.3%). The above three factors contributed to over 85.3% (i.e., ±6.82 years) of the peaking time of carbon emissions in the commercial building sector. The above sensitivity analysis reveals that the emission factors, GDP per capita, and the industrial structure are determining to the uncertainty of emission peak. Therefore, they should be further discussed for the identification of a low carbon pathway for the commercial building sector in China.

## 4 Conclusions

This study proposed a historical carbon mitigation assessment and a dynamic emission scenario analysis to identify the low carbon pathways of the commercial building sector in China up to 2070. For that, a building emission model was established and the carbon mitigation during 2000–2016 was assessed by the decomposition analysis. Then, the prospective energy and emission scenarios up to 2070 were analyzed via Monte Carlo simulation. Based on the simulation result, the strict target schemes on the energy demand of the commercial building sector were proposed. Moreover, the emission benchmarks for the future commercial building sector in the 2030 peaking scheme were discussed. Furthermore, the uncertainty sources with their impacts were discussed to reinvent the future commercial building sector for it to achieve its earliest emission peak. The most important new findings are organized as follows.

**Carbon mitigation of the China's commercial building sector during 2000–2016 is 1221.50 (±486.89) MtCO$_2$.** This study established the emission model via Kaya identity to characterize the carbon emissions of the commercial building sector. Then, the decomposition analysis was applied to investigate the contribution of each factor to the carbon emissions of commercial buildings. Furthermore, the carbon mitigation of the commercial building sector was assessed based on the decomposition result. Considering the uncertainty impacts, the total carbon mitigation of the commercial building sector in 2000–2016 is 1221.50 (±486.89) MtCO$_2$. Specifically, the emission mitigation values during the following three periods are: 211.13 (±152.15, 2000–2005), 391.40 (±152.15, 2006–2010), and 618.97 (±182.58, 2011–2016) MtCO$_2$, respectively. It is believed that the continuous increase of carbon mitigation at different scales is significant.

**The commercial building sector is expected to achieve its carbon emission peak in 2039 (±8) with a peak of 1154.88 (±191.05) MtCO$_2$.** This study set the static emission scenario of the commercial building sector at the BAU level. Then, the BAU scenario was modeled using 100,000 Monte Carlo simulations. Considering the uncertainty impacts, it was observed that the commercial building sector would achieve its emission peak in 2039 (±8) with a peak of 1154.88 (±191.05) MtCO$_2$. Furthermore, the sensitivity analysis revealed that the emission factor, GDP per capita, and the industrial structure are determining to the uncertainty of emission peak and peaking time of the commercial building sector. For the earliest peaking time, if the commercial building sector aims to achieve its emission peak in 2030, the emission peak should be controlled at 960 MtCO$_2$ and the carbon mitigation will be over 195 MtCO$_2$ nationwide compared to the BAU scenario.

Due to the space limitation, several gaps that exist in this study can be addressed in the future. For example, although this study has successfully performed a dynamic emission scenario analysis to identify low carbon roadmaps of the commercial building sector in China to 2070, the specific contribution of theses mitigations to the global 1.5–2 °C goals have yet to be clarified and assessed. Therefore, there should be a considerable effort by upcoming studies to illustrate a low carbon pathway of the global commercial buildings in 1.5–2 °C scenarios.

## References

1. UN, IEA.: 2019 Global status report for buildings and construction (2019). Available at https://www.gbpn.org/china/newsroom/2019-global-status-report-buildings-and-construction. Access date 15 Aug 2020
2. Ma, M., Cai, W., Cai, W.: Carbon abatement in China's commercial building sector: a bottom-up measurement model based on Kaya-LMDI methods. Energy **165**, 350–368 (2018)

3. Röck, M., Saade, M.R.M., Balouktsi, M., Rasmussen, F.N., Birgisdottir, H., Frischknecht, R., et al.: Embodied GHG emissions of buildings—the hidden challenge for effective climate change mitigation. Appl. Energy **258**, 114107 (2020)
4. Chen, L., Cai, W., Ma, M.: Decoupling or delusion? Mapping carbon emission per capita based on the human development index in Southwest China. Sci. Total Environ. 138722 (2020)
5. Ma, M., Ma, X., Cai, W., Cai, W.: Carbon-dioxide mitigation in the residential building sector: a household scale-based assessment. Energy Convers. Manag. **198**, 111915 (2019)
6. Schäuble, D., Marian, A., Cremonese, L.: Conditions for a cost-effective application of smart thermostat systems in residential buildings. Appl. Energy **262**, 114526 (2020)
7. Yan, J., Yang, Y., Elia Campana, P., He, J.: City-level analysis of subsidy-free solar photovoltaic electricity price, profits and grid parity in China. Nat. Energy **4**, 709–717 (2019)
8. Zhou, N., Khanna, N., Feng, W., Ke, J., Levine, M.: Scenarios of energy efficiency and $CO_2$ emissions reduction potential in the buildings sector in China to year 2050. Nat. Energy **3**, 978–984 (2018)
9. Ang, B.: LMDI decomposition approach: a guide for implementation. Energy Policy **86**, 233–238 (2015)
10. Wang, H., Lu, X., Deng, Y., Sun, Y., Nielsen, C.P., Liu, Y., et al.: China's $CO_2$ peak before 2030 implied from characteristics and growth of cities. Nat. Sustain. **2**, 748–754 (2019)

# Impact of Waste Import Restriction on Carbon Emission: Evidence from East Asia

Yi Liu and Wenqian Yao

**Abstract** Evidence shows that China's implementation of waste import restriction since the beginning of 2018 provoked the waste import surge of other Asian countries. In order to avoid being the new "pollution haven" of the "North", many Asian countries started to follow China's waste import restriction policy (WIRP). WIRP has great impact on the world waste trade network, however, whether it mitigate environment pollution of these countries has not been answered. The objective of this research is to investigate the impact of WIRP on $CO_2$ emission. Using time-varying Difference-in-Difference (DID) model and seasonal data from 2015 to 2019, we find that WIRP in Asian countries has significant effects on $CO_2$ emission embodied in four major categories of waste import. Moreover, the marginal effect of WIRP is increasing. Through the analysis of mediation effects, it is concluded that policy act on the reduction of $CO_2$ emission through three decomposition channels, namely, economic scale, technological improvement, and structural change. The results are robust using the Difference in Difference in Difference model. This finding suggests trade policy on restricting waste import has great impact on world environmental pollution.

**Keywords** Time-varying difference in difference model · Waste import · $CO_2$ emission · Mediation test

## 1 Introduction

The "South" countries are natural "Pollution Haven" of the "North", since the trade and industrial structure of the "South" are more pollution intensive [1]. Advantages in weaker environmental regulations, resource endowments and pollution intensive FDI inflow from the "North" reinforced the "pollution haven effects". Massive

Y. Liu (✉) · W. Yao
School of International Trade and Economics, Jiangxi University of Finance and Economics, Nanchang 300013, China
e-mail: louisones@qq.com



F. K. S. Chan et al. (eds.), *Proceedings of the 2020 International Conference on Resource Sustainability: Sustainable Urbanisation in the BRI Era (icRS Urbanisation 2020)*, Environmental Science and Engineering,
https://doi.org/10.1007/978-981-15-9605-6_30

trade surplus of the developing countries is built upon even greater trade deficit in carbon emission [2]. The developing countries need green growth policy that balance economic growth and the environmental protection. As a result, changing trade structure could be an important channel for $CO_2$ emission mitigation in the developing countries, such as China and India. Others suggest the changing of the industrial structure may help to mitigate environmental pollution. Although, it is worried that unemployment constraint might limit the impact of industrial or trade structure adjustment efforts on emission reduction of the developing countries [3]. Regardless to say, little evidence explained how adjustment of domestic trade structure will impact on carbon emission of the developing countries. The objective of this study is to investigate the impact of trade adjustment policy on carbon dioxide emission ($CO_2$) using the case of renewable resources (recyclable waste). Using time-varying Difference-in-Difference (DID) model and seasonal data from 2015 to 2019, we find that waste import restriction policy in targeted Asian countries reduced $CO_2$ emission embodied in four major categories of waste import. The result is robust using Difference-in-Difference model, etc. We also find 1% $CO_2$ emission mitigation is accompanied with the increase of unemployment rate by 1%, nevertheless, the increase speed of marginal effect of restriction policy is faster. Implying although structural adjustment in trade and industry hurts, its effects on pollution mitigation worth it in a long run.

China's implementation of waste import restriction policy in 2018 provoked the waste import surge of other Asian countries. In order to avoid being the new "pollution haven" of the "North", many Asian countries started to follow China's waste import restriction policy (WIRP). WIRP has great impact on the world waste trade network, however, whether this trade structural adjustment policy mitigate environment pollution of these countries was not answered. Due to advantages of resource endowments of the developing countries in cheap labor and low abatement cost, most of the international trade of waste were exported from the "North" to the "South". Majority of the waste was exported to Asia, especially to China. As environmental problems mount, southeast Asian countries are coming up with solutions. In order to alleviate serious environmental problems due to waste import, India, Malaysia, Thailand, Indonesia, and other countries took measures one after another. Malaysia and Vietnam are imposing more restrictions on waste imports, Thailand banned plastic waste outright. Both Malaysia and the Philippines repatriated large amounts of waste back to their countries of origin.

Waste is a low value product with heavy weight. It is imported and recycled as an important type of recyclable resource for manufacturing input. A series of research proves that environmental policies reduce the domestic comparative advantage of domestic pollution intensive sectors, thus stimulates the pollution haven transfer to other countries and regions with lower regulation stringencies [4, 5]. Does trade policy have the same effects? For the waste importing countries, can the implementation of strict waste import management policies achieve environmental improvement? These question not only helps to demonstrate the relationship between trade policy and carbon emission intensity, but also significant for the import waste management system and environmental issues in developing countries.

In this paper, time-varying DID Model is adopted to analyze this problem. Since all countries adopted WIRP are developing countries, we randomly selected 50 developing countries as the research targets, using time-varying DID model, we evaluate the impact of WRIP on the environment of different countries in different periods. Relevant studies provide theoretical and empirical support for us on carbon emission impact factors. Using mediation test, the influence mechanism of the policy of restricting waste import on carbon emission is studied at the product level in this paper. Mediation test shows 68% of the carbon emission mitigation is coming from scale effects and 25% is from technology effects. Implying that the industrial structural change and technological innovation due to trade structural adjustment policy (restriction on recyclable resources) helps them to avoid being the "pollution haven". It is helpful to enrich the existing theoretical research on trade and the environment. Meanwhile, the research content of this paper is conducive to the further expansion of environmental impact model under pressure by Population, Affluence, and Technology (Li 2011).

## 2 Literature Review

It is important for the policy makers of the developing countries to make good environmental policy that creates more benefits and less cost. In the past few years, some researchers started to answer what environmental policy is a good one, such as, impact of pollution and income on life expectancy [6], impact of pollution reduction on fertility rate during Olympic Games [7], and the impact of environmental regulations of water pollution on enterprise productivity [8]. Policy interactions impact $CO_2$ emission in cement production, policy such as optimized production scheme can promote the use of alternative resources and energy, thus to mitigate carbon emission.

Impact of trade liberalization and the environment is also an import strand of research. Researchers argue that trade liberalization, measured by the ratio of trade to income can mitigate pollution such as $SO_2$, $NO_x$ and PM2.5 [9]. However, the causality impact between them was not found. In recent years, more and more researchers found contrary evidence. For example, trade openness can increase China's air pollution such as $SO_2$, $NO_2$ and Aerosol concentration [10]. Trade liberalization of ethylene glycol doesn't not necessarily conducive to the mitigation of global carbon emission, since the variety of ends usages with conventional energy may even damage the environment [11]. As another example, trade openness enlarges pollution of Hong Kong [12]. Most of these research focus on trade liberalization policy. However, nobody answers that what happens to the environment if a de-liberalization trade policy is adopted. We will try to fill this gap by answering this question.

Carbon emission embodied in copper trade shows that global economic event is the major cause of the fluctuations in copper mine import [13]. However, there is little empirical evidence about the impacts of such changes of resource import policy on the environment. We will try to answer this question using several representative recyclable resources. This research helps us to understand impact of trade policy on the environment of the developing countries.

Most of the studies about impact of imported waste are case studies focusing on the negative impact of hazardous elements from imported waste on human body and the environmental through biological and chemical test. Most of the study focus on small area or individuals, there is few studies about the impact of imported waste on the environment at the national level. Another difference between the previous studies and ours is that there are few studies investigate the $CO_2$ emission in recyclable resources.

Restricting or even banning the import of wastes might be helpful to promote industrial and economic transformation. However, some existing literature is skeptical about the impact of restrictions on waste import mainly because different countries have different needs of waste for industrial production. Researchers argues that, using scenario analysis, the ban on plastic waste import creates more pollution since it may induce the increase the use of virgin feedstock of Polyethylene terephthalate (PET) fiber. However, the causality relation between plastic import ban and the increase of pollution still needs to be approved. Using DID model, our research is helpful to comprehensively evaluate the effect of restricting waste import.

## 3 Data and Model Specification

### 3.1 Data

Waste import restriction of a nation will induce the increase of virgin resources. The carbon emission intensity of the virgin resource and recyclable resources is different. In order to describe the whole picture of pollution emission caused by waste import, we use specific energy carbon method to quantify $CO_2$ emission embodied in waste import. It enables us to quantify pollution from imported waste (Dong 2018). The model specification is as follow.

$$eco2i = E\mathrm{i} = Mi \times fi \tag{1}$$

$E_\mathrm{i}$ is the embodied $CO_2$ emission of imported waste, $M_i$ is the amount of imported recyclable waste resources, and f is the coefficient of carbon of producing one unit of recycled materials. $E_\mathrm{i}$ can be calculated according to waste import data from TRADEMAP statistics, efficient of $CO_2$ comes from ICE database of Base University.

FDI to the developing countries coming from the developed countries promote the transfer of pollution haven (Essandoh 2020). For example, inward FDI increases China's $CO_2$ emission [2], Other factors such as population density, GDP also can

**Table 1** Descriptive statistics

| | (1) | (2) | (3) | (4) | (5) |
|---|---|---|---|---|---|
| Vars | Definition | Median | Std. error | Min | Max |
| lneco2[a] | ln$CO_2$ | 15.46 | 4.120 | −0.0834 | 23.46 |
| Lnte[b] | Tech | 9.382 | 4.778 | 0 | 18.42 |
| Lnes[b] | Economic of scale | 7.553 | 4.051 | 0 | 15.54 |
| lnse[b] | Structural | 6.668 | 2.785 | −2.996 | 9.210 |
| lngdp[b] | Log of GDP | 7.496 | 2.107 | 3.469 | 12.79 |
| lnrunem[b] | Log of unemployment | 0.199 | 0.121 | 0.0200 | 0.710 |

*Notes* [a]TRADEMAP database; [b]WEO report

positively impact $CO_2$ emission embodied in trade of Hong Kong [12]. FDI in the recent two years are not available, thus we use

GDP, industrial structure and technological level are the main factors influencing China's $CO_2$ emission. Country size has direct effect on import of hazardous waste [14]. The developed countries import less waste than the developing countries [15]. Fast economic growth of the developing countries in Asia caused serious resource depletion and environmental damage. Sustained population growth has exacerbated their environment problems [16]. The traditional linear development model includes mining, processing, using, and discarding is unsustainable [17, 18]. Thus, the scares of resources stimulate the using of recyclable resources [19]. This research expands the decomposition model by including control variables such as, GDP level, population size and total unemployment. GDP and population size represent the level of economic activity of a nation. Descriptive statistic of the variables is written in Table 1.

A series of studies approves that stringent environmental regulation reduces import of waste, thus transferring it to other countries and regions [20, 21]. This shows that policies and regulations have great impact on the flow of waste trade. First, most of the wastes that are restricted to be imported are of high pollutants, and thus belongs to the more polluted part of the industry. Second, controlling the input of intermediate products in the industrial chain, import restriction effectively controls the scale of pollution industry, to achieve the purpose of reducing production and emission reduction.

Based on the above analysis, the following hypotheses are proposed. First, the policy of restricting waste import reduces the flow of waste trade, controls the overall economic scale of the waste industry (scale effect), and thus reduces carbon emissions. Secondly, the policy of restricting the import of waste products can reduce the import of low-priced intermediate products (structural effect), as a result, the internal industrial structure changes and cleaner industries are developed. Thirdly, the policy of restricting the import of waste makes a country's environmental standards more stringent, and it loses the original comparative advantage of environmental policies, inducing the technological upgrading of enterprises (technological effect). All three effects indirectly help to achieve the purpose of improving the environment.

# 4 Basic Model Specification and Result

## 4.1 Basic Model Specification

In recent years, the DID model has been widely used as an econometric tool to evaluate the effect of public policy implementation, the basic setting of the model is as follows:

$$y_{it} = \alpha + \theta\, treat_i \times post_t + \beta X_{it} + \mu_i + \sigma_t + \epsilon_{it} \tag{2}$$

$y_{it}$ is explained variable, i is countries. t stands for time, $\mu_i$ indicates country fixed effects. $\sigma_t$ is year fixed effect. $X_{it}$ are control variables. $\beta$ is the coefficient of control variable, $\epsilon it$ indicates random error. $treat_i$ is a country dummy variable, or the treatment group. If country i adopts waste import restriction relevant policies or measures, the value of $treat_i$ is 1; otherwise, the value is 0. $post_t$ is a year dummy variable, indicating whether the WIRP has been implemented or not. If country i enters the implementing period, the value is 1; otherwise, the value is 0. $\Theta$ represents the average treatment effect of the policy implementation, it is the key variable that measures the pre and post changes of the treatment group and the control group.

Traditional DID model assumes that treatment group receives policy shocks at the same, however, in our study, countries adopted WIRP policies at different time. Time-varying DID deals with the problem when the starting time of policy shock on treatment group is not completely consistent.

We replace multiplicative term $treat_i \times post_t$ with. $D_{it}$ for simplicity. If a country i started to implement WIRP policy in year t, we mark it as 1, otherwise, we mark it as zero. The time-varying difference in difference model is set as follow.

$$y_{it} = \alpha + \theta D_{it} + \beta X_{it} + \mu_i + \sigma_t + \epsilon_{it} \tag{3}$$

According to the basic framework of time-varying DID model, the left side is carbon emission referring to the specific energy carbon method, which comprehensively evaluate the environmental pollution caused by a country's waste import. The formula is as follows.

A strand of research implements Stochastic Impacts by Regression on Population, Affluence, and Technology (STIRIPAT) model to investigate the impact of the above-mentioned variables, including economic development level, industrial structure, and R&D on carbon emission in urbans. For example, Shahbaz et al. [22] investigate the impact of urbanization and economic development on $CO_2$ emission of Malaysia. STIRIPAT model is the extension and comprehensive evaluation of the environmental pressure. We include the following variables as our independent variable, including GDP per capita and unemployment rate.

$$lneco2_{it} = \alpha + \theta D_{it} + \beta_1 lnrunem_{it} + \beta_2 lngdp_{it} + \beta_3 lnes_{it} + \beta_4 lnte_{it} + \beta_5 lnse_{it} + \mu_i + \sigma_t + \epsilon_{it} \quad (4)$$

$lneco2_{it}$ represents carbon emissions of country i in quarter t. $D_{it}$ is the treatment effect. $\mu_i$ indicates country fixed effects. $\sigma_t$ is year fixed effect. $X_{it}$ are control variables. $lnrunem_{it}$ and $\beta_2 lngdp_{it}$ indicates unemployment and GDP per capita of country i in year t. In order to test the direct impact of scale effect, technical effect and structural effect on carbon emissions, $lnes_{it}$, $lnte_{it}$ and $lnse_{it}$ are introduced based on Eq. 5, indicates scale effect, technical effect and structural effect of country i in year t.

## 4.2 Basic Model Results

The policy of waste import restriction mainly restricts the import volume of wastes from various countries. Therefore, it has a direct impact on the import volume of wastes, while its effect on carbon emissions still needs to be investigated. Therefore, this paper adopts the multi-stage dual difference model to analyze the impact of the policy on carbon emissions of various countries (Table 2). At the same time, the policy interaction terms of economic effect, technological effect and structural effect are introduced to test the mechanism of policy's effect on carbon emission. The parallel trend test results show that the policy of restricting the import of waste has a lag effect on carbon emissions, so the carbon emissions are treated with a lag phase in this paper.

Regression results show that for the panel data set with carbon emissions as the explained variable, the regression coefficient of interaction term D is always negative at the significance level of 10% after all the control variables such as economic development level and unemployment rate are included in the regression equation (Table 2, C1 and C2). This fact shows that the policy of restricting the import of waste can significantly reduce the carbon emission of each country and thus improve the environment. At the same time (Table 2, C3), from the regression results of economic effect and technology effect, economic scale expansion will increase carbon emissions, technology effect upgrade and carbon emissions show an inverse relationship, and structural effect regression coefficient is not significant. This result basically conforms to the empirical research conclusion, that is, the structural effect of trade is often different from country to country, and the effect is not significant (Antweiler 2001; [23]). It is reasonable to assume that the indirect effect of the policy exists, that is, the intermediary effect may exist, from the result that the impact of the policy on carbon emission is always significant.

Table 2 Time-varying DID test

| | (1) | (2) | (3) |
|---|---|---|---|
| Variables | Carbon emission | Carbon emission | Carbon emission |
| D | −0.963*** | −0.987*** | −1.015*** |
| | (−3.28) | (−3.36) | (−3.49) |
| Unemployment | | −0.930** | −1.007** |
| | | (−2.20) | (−2.40) |
| GDP | | −0.325 | −0.418 |
| | | (−0.69) | (−0.89) |
| Economic scale | | | 0.096*** |
| | | | (3.00) |
| Tech | | | −0.036* |
| | | | (−1.73) |
| Structure | | | −0.054 |
| | | | (−1.19) |
| Constant | 15.348*** | 16.153*** | 16.673*** |
| | (83.12) | (4.63) | (4.76) |
| Obs | 855 | 855 | 855 |
| Prob > F | 0.000 | 0.000 | 0.000 |
| Num of nations | 44 | 44 | 44 |
| Country fixed | Y | Y | Y |
| Year fixed | Y | Y | Y |

*Note* ***$p < 0.01$, **$p < 0.05$, *$p < 0.1$. It stands for statistically significant level at 1, 5 and 10%

# 5 Mediating Effect

In recent years, mediating effect has been widely used to study the mechanism of relationship between policy influence and independent variables. For example, in the relationship between trade liberalization and haze pollution, through the mediating effect method, it is found that the positive effect of technology on reducing haze pollution has exceeded the scale effect and component effect [24].

Considering the impact of independent variables X on dependent variable Y, if X exerts a shadow on Y by influencing variable M, then M is called the intervening variable (mediator variable). In this paper, the model of indirect influence of WIRP is set as follows.

$$\begin{cases} lneco2_{it} = \theta D_{it} + \beta X_{it} + \mu_i + \sigma_t + \epsilon_{it} \\ M_{it} = \alpha_i D_{it} + \mu_i + \sigma_t + \epsilon_{it} \\ lneco2_{it} = \theta' D_{it} + \sum_{i=1}^{3} \gamma_i M_{it} + \beta X_{it} + \mu_i + \sigma_t + \epsilon_{it} \end{cases} \quad (5)$$

where coefficient $\theta$ is the total effect of policy $D_{it}$ on carbon emissions. $M_{it}$ is the mediation variable. It represents the scale effect, technical effect, and structural effect. $\alpha_i$ is the effect of policy $D_{it}$ on the mediating variable $M_{it}$. Coefficient $\gamma_i$ is the effect of the mediating variable M on carbon emissions after controlling for the influence of policy $D_{it}$. Coefficient $\theta'$ is the effect of policy $D_{it}$ on $CO_2$ emission when controlling for the mediating variable M. $\epsilon_{it}$ is the residual. If the parameter $\beta$ is significant, and if the $\alpha_i$ and $\gamma_i$ are statistically significant, it indicates WIRP affect $CO_2$ emission, through mediator M.

In this mediation model, the mediating effect is equal to the indirect effect, which is the coefficient of product $\alpha_i * \gamma_i$. It has the following relation with total effect and direct effect.

$$\theta = \theta' + \alpha_i * \gamma_i \quad (6)$$

We use stepwise regression and bootstrap method to analyze the mediating effects according to Eq. (5). A common way to test the mediating effect is to test the regression coefficient step by step, which is also known as the stepwise method. In the first step, the mediating effects hypothesis hold if the coefficient $\theta$ is significant. Otherwise, the shadowing effects hold. In the second step, we test the coefficients of the equation $\alpha_i$ and $\gamma_i$. If both coefficients are significant, the indirect effects hold. If at least one of them is not significant, proceed to step 3 [25].

The third step, we use bootstrap to test hypothesis that $H0 : \alpha_i * \gamma_i = 0$. If the null hypothesis is significant, indirect effect hold. Otherwise, the whole process ends. First, according to C1 and C2 of Table 3, the impact of WIRP on scale effect ($\alpha_i$) and technology effects ($\gamma_i$) are statistically negative. Indicating they are playing a partial mediating role in restricting. It shows that scale and technology effect are mediating the WIRP in the inhibition of carbon emissions. Since the regression coefficients of

**Table 3** Direct effects

| | (1) | (2) | (3) |
|---|---|---|---|
| Vars | Economic scale | Tech | Structural |
| D | 4.879*** | −2.984*** | 0.632 |
| | (0.746) | (0.900) | (0.544) |
| Constant | −0.0658 | 19.01*** | 7.548*** |
| | (0.905) | (1.092) | (0.659) |
| Obs | 855 | 855 | 855 |
| Prob > F | 0.000 | 0.000 | 0.000 |
| Num of nations | 44 | 44 | 44 |
| Country fixed | Y | Y | Y |
| Year fixed | Y | Y | Y |
| Controls | Y | Y | Y |

*Note* $\alpha_i$; ***$p < 0.01$, **$p < 0.05$ and *$p < 0.1$ stand for statistically significant level at 1, 5, and 10%

**Table 4** Mediation effects

| | Coef. | Std. Err. | z | P > z | [95%Conf. | Interval] |
|---|---|---|---|---|---|---|
| Economic of scale# | 3.056 | 0.356 | 8.580 | 0.000 | 2.358 | 3.755 |
| Tech# | 1.136 | 0.195 | 5.820 | 0.000 | 0.753 | 1.519 |
| Structural# | 0.299 | 0.083 | 3.620 | 0.000 | 0.137 | 0.462 |
| Total effects# | 4.492 | 0.465 | 9.670 | 0.000 | 3.581 | 5.402 |

*Notes* The results of this table are derived from using bootstrap 5000 times. #stands for the mediation effects of WIPR on carbon emission, $\alpha_i * \gamma_i$. $p < 0.01$, $p < 0.05$, $p < 0.1$, stand for statistically significant value at 1, 5, and 10%

structural effects in Tables 2 and 3 are insignificant, according to the procedure of mediating effect test [26], we need to bootstrap test to confirm whether the structural effect is the mediating variable of WIRP on carbon emissions. According to Table 3, structural effect is significant, indicating that structural effect is also a mediating variable.

The results of Table 4 show that restricted imports of waste policy indirectly affect three hypotheses can be established, but the scale effect, technological effect and structure effect to explain the impact strength is waning, scale effect explain restricted imports of waste policy inhibition of carbon emissions is one of the most effective (accounting for 68% of the overall effect of mediation), control of waste industry economies of scale can effectively control the overall carbon emissions. The mediating effect of improving the overall carbon emission intensity through stimulus technology is not significant in the sample range, accounting for 25% of the overall mediating effect. When the two groups of direct effects of policies are not significant, the mediating effect is significant at the significance level of 1%, accounting for 7% of the overall mediating effect. This indicates that it is still difficult for developing countries to directly change the overall industry composition through policies, but some objective facts of mediating effect still exist.

## 6 Conclusion

In the context of severe environmental pollution, can the developing countries mitigate carbon emission by adopting waste import restriction policy? In this paper, the impact of WIRP on carbon emission was empirically by constructing a time-varying DID model to control for the potential endogenous problems. The Preliminary estimates of quarterly data show that the policy has had a significant impact on carbon emissions reduction. Marginal effects of WIRP was 2% a year in Asia quarterly. The carbon emission embodied in waste import has been reduced by 8% until the end of 2019. The results of mediation mechanism analysis show that scale effect, technology effect and structure indirectly affect the carbon emission inhibition effect of waste import restriction policy. The research results showed that the effect of WIRP on carbon emission in various countries would not be affected by China's waste ban. Based on this research, this paper proposes the following policy recommendations.

Controlling recycling industry by the restriction the size and types of the waste import may help to reduce pollution intensity of the domestic industries. In the process of waste importing of the developing countries, a large amount of hazardous wastes is often carried along with them, which has a profound impact on the industrial structure and pollution intensity of developing countries. As a result, cooperative surveillance system and harmless waste treatment technology transfer between the "north" and the "south" is very important to mitigate the negative effects of waste import of the developing countries. Cutting the import of recyclable resources will generate negative impact on domestic employment. Empirical evidence shows the magnitude of carbon emission reduction is two times higher than the employment due to the multiplication of mediating effects. What impact will WIRP has on employment and long-term carbon emission and economic growth is unclear. Further study awaits the availability of long-term data.

**Acknowledgements** This research is funded under National Social Science Foundation of China <Grant NO: 16CJY025>.

## References

1. Wu, X.D., Guo, J.L., Li, C., Chen, G.Q., Ji, X.: Carbon emissions embodied in the global supply chain: intermediate and final trade imbalances. Sci. Total Environ. **707**, 134670 (2019)
2. Ren, S., Yuan, B., Ma, X., Chen, X.: The impact of international trade on China's industrial carbon emissions since its entry into WTO. Energy Policy **69**, 624–634 (2014)
3. Tang, X., Jin, Y., Wang, X., Wang, J., McLellan, B.C.: Will China's trade restructuring reduce $CO_2$ emissions embodied in international exports? J. Clean. Prod. **161**, 1094–1103 (2017)
4. Kellenberg, D.: An empirical investigation of the pollution haven effect with strategic environment and trade policy. J. Int. Econ. **78**, 242–255 (2009)
5. Levinson, A: Taylor MS. Unmasking the pollution haven effect. Int. Econ. Rev. **49**, 223–254 (2008)
6. Ebenstein, A., Fan, M., Greenstone, M., He, G., Yin, P., Zhou, M.: Growth, pollution, and life expectancy: China from 1991–2012. Am. Econ. Rev. **105**(5), 226–231 (2015)
7. He, G., Fan, M., Zhou, M.: The effect of air pollution on mortality in China: evidence from the 2008 Beijing Olympic Games. J. Environ. Econ. Manag. **79**, 18–39 (2016)
8. He, G., Wang, S., Zhang, B.: Watering down environmental regulation in China. Q. J. Econ. **135**(4), 2135–2185 (2020)
9. Frankel, J.A., Rose, A.K.: Trade good or bad for the environment? Sorting out the causality. NBER. Working Paper No. 9021 (2002)
10. Lin, F.: Trade openness and air pollution: city-level empirical evidence from China. China Econ. Rev. **45**, 78–88 (2017)
11. Hu, X., Pollitt, H., Pirie, J., Mercure, J.-F., Liu, J., Meng, J., Tao, S.: The impacts of the trade liberalization of environmental goods on power system and $CO_2$ emissions. Energy Policy **140**, 111173 (2020)
12. Huang, R., Lv, G., Chen, M., Zhu, Z.: $CO_2$ emissions embodied in trade: evidence for Hong Kong SAR. J. Clean. Prod. **239**(1), 117918 (2019)

13. Dong, D., An, H., Huang, S.: The transfer of embodied carbon in copper international trade: An industry chain perspective. Res. Policy **52**, 173–180 (2017)
14. Keisaku, H., Managi, S.: Determinants of trade in recyclable wastes: evidence from commodity-based trade of waste and scrap. Environ. Dev. Econ. (19), 250–270 (2014)
15. Mazzanti, M., Zoboli, R.: International waste trade: impacts and drivers. Waste Manag. Spatial Environ., 99–136 (2013)
16. Ekins, P.: The Kuznets curve for the environment and economic growth: examining the evidence. Environ. Plan. A **29**(5), 805–830 (1997)
17. Korhonen, J., Honkasalo, A., Seppälä, J.: Circular economy: the concept and its limitations. Ecol. Econ. **143**, 37–46 (2018)
18. Preston, F.: A Global Redesign? Shaping the Circular Economy. Chatham House, London (2012)
19. García-Barragán, J.F., Eyckmans, J., Rousseau, S.: Defining and measuring the circular economy: a mathematical approach. Ecol. Econ. **157**, 369–372 (2019)
20. Dubois, M.: Towards a coherent European approach for taxation of combustible waste. Waste Manag. **33**, 1776–1783 (2013)
21. Fikru, M.G.: Trans boundary movement of hazardous waste: evidence from a new micro-data in the European Union. Rev. Eur. Stud. **4**(1), 3–14 (2012)
22. Shahbaz, M., Loganathan, N., Muzaffar, A.T., Ahmed, K.: How UrbaNization affects $CO_2$ emissions in Malaysia? The application of STIRPAT. Model Renew. Sustain. Energy Rev. **57**, 83–93 (2016)
23. Brunel, C.: Pollution offshoring and emission reductions in EU and US manufacturing. Environ. Res. Econ. **68**(3), 621–641 (2016). https://doi.org/10.1007/s10640-016-0035-1
24. Xu, Y., Fan, X., Zhang, Z., Zhang, R.: Trade liberalization and haze pollution: evidence from China. Ecol. Ind. **109**, 105825 (2020)
25. Baron, R.M., Kenny, D.A.: The moderator-mediator variable distinction in social psychological research: conceptual, strategic and statistical considerations. J. Pers. Soc. Psychol. **51**, 1173–1182 (1986)
26. Efron, B., Tibshirani, R.J.: An Introduction to the Bootstra. Chapman & Hall, CRC, Boca Raton (1993)

# Circular Economy Legislation and Environmental Pollution: Evidence from Urban Mining Pilot Cities in China

**Hongcheng Shen and Yi Liu**

**Abstract** Urban Mining Pilot Policy (UMP) is an important implementation of the Circular Economy Promotion Law of China. It promotes the building of a circular economy industry and environment-friendly society. This study applies time-varying difference-in-difference estimation on city-level panel data to estimate the impacts of gradually adoption of UMP on pollution reduction of China. To avoid model dependence and distribution imbalance of control and treatment groups, Propensity Score Matching-Difference in Difference (PSM-DID) is also adopted. The results show us, gradual adoption of UMP among municipal cities significantly improves waste reuse and reduces pollutant emission. We use various types of data and robustness check. The impact of UMP policy on improving China's environmental condition is positive and significant.

**Keywords** Urban mining · Environmental pollution · Industrial solid wastes · Time-varying DID

## 1 Introduction

Waste pollution is the major conflict in the harmonious coexistence between economic growth and good environment. If improperly managed, it leads to severe damage to the environment [5] and human health [7, 19, 35], (Liu and Salvo 2018). To encounter such problem, governments around the world spare no effort in reducing the negative impacts from industrial and urban waste. In the meanwhile, the world is increasingly dependent on material recycling due to environmental reasons and primary resource scarcity [21]. Under this circumstance, major economies like USA, EU, Japan and China enacted Circular Economy Promotion Law (CEPL) in

H. Shen
Taizhou University, Taizhou 318000, China

Y. Liu (✉)
Jiangxi University of Finance and Economics, Nanchang 300013, China
e-mail: louisones@qq.com



F. K. S. Chan et al. (eds.), *Proceedings of the 2020 International Conference on Resource Sustainability: Sustainable Urbanisation in the BRI Era (icRS Urbanisation 2020)*, Environmental Science and Engineering,
https://doi.org/10.1007/978-981-15-9605-6_31

order to promote the development of resources recycling industry. The development of circular economic industry helps to maintain the harmonized relation between sustainable economic growth and environmental safety [9]. The objective of this research is to investigate the impact of CEPL on the environmental performance of municipalities in China. Using several models including, difference-in-difference model, difference-in-difference-indifferences mode, propensity score matching and time-varying difference in difference model (PSM-DID), we found adoption of Urban Mining Pilot Bases (UMPB) significantly improves the environment performance of that cities. That is the environmental performance indicators (measured by industrial solid waste recycling rate, harmless treatment rate of municipal solid waste, $SO_2$ emission and $PM_{2.5}$ concentration) have been greatly improved in the cities that were permitted to establish UMPB.

The rapid industrialization and urbanization have made China the second largest economy in the world, but it led to resource exhaustion and serious environmental pollution. Urban Mining (UM) has attracted great attention from both academic and industrial sectors since it recovers rare and precious metals (RPMs) and it is compared to on-surface stock of natural resource in technosphere because it contains abundant of renewable resources including copper, iron, aluminum while primary mining is compared to underground stock in geo-sphere [24, 36]. In order to mitigate the environmental pollution, the Chinese central government has promulgated UMPB to balance economic growth, save resources and protect the environment. In addition, UMPB program aims at fostering comprehensive recovering capability of varies types of wastes such as municipal, urban construction, general industrial solid and hazardous solid waste [28]. Because of fast economic growth and rapid urbanization. China's collected waste increased tremendous since 2000. In 2016, China collected a total volume of 0.256 billion tons of waste material, increased by 3.7% than year 2015 (see Fig. 1).

Since 2010, China has introduced UMPB program and tested it in a range of cities to promote resource recycling industrial development and improve resource utilization [33]. UM has been the key strategic industry of China's 12th Five-year Plan (FYP) and improved the efficiency in combating municipal solid waste by facilitating recycling network and large-scale handling of waste through financial aid and fiscal policies. The adoption of UMPB forms a circular economy development model of

**Fig. 1** UM collection amount of waste materials in China between 2006 and 2016

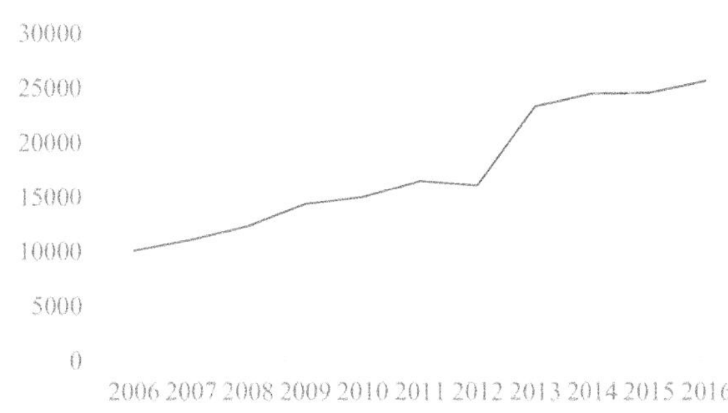

**Table 1** The implementation of UMP and candidate cities from 2010 to 2015

| Year | Batch | Cities | UMP candidate city |
|---|---|---|---|
| 2010 | First batch | 7 cities | Tianjin, Ningbo, Yueyang, Qingyuan, Fuyang, Qingdao, Neijiang |
| 2011 | Second batch | 15 cities | Shanghai, Wuzhou, Xuzhou, Linyi, Chongqing, Hangzhou, Xiangfan, Dalian, Xinyu, Tangshan, Xuchang, Fuzhou, Yinchuan, Beijing, Dandong |
| 2012 | Third batch | 6 cities | Foshan, Chuzhou, Bazhou, Yangquan, Qitaihe, Chenzhou |
| 2013 | Fourth batch | 10 cities | Jingmen, Yingtan, Nantong, Taizhou, Xingtai, Mianyang, Luoyang, Guiyang, Quanzhou, Xiamen |
| 2014 | Fifth batch | 6 cities | Yantai, Baotou, Lanzhou, Kelamayi, Harbin, Yulin |
| 2015 | Sixth batch | 5 cities | Taaizhou, Yichun, Huangshi, Baoding, Xianyang |
| Total | | 49 cities | |

"Reduce, Reuse, and Resource" (3Rs) [27]. UM plays an important role in easing resource scarcity and it is a viable way to reduce natural resource exploitation and protect resource supplies [33]. Annual ISW recycled raised from a merely less than 0.5 billion tons in 2000 to about 1.8 billion tons in 2017.

During 2010 and 2015, NDRC of China approved 49 pilot cities. Figure 1 shows UM collection of waste each year during 2010 to 2015. The trend has obvious characteristics of batching and gradual promotion, which is the reason for designing time-varying DID as empirical analysis methods (see Table 1). Five provinces, namely Tibet, Yunnan, Qinghai, Hainan, Jilin do not have UMPB. A total of ten provinces set up one UMPB, and eight provinces set up two UMPBs, eight provinces set up three UMPBs. Most of the provinces that have UMPBs locate in the eastern area (see Fig. 2). Therefore, this study believes that the establishment of UMPB has regional heterogeneity and following the standard of National Bureau of Statistics of China, the authors divide the 31 provinces into eastern, central, western, and northeast China [20].

Our research object conforms to the premise of natural experiments. First, the experimental group and the treatment group are randomly chosen. In this paper, the policy maker and the experimental group do not know which cities will be selected as treatment group. Second, the UM policy is to improve the recovery rate of industrial and municipal solid wastes. The preconditions for cities applying for UMPB are to have a complete circular economy industrial system and sound technologies for recycling, which are independent of the environmental conditions of a city. In other words, the environmental conditions of the UMPB and the UM policy are independent. The government's choice of UMPB and the cities applying for UMPB are not determined by the degree of environmental pollution. Therefore, the impact of the establishment of the UMPB on the environmental pollution of the city conforms to the premise of the natural experiment. This study innovates by treating UMPB

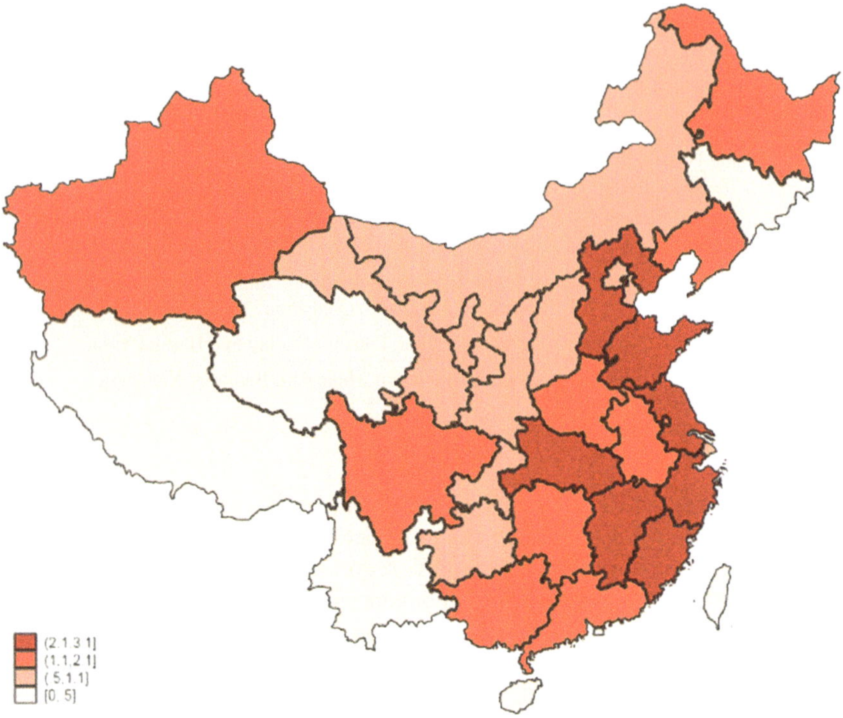


**Fig. 2** Geographical distribution of UMPs in China

as a quasi-natural experiment of circular economy legislation in mitigating environmental pollution and promoting circular economy and evaluates the policy impact of UMPB and thus contributes to the existing body of literature on UM.

## 2 Literature Review

Different levels of data have been applied in the studies addressing the roles of government regulation and policies in mitigating environmental pollution. These studies include national level, province-level, and city level. For examples, Auffhammer [3] used province-level data to forecast the path of China's $CO_2$ emission. County level data was used by [5] to study the consequences of China's central government's water pollution reduction mandates. Chen et al. [6] used single city level to study the impact of the 2008 Olympic Games on air quality. Fu et al. [14] adopts multiple city level data to investigate the relationship between highway toll and air pollution.

Different empirical models have been used to study the effect of environmental regulation and policy on environmental protection in China, such as panel data model [3], case studies [36], Difference-in-difference-in-differences (DDD) model

[5, 25], regression discontinuity design (RDD), differences-in-differences method (DID) [14], and instrumental variables. As other examples, Li et al. [9] constructs a two-stage DEA model with undesirable inputs and evaluates the industrial waste recycling and reuse efficacy in China during 2006 and 2015. Many modeling techniques have been adopted in studying solid waste generation and recycling, including regression models [29], time series analysis [10, 34], system dynamics [23], spatial panel model [6, 8] and input-output analysis [22].

UMPB brings significant resource effects [15] and obvious environmental effects [4, 31]. Eygen et al. [13] derives UM reduces resource consumption by 80% and 87% through recovery of desktop computers and notebooks comparing to primary mining. UM recyclers play an important role in recovering resources and redirecting them into productive cycles, performing a vital environmental service to local community [16]. The main tasks of establishing UMPB are to form recycled copper, aluminum, lead, and recover waste plastics, reduce the external dependence on important strategic resources such as iron ore, petroleum, ensure national economic security, and solve environmental pollution problems.

Existing research on UM has mainly centered on investigating the quantities, scale and spatial location of metal stocks and on economic and environmental motives of cable recovery [21]. Few have paid much attention to the effects and consequences of the implementation of UMPB. The impact of UM implementation policy on improving environmental pollution through increasing industrial solid wastes recycling have not been thoroughly researched. Most existing literatures on UM are based on qualitative analysis, few adopts empirical methods to study the impact of UM on environment protection or economic development.

# 3 Data and Methodology

## *3.1 Data*

This article applies a panel dataset of 285 municipal cities in 31 provinces from 2003 to 2016, including 4 municipalities directly under the central government's control. They are Shanghai, Beijing, Chongqing, and Tianjin. The data used in this study is obtained from the China City Statistical Yearbook. The PM2.5 data of the cities are from Beijing City Lab (BCL). In the process of selecting the treatment group and control group, this article made several adjustments of the raw data because a series of adjustments have been made to the setting of China's prefecture-level cities. Lhasa in Tibet was not included in the study because of lack of corresponding data. The dependent and independent variables of this paper are as follow.

**$SO_2$Emission**: Following Dong et al. [11], this paper uses industrial $SO_2$ emission volume and $PM_{2.5}$ concentration density level to measure air pollution in China. Industrial $SO_2$ has been gradually alleviated from 16 Mio. Tons in 2010 to 7.1 Mio. Tons in 2017.

**ISW**: Following Lamia and Sami [1], the Industrial Solid Waste regeneration rate (ISW) is chosen since it implies recycling ability of pilot cities. The higher the ISW reuse rate, the better chance to reduce ISW pollution.

**MSW**: MSW is measured by the ratio of harmless treatment of municipal solid waste of each city in China. It reflects local government's effort in protecting the environment. In addition, to robustly test the impact of UMPB pilot program on environmental alleviation, we also select industrial $SO_2$ emission and pm 2.5 density index to reflect environmental pollution level. Definition of each dependent variable is as follow.

**GDP**: is adopted to reflect the scale effect of real GDP of each city. To eliminate the price effect, GDP is deflated into 2003 constant price.

**Composition Effect (CMP)**: We use the interaction between the share of secondary industrial carbon emission as total carbon emission and the ratio of second industry to total GDP output to measure the composition effect.

**Technology Effect (Tech)**: The interaction between the share of secondary industrial pollution as total pollution and the average pollution of secondary generate by each unit of GDP.

**Industrial Structure (Sigdprate)**: Sigdprate is represented by the ratio of secondary industry to total GDP. Generally, second industry generates ISW and leads to pollutant emission including $SO_2$ and $PM_{2.5}$.

**Population Density (Popudens)**: Urbanization leads to increasing consumption of various goods and residents.

**Real Foreign Direct Investment (RFDI)**: For the developing countries, inward FDI especially actually utilized foreign capital enlarges energy intensive and pollution intensive manufacturing sectors. RFDI tends to increase pollution.

Table 2 presents the descriptive statistics for the variables used in this research.

**Table 2.** Descriptive statistics

| Variable | Definitions (unit) | Obs | Mean | Min | Max |
|---|---|---|---|---|---|
| ISW | ISW regeneration rate (%) | 3990 | 76.935 | .49 | 143.24 |
| MSW | MSW harmless treatment rate (%) | 3990 | 73.394 | .44 | 157.94 |
| $SO_2$ | Industrial $SO_2$ emission (ton) | 3990 | 10.413 | .693 | 13.434 |
| $PM_{2.5}$ | Annual $PM_{2.5}$ concentration (ug/m$^3$) | 3990 | 36.523 | 4.517 | 90.856 |
| lnPopudens | Population density (person/sq. km) | 3990 | 5.718 | 1.548 | 7.887 |
| Sigdprate | Proportion of 2nd industry to GDP (%) | 3990 | 48.636 | 9 | 90.97 |
| lnRfdi | Foreign capital actually utilized ($10,000) | 3990 | 8.999 | −2.408 | 14.941 |
| LnGDP | Gross regional product (yuan) | 3856 | 15.977 | 10.34 | 19.191 |
| lnCmpnt | Cargo trans. Ratio*2nd ind. GDP ratio (%) | 3854 | 3.471 | 1.441 | 4.291 |
| lnTech | Tech development in cargo trans. (%) | 3854 | −12.34 | −17.601 | −6.862 |

### 3.2 Model Specification

The research about the impact of the UMPB is as a quasi-natural experiment. Many factors may affect the comprehensive utilization rate of waste of pilot cities. Following Wang [32], by controlling other factors, the differences between the treated group and the control group before and after the policy occur are compared to evaluate the effect of the policy. A sample of pilot cities established between 2010 and 2015. If a city is batched as a pilot city in year t, it is marked as 1, otherwise, it is marked as 0. Following Heckman [18] and Leuven [26], DID method is adopted to avoid bias general factors that affect the dependent variable.

We are going to use several factors to indicate the changing of municipal environment conditions, they are utilization of ISW, harmless treatment rate of MSW. ISW and MSW utilization rates are adopted to reflect urban environmental governance performance while SO2 emission and PM2.5 density.This setting automatically generates a "treated group" and a "control group". Time-varying DID method is based on the regression as follow:

$$\text{Pollution}_{ct} = \alpha + \beta D_{ct} + \gamma X_{ct} + A_c + B_t + \varepsilon_{ct} \quad (1)$$

Based on Eq. (1), PSM-DID is used to conduct robust estimation based on regression model. First, PSM is used to find the control group that is closest to the characteristics of the treatment group; Second, we use the DID regression. The specific model is as follow:

$$\text{Pollution}_{ct} PSM = \alpha + \beta D_{ct} + \gamma X_{ct} + A_c + B_t + \varepsilon_{ct} \quad (2)$$

In Eqs. (1) and (2), dependent variables including several indicators, such as ISW reutilization rate, harmless treatment of MSW, industrial $SO_2$ emission, and PM2.5 concentration in city $c$ in year $t$. $A_c$ and $B_t$ are vectors of city and year dummy variables that account for city and year fixed effects. $X_{ct}$ is a set of time varying city-level control variables. $\varepsilon_{ct}$ is the error term. The key independent variable is $D_{ct}$, which is a dummy variable that equals 1 in the years after city $c$ becomes an UMPB and 0 otherwise. Coefficient, $\beta$, examines the effect of UMPB on dependent variables.

UMPB is set up in industrial agglomeration and development zones with a certain scale, which can form a more complete recycling network and help to extend the industrial chain. Moreover, recycling companies in the park where UMPB is located can share infrastructure, logistics facilities and information service facilities, and the by-products between industrial enterprises can be exchanged and used more efficiently, which reduces the freight cost and improves the efficiency of industrial waste treatment. Furthermore, this paper uses the difference-in-difference-in-difference (DDD) method to verify the impact of UMPB on pollution reduction. Using DDD requires finding another pair of "treatment group" and "control group". The authors select provinces that have established UMPB as the treatment group, and provinces that have not established UMPB as the control group. The DDD model is designed as follows:

$$\begin{aligned}\text{Pollution}_{ct} = {} & \alpha + \beta 1 time * treat * group + \beta 2 time * treat \\ & + \beta 3 time * \text{group} + \beta 4 treat * group \\ & + \gamma X_{ct} + A_c + B_t + \varepsilon_{ct}\end{aligned} \tag{3}$$

Group is a dummy variable. When the sample city is in a province where UMPB is implemented, it is assigned a value of 1, otherwise it is 0. The definition of other variables is the same as Eq. (2). This article is interested in the estimated coefficient $\beta 1$ of time*treated*group. In order to test the mechanism of UMPB affecting pollution alleviation, this research further examines the channels based on the mediation model which is usually adopted to offer a mechanism test to clarify how the explanatory variables affect the explained variables via mediating variables [12]. Following Antweiler et al. [2], this paper divides the effects of industrial development on environmental pollution into three effects. First, scale effect, which means that the economic development will consume more industrial resources and lead to more industrial pollution; secondly, composition effect, which refers to domestic industrial structure, the logic behind composition effect is that different industrial structure, namely agriculture, manufacturing industry and tertiary industry will generate different levels of pollution and affect environment verily; and thirdly, technology effect, which means that newly introduced technology will positively reduce environmental pollution by lowering pollutant emissions. The recursive equations to test the mediation effect are designed as follows:

$$\text{Yit} = \alpha_1 + \alpha_2 * DID + \alpha_3 * X_{it} + \eta t + \mu_i + \varepsilon it \tag{4}$$

$$\text{Mit} = \alpha_4 + \alpha_5 * DID + \alpha_5 * X_{it} + \text{vt} + \delta_i + \theta it \tag{5}$$

where Mit is the mediation variable, including scale effect (GDP), composition effect (COMP), and technology effect (TECHNO). Yit is a set of pollution variables, including industrial sulfur dioxide discharge volume and PM2.5 density level. If parameter $\alpha_2$ in Eq. (4) is significant and if the parameter $\alpha_5$ in Eq. (5) is statistically significant, it means DID affects Yit through mediator Mit.

# 4 Empirical Results

## 4.1 Impact of UMP Construction on Environmental Pollution

According to the baseline regression results, UMP program has a significant impact on improving the comprehensive utilization rate of ISW, the harmless treatment rate of MSW and reducing industrial $SO_2$ discharge and PM2.5 concentration, whether control variables are added. UMP construction has significantly reduced urban environmental pollution. Among them, UMP has significantly improved the utilization rate of ISW by about 2.68%, the harmless treatment rate of MSW by 13.60, and 0.39 Industrial $SO_2$ emissions and PM2.5 concentration by 0.80 (see Table 3).

### 4.2 PSM-DID Method

Due to the heterogeneity among cities in China, it is difficult to meet the conditions of consistent time effect. Therefore, we also apply PSM approach raised by Heckman [17] and Rosenbaum and Rubin [30] to avoid potential bias by paring treatment cities with cities that have similar observed attributes from the control group.

The results of the common support hypothesis test show that there is no significant difference between the covariates after matching. The kernel density function curve is drawn after matching the propensity scores of the treatment group and the control group (see Fig. 3). According to Fig. 3, the probability density distribution of the propensity score values of the two groups of samples is obviously different. The difference between the sample cities of the group will inevitably produce biased estimation. After the matching is completed, the characteristics of the two groups of sample cities are very close, and the selection bias of the samples is basically eliminated.

The results show that the core explanatory variables (Treated*Post) that capture the effects of different periods have significant coefficients in each model, indicating that UMPB has significantly improved the level of ISW utilization and harmless treatment of MSW in UMPB. In addition, UMPB decreased both industrial $SO_2$ emission and PM2.5 density level (see Table 4).

### 4.3 DDD Method

Table 5 reports the average treatment effect estimated by the DDD method, and the result is basically consistent with DID and PSM-DID results, indicating that the UMPB policy has significantly improved the ISW utilization rate and the harmless treatment rate of MSW. At the same time, it significantly reduces industrial $SO_2$ emissions and PM2.5 concentration values.

## 5 Conclusion and Policy Implications

In order to test the impact of UMPB on environmental pollution. We have done the following tests, firstly, the relationship between UMPB construction and ISW

**Table 3** Baseline results

| | (1) | (2) | (3) | (4) | (5) | (6) | (7) | (8) |
|---|---|---|---|---|---|---|---|---|
| | ISW | MSW | SO2 | PM2.5 | ISW | MSW | $SO_2$ | PM2.5 |
| $D_{it}$ | 3.3590** | 15.6997*** | −0.3927*** | −0.8019** | 2.6768* | 13.6012*** | −0.3599*** | −0.8757** |
| | (1.48) | (2.40) | (0.09) | (0.39) | (1.47) | (2.36) | (0.09) | (0.39) |
| lnrfdi | | | | | 0.7433*** | 2.7308*** | −0.013 | 0.1226** |
| | | | | | (0.19) | (0.30) | (0.01) | (0.05) |
| sigdprate | | | | | 0.3732*** | 0.7663*** | 0.0215*** | 0.0045 |
| | | | | | (0.06) | (0.09) | (0.00) | (0.02) |
| lnpopudens | | | | | 10.5695*** | 15.3261*** | 0.1027 | −0.4756 |
| | | | | | (2.41) | (3.86) | (0.15) | (0.64) |
| _cons | 76.7102*** | 72.4813*** | 10.4354*** | 36.5693*** | -8.5277 | −76.8754*** | 8.9155*** | 37.9686*** |
| | (0.29) | (0.48) | (0.02) | (0.08) | (13.98) | (22.33) | (0.86) | (3.72) |
| N | 3990 | 3990 | 3990 | 3990 | 3990 | 3990 | 3990 | 3990 |

Standard errors in parentheses
$*p < 0.1$, $**p < 0.05$, $***p < 0.01$

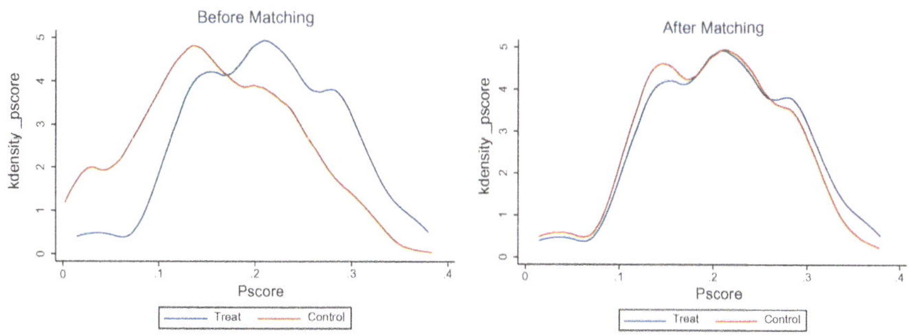


**Fig. 3** Before and after PSM

and MSW is investigated through PSM-DID method, UMPB significantly improves ISW and MSW. Secondly, the effects of UMPB on alleviating pollution are analyzed followed by robustness tests including DDD method. Finally, the impact of UMPB on improving ISW and MSW regeneration rates and mitigating pollutant emissions is examined by adopting mediating effect methods. The main findings of this study are summarized as follows. The empirical analysis results show that UMPB construction can reduce the pollution of urban ecological environment by increasing the comprehensive utilization rate of ISW and MSW and reducing industrial SO2 emissions and PM2.5 concentration level.

China has experienced rapid economic growth since Opening-up Reform in 1978 under linear growth model which generated tremendous amount of pollutants. Huge amount of industrial solid wastes lead to severe pollution. In addition, municipal solid waste grew dramatically along with fast urbanization and economic growth. The positive effect of UMPB has improved waste recovering rate and significantly mitigated air pollution. Therefore, UMPB program is a success and should be further enacted step by step among other cities. Meanwhile, the central government should regulate to avoid repeated investment and vicious competition among existing and potential UMPBs. More attention should be paid to adopting advanced recovering technique while processing solid wastes.

Generally, UMPB is composed of multiple renewable resource processing enterprises. However, because it belongs to a low-profit industry, the government should implement a tax incentive mechanism and tax preferential policies for recycling entities to support the development of the circular economy industry. In addition, due to the limited financial resources of local governments in small and medium-sized cities, the central government should consider increasing general transfer payments to UMPB in these cities and the Northeast region.

**Table 4** UMPB on environmental pollution alleviation under PSM-DID

| | (1) | (2) | (3) | (4) | (5) | (6) | (7) | (8) |
|---|---|---|---|---|---|---|---|---|
| | ISW | ISW | MSW | MSW | $SO_2$ | $SO_2$ | PM2.5 | PM2.5 |
| Dit | 3.3590* | 1.6337 | 15.6997*** | 13.1709*** | −0.3927*** | −0.4291*** | −0.8019* | −1.0203** |
| | (1.83) | (2.15) | (3.40) | (3.04) | (0.10) | (0.11) | (0.43) | (0.41) |
| lnrfdi | | 1.7427** | | 3.0083*** | | 0.0225 | | 0.2435* |
| | | (0.71) | | (1.08) | | (0.05) | | (0.13) |
| sigdprate | | 0.1654 | | 0.4628 | | 0.0227*** | | −0.0019 |
| | | (0.15) | | (0.29) | | (0.01) | | (0.04) |
| lnpopudens | | 9.3889 | | 11.6225 | | 0.9237 | | 0.223 |
| | | (8.25) | | (9.96) | | (0.90) | | (0.78) |
| _cons | 80.0847*** | −2.2308 | 75.7740*** | −47.8334 | 10.8446*** | 3.9005 | 40.5752*** | 36.9080*** |
| | (0.35) | (49.56) | (0.65) | (61.86) | (0.02) | (5.36) | (0.08) | (5.39) |
| N | 1212 | 1212 | 1212 | 1212 | 1212 | 1212 | 1212 | 1212 |

Standard errors in parentheses
$^{*}p < 0.1$, $^{**}p < 0.05$, $^{***}p < 0.01$

**Table 5** UMPB on environmental pollution alleviation under DDD

| | (1) | (2) | (3) | (4) | (5) | (6) | (7) | (8) |
|---|---|---|---|---|---|---|---|---|
| | ISW | ISW | MSW | MSW | $SO_2$ | $SO_2$ | PM2.5 | PM2.5 |
| Dit*group | 3.4284* | 2.532 | 15.4437*** | 14.0629*** | −0.3894*** | −0.3679*** | −0.7930* | −1.1545*** |
| | (1.87) | (1.30) | (4.68) | (4.54) | (−4.157) | (−3.915) | (−1.862) | (−2.668) |
| treats | 3.1781 | −1.2918 | 2.427 | −2.6377 | 0.7316*** | 0.6281*** | 2.9721 | 1.1742 |
| | (1.16) | (-0.514) | (0.66) | (-0.790) | (5.55) | (4.82) | (1.31) | (0.61) |
| group | 7.5690* | 0.8385 | 7.8987 | 2.6309 | 0.9258** | 0.7440* | 13.5274*** | 10.6173*** |
| | (1.71) | (0.19) | (1.37) | (0.44) | (2.24) | (1.94) | (5.14) | (4.24) |
| lnrfdi | | 0.7902*** | | 2.4215*** | | 0.0106 | | 0.1640*** |
| | | (2.64) | | (5.57) | | (0.69) | | (3.71) |
| sigdprate | | 0.2032** | | 0.6331*** | | 0.0254*** | | 0.007 |
| | | (2.44) | | (5.26) | | (5.94) | | (0.38) |
| lnpopudens | | 8.9593*** | | 1.355 | | 0.0664 | | 4.7464*** |
| | | (6.53) | | (0.98) | | (0.78) | | (7.49) |
| _cons | 69.1065*** | 7.9681 | 64.7153*** | 10.237 | 9.4478*** | 7.9242*** | 23.4427*** | −2.4732 |
| | (16.31) | (0.96) | (11.51) | (1.04) | (23.09) | (15.14) | (9.76) | (−0.644) |
| N | 3990 | 3990 | 3990 | 3990 | 3990 | 3990 | 3990 | 3990 |

t statistics in parentheses
$^{*}p < 0.1$, $^{**}p < 0.05$, $^{***}p < 0.01$

**Acknowledgements** This research is funded by National Social Science Foundation of China <Grant NO: 16CJY025>.

# References

1. Amor, L.B., Hammami, S.: The determinant factors for the rate of recycling: the example of used lube oils in Tunisia. Int. J. Waste Resour. **7**, 1–6 (2017)
2. Antweiler, W., Copeland, B.R., Taylor, M.S.: Is free trade good for the environment? Am. Econ. Rev **91**, 877–908 (2001)
3. Auffhammer, M., Carson, R.T.: Forecasting the path of China's $CO_2$ emissions using province-level information. J. Environ. Econ. Manage. **55**(3), 229–247 (2008)
4. Brunner, P.H.: Urban mining A: contribution to reindustrializing the city. J. Ind. Ecol. **15**(3), 339–341 (2011)
5. Cai, H., Chen, Y., Gong, Q.: Polluting thy neighbor: Unintended consequences of China's pollution reduction mandates. J. Environ. Econ. Manage. **76**, 86–104 (2016)
6. Chen, Y., Ginger, Z., Kumar, N., Shi, G.: The promise of Beijing: evaluating the impact of the 2008 Olympic games on air quality. J. Environ. Econ. Manage. **66**(3), 424–443 (2013)
7. Chen, S., Guo, C., Huang, X.: Air pollution, student health, and school absences: evidence from China. J. Environ. Econ. Manage. **92**, 465–497 (2018)
8. Chen, Z., Kahn, M.E., Liu, Yu., Wang, Z.: The consequences of spatially differentiated water pollution regulation in China. J. Environ. Econ. Manage. **88**, 468–485 (2018)
9. Dan Li, Mei-Qiang Wang, Chieh Lee: The waste treatment and recycling efficiency of industrial waste processing based on two-stage data envelopment analysis with undesirable inputs, J. Cleaner Prod. 242, (2020)
10. Denafas, G., Ruzgas, T., Martuzevicius, D., Shmarin, S., Hoffmann, M., Mykhaylenko, V., Ogorodnik, S.: Seasonal variation of municipal solid waste generation and composition in four East European cities. Resour. Conserv. Recycl. **89**, 22–30 (2014)
11. Dong, F., Zhang, S., Long, R., Zhang, X., Sun, Z.: Determinants of haze pollution: an analysis from the perspective of spatiotemporal heterogeneity. J. Clean. Prod. (2019).
12. Dou, J., Han, X.: How does the industry mobility affect pollution industry transfer in China: empirical test on pollution haven hypothesis and porter hypothesis. J. Clean. Prod. **217**, 105–115 (2019)
13. Van Eygen, E., De Meester, S., Tran, H.P., Dewulf, J.: Resource savings by urban mining: the case of desktop and laptop computers in Belgium. Resour. Conserv. Recycl. **107**, 53–64 (2016)
14. Fu, S., Gu, Y.: Highway toll and air pollution: evidence from Chinese cities. J. Environ. Econ. Manage. **83**, 32–49 (2017)
15. Gu Yifan, Wu Yufeng, Mu Xianzhong: Coupling allocation of primary and secondary resources. China Ind. Econ. (5):22–39, 2016
16. Gutberlet, J.: Cooperative urban mining in Brazil: collective practices in selective household waste collection and recycling. Waste Manage. **45**, (2015)
17. Heckman, J.J.: The common structure of statistical models of truncation, sample selection and limited dependent variables and a simple estimator for such models. Annals Econ. Soc. Meas. **5**, 475–492 (1976)
18. Heckman, J., Ichimura, H., Todd, P.: Matching as an econometric evaluation estimator. Rev. Econ. Stud. **65**(2), 261–294 (1998)
19. Heyes, A., Zhu, M.: Air pollution as a cause of sleeplessness: social media evidence from a panel of Chinese cities. J. Environ. Econ. Manage. **98**, 102247 (2019)
20. Huang, Daiyue: A study of new industrialization and foreign direct investment (FDI) based on China's East, Middle and West regions. African J. Bus. Manage. **6**(27), (2012)

21. Joakim, K.: Urban infrastructure mines: on the economic and environmental motives of cable recovery from subsurface power grids. J. Cleaner Prod. **104**, 353–363 (2015)
22. Joosten, L.A.J., Hekkert, M.P., Worrell, E.: Assessment of the plastic flows in the Netherlands using STREAMS. Resour. Conserv. Recycl. **30**, 135–161 (2000)
23. Kollikkathara, N., Feng, H., Yu, D.: A system dynamic modeling approach for evaluating municipal solid waste generation, landfill capacity and related cost management issues. Waste Manage. **30**(11), 2194–2203 (2010)
24. Krook, J., Baas, L.: Getting serious about mining the technosphere: a review of recent landfill mining and urban mining research. J. Cleaner Prod. **55**(15), 1–9 (2013)
25. Lai, W.: Pesticide use and health outcomes: evidence from agricultural water pollution in China. J. Environ. Econ. Manage. **86**, 93–120 (2017)
26. Leuven, E., Sianesi, B.: PSMATCH2: Stata module to perform full Mahalanobis and propensity score matching, common support graphing, and covariate imbalance testing. Statistical Software Components, Boston College Department of Economics (2014)
27. Liu, L., Liang, Y., Song, Q.: A review of waste prevention through 3R under the concept of circular economy in China. J. Mater. Cycles Waste Manage. **19**(4), 1314–1323 (2017)
28. NDRC, 2010. On Implementing Urban Mining Base.
29. Rimaityte, I., Ruzgas, T., Denafas, G., Racys, V., Martuzevicius, D.: Application and evaluation of forecasting methods for municipal solid waste generation in an eastern-European city. Waste Manage. Res **30**(1), 89–98 (2012)
30. Rosenbaum, P.R., Rubin, D.B.: The central role of the propensity score in observational studies for causal effects. Biometrika **70**, 41–55 (1983)
31. Simoni, M., Kuhn, E.P., Morf, L.S., et al.: Urban mining as a contribution to the resource strategy of the Canton of Zurich. Waste Manage. **45**, 10–21 (2015)
32. Wang, J., Li, Y., Huang, K., Chen, H.: Treatment situation and determinants of rural domestic waste. China Popul. Resour. Environ. **21**(6), 71–78 (2011)
33. Wen, Z., Zhang, C., Ji, X., et al.: Urban mining's potential to relieve China's coming resource crisis. J. Ind. Ecol. **19**(6), 1091–1102 (2015)
34. Zaman, A.U., Lehmann, S.: The zero-waste index: a performance measurement tool for waste management systems in a "zero waste city". J. Cleaner Prod. **50**, 123–132 (2013)
35. Zhang, X., Zhang, X., Chen, Xi.: Happiness in the air: how does a dirty sky affect mental health and subjective well-being? J. Environ. Econ. Manage. **85**, 81–94 (2017)
36. Zhou, X.L.: Development and utilization of circular economy and urban mining—Chengdu city based renewable resource industry survey. Appl. Mech. Mater. **768**, 644–651 (2015)

# Revealing the Psychological Basis of Green Hotel Visiting Intention with the Extended Theory of Planned Behavior: An Empirical Study in Shenzhen, China

**Yu-Tong Gao**

**Abstract** The lodging industry has been a critical energy consumer and greenhouse gas emitter in China. Customers present increasing demands for green hotels. However, the psychological factors affecting the green hotel visiting intention of consumers is still unclear. This study aims to investigate the psychological predictors of green hotel visiting intention of customers. To further explain the psychological basis of green hotel visiting behaviour, this study extends the theory of planned behaviour (TPB) by adding two factors: moral norms and past behaviour. The study conducted a questionnaire-based survey with a sample size of 461 respondents in Shenzhen, a typical city in southeast China. The survey was conducted in 2019. The study employed structural equation modelling for data analysis. The results indicate that, in addition to two TPB factors (attitude and perceived behavior control), past behaviour and moral norms are also significantly positively related to intention. However, the impact of subjective norms on green hotel visiting intention is less significant. The study provides a novel psychological explanation of green hotel visiting intention. The results might help the development policy interventions towards "sustainable lodging". Furthermore, the findings would contribute to better marketing and service strategies of green hotels through more effective management of customers' normative processes of pro-environmental decision-making.

**Keywords** Green hotel · Pro-environment · Theory of planned behavior · Past behavior · Moral norms · China

## 1 Introduction

With the social and economic development, the demand for energy maintained an increasing trend in the past decades. The building sector has become one of the largest energy consumers [1, 2], as well as greenhouse gas emitters [3, 4]. Especially, in

Y.-T. Gao (✉)
New-Channel International Education Group Ltd, Qingdao 266555, Shandong, China
e-mail: yu-tong.gao@outlook.com



F. K. S. Chan et al. (eds.), *Proceedings of the 2020 International Conference on Resource Sustainability: Sustainable Urbanisation in the BRI Era (icRS Urbanisation 2020)*, Environmental Science and Engineering,
https://doi.org/10.1007/978-981-15-9605-6_32

China, hotel buildings have an important impact on the environment. Although hotel buildings spend much more on high-efficiency building service systems and green materials than residential or even office buildings, inappropriate operating model and wasteful behavior of hotel guests make it difficult for the hotel industry to achieve a satisfactory level of energy efficiency [5].

Energy shortages and environmental pollution have triggered people's environmental concerns [6]. In recent years, more and more people choose environmentally friendly products and services [7]. In the lodging industry, some scholars have put forward the concept of green hotels. Green hotels refer to those hotels and B&Bs who reduce or even minimize the impact on the environment where they are located in their operations [8]. Green hotels are profitable by reducing operating costs and attracting tenants who support environmental protection. The market for green hotels has continued to expand since 2000s [7]. A growing number of hotels and B&Bs are joining the green movement to reduce their negative environmental impact. Green hotels have become a promising sustainable hotel industry development model [9].

Understanding the decision-making process of hotel guests would help hotel operators to find target customers more accurately and meet the needs of guests. Some previous studies explained the guest's green hotel visiting intention with classical psychology models. For example, Han and Kim [7] and Chen and Tung [10] used Theory of planned behavior (TPB) to explain the willingness of American and Taiwanese residents to choose environmentally friendly hotels respectively. Besides, Teng, Wu and Liu [11] explored the role of altruism on the basis of TPB in green hotel visiting behavioral process. Han [12] then combined TPB with value-belief-norm to further explain the psychological model in the green hotel visiting process, and further extend the model in 2017 [13]. These studies have revealed many important psychological factors affecting green hotel visiting intention of guests. However, few studies have paid attention to the role of past behavior and moral norm in the behavioral process of green hotel visits.

In order to further explain the psychological basis of green hotel visits, this study expands the planned behavior theory (TPB) by adding two factors: moral norms and past behavior. This article first reviews relevant literature in the field of green hotels and proposes a series of hypotheses. After that, the paper uses a survey based on a questionnaire survey conducted in Shenzhen, a typical city in southeast China in 2019, to verify the hypotheses. The specific research process is described in detail in the article. The study employs structural equation model for data analysis and then further discusses the analysis results. Finally, the article also shows the limitations and implication of the research results. This research provides a novel psychological explanation for the visit intention of the green hotel. The results may help development policy interventions in green hotels for "sustainable accommodation". In addition, the research results will help to better formulate the marketing and service strategies of green hotels through more effective management of customers' environmental decision-making procedures.

## 2 Literature Review

In 1975, Fishbein and Ajzen [14] jointly proposed the Theory of Reasoned Action (TRA). TRA suggests that human behavior depends on their intention: this intention is the result of people's rational consideration of their own situation and environment [14]. In 1985, Ajzen [15] made modifications on the basis of TRA and proposed TPB. TPB has become one of the most commonly-used psychological theories to explain pro-environmental behavior. Previous studies have used TPB to explain household energy-saving behaviors [16], resource recycling [17], green investment [18] and purchasing green products [19].

Similar to TRA, individual behavior is a function of intention in the TPB model. TPB uses three factors to explain an individual's intentions for a particular behavior: attitude, subjective norms and perceived behavior control (PBC). Attitude is an individual's subjective assessment of the results and the outcomes of one particular behavior. This subjective judgment may be positive or negative [20]. A positive attitude can promote the occurrence of certain behaviors. For example, hotel guests may choose green hotels because of their emphasis on environmental protection [7]. Subjective norms mainly reflect the impact of pressure from society and the surrounding environment on decision-making and behavior. This influence often comes from people who have important relationships with the individual, such as bosses, friends, and families [20]. Consumers may reject green hotels because they consider the feelings of their families [10]. On the other hand, hotel guests may also choose green hotels because they consider the expectations of their superiors or colleagues. Human behavior is often affected by the surrounding environment and personal abilities. Therefore, TPB also introduced a new concept: PBC. PBC reflects the individual's subjective judgments about the difficulty or convenience of one particular behavior. For example, some customers abandon green hotels due to the complex process of evaluating hotel sustainability.

TPB has been widely used in both psychology practice and application. In previous studies, TPB has been proven to have satisfactory explanatory power. However, some studies have also found different results. For example, Lim et al. [21] found that the impact of attitude on intention is less significant. Moser [22, 23] also found the linkage between attitude and intention less significant. Chau and Hu [24] noted that subjective have no significant relationship with acceptance of telemedicine technology. Husin and Rahman [25] and Liu et al. [16] also found similar results. There are also some empirical evidence showing that the contribution of PBC on intention is small or less significant (e.g., [26]). Considering the previous studies and above evidence, this study makes the following hypotheses:

**$H_1$** Attitude affects green hotel visiting intention positively;

**$H_2$** Subjective Norms affect green hotel visiting intention positively;

**$H_3$** PBC affects green hotel visiting intention positively;

Many studies have also expanded the TPB model by adding other psychological elements to achieve better explanatory power. Moral Norms is a factor widely used in extended TPB. Moral norms reflects the evaluation of a particular behavior by personal moral perception [27]. Moral norms are closely related to personal responsibility [28]. Different from subjective norms, moral norms emphasize the influence of individual internalization and emotion. Previous research has shown that moral norms play a vital role in pro-environmental behavior. Ru et al. [29], for example, found that moral norms are directly related to energy-saving behavior. Wang et al. [28] also employ moral norms to explore people's electronic vehicle adoption intention. Shi et al. [30] believe that moral norms significantly affect the environmental protection behavior of Chinese residents. There have been many previous studies showing that adding moral norms can significantly improve the variance of TPB's interpretation of pro-environmental behavior (e.g., [29, 31]). Therefore, the study put forward the hypothesis:

**$H_4$** Moral Norms affect green hotel visiting intention positively;

Past behavior is another proposed addition to the TPB. Some studies found that with repeated performance, behavior might be determined by a person's past behavior [32]. There is sufficient evidence supporting the view that past behavior explains unique differences in intention and behavior [33–35]. Knowles et al. [36] for example, found that past behaviors is highly related to financial donation behavior. Beside, Lee and Choi [37] also found that past experience was an important predictor of behavioral intention in the hospitality and tourist sectors. The green hotel visiting behavior of guests might be influenced by their past experience. In this study, therefore, the impact of past behavior on green hotel visiting intention and behavior was examined. The study put forward the hypothesis:

**$H_5$** Past Behaviors affect green hotel visiting intention positively.

The theoretical framework with hypotheses is presented in Fig. 1.

# 3 Methodology

## *3.1 Questionnaire Design and Sampling*

The researchers collected data through a questionnaire survey conducted in Shenzhen, China in 2019. The researcher designed a questionnaire with two parts. In the first part, the questionnaire tested the respondents' basic psychological factors (including attitude, subjective norms, PBC, moral norms, past behavior and green hotel visiting intention) through 17 items. Some items was employed from Han, Hsu and Sheu [38]. The items in the section is presented in Table 1. The second part focuses on the demographic information (gender, age, income, education) of the interviewee. In the cover letter of the questionnaire, the researchers defined the green hotel. In order to reduce the impact on the respondents, the respondents were told that the questionnaire did not have the correct answer.

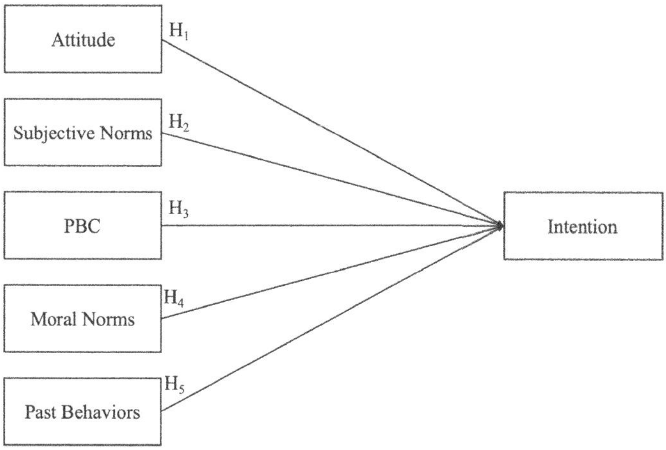


**Fig. 1** The theoretical framework with hypotheses. PBC refers to perceived behavior control

**Table 1** Items in the questionnaire

| Construct (reference) | Item | Code |
|---|---|---|
| Attitude | For me, staying at a green hotel when travelling is good | ATT-1 |
| | For me, staying at a green hotel when travelling is pleasant | ATT-2 |
| | For me, staying at a green hotel when travelling is positive | ATT-3 |
| Subjective norms | My families (or relatives) think I should stay at a green hotel when traveling | SN-1 |
| | My friends think I should stay at a green hotel when traveling | SN-2 |
| | My colleagues (or co-works) thinks I should stay at a green hotel when traveling | SN-3 |
| PBC | Whether or not I stay at a green hotel when traveling is completely up to me | PBC-1 |
| | I am confident that if I want, I can stay at a green hotel when traveling | PBC-2 |
| | I have resources, time, and opportunities to stay at a green hotel when traveling | PBC-3 |
| Moral norms | I have a moral responsibility to choose green hotels when traveling | MN-1 |
| | Visiting green hotels is depending on my own moral obligation | MN-2 |
| | I would feel unhappy if I do not stay in green hotels when traveling | MN-3 |
| Past behavior | I used to visit green hotel when traveling | PB-1 |
| | I am used to visiting green hotel when traveling | PB-2 |
| Behavioral intention | I am willing to stay at a green hotel when traveling | BI-1 |
| | I plan to stay at a green hotel when traveling | BI-2 |
| | I will make an effort to stay at a green hotel when traveling | BI-3 |

The researchers distributed 1500 questionnaires outside the local high-speed railway station with the largest passenger traffic in Shenzhen and received 880 responses. Among them, 393 respondents do not need hotel services in Shenzhen. Besides, there are 26 invalid responses of the remaining 487 feedbacks. This research finally received 461 valid responses. There were 242 male (i.e., 52.49%) and 219 female respondents (i.e., 47.51%). Respondents ranged in age from 16 to 72 and the average age of respondents is 30.65. There are 263 (57.05%) respondents with bachelor's or polytechnics degrees. In general, the demographic characteristics of the respondents are in line with the general situation in China.

### 3.2 *Data Analysis*

Structural equation model (SEM) is a statistical method to analyze the relationship based on the covariance matrix of variables [39]. It is an important method for multivariate data analysis. SEM is an important method of quantitative research in contemporary behavior and social fields and is widely used in pro-environmental behavior research modelling [7, 16, 29, 40]. The application of structural equation modeling (SEM) can not only estimate the unknown coefficients of the causal relationship between latent variables but also can specify how the observation variables represent the hypothetical structure [41]. The study followed the two-step method of Anderson and Gerbing [42]. Firstly, the study used confirmatory factor analysis (CFA) to evaluate the measurement model. Secondly, SEM analysis methods were used to evaluate the model after evaluating the appropriateness of the measurement model. The purpose of evaluating the structural model is to determine whether the data supports the specified theoretical relationship. Since Amos is widely recognized in behavioral science research, this study employs Amos for SEM analysis.

## 4 Results

In this study, a two-step structural equation modeling analysis was conducted based on [41] to ensure all measurement constructs are statistically reliable. The measurement modeling results indicate that the factor loadings of all 17 items (minimum at 0.625) and the average variance extracted values of the constructs (minimum at 0.722) in the section one are higher than 0.5. Besides, the composite reliability values and the Cronbach's alphas of constructs of the constructs in the first section are higher than 0.7. The results suggest that all constructs in the current study well fit the convergent validity requirement.

Then, the study presents the Heterotrait-Monotrait (HTMT) Ratio of the constructs in the section one (see Table 2), which reflects discriminant validity assessment results. The presented values are expected to be less than 0.9. The results suggest that the constructs in the questionnaires well satisfy the discriminant validity requirement.

**Table 2** Heterotrait-Monotrait (HTMT) ratio results

| Test name | ATT | SN | PBC | MN | PB | BI |
|---|---|---|---|---|---|---|
| ATT | | | | | | |
| SN | 0.313 | | | | | |
| PBC | 0.421 | 0.366 | | | | |
| MN | 0.450 | 0.317 | 0.712 | | | |
| PB | 0.507 | 0.445 | 0.707 | 0.555 | | |
| BI | 0.687 | 0.396 | 0.522 | 0.435 | 0.428 | |

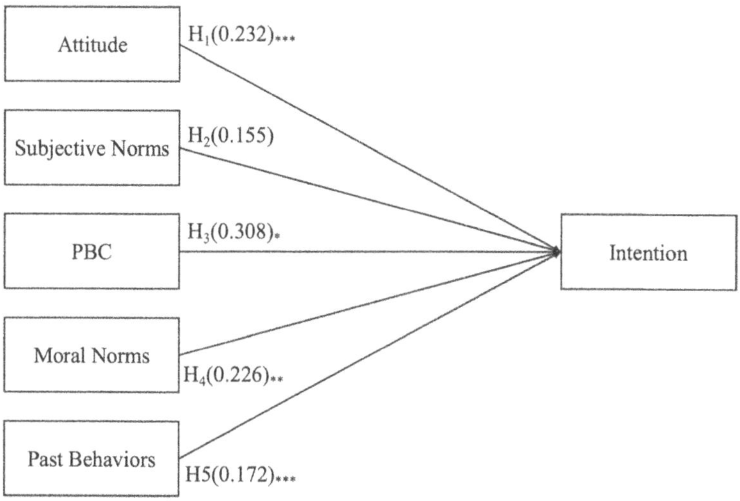


**Fig. 2** Structural modeling analysis results. $^{*}p < 0.05$; $^{***}p < 0.01$;$^{***}p < 0.001$

The structural analysis results show that two TPB factors (attitude and PBC) are significantly related to green hotel visiting intention. Both attitude (Beta = 0.232; $p < 0.001$) and PBC (Beta = 0.308; $p < 0.01$) directly and positively contributes to intention. However, the impact of subjective norms on green hotel visiting intention is less significant (Beta = 0.155; $p = 0.255$). Besides, all two additional factors, moral norms (Beta = 0.226; $p = 0.007 < 0.01$) and past behaviors (Beta = 0.172; $p < 0.001$) are significantly contribute to green hotel visiting intention. The structural analysis results are presented in Fig. 2 and Table 3.

## 5 Discussion

The research shows that TPB is a reasonable theory to explain green hotel visiting behavior. The result indicates that attitude, perceived behavior control, past

**Table 3** Standard test condition applied in the PV performance experimental study

| Parameter | Beta | $p$-value | Result |
|---|---|---|---|
| $H_1$: Attitude > intention | 0.232 | <0.001*** | Do not reject |
| $H_2$: Subjective norms > intention | 0.155 | 0.255 | Reject |
| $H_3$: PBC > intention | 0.308 | 0.022* | Do not reject |
| $H_4$: Moral norms > intention | 0.226 | 0.007** | Do not reject |
| $H_5$: Past Behaviors > intention | 0.172 | <0.001*** | Do not reject |

*Note*: $*p < 0.05$; $***p < 0.01$; $***p < 0.001$

behavior and moral norm are significantly affecting green hotel visiting intention. Firstly, personal attitudes affect green hotel visiting behavior. People's preference for environment protection will promote their choice of visiting a green hotel. Secondly, the obvious intention to choose an environmentally friendly environment is likely to encourage people's decision on living in a green hotel when traveling. Thirdly, individual's past energy-saving behavior will cultivate their environmental awareness. People may be consistent in their pro-environmental behaviors which will impact on their decision to visit a green hotel positively. In addition, if people's moral norm is higher, they may pay more attention to social-environmental protection and other related environmental issues, which will also have an optimistic impact on their green hotel visiting behavior.

On the contrary, the research shows that subjective norm has no significant effect on green hotel visiting behavior. There might have some reasons. Firstly, the study researched in individual behavior. The expectations of family members, colleagues and other people will affect a person's intention of hotel visiting. However, some other factors might impact on individual decision-making process. Secondly, the research was conducted in Shenzhen, a well-developed city in China. The average age of the population is lower and the education level is higher in Shenzhen. Previous research has found that people with higher education levels are likely to pay more attention to environmental behaviors rather than others' expectations. Therefore, the role of subjective norm in the green hotel visiting behavior in this study is not apparent. The governments have paid much attention on green building [43]. The paper points out a psychological way to promote hotel building energy efficiency. This research provides a novel psychological explanation for the visit intention of the green hotel. The results may help development policy interventions in green hotels for "sustainable accommodation". In addition, the research results will help to better formulate the marketing and service strategies of green hotels through more effective management of customers' environmental decision-making procedures.

This study also has some certain limitations. First of all, the data of this research comes from self-report questionnaire. Although the questionnaire description has suggested objective standards, some previous research has indicated that self-report data is subjective and might not necessarily completely objective. Secondly, this research was conducted in one of the most economically developed cities in China. The demographic data has specific regional characteristics and may not be fully applicable to all cities of China. Third, the questionnaire of the study was mainly distributed around the high-speed railway station. Although the high-speed railway is one of the main ways for out-of-town visitors, some people may also arrive in the city by train or plane. The questionnaires concentrated on high-speed railway stations may not necessarily represent the green hotel visiting behavior of all travelers. Moreover, the sample size of this study is 461. A total of 1500 questionnaires were distributed in the study, but only 461 valid questionnaires were collected. Smaller sample size questionnaire might not fully represent the choices and preferences of visitors throughout the city.

## References

1. Huo, T., et al.: China's building stock estimation and energy intensity analysis. J. Cleaner Prod. **207**, 801–813 (2019)
2. Chang, S., Wang, Q., Hu, H., Ding, Z., Guo, H.: An NNwC MPPT-based energy supply solution for sensor nodes in buildings and its feasibility study. Energies **12**(1), 101 (2019)
3. Chang, R., Wang, Q., Ding, Z.: How is the energy performance of buildings assessed in Australia? A comparison between four evaluation systems (2019)
4. Ma, M., Yan, R., Cai, W.: An extended STIRPAT model-based methodology for evaluating the driving forces affecting carbon emissions in existing public building sector: evidence from China in 2000–2015. Nat. Hazards **89**(2), 741–756 (2017)
5. Long, X., Wang, X., Mensah, C.N., Wang, M., Zhang, J.: Spatial and temporal heterogeneity of environmental efficiency for China's hotel sector: new evidence through metafrontier global Malmquist-Luenberger. Environ. Sci. Pollut. Res. **26**(26), 27534–27541 (2019)
6. Wang, Q., Wei, H.H., Xu, Q.: A solid oxide fuel cell (SOFC)-based biogas-from-waste generation system for residential buildings in China: a feasibility study. Sustainability **10**(7), 2395 (2018)
7. Han, H., Kim, Y.: An investigation of green hotel customers' decision formation: developing an extended model of the theory of planned behavior. Int. J. Hosp. Manage. **29**(4), 659–668 (2010)
8. Verma, V.K., Chandra, B.: An application of theory of planned behavior to predict young Indian consumers' green hotel visit intention. J. Cleaner Prod. **172**, 1152–1162 (2018)
9. Wang, J., Wang, S., Xue, H., Wang, Y., Li, J.: Green image and consumers' word-of-mouth intention in the green hotel industry: the moderating effect of Millennials. J. Cleaner Prod. **181**, 426–436 (2018)
10. Chen, M.F., Tung, P.J.: Developing an extended theory of planned behavior model to predict consumers' intention to visit green hotels. Int. J. Hosp. Manage. **36**, 221–230 (2014)
11. Teng, Y.M., Wu, K.S., Liu, H.H.: Integrating altruism and the theory of planned behavior to predict patronage intention of a green hotel. J. Hosp. Tourism Res. **39**(3), 299–315 (2015)
12. Han, H.: Travelers' pro-environmental behavior in a green lodging context: converging value-belief-norm theory and the theory of planned behavior. Tourism Manage. **47**, 164–177 (2015)
13. Han, H., Hwang, J., Lee, M.J.: The value–belief–emotion–norm model: investigating customers' eco-friendly behavior. J. Travel Tourism Mark. **34**(5), 590–607 (2017)
14. Fishbein, M., Ajzen, I.: Belief, attitude, intention, and behavior: an introduction to theory and research (1977).
15. Ajzen, I.: From intentions to actions: a theory of planned behavior In: Action Control, pp. 11–39, Springer (1985).
16. Liu, X., Wang, Q., Wei, H.H., Chi, H.L., Ma, Y., Jian, I.Y.: Psychological and demographic factors affecting household energy-saving intentions: a TPB-based study in Northwest China. Sustainability **12**(3), 836 (2020)

17. Tonglet, M., Phillips, P.S., Read, A.D.: Using the theory of planned behaviour to investigate the determinants of recycling behaviour: a case study from Brixworth, UK. Resour. Conserv. Recycl. **41**(3), 191–214 (2004)
18. Husni, N.S.A., Alwi, S.F.S., Hassan, F.: Determinants of behavioural intention towards green investments: the perspectives of muslims* Ismah Osman (corresponding author) Masturah Ma'in Ruhaini Muda. Int. J. **4**(1), 16–38 (2019)
19. Paul, J., Modi, A., Patel, J.: Predicting green product consumption using theory of planned behavior and reasoned action. J. Retail. Consum. Serv. **29**, 123–134 (2016)
20. Conner, M.: Theory of planned behavior. Handbook of Sport Psychology, pp. 3, (2020)
21. Lim, X.J., Radzol, A.M., Cheah, J., Wong, M.: The impact of social media influencers on purchase intention and the mediation effect of customer attitude. Asian J. Bus. Res. **7**(2), 19–36 (2017)
22. Moser, A.K.: Thinking green, buying green? Drivers of pro-environmental purchasing behavior J. Cons. Market. (2015)
23. Moser, A.K.: Consumers' purchasing decisions regarding environmentally friendly products: an empirical analysis of German consumers. J. Retail. Consum. Serv. **31**, 389–397 (2016)
24. Chau, P.Y., Hu, P.J.H.: Investigating healthcare professionals' decisions to accept telemedicine technology: an empirical test of competing theories. Inform. Manage. **39**(4), 297–311 (2002)
25. Husin, M.M., Ab Rahman, A.: Predicting intention to participate in family takaful scheme using decomposed theory of planned behavior. Int. J. Soc. Econ. (2016)
26. Conner, M., Armitage, C.J.: Extending the theory of planned behavior: a review and avenues for further research. J. Appl. Soc. Psychol. **28**(15), 1429–1464 (1998)
27. Schwartz, S.H.: Normative influences on altruism. Adv. Exp. Soc. Psychol. **10**(1), 221–279 (1977)
28. Wang, S., Fan, J., Zhao, D., Yang, S., Fu, Y.: Predicting consumers' intention to adopt hybrid electric vehicles: using an extended version of the theory of planned behavior model. Transportation **43**(1), 123–143 (2016)
29. Ru, X., Wang, S., Yan, S.: Exploring the effects of normative factors and perceived behavioral control on individual's energy-saving intention: an empirical study in eastern China. Resour. Conserv. Recycl. **134**, 91–99 (2018)
30. Shi, H., Fan, J., Zhao, D.: Predicting household PM2. 5-reduction behavior in Chinese urban areas: an integrative model of theory of planned behavior and norm activation theory. J. Cleaner Prod. **145**, 64–73 (2017)
31. Shi, H., Wang, S., Zhao, D.: Exploring urban resident's vehicular PM2. 5 reduction behavior intention: an application of the extended theory of planned behavior. J. Cleaner Prod. **147**, 603–613 (2017)
32. Sutton, S.: The past predicts the future: interpreting behaviour–behaviour relationships in social psychological models of health behaviour (1994)
33. Wang, X.: The role of anticipated negative emotions and past behavior in individuals' physical activity intentions and behaviors. Psychol. Sport Exerc. **12**(3), 300–305 (2011)
34. Kidwell, B., Jewell, R.D.: The influence of past behavior on behavioral intent: an information-processing explanation. Psychol. Mark. **25**(12), 1151–1166 (2008)
35. Kokkinaki, F.: Predicting product purchase and usage: the role of perceived control, past behavior and product involvement. ACR North Am. Adv. (1999)
36. Knowles, S.R., Hyde, M.K., White, K.M.: Predictors of young people's charitable intentions to donate money: an extended theory of planned behavior perspective. J. Appl. Soc. Psychol. **42**(9), 2096–2110 (2012)
37. Lee, W.J., Choi, H.S.C.: Understanding meeting planners' internet use behavior: an extension to the theory of planned behavior. Int. J. Hosp. Tourism Adm. **10**(2), 109–128 (2009)
38. Han, H., Hsu, L.T.J., Sheu, C.: Application of the theory of planned behavior to green hotel choice: testing the effect of environmental friendly activities. Tourism Manage. **31**(3), 325–334 (2010)
39. Hair, J.F., Hult, G.T.M., Ringle, C.M., Sarstedt, M., Thiele, K.O.: Mirror, mirror on the wall: a comparative evaluation of composite-based structural equation modeling methods. J. Acad. Mark. Sci. **45**(5), 616–632 (2017)

40. Wang, Q.C., Wang, Y.X., Jian, I.Y., Wei, H.H., Liu, X., Ma, Y.T.: Exploring the "Energy-saving personality traits" in the office and household situation: an empirical study. Energies **13**(14), 3535 (2020)
41. Jöreskog, K., Sörbom, D.: PRELIS 2.20 for Windows. Scientific Software International, Chicago (1998)
42. Anderson, J.C., Gerbing, D.W.: Assumptions and comparative strengths of the two-step approach: comment on Fornell and Yi. Sociol. Methods Res. **20**(3), 321–333 (1992)
43. Yu, C.R., Guo, H.S., Wang, Q.C., Chang, R.D.: Revealing the impacts of passive cooling techniques on building energy performance: a residential case in Hong Kong. Appl. Sci. **10**(12), 4188 (2020)

The manufacturer's authorised representative in the EU is Springer Nature Customer Service Centre GmbH, Europaplatz 3, 69115 Heidelberg, Germany. If you have any concerns regarding our products, please contact ProductSafety@springernature.com

Printed and bound by CPI Group (UK) Ltd, Croydon, CR0 4YY
30/09/2026
02225289-0014